Michael R. Chernick
15 Quail Drive
Holland PA 18966

Probability and Mathematical Statistics (Continued)

PURI and SEN • Nonparametric Methods in Multivariate Analysis
RANDLES and WOLFE • Introduction to the Theory of Nonparametric Statistics
RAO • Linear Statistical Inference and Its Applications, *Second Edition*
RAO • Real and Stochastic Analysis
RAO and SEDRANSK • W.G. Cochran's Impact on Statistics
RAO • Asymptotic Theory of Statistical Inference
ROHATGI • An Introduction to Probability Theory and Mathematical Statistics
ROHATGI • Statistical Inference
ROSS • Stochastic Processes
RUBINSTEIN • Simulation and The Monte Carlo Method
SCHEFFE • The Analysis of Variance
SEBER • Linear Regression Analysis
SEBER • Multivariate Observations
SEN • Sequential Nonparametrics: Invariance Principles and Statistical Inference
SERFLING • Approximation Theorems of Mathematical Statistics
SHORACK and WELLNER • Empirical Processes with Applications to Statistics
TJUR • Probability Based on Radon Measures

Applied Probability and Statistics

ABRAHAM and LEDOLTER • Statistical Methods for Forecasting
AGRESTI • Analysis of Ordinal Categorical Data
AICKIN • Linear Statistical Analysis of Discrete Data
ANDERSON, AUQUIER, HAUCK, OAKES, VANDAELE, and WEISBERG • Statistical Methods for Comparative Studies
ARTHANARI and DODGE • Mathematical Programming in Statistics
BAILEY • The Elements of Stochastic Processes with Applications to the Natural Sciences
BAILEY • Mathematics, Statistics and Systems for Health
BARNETT • Interpreting Multivariate Data
BARNETT and LEWIS • Outliers in Statistical Data, *Second Edition*
BARTHOLOMEW • Stochastic Models for Social Processes, *Third Edition*
BARTHOLOMEW and FORBES • Statistical Techniques for Manpower Planning
BECK and ARNOLD • Parameter Estimation in Engineering and Science
BELSLEY, KUH, and WELSCH • Regression Diagnostics: Identifying Influential Data and Sources of Collinearity
BHAT • Elements of Applied Stochastic Processes, *Second Edition*
BLOOMFIELD • Fourier Analysis of Time Series: An Introduction
BOX • R. A. Fisher, The Life of a Scientist
BOX and DRAPER • Empirical Model-Building and Response Surfaces
BOX and DRAPER • Evolutionary Operation: A Statistical Method for Process Improvement
BOX, HUNTER, and HUNTER • Statistics for Experimenters: An Introduction to Design, Data Analysis, and Model Building
BROWN and HOLLANDER • Statistics: A Biomedical Introduction
BUNKE and BUNKE • Statistical Inference in Linear Models, Volume I
CHAMBERS • Computational Methods for Data Analysis
CHATTERJEE and PRICE • Regression Analysis by Example
CHOW • Econometric Analysis by Control Methods
CLARKE and DISNEY • Probability and Random Processes: A First Course with Applications, *Second Edition*
COCHRAN • Sampling Techniques, *Third Edition*
COCHRAN and COX • Experimental Designs, *Second Edition*
CONOVER • Practical Nonparametric Statistics, *Second Edition*
CONOVER and IMAN • Introduction to Modern Business Statistics
CORNELL • Experiments with Mixtures: Designs, Models and The Analysis of Mixture Data

Applied Probability and Statistics (Continued)

COX • Planning of Experiments
COX • A Handbook of Introductory Statistical Methods
DANIEL • Biostatistics: A Foundation for Analysis in the Health Sciences, *Third Edition*
DANIEL • Applications of Statistics to Industrial Experimentation
DANIEL and WOOD • Fitting Equations to Data: Computer Analysis of Multifactor Data, *Second Edition*
DAVID • Order Statistics, *Second Edition*
DAVISON • Multidimensional Scaling
DEGROOT, FIENBERG and KADANE • Statistics and the Law
DEMING • Sample Design in Business Research
DILLON and GOLDSTEIN • Multivariate Analysis: Methods and Applications
DODGE • Analysis of Experiments with Missing Data
DODGE and ROMIG • Sampling Inspection Tables, *Second Edition*
DOWDY and WEARDEN • Statistics for Research
DRAPER and SMITH • Applied Regression Analysis, *Second Edition*
DUNN • Basic Statistics: A Primer for the Biomedical Sciences, *Second Edition*
DUNN and CLARK • Applied Statistics: Analysis of Variance and Regression
ELANDT-JOHNSON and JOHNSON • Survival Models and Data Analysis
FLEISS • Statistical Methods for Rates and Proportions, *Second Edition*
FLEISS • The Design and Analysis of Clinical Experiments
FOX • Linear Statistical Models and Related Methods
FRANKEN, KÖNIG, ARNDT, and SCHMIDT • Queues and Point Processes
GALLANT • Nonlinear Statistical Models
GIBBONS, OLKIN, and SOBEL • Selecting and Ordering Populations: A New Statistical Methodology
GNANADESIKAN • Methods for Statistical Data Analysis of Multivariate Observations
GREENBERG and WEBSTER • Advanced Econometrics: A Bridge to the Literature
GROSS and HARRIS • Fundamentals of Queueing Theory, *Second Edition*
GUPTA and PANCHAPAKESAN • Multiple Decision Procedures: Theory and Methodology of Selecting and Ranking Populations
GUTTMAN, WILKS, and HUNTER • Introductory Engineering Statistics, *Third Edition*
HAHN and SHAPIRO • Statistical Models in Engineering
HALD • Statistical Tables and Formulas
HALD • Statistical Theory with Engineering Applications
HAND • Discrimination and Classification
HOAGLIN, MOSTELLER and TUKEY • Exploring Data Tables, Trends and Shapes
HOAGLIN, MOSTELLER, and TUKEY • Understanding Robust and Exploratory Data Analysis
HOEL • Elementary Statistics, *Fourth Edition*
HOEL and JESSEN • Basic Statistics for Business and Economics, *Third Edition*
HOGG and KLUGMAN • Loss Distributions
HOLLANDER and WOLFE • Nonparametric Statistical Methods
IMAN and CONOVER • Modern Business Statistics
JAGERS • Branching Processes with Biological Applications
JESSEN • Statistical Survey Techniques
JOHNSON • Multivariate Statistical Simulation: A Guide to Selecting and Generating Continuous Multivariate Distributions
JOHNSON and KOTZ • Distributions in Statistics
Discrete Distributions
Continuous Univariate Distributions—1
Continuous Univariate Distributions—2
Continuous Multivariate Distributions

(*continued on back*)

Asymptotic Theory of Statistical Inference

Asymptotic Theory of Statistical Inference

B. L. S. PRAKASA RAO

Indian Statistical Institute, New Delhi

JOHN WILEY & SONS

New York • Chichester • Brisbane • Toronto • Singapore

Library of Congress Cataloging-in-Publication Data:
Prakasa Rao, B. L. S.
Asymptotic theory of statistical inference.
(Wiley series in probability and mathematical statistics. Probability and statistics, ISSN 0271-2728)
Includes bibliographies and index.
1. Mathematical statistics–Asymptotic theory.
I. Title. II. Series.
QA276.P67 1986 519.5′4 86-15735
ISBN 0-471-84335-0

Printed in the United States of America

10 9 8 7 6 5 4 3 2 1

To
Vasanta
and
Gopi, Vamsi, Venu

Preface

Asymptotic theory of statistical inference encompasses such a vast area of study, including most of theoretical statistics, that it is impossible to cover all the topics in a single volume. The aim in writing this book is to bring together the interplay between the recent advances in probability and the asymptotic theory of inference and to present a modern approach. Attention is restricted mostly to the classical independent and identically distributed (i.i.d.) case even though it is now well known that statistical inference for stochastic processes presents special problems not present in the classical i.i.d. situation. The only exception to this coverage is the material presented in Section 3.11, where we discuss maximum likelihood estimation for stochastic processes. Further, we discuss mostly parametric inference throughout the text except in Chapter 7, where the asymptotic theory for von Mises functionals is presented. We have not incorporated the recent work on, for example, the Bootstrap, Jackknife, and Projection pursuit methods, in an attempt to keep the book to a reasonable size and because there are other works in the literature dealing with these topics.

An introduction to important results in the theory of probability and stochastic processes which are useful in the asymptotic theory of inference is given in Chapter 1. No proofs are presented. References at the end of each chapter pertain to the topics discussed in that chapter. Topics from the book may be used for a one-semester graduate course in the asymptotic theory of statistical inference.

It is a pleasure to thank the Indian Statistical Institute for its support and Mr. V. P. Sharma for his excellent typing of the manuscript.

B. L. S. Prakasa Rao

New Delhi, India
October 1986

Contents

Notation

$\xrightarrow{\text{a.s}}$	Convergence almost surely
$\xrightarrow{\text{c}}$	Convergence completely
$\xrightarrow{\mathscr{L}}$	Convergence in law
$\xrightarrow{\text{q.m}}$	Convergence in quadratic mean
$\xrightarrow{p}$	Convergence in probability
$\xrightarrow{r\text{th-m}}$	Convergence in rth mean
$\xrightarrow{\text{w}}$	Convergence weakly
$\xrightarrow{\text{w}}_u$	Convergence weakly uniformly
$\xrightarrow{\mathscr{L}}_u$	Uniform weak convergence
$\xrightarrow{p}_u$	Uniform convergence in probability
$\xrightarrow{\mathscr{L}^*}$	Weak convergence of finite dimensional distributions
$\mu \ll \nu$	μ is absolutely continuous with respect to ν
$\mu \perp \nu$	μ is singular with respect to ν
$\rho(\mu, \nu)$	Affinity between the measures μ and ν
$H(\mu, \nu)$	Hellinger distance between the measures μ and ν
$o_p(1)$	Converges to zero in probability
$O_p(1)$	Bounded in probability
$\mathscr{P}(S)$	Set S occurs in probability
$\langle M, M \rangle$	Quadratic characteristic of a process M
$\langle M_1, M_2 \rangle$	Predictable covariation process of M_1 and M_2
$(Q_n) \lhd (P_n)$	$\{Q_n\}$ is contiguous to $\{P_n\}$
$(Q_n) \triangle (P_n)$	$\{Q_n\}$ and $\{P_n\}$ are completely separable
$\mathscr{L}(\mathbf{X})$	Law of the random vector $\mathbf{X}$
$\mathscr{L}(\mathbf{X} \mid P)$	Law of the random vector $\mathbf{X}$ under the probability measure P

CLT	Central limit theorem
IC	Influence curve
IEE	Integrated error estimator
ISEE	Integrated squared error estimator
GLSE	Generalized least squares estimator
LSE	Least-squares estimator
WLSE	Weighted least-squares estimator
LAN	Local asymptotic normality
AMLE	Approximate maximum likelihood estimator
MLE	Maximum likelihood estimator
QMLE	Quasi-maximum likelihood estimator
SLLN	Strong law of large numbers
SLLR	Similar to logarithm of likelihood ratio
WLLN	Weak law of large numbers
$a \vee b$	max (a, b)
$a \wedge b$	min (a, b)
$\overline{\lim}$	Limit superior
$\underline{\lim}$	Limit inferior
Var	Variance
Cov	Covariance
tr	Trace
∂A	Boundary of a set A
$\bar{A}$	Closure of set A
A^c	Complement of set A
$A \backslash B$	Set of elements of A not in B
$\nabla \mathbf{g}$	Gradient of $\mathbf{g}$
$g^{(i)}(x)$	ith derivative of $g(\cdot)$
$\mathbf{x}'$	Transpose of column vector $\mathbf{x}$
$\in$	Belongs to
$\notin$	Does not belong to
$\not\to$	Does not tend to
#	Number
$\Rightarrow$	Implies
$\Leftrightarrow$	Implies and is implied by
$\mathbf{N}_k(\boldsymbol{\mu}, \boldsymbol{\Gamma})$	k-variate normal distribution with mean vector $\boldsymbol{\mu}$ and covariance matrix $\boldsymbol{\Gamma}$

Asymptotic Theory
of Statistical Inference

CHAPTER 1

Probability and Stochastic Processes

1.1 INTRODUCTION

Let Ω be an abstract space and $\mathscr{B}$ be a collection of subsets of Ω satisfying the following conditions:

(1.1.0) (i) $\Omega \in \mathscr{B}$,

(ii) $A \in \mathscr{B} \Rightarrow A^c \in \mathscr{B}$,

(iii) $A_i \in \mathscr{B}, i \geqslant 1 \Rightarrow \bigcup_{i=1}^{\infty} A_i \in \mathscr{B}$,

where A^c denotes the complement of the set A. $\mathscr{B}$ is said to be a *σ-algebra* of subsets of Ω and $(\Omega, \mathscr{B})$ is called a *measurable space.*

Let P be a set function defined on $(\Omega, \mathscr{B})$ with the following properties:

(1.1.1) (i) $0 \leqslant P(A) \leqslant 1$ for every $A \in \mathscr{B}$,

(ii) $P(\Omega) = 1$,

(iii) $A_i \in \mathscr{B}, A_i \cap A_j = \phi, \quad i, j \geqslant 1$

$$\Rightarrow P\left(\bigcup_{i=1}^{\infty} A_i\right) = \sum_{i=1}^{\infty} P(A_i).$$

P is called a *probability measure* defined on $(\Omega, \mathscr{B})$, and $(\Omega, \mathscr{B}, P)$ is called a *probability space.* The following implications are immediate from the definition of a probability measure P on $(\Omega, \mathscr{B})$:

(1.1.2) (i) $A \in \mathscr{B} \Rightarrow P(A^c) = 1 - P(A)$,

(ii) $A \in \mathscr{B}, B \in \mathscr{B} \Rightarrow P(A \cup B) = P(A) + P(B) - P(A \cap B)$,

(iii) $A_i \in \mathscr{B}, i \geqslant 1 \Rightarrow P\left(\bigcup_{i=1}^{\infty} A_i\right) \leqslant \sum_{i=1}^{\infty} P(A_i)$,

(iv) $A \subset B, A, B \in \mathscr{B} \Rightarrow P(A) \leqslant P(B)$,

(v) $A_n \uparrow A, A_n, A \in \mathscr{B} \Rightarrow P(A_n) \uparrow P(A)$ as $n \to \infty$,

(vi) $A_n \downarrow A, A_n, A \in \mathscr{B} \Rightarrow P(A_n) \downarrow P(A)$ as $n \to \infty$.

Consider a measurable space $(\Omega, \mathscr{B})$. Let X be a real-valued function defined on Ω such that for any Borel set $B \subset R^k$, $\{\omega \in \Omega : X(\omega) \in B\} \in \mathscr{B}$. X is called a *measurable function* or a *random vector*. If $k = 1$, then X is called a *random variable*.

Let $(\Omega, \mathscr{B}, P)$ be a probability space and X be a random variable defined on it. For any real number x, define

(1.1.3) $$F(x) = P[\{\omega : X(\omega) \leqslant x\}] \equiv P[X \leqslant x].$$

$F(\cdot)$ is called the *distribution function* of the random variable X. If $\mathbf{X} = (X_1, \ldots, X_k)$ is a k-dimensional random vector, define

(1.1.4) $$F(\mathbf{x}) = P\{[\omega : X_1(\omega) \leqslant x_1, X_2(\omega) \leqslant x_2, \ldots, X_k(\omega) \leqslant x_k]\}$$
$$= P[X_1 \leqslant x_1, \ldots, X_k \leqslant x_k].$$

$F(\cdot)$ is called the *joint distribution* function of the random vector $\mathbf{X}$. The following properties are easy consequences from the definition of a distribution function of a random variable X in view of (1.1.2):

(1.1.5) (i) $0 \leqslant F(x) \leqslant 1$,

(ii) $F(x) \to 1$ as $x \to +\infty$,

(iii) $F(x) \to 0$ as $x \to -\infty$,

(iv) $F(\cdot)$ is nondecreasing, that is, $F(x) \leqslant F(y)$ if $x \leqslant y$,

(v) $F(\cdot)$ is right continuous that is, $F(y) \to F(x)$ as $y \to x + 0$.

It can be shown that the number of discontinuities of F is at most countable.

Let $\mathscr{F}$ be the family of all distribution functions. It is known that there is one-to-one correspondence between the family $\mathscr{F}$ and the family $\mathscr{M}$ of all probability measures on $(\Omega, \mathscr{B})$. Furthermore, given any function $F(\cdot)$ satisfying properties (i)–(v) listed above, one can construct a probability space $(\Omega, \mathscr{B}, P)$ and a random variable X defined on it such that F is the distribution function of X.

Given a random variable X defined on $(\Omega, \mathscr{B}, P)$, one can study either

$$\int_\Omega X \, dP \tag{1.1.6}$$

or equivalently

$$\int_{-\infty}^{\infty} x \, dF(x), \tag{1.1.7}$$

where the former is the integral of the measurable function X with respect to the measure P and the latter is interpreted as a Lebesgue–Stieljes integral [cf. Halmos (1950)]. It is again known that X is integrable with respect to P if and only if $|X|$ is integrable with respect to P. Denote

$$\int_\Omega X \, dP$$

by EX when it exists. EX is called the *mean*, and EX^r is called the rth *raw moment* of X when it exists. It is easy to see that if EX^2 exists, then $\mu \equiv EX$ exists; $E(X-\mu)^2 = EX^2 - (EX)^2$ is called the *variance* of the random variable X. It is easy to check that if $X \geqslant 0$ and $EX < \infty$, then

$$EX = \int_0^\infty [1 - F(x)] \, dx$$

and, in general, for any random variable X with $E|X| < \infty$,

$$EX = \int_0^\infty [1 - F(x) - F(-x)] \, dx.$$

Suppose $\mathbf{X}$ is a k-dimensional random vector $(X_1, \ldots, X_k)$ defined on $(\Omega, \mathscr{B}, P)$ with joint distribution function $F_{\mathbf{X}}(x_1, \ldots, x_k)$. Clearly

$$F_{X_j}(x) = P(X_j \leqslant x) = \int_{-\infty}^{\infty} \cdots \int_{-\infty}^{x} \cdots \int_{-\infty}^{\infty} dF(x_1, \ldots, x_j, \ldots, x_k) \tag{1.1.8}$$

for any $1 \leqslant j \leqslant k$. $F_{X_j}(x)$ is called the *marginal distribution* function of X_j, and, in general, for any $j_1, \ldots, j_l$,

$$F_{X_{j_1},\ldots,X_{j_l}}(x_1,\ldots,x_l) = P(X_{j_1} \leqslant x_1, \ldots, X_{j_l} \leqslant x_l) \tag{1.1.9}$$

gives the l-dimensional marginal distribution function of $(X_{j_1}, \ldots, X_{j_l})$. Suppose $E|X_j X_l| < \infty$ for $1 \leqslant j, l \leqslant k$. The matrix $\Sigma = ((\sigma_{jl}))$, where

$$\sigma_{jl} = E(X_j X_l) - E(X_j)E(X_l), \qquad 1 \leqslant j, l \leqslant k, \tag{1.1.10}$$

is called the *covariance matrix* (or *dispersion matrix*) of the random vector **X**. It is easy to see that Σ is a nonnegative definite symmetric matrix when it exists.

A distribution function $F(\cdot)$ of a random variable X is called *discrete* if

$$F(x) = \sum_{j: x_j \leqslant x} p_j, \qquad -\infty < x < \infty, \tag{1.1.11}$$

where $p_j > 0$ for all j, $\sum_j p_j = 1$ and $S = \{x_j, j \geqslant 1\}$ is a subset of R. $F(\cdot)$ is called *absolutely continuous* if there exists a measurable function f such that

$$F(x) = \int_{-\infty}^{x} f(y)\,dy, \qquad -\infty < x < \infty. \tag{1.1.12}$$

Clearly $f \geqslant 0$ almost everywhere and

$$\int_{-\infty}^{\infty} f(x)\,dx = 1.$$

Here *almost everywhere* means that the relation holds except possibly for a set of Lebesgue measure zero. $F(\cdot)$ is said to be *singular* if it is continuous and the probability measure corresponding to it is singular with respect to the Lebesgue measure. It is known that every distribution function $F(\cdot)$ can be written $\alpha_1 F_1 + \alpha_2 F_2 + \alpha_3 F_3$, where $\sum_{i=1}^{3} \alpha_i = 1$, $\alpha_i \geqslant 0$, $i = 1, 2, 3$, and F_1, F_2, F_3 are the discrete, absolutely continuous, and singular distribution functions.

It is easy to see that if two random vectors **X** and **Y** defined on a probability space $(\Omega, \mathscr{B}, P)$ have identical distributions, then $g(\mathbf{X})$ and $g(\mathbf{Y})$ will have the same distribution for any Borel-measurable function $g(\cdot)$.

Consider a probability space $(\Omega, \mathscr{B}, P)$. Any set $A \in \mathscr{B}$ is called an *event*. Two events A and B in $\mathscr{B}$ are said to be *independent* if

$$P(A \cap B) = P(A)P(B). \tag{1.1.13}$$

Three events A, B, C in $\mathscr{B}$ are said to be (*mutually*) *independent* if

$$P(A \cap B) = P(A)P(B),$$
$$P(B \cap C) = P(B)P(C),$$
$$P(C \cap A) = P(C)P(A),$$

and

$$P(A \cap B \cap C) = P(A)P(B)P(C).$$

In general, a collection of events A_i, $1 \leqslant i \leqslant k$, are said to be *independent* if, for every finite subset $1 \leqslant i_j \leqslant k, j = 1, \ldots, r$,

$$P(A_{i_1} \cap A_{i_2} \cap \cdots \cap A_{i_r}) = \prod_{j=1}^{r} P(A_{i_j}).$$

A family of events $\{A_\alpha, \alpha \in I\}$ is said be *independent* if every finite subset of the family is independent. Let $\mathscr{B}_i, i = 1, 2$, be sub-σ-algebras of $\mathscr{B}$, that is, $\mathscr{B}_i \subset \mathscr{B}$ and $\mathscr{B}_i$ is a σ-algebra for $i = 1, 2$. $\mathscr{B}_1$ and $\mathscr{B}_2$ are said to be *independent* if

$$P(A \cap B) = P(A)P(B)$$

for any $A \in \mathscr{B}_1$ and $B \in \mathscr{B}_2$. The notion of independence can be extended to any family of σ-algebras in an obvious manner.

Let X_i, $1 \leqslant i \leqslant n$, be a set of random variables defined on a probability space $(\Omega, \mathscr{B}, P)$. They are said to be *independent* if, for every real x_i, $1 \leqslant i \leqslant n$,

$$P(X_1 \leqslant x_1, X_2 \leqslant x_2, \ldots, X_n \leqslant x_n) = \prod_{i=1}^{n} P(X_i \leqslant x_i).$$

In general, a family of random variables $\{X_\alpha, \alpha \in J\}$ is said to form an independent family if *every* finite subfamily constitutes an independent set of random variables. It is clear from the definition given above that, if X_i, $1 \leqslant i \leqslant n$, form an independent family, then

$$F_{\mathbf{X}}(x_1, \ldots, x_n) = F_{X_1}(x_1) \cdots F_{X_n}(x_n), (x_1, \ldots, x_n) \in R^n,$$

where F_{X_i} denotes the distribution function of X_i and $F_{\mathbf{X}}$ denotes the distribution function of $\mathbf{X} = (X_1, \ldots, X_n)$. One can define independence of a set of random vectors in a similar fashion.

A collection of random vectors $\{\mathbf{X}_1, \ldots, \mathbf{X}_n\}$ is said to be *independent and identically distributed* (i.i.d.) if each $\mathbf{X}_i$ has the same distribution and the set $\{\mathbf{X}_1, \ldots, \mathbf{X}_n\}$ are independent as defined above.

Let $\mathbf{X} = (X_1, \ldots, X_k)$ be a random vector defined on a probability space

$(\Omega, \mathscr{B}, P)$. Let

$$\phi(\mathbf{t}) = E[e^{i\mathbf{t}'\mathbf{X}}],$$

where $\mathbf{t} \in R^k$ (here prime denotes the transpose of vector). It is clear that $|\phi(\mathbf{t})| \leqslant 1$ and $\phi(\mathbf{0}) = 1 \cdot \phi(\mathbf{t})$ is called the *characteristic function* of the random vector $\mathbf{X}$. It is well known that there is a one-to-one correspondence between the family of characteristic functions and the family of distribution functions on R^k. Note that $\phi(\mathbf{t})$ exists for any $\mathbf{t} \in R^k$. Furthermore, $\phi(\mathbf{t}) = \overline{\phi(-\mathbf{t})}$ (where $\bar{z}$ denotes the complex conjugate of z) and $\phi(\cdot)$ is uniformly continuous. Let

$$\psi(\mathbf{t}) = E[e^{\mathbf{t}'\mathbf{X}}].$$

$\psi(\mathbf{t})$ is called the *moment generating function* of $\mathbf{X}$. $\psi(\mathbf{t})$ need not be finite for all $\mathbf{t}$ in general. Properties of characteristic functions and their applications are given, for instance, in Chow and Teicher (1978). It is easy to see that, if $\mathbf{X}_i, 1 \leqslant i \leqslant n$, are independent random vectors with characteristic functions $\phi_i(\mathbf{t})$, then $\sum_{i=1}^n \mathbf{X}_i$ has the characteristic function $\prod_{i=1}^n \phi_i(\mathbf{t})$.

Before we conclude this section, we state a result known as an *inversion formula*, which gives the correspondence between characteristic functions and distribution functions. Even though this result is not of use for computational purposes, it is useful in the study of limit theorems in probability.

Proposition 1.1.1 (Inversion Formula). If X is a random variable with characteristic function ϕ, then

$$\lim_{k\to\infty} \frac{1}{2\pi}\int_{-k}^{k} \frac{e^{-ita} - e^{-itb}}{it}\phi(t)\,dt = P[a < X < b] + \frac{P(X = a) + P(X = b)}{2}$$

for $-\infty < a < b < \infty$.

As a corollary, we have the following:

Proposition 1.1.2. If the characteristic function ϕ of a random variable X is absolutely integrable, then the distribution function F of X is absolutely continuous with a bounded continuous density f given by

$$f(x) = \frac{1}{2\pi}\int_{-\infty}^{\infty} e^{-itx}\phi(t)\,dt.$$

1.2 CONVERGENCE OF RANDOM VARIABLES

Let $(\Omega, \mathscr{B}, P)$ be a probability space and $\{X_n, n \geqslant 1\}$ be a sequence of random variables defined on it. We now define several types of convergence for the sequence $\{X_n\}$ and study their interrelations.

Definition 1.2.1. A sequence $\{X_n\}$ is said to *converge in probability* to a random variable X if, for every $\varepsilon > 0$,

$$P(|X_n - X| > \varepsilon) \to 0 \quad \text{as} \quad n \to \infty. \tag{1.2.1}$$

We denote this convergence as $X_n \xrightarrow{p} X$ as $n \to \infty$.

Definition 1.2.2. A sequence $\{X_n\}$ is said to *converge almost surely* to a random variable X if there exists a measurable set $N \in \mathscr{B}$ such that $P(N) = 0$ and

$$X_n(\omega) \to X(\omega) \quad \text{as} \quad n \to \infty \tag{1.2.2}$$

for every $\omega \notin N$.

We denote this convergence as $X_n \xrightarrow{\text{a.s.}} X$ as $n \to \infty$.

Definition 1.2.3. A sequence $\{X_n\}$ is said to *converge completely* to a random variable X if, for every $\varepsilon > 0$,

$$\sum_{n=1}^{\infty} P(|X_n - X| > \varepsilon) < \infty. \tag{1.2.3}$$

We denote this convergence as $X_n \xrightarrow{c} X$ as $n \to \infty$.

Let $A_n = [|X_n - X| \leqslant \varepsilon]$. One can define the convergence concepts discussed above in an alternative fashion:

(1.2.4) (i) $X_n \xrightarrow{p} X$ if $P(A_n) \to 1$ as $n \to \infty$ for every $\varepsilon > 0$,

(ii) $X_n \xrightarrow{\text{a.s.}} X$ if $P\left[\bigcap_{m=n}^{\infty} A_m\right] \to 1$ as $n \to \infty$ for every $\varepsilon > 0$,

(iii) $X_n \xrightarrow{c} X$ if $\sum_{n=1}^{\infty} P[A_n^c] < \infty$ for every $\varepsilon > 0$.

It is clear from the alternative definitions given above that the following implications hold:

$$X_n \xrightarrow{c} X \Rightarrow X_n \xrightarrow{\text{a.s.}} X \Rightarrow X_n \xrightarrow{p} X. \tag{1.2.5}$$

The converse implications are not true. For counterexamples, see Chung (1974), Loeve (1963), and Lukacs (1975).

Suppose $\{X_n\}$ is a sequence of random variables defined on $(\Omega, \mathscr{B}, P)$ such that $E|X_n|^r < \infty$ for some $r > 0$.

Definition 1.2.4. A sequence $\{X_n\}$ is said to *converge in rth mean* to a random variable X if

$$E|X_n - X|^r \to 0 \quad \text{as} \quad n \to \infty. \tag{1.2.6}$$

We denote this convergence as $X_n \xrightarrow{r\text{th}-\text{m.}} X$ as $n \to \infty$. If $r = 2$, we say that $X_n \xrightarrow{\text{q.m.}} X$ as $n \to \infty$.

We remark that if $X_n \xrightarrow{r\text{th}-\text{m.}} X$, it implies a priori that $E|X|^r < \infty$. It can be shown that

$$X_n \xrightarrow{r\text{th}-\text{m.}} X \Rightarrow X_n \xrightarrow{p} X. \tag{1.2.7}$$

This follows from an application of the Chebyshev inequality. The converse implication is not true in general, however. In fact, even if $X_n \xrightarrow{\text{a.s.}} X$, it does not follow that $X_n \xrightarrow{r\text{th}-\text{m.}} X$. For a counterexample, see Lukacs (1975). Furthermore, $X_n \xrightarrow{\text{q.m.}} X \not\Rightarrow X_n \xrightarrow{\text{a.s.}} X$ [see Lukacs (1975)]. However, the following result is true.

If $\sum_{n=1}^{\infty} E|X_n - X|^r < \infty$, then $X_n \xrightarrow{\text{a.s.}} X$. For proof, see Lukacs (1975). If $X_n \xrightarrow{p} X$ and the X_n are dominated by a random variable Y such that $E|Y|^r < \infty$, then $X_n \xrightarrow{r\text{th}-\text{m.}} X$. This follows from the dominated convergence theorem [cf. Halmos (1950)].

The convergence concepts discussed up to now deal with random variables directly. Let $\{\mathbf{X}_n\}$ be a sequence of random vectors with the corresponding distribution functions $\{F_n\}$.

Definition 1.2.5. A sequence $\{\mathbf{X}_n\}$ is said to *converge in law* (*distribution*) to $\mathbf{X}$

with distribution function F if

$$\lim_{n\to\infty} F_n(\mathbf{x}) = F(\mathbf{x})$$

at all continuity points $\mathbf{x}$ of $F(\cdot)$.

We denote this convergence by $\mathbf{X}_n \xrightarrow{\mathscr{L}} \mathbf{X}$. In such an event, F_n is said to *converge weakly* to F $(F_n \xrightarrow{w} F)$.

It is easy to see that Definitions 1.2.1–1.2.4 can be extended easily to a sequence $\{\mathbf{X}_n\}$ of random vectors by interpreting $|\cdot|$ as the norm in the appropriate Euclidean space under consideration.

We now state some propositions concerning convergence of sequences of random variables. For proofs, see Chow and Teicher (1978).

Proposition 1.2.1. $X_n \xrightarrow{\text{a.s.}} X$ iff $\sup_{j\geqslant n}|X_j - X| \xrightarrow{p} 0$ as $n\to\infty$.

Proposition 1.2.2. If $X_n \xrightarrow{\text{a.s.}} X$, then $g(X_n) \xrightarrow{\text{a.s.}} g(X)$ as $n\to\infty$ for any continuous function $g(\cdot)$.

Proposition 1.2.3. A sequence X_n converges in probability to some random variable X iff for every $\varepsilon > 0$

$$P(|X_m - X_n| > \varepsilon) \to 0 \quad \text{as} \quad n\to\infty \quad \text{and} \quad m\to\infty.$$

Proposition 1.2.4. $X_n \xrightarrow{p} X$ if and only if every subsequence of $\{X_n\}$ has another subsequence converging almost surely to X.

Proposition 1.2.5. If $X_n \xrightarrow{p} X$, then $g(X_n) \xrightarrow{p} g(X)$ for any continuous function $g(\cdot)$.

The next definition is useful in the study of convergence of moments of random variables.

Definition 1.2.6. A sequence of random variables $\{X_n\}$ is said to be *uniformly integrable* if, for every $\varepsilon > 0$, there exists a $\delta > 0$ such that

$$P(A) < \delta \Rightarrow \sup_{n\geqslant 1} \int_A |X_n|\, dP < \varepsilon$$

and

$$\sup_{n\geqslant 1} E|X_n| < \infty.$$

It is easy to see that if $|X_n| \leqslant Y$ and $EY < \infty$, then $\{X_n\}$ is uniformly integrable.

Proposition 1.2.6. If $X_n \xrightarrow{p} X$ and $\{|X_n|^r, n \geqslant 1\}$ is uniformly integrable, then $X_n \xrightarrow{r\text{th}-\text{m.}} X$. Conversely, if $E|X_n|^r < \infty, n \geqslant 1$, and $X_n \xrightarrow{r\text{th}-\text{m.}} X$, then $E|X|^r < \infty$, $X_n \xrightarrow{p} X$, and $\{|X_n|^r, n \geqslant 1\}$ is uniformly integrable.

Proposition 1.2.7 (Monotone Convergence Theorem). If $X_n \geqslant 0$ and $X_n \uparrow X$ a.s., then $EX_n \uparrow EX$.

Proposition 1.2.8 (Fatou's Lemma). If $X_n \geqslant Y$ and $E|Y| < \infty$ then

$$\varliminf_n \int X_n \, dP \geqslant \int \varliminf_n X_n \, dP.$$

Proposition 1.2.9 (Dominated Convergence Theorem). If $|X_n| \leqslant Y$ where $E|Y| < \infty$, $X_n \to X$, then $E|X_n - X| \to 0$ and, in particular, $EX_n \to EX$.

Proposition 1.2.10. If $\{X_n, n \geqslant 1\}$ and X are nonnegative random variables with $E(X_n) < \infty$ and $E(X) < \infty$ and if $X_n \xrightarrow{p} X$, then $EX_n \to EX$ iff $E|X_n - X| \to 0$, which in turn holds iff $\{X_n, n \geqslant 1\}$ is uniformly integrable.

Proposition 1.2.11. If $\{X_n, n \geqslant 1\}$ is a sequence of random variables with $E|X_n|^r < \infty$ and $X_n \xrightarrow{r\text{th}-\text{m.}} X$ for $r > 0$, then $E|X|^r < \infty$ and $E|X_n|^r \to E|X|^r$.

The next step is to study the relationship between the concepts of convergence in probability and convergence in law (distribution). A fundamental result that we shall have several occasions to use in this context is due to Slutsky (1925).

Proposition 1.2.12 (Slutsky's Lemma). Let $\{X_n\}$ and $\{Y_n\}$ be sequences of random variables defined on a probability space $(\Omega, \mathscr{B}, P)$ such that $X_n \xrightarrow{\mathscr{L}} X$ and $Y_n \xrightarrow{p} c$, where $-\infty < c < \infty$. Then

(i) $X_n + Y_n \xrightarrow{\mathscr{L}} X + c$,

(ii) $X_n Y_n \xrightarrow{\mathscr{L}} cX$,

(iii) $X_n / Y_n \xrightarrow{\mathscr{L}} X/c \quad \text{if} \quad c \neq 0$.

Proposition 1.2.13. $X_n \xrightarrow{p} X \Rightarrow X_n \xrightarrow{\mathscr{L}} X$ and conversely, if X is a degenerate random variable, then $X_n \xrightarrow{\mathscr{L}} X \Rightarrow X_n \xrightarrow{p} X$.

Proposition 1.2.14. If $X_n - Y_n \xrightarrow{p} 0$ and $Y_n \xrightarrow{\mathscr{L}} Y$, then $X_n \xrightarrow{\mathscr{L}} Y$.

For proofs of Propositions 1.2.12–1.2.14, see, for instance, Chow and Teicher (1978).

The next proposition is useful in connecting the concepts of convergence in law and almost sure convergence.

Proposition 1.2.15. Suppose $\{X_n\}$ and X are random variables defined on a probability space $(\Omega, \mathscr{B}, P)$ such that $X_n \xrightarrow{\mathscr{L}} X$ as $n \to \infty$. Then there exists a probability space $(\Omega', \mathscr{B}', P')$ and a sequence of random variables $\{Y_n\}$ and $\{Y\}$ defined on it such that X_n and Y_n have the same distribution and X and Y have identical distributions. Furthermore,

$$Y_n \xrightarrow{\text{a.s.}} Y \quad \text{as} \quad n \to \infty.$$

Proof. Let $\Omega' = [0, 1]$, $\mathscr{B}'$ be the σ-algebra of Borel subsets of $[0, 1]$, and P' be the Lebesgue measure on $[0, 1]$. For any distribution function $F(\cdot)$, define

$$F^{-1}(x) = \inf\{y : F(y) \geqslant x\}.$$

It is easy to see that $F(Y) \geqslant x$ iff $y \geqslant F^{-1}(x)$. For any $t \in [0, 1]$, define

$$Y_n(t) = F_{X_n}^{-1}(t) \quad \text{and} \quad Y(t) = F_X^{-1}(t).$$

Clearly

$$\begin{aligned} F_Y(y) &= P'[\{t : Y(t) \leqslant y\}] \\ &= P'[\{t : t \leqslant F_X(y)\}] \\ &= F_X(y), \end{aligned}$$

which shows that X and Y are identically distributed. Similarly we obtain that X_n and Y_n have the same distribution. Furthermore, $\{t : Y_n(t) \nrightarrow Y(t)\}$ is utmost countable [see Serfling (1980, p. 21)]. Suppose $0 < t_0 < 1$ such that $Y_n(t_0) \equiv F_{X_n}^{-1}(t_0)$ does not converge to $F_X^{-1}(t_0) \equiv Y(t_0)$ as $n \to \infty$. Then it can be checked that, for any $\varepsilon > 0$, $F_{X_n}^{-1}(t_0) > F_X^{-1}(t_0) + \varepsilon$ infinitely often. Therefore

$F_{X_n}(F_X^{-1}(t_0)+\varepsilon)<t_0$ infinitely often. But $F_{X_n}\xrightarrow{w}F_X$. Hence $F_X(F_X^{-1}(t_0)+\varepsilon)\leqslant t_0$. But $F_X(F_X^{-1}(t_0))\geqslant t_0$. Therefore $t_0=F_X(F_X^{-1}(t_0))$ and $F(x)=t_0$ for $x\in[F_X^{-1}(t_0),\ F_X^{-1}(t_0)+\varepsilon]$. In other words F_X does not have a jump over $[F_X^{-1}(t_0),\ F_X^{-1}(t_0)+\varepsilon]$. Since there can be utmost countably many such intervals, it follows that the set $\{t: Y_n(t)\nrightarrow Y(t)\}$ is countable. Hence

$$Y_n\xrightarrow{\text{a.s.}}Y \quad \text{as} \quad n\to\infty.$$

An immediate application of this result is the following proposition.

Proposition 1.2.16. Let $\{X_n\}$ and X be random variables defined on a probability space $(\Omega,\mathscr{B},P)$. Suppose $g(\cdot)$ is a Borel-measurable function and the set of discontinuities of $g(\cdot)$ has probability zero with respect to the distribution of X. Then

$$\text{(i)}\quad X_n\xrightarrow{\text{a.s.}}X\Rightarrow g(X_n)\xrightarrow{\text{a.s.}}g(X),$$

$$\text{(ii)}\quad X_n\xrightarrow{p}X\Rightarrow g(X_n)\xrightarrow{p}g(X),$$

$$\text{(iii)}\quad X_n\xrightarrow{\mathscr{L}}X\Rightarrow g(X_n)\xrightarrow{\mathscr{L}}g(X).$$

Proof. We prove part (iii). (i) and (ii) are left as exercises for the reader. Since $X_n\xrightarrow{\mathscr{L}}X$, there exist random variables $\{Y_n\}$ and Y defined on possibly another probability space $(\Omega',\mathscr{B}',P')$ such that $\mathscr{L}(X_n)=\mathscr{L}(Y_n)$, $\mathscr{L}(X)=\mathscr{L}(Y)$, and $Y_n\xrightarrow{\text{a.s.}}Y$ as $n\to\infty$, where $\mathscr{L}(X)$ denotes the distribution of X. Hence, it is clear that $g(Y_n)\xrightarrow{\text{a.s.}}g(Y)$. This in turn implies that $g(Y_n)\xrightarrow{\mathscr{L}}g(Y)$ by Proposition 1.2.12. Hence $g(X_n)\xrightarrow{\mathscr{L}}g(X)$.

1.3 LIMIT THEOREMS

We shall now state some limit theorems in probability useful in the asymptotic theory of statistics. For more details and proofs, see Loeve (1963) or Chow and Teicher (1978).

Proposition 1.3.1 (Borel–Cantelli Lemma). Let $\{A_n, n\geqslant 1\}$ be a sequence of events with $\sum_{n=1}^{\infty}P(A_n)<\infty$. Then $P(A_n$ occurs infinitely often$)=0$. Conversely, if the events $\{A_n, n \geqslant 1\}$ are independent and $\sum_{n=1}^{\infty} P(A_n) = \infty$, then $P(A_n$ occurs infinitely often$) = 1$.

As a consequence of this result, it follows that if $\{A_n, n \geqslant 1\}$ are independent events, then $P(A_n$ occurs infinitely often$) = 0$ or 1 according as $\sum_{n=1}^{\infty} P(A_n) < \infty$ or $\sum_{n=1}^{\infty} P(A_n) = \infty$.

Let $\{X_n, n \geqslant 1\}$ be a sequence of random variables defined on a probability space $(\Omega, \mathscr{F}, P)$. Let $\mathscr{F}_n$ denote the smallest σ-algebra generated by the sequence $\{X_j, j \geqslant n\}$. The σ-algebra $\bigcap_{n=1}^{\infty} \mathscr{F}_n$ is called the *tail σ-algebra* for the random sequence $\{X_n, n \geqslant 1\}$. It is said to be *trivial* if it consists of sets E with $P(E) = 0$ or 1.

Proposition 1.3.2 (Kolmogorov's Zero–One Law). If $\{X_n, n \geqslant 1\}$ is an independent sequence of random variables, then $P(E) = 0$ or 1 for every $E \in \bigcap_{n=1}^{\infty} \mathscr{F}_n$.

One immediate application of this result is to the study of convergence of infinite series of independent random variables. Let

$$S = \left\{ \omega \colon \sum_{n=1}^{\infty} X_n(\omega) < \infty \right\},$$

where $\{X_n, n \geqslant 1\}$ are independent random variables. Clearly $S \in \bigcap_{n=1}^{\infty} \mathscr{F}_n$, and hence $P(S) = 0$ or 1 by Proposition 1.3.2. The next result gives necessary and sufficient conditions so that $P(S) = 1$.

Proposition 1.3.3 (Kolmogorov's Three-Series Theorem). Let $\{X_n, n \geqslant 1\}$ be a sequence of independent random variables defined on a probability space $(\Omega, \mathscr{B}, P)$. Then $\sum_{n=1}^{\infty} X_n < \infty$ a.s. iff there exists $c > 0$ such that

$$\text{(i)} \quad \sum_{n=1}^{\infty} P(|X_n| \geqslant c) < \infty,$$

$$\text{(ii)} \quad \sum_{n=1}^{\infty} E(X_n^c) < \infty,$$

$$\text{(iii)} \quad \sum_{n=1}^{\infty} \operatorname{Var}(X_n^c) < \infty,$$

where

$$X_n^c = \begin{cases} X_n & \text{if } |X_n| \leqslant c \\ 0 & \text{if } |X_n| > c. \end{cases}$$

Remark. Conditions (i)–(iii) of Proposition 1.3.3 hold for every $c > 0$ in case $\sum_{n=1}^{\infty} X_n < \infty$ a.s.

As an immediate consequence of Proposition 1.3.3, we have the following result.

Proposition 1.3.4. Suppose $\{X_n, n \geqslant 1\}$ are independent random variables such that $\sum_{n=1}^{\infty} E|X_n|^{\alpha_n} < \infty$ for some $0 < \alpha_n \leqslant 2$ with $EX_n = 0$ if $1 \leqslant \alpha_n \leqslant 2$. Then $\sum_{n=1}^{\infty} X_n$ converges a.s.

Laws of large numbers play an important role in the study of asymptotic behavior of statistics. The following results are well known for i.i.d. sequences of random variables $\{X_n\}$.

Proposition 1.3.5 (Strong Law of Large numbers) (SLLN). Let $\{X_n, n \geqslant 1\}$ be a sequence of i.i.d. random variables with $E(X_1) < \infty$. Then

$$\frac{S_n}{n} = \frac{X_1 + \cdots + X_n}{n} \xrightarrow{\text{a.s.}} EX_1.$$

A sequence $\{X_n, n \geqslant 1\}$ of random variables is said to obey the *weak law of large numbers* (WLLN) if there exists sequence of constants a_n and b_n, $0 < b_n \uparrow \infty$, such that

$$\frac{S_n - a_n}{b_n} \xrightarrow{p} 0.$$

It is clear from Proposition 1.3.5 that $\{X_n, n \geqslant 1\}$ obeys WLLN with $a_n = nEX_1$ and $b_n = n$ in the case of i.i.d. random variables with finite expectation.

A consequence of the three-series theorem and hence of Proposition 1.3.4 is the following SLLN.

Proposition 1.3.6. If $\{X_n\}$ are independent random variables satisfying

$$\sum_{n=1}^{\infty} \frac{E|X_n|^{\alpha_n}}{n^{\alpha_n}} < \infty$$

for some $0 < \alpha_n \leqslant 2$, where $EX_n = 0$ if $1 \leqslant \alpha_n \leqslant 2$, then

$$\frac{S_n}{n} \xrightarrow{\text{a.s.}} 0.$$

A more precise behavior of the partial sums $S_n = X_1 + \cdots + X_n$ of independent random variables is given by the law of iterated logarithm stated below.

Proposition 1.3.7. Let $X_1, X_2, \ldots$ be i.i.d. random variables with $E(X_1) = 0$ and $0 < E(X_1^2) = \sigma^2 < \infty$. Let $S_n = X_1 + \cdots + X_n$. Then

$$\overline{\lim_{n\to\infty}} \frac{S_n}{\sigma(2n \log \log n)^{1/2}} = 1 \quad \text{a.s}$$

and

$$\underline{\lim_{n\to\infty}} \frac{S_n}{\sigma(2n \log \log n)^{1/2}} = -1 \quad \text{a.s.}$$

In particular, the relations given above imply that

$$P[|S_n| \geqslant (1 - \varepsilon)\sigma(2n \log \log n)^{1/2} \text{ infinitely often}] = 1$$

and

$$P[|S_n| \geqslant (1 + \varepsilon)\sigma(2n \log \log n)^{1/2} \text{ infinitely often}] = 0$$

for every $\varepsilon > 0$.

Proposition 1.3.8. Let $X_1, X_2, \ldots$ be independent random variables with $E(X_i) = 0$ and $\text{Var}(X_i) = \sigma_i^2 < \infty$. Let $S_n = X_1 + \cdots + X_n$ and $s_n^2 = \sigma_1^2 + \cdots + \sigma_n^2$. Suppose $s_n \to \infty$ as $n \to \infty$ and

$$|X_n| = o\left(\frac{s_n}{(\log \log s_n)^{1/2}}\right) \quad \text{a.s.}$$

Then

$$\overline{\lim_{n\to\infty}} \frac{S_n}{s_n(2 \log \log s_n)^{1/2}} = 1 \quad \text{a.s.}$$

and

$$\underline{\lim_{n\to\infty}} \frac{S_n}{s_n(2 \log \log s_n)^{1/2}} = -1 \quad \text{a.s.}$$

For proofs of Proposition 1.3.7 and 1.3.8, see Loeve (1963).

The following theorem deals with almost complete convergence for the sum of a random number of independent and identically distributed random variables.

Proposition 1.3.9 (Szynal, 1972). Let $\{X_n, n \geqslant 1\}$ be a sequence of i.i.d. random variables with $E(X_1) = \mu$ and $\text{Var } X_1 = \sigma^2 < \infty$. Let $\{N_n, n \geqslant 1\}$ be a sequence of positive integer-valued random variables such that

$$\sum_{n=1}^{\infty} P\left(\left|\frac{N_n}{n} - L\right| \geqslant \varepsilon\right) < \infty$$

for every $\varepsilon > 0$, where L is a random variables with values from an interval (c, d) with $0 < c < d < \infty$. Then

$$\sum_{n=1}^{\infty} P\left(\left|\frac{S_{N_n}}{N_n} - \mu\right| \geqslant \varepsilon\right) < \infty,$$

where $S_n = X_1 + \cdots + X_n$.

In contrast to the laws of large numbers, which deal with asymptotic behavior in the sense of either convergence in probability or almost sure convergence of averages of sequences of random variables, we now present some results dealing with convergence in law of normalized averages of sequences of random variables. Before we present these results, we first state some theorems concerning weak convergence of distribution functions.

Proposition 1.3.10 (Levy–Cramer Continuity Theorem). Let $\{F_n, n \geqslant 1\}$ be a sequence of distribution functions with the corresponding characteristic functions $\{\phi_n, n \geqslant 1\}$. If F_n converges weakly to F, then $\phi_n(t)$ tends to $\phi(t)$ for every real t, where $\phi(\cdot)$ is the characteristic function of F. Conversely, if $\phi_n(t) \to \phi(t)$ for every real t, where $\phi(\cdot)$ is a function continuous at 0, then $\phi(\cdot)$ is a characteristic function of a distribution function F and F_n converges to F weakly.

Proposition 1.3.11 (Helly–Bray Theorem). Let $\{F_n, n \geqslant 1\}$ and F be distribution functions. Then F_n converges weakly to F if and only if for every bounded continuous function $g(\cdot)$,

$$\int_{-\infty}^{\infty} g \, dF_n \to \int_{-\infty}^{\infty} g \, dF.$$

Proposition 1.3.12. Suppose $F_n \xrightarrow{w} F$ or equivalently $X_n \xrightarrow{\mathscr{L}} X$. Then, for any $\alpha > 0$, $E|X_n|^\alpha \to E|X|^\alpha$ iff $\{|X_n|^\alpha\}$ are uniformly integrable.

Proposition 1.3.10 gives sufficient conditions for the weak convergence of F_n to F via the characteristic functions. The next theorem gives sufficient conditions for $F_n \xrightarrow{w} F$ via their moments.

Proposition 1.3.13 (Frechet–Shohat). If F_n is a sequence of distribution functions such that

$$\mu_{n,k} = \int_{-\infty}^{\infty} x^k \, dF_n(x) \to \mu_k < \infty, \qquad k \geqslant 1,$$

and, further, if $\{\mu_k\}$ are the moments of a *uniquely* determined distribution function F, then

$$F_n \xrightarrow{\text{w}} F.$$

We have discussed up to now convergence in law (weak convergence) of one-dimensional random variables (distribution functions). One can study the convergence in law for multivariate random vectors via the univariate case in the light of the following proposition.

Proposition 1.3.14 (Cramer–Wold). A sequence of k-dimensional random vectors $\mathbf{X}_n = (X_{n1}, \ldots, X_{nk})$ converges in law to a k-dimensional random vector $\mathbf{X} = (X_1, \ldots, X_k)$ if and only if for every $\boldsymbol{\lambda}' = (\lambda_1, \ldots, \lambda_k) \in R^k$,

$$\boldsymbol{\lambda}' X_n = \sum_{i=1}^{k} \lambda_i X_{ni} \xrightarrow{\mathscr{L}} \sum_{i=1}^{k} \lambda_i X_i = \boldsymbol{\lambda}' \mathbf{X}.$$

The next theorem deals with the uniform convergence for a sequence of distribution functions.

Proposition 1.3.15 (Polya). If $\{F_n\}$ is a sequence of distributions functions such that $F_n \xrightarrow{\text{w}} F$ and F is a continuous distribution function, then

$$\sup_x |F_n(x) - F(x)| \to 0 \quad \text{as} \quad n \to \infty.$$

Let $X_1, X_2, \ldots, X_n$ be i.i.d. with distribution function F. Define

$$I(X_i, x) = \begin{cases} 1 & \text{if} \quad X_i \leqslant x \\ 0 & \text{if} \quad X_i > x \end{cases}$$

and

$$F_n(x) = n^{-1} \sum_{i=1}^{n} I(X_i, x)$$

for any fixed $x \in R$. F_n is called the *empirical distribution function* based on the sample $(X_1,\ldots,X_n)$. It is clear that

$$F_n(x) \xrightarrow{\text{a.s.}} F(x)$$

by the SLLN (Proposition 1.3.5).

Proposition 1.3.16 (Glivenko–Cantelli). If $\{X_n, n \geqslant 1\}$ are i.i.d. random variables and F_n is the empirical distribution function based on $X_1,\ldots,X_n$, then

$$\sup_x |F_n(x) - F(x)| \xrightarrow{\text{a.s.}} 0.$$

The following proposition deals with the L_1 convergence of densities when they exist.

Proposition 1.3.17 (Scheffé). Suppose $f_n(x)$ is a sequence of densities with respect to a σ-finite measure μ converging to a density $f(x)$ a.e. Then

$$\lim_{n\to\infty} \int_{-\infty}^{\infty} |f_n(x) - f(x)|\,\mu(dx) = 0.$$

For any two probability measures P_1 and P_2 on $(\Omega, \mathscr{B})$, define

$$\|P_1 - P_2\| = \int_\Omega \left|\frac{dP_1}{d\mu} - \frac{dP_2}{d\mu}\right| d\mu, \tag{1.3.0}$$

where $P_i \ll \mu$, $i = 1, 2$ and μ is a σ-finite measure on $(\Omega, \mathscr{B})$. It is easy to see that

$$\|P_1 - P_2\| = \int_\Omega \left|\frac{dP_1}{d\mu} - \frac{dP_2}{d\mu}\right| d\mu = 2 \sup_{B\in\mathscr{B}} |P_1(B) - P_2(B)|. \tag{1.3.1}$$

It follows now from Proposition 1.3.17 that

$$\lim_{n\to\infty} \sup_{B\in\mathscr{B}} |Q_n(B) - Q(B)| = 0$$

where Q_n and Q are the probability measures corresponding to the densities f_n and f, respectively.

For any two probability measures P_1 and P_2 on $(\Omega, \mathscr{B})$, define

(1.3.2) $$\rho(P_1, P_2) = \int_\Omega \left(\frac{dP_1}{d\mu} \frac{dP_2}{d\mu} \right)^{1/2} d\mu$$

where $P_i \ll \mu, i = 1, 2$. $\rho(\cdot, \cdot)$ is called the *affinity* between the probability measures P_1 and P_2. Note that $\rho(P_1, P_2)$ does not depend on the measure μ dominating P_1 and P_2.

Proposition 1.3.18. (i) $0 \leqslant \rho(P_1, P_2) \leqslant 1$.

(ii) $\rho(P_1, P_2) = 1$ if and only if $P_1 = P_2$.

(iii) $\rho(P_1, P_2) = 0$ if and only if $P_1 \perp P_2$.

Proof. Applying the Cauchy–Schwarz inequality, it follows that

$$\begin{aligned}
\rho^2(P_1, P_2) &= \left\{ \int_\Omega \left(\frac{dP_1}{d\mu} \frac{dP_2}{d\mu} \right)^{1/2} d\mu \right\}^2 \\
&\leqslant \int_\Omega \frac{dP_1}{d\mu} d\mu \int_\Omega \frac{dP_2}{d\mu} d\mu = 1,
\end{aligned}$$

equality occurring if and only if $(dP_1/d\mu)^{1/2}$ and $(dP_2/d\mu)^{1/2}$ are linearly related. Since $dP_1/d\mu$ and $dP_2/d\mu$ are both densities, equality occurs if and only if $dP_1/d\mu = dP_2/d\mu$ a.e.$[\mu]$, which in turn holds iff $P_1 = P_2$. It is clear that $\rho(P_1, P_2) \geqslant 0$. Suppose $P_1 \perp P_2$. Then there exists a set $M \in \mathscr{B}$ such that $P_2(M) = 0$ and $P_1(M) = 1$. Hence

$$\begin{aligned}
0 \leqslant \rho(P_1, P_2) &= \int_M \left(\frac{dP_1}{d\mu} \frac{dP_2}{d\mu} \right)^{1/2} d\mu + \int_{M^c} \left(\frac{dP_1}{d\mu} \frac{dP_2}{d\mu} \right)^{1/2} d\mu \\
&= \int_\Omega I_M \left(\frac{dP_1}{d\mu} \frac{dP_2}{d\mu} \right)^{1/2} d\mu + \int_\Omega I_{M^c} \left(\frac{dP_1}{d\mu} \frac{dP_2}{d\mu} \right)^{1/2} d\mu \\
&\leqslant \left(\int_\Omega \frac{dP_1}{d\mu} d\mu \right)^{1/2} \left(\int_\Omega I_M \frac{dP_2}{d\mu} d\mu \right)^{1/2} \\
&\quad + \left(\int_\Omega \frac{dP_2}{d\mu} d\mu \right)^{1/2} \left(\int_\Omega I_{M^c} \frac{dP_1}{d\mu} d\mu \right)^{1/2} \\
&= 0,
\end{aligned}$$

where I_M is the indicator function of the set M. Therefore $\rho(P_1, P_2) = 0$ if $P_1 \perp P_2$. Conversely, suppose $\rho(P_1, P_2) = 0$. Clearly $(dP_1/d\mu \, dP_2/d\mu) = 0$ a.e.$[\mu]$ from the definition of $\rho(\cdot,\cdot)$. Hence

$$\frac{dP_1}{d\mu} > 0 \Rightarrow \frac{dP_2}{d\mu} = 0 \quad \text{a.e.}[\mu]$$

and

$$\frac{dP_2}{d\mu} > 0 \Rightarrow \frac{dP_1}{d\mu} = 0 \quad \text{a.e.}[\mu].$$

Therefore $P_1 \perp P_2$.

We shall see later additional applications of these concepts.

The *Hellinger distance* $H(P_1, P_2)$ between two probability measures P_1 and P_2 is defined by

$$H^2(P_1, P_2) = 2[1 - \rho(P_1, P_2)] = \int \left[\left(\frac{dP_1}{d\mu} \right)^{1/2} - \left(\frac{dP_2}{d\mu} \right)^{1/2} \right]^2 d\mu. \tag{1.3.3}$$

We rewrite $H^2(P_1, P_2)$

$$\int ((dP_1)^{1/2} - (dP_2)^{1/2})^2$$

symbolically as it does not depend on the dominating measure μ. It can be checked that

$$\| P_1 - P_2 \| = 2[1 - \| P_1 \wedge P_2 \|] \tag{1.3.4}$$

where $\| P_1 - P_2 \|$ is the L_1 norm of $P_1 - P_2$ and $P_1 \wedge P_2$ denotes the measure whose density is $\min(dP_1/d\mu, dP_2/d\mu)$ with respect to μ. Furthermore,

$$\begin{aligned} H^2(P_1, P_2) \leqslant \| P_1 - P_2 \| &\leqslant H(P_1, P_2)(4 - H^2(P_1, P_2))^{1/2} \\ &\leqslant 2H(P_1, P_2). \end{aligned} \tag{1.3.5}$$

If $P = \prod_j P_j$ and $Q = \prod_j Q_j$, then

$$H^2(P, Q) = 2\{1 - \textstyle\prod_j [1 - \frac{1}{2} H^2(P_j, Q_j)\}. \tag{1.3.6}$$

Note that $P_1 \perp P_2$ if and only if $\| P_1 - P_2 \| = 2$, which holds if and only if $H^2(P_1, P_2) = 2$.

Before we conclude this section, we introduce another function π defined on sets of probability measures over a measurable space $(\Omega, \mathscr{B})$. An application of this function is given in Section 3.2. Let Φ denote the space of all test functions ϕ defined on $(\Omega, \mathscr{B})$, that is, ϕ is measurable and $0 \leqslant \phi \leqslant 1$ for all $\omega \in \Omega$. For any test function ϕ, define

$$\pi(A, B; \phi) = \sup_{P \in A, Q \in B} \left[\int (1 - \phi)\, dP + \int \phi\, dQ \right] \tag{1.3.7}$$

for any two sets A, B of probability measures over $(\Omega, \mathscr{B})$. In statistical language, this number represents the maximum of the sum of the probabilities of errors corresponding to a test function ϕ. Let

$$\pi(A, B) = \inf_{\phi} \pi(A, B; \phi). \tag{1.3.8}$$

If A and B are singletons consisting of P and Q, denote $\pi(A, B)$ by $\pi(P, Q)$. It can be checked that $\pi(P, Q)$ is the L_1 norm of $P \wedge Q$ and

$$\pi^2(P, Q) \leqslant \rho^2(P, Q) \leqslant 1 - (1 - \pi(P, Q))^2 = \pi(P, Q)(2 - \pi(P, Q)). \tag{1.3.9}$$

1.4. CENTRAL LIMIT THEOREMS

In this section, we briefly discuss central limit theorems concerning sums of independent random variables, which are useful in the asymptotic theory of statistics.

The most useful and widely applied result is the following basic central limit theorem (CLT).

Proposition 1.4.1. Let $\{X_n, n \geqslant 1\}$ be i.i.d. random variables with mean μ and finite positive variance σ^2. Let

$$S_n = X_1 + X_2 + \cdots + X_n.$$

Then

$$\frac{S_n - n\mu}{\sigma\sqrt{n}} \xrightarrow{\mathscr{L}} N(0, 1), \tag{1.4.0}$$

where $N(0, 1)$ denotes the standard normal distinction.

The following result gives sufficient conditions for the asymptotic normality for sums of independent random variables.

Proposition 1.4.2 (Lindeberg). Let $\{X_n, n \geqslant 1\}$ be independent random variables with mean 0 and variance $EX_n^2 = \sigma_n^2 < \infty$. Furthermore, suppose that

$$\frac{1}{s_n^2} \sum_{j=1}^{n} \int_{[|x| > \varepsilon s_n]} x^2 \, dF_j(x) \to 0 \quad \text{as} \quad n \to \infty \tag{1.4.1}$$

for every $\varepsilon > 0$ where $s_n^2 = \sigma_1^2 + \cdots + \sigma_n^2$. Then

$$\frac{S_n}{s_n} \xrightarrow{\mathscr{L}} N(0, 1). \tag{1.4.2}$$

Chow and Teicher (1978) show that the condition (1.4.1), which is known as the Lindeberg condition, is equivalent to the condition

$$\frac{1}{s_n^2} \sum_{j=1}^{n} \int_{[|x| > \varepsilon s_j]} x^2 \, dF_j(x) \to 0 \quad \text{as} \quad n \to \infty \tag{1.4.1}$$

for every $\varepsilon > 0$. An important and useful consequence of Proposition 1.4.2 is the following result.

Proposition 1.4.3 (Liapunov). If $\{X_n, n \geqslant 1\}$ are independent random variables with $EX_n = 0$ and if

$$\frac{1}{s_n^{2+\delta}} \sum_{j=1}^{n} E|X_j|^{2+\delta} \to 0 \tag{1.4.3}$$

for some $\delta > 0$, then

$$\frac{S_n}{s_n} \xrightarrow{\mathscr{L}} N(0, 1). \tag{1.4.4}$$

In particular (1.4.3) holds provided

$$\sup_j E|X_j|^{2+\delta} < \infty \quad \text{and} \quad s_n \to \infty. \tag{1.4.5}$$

All the theorems given above can be stated in case $EX_n = \mu_n$ by defining $Y_n = X_n - \mu_n$ and applying the results for the sequence of random variables $\{Y_n\}$.

The next theorem deals with the convergence of moments of S_n/s_n to the corresponding moments of a normal distribution.

Proposition 1.4.4. Let $\{X_n, n \geqslant 1\}$ be independent random variables with $EX_n = 0$, $EX_n^2 = \sigma_n^2$. Suppose $\{X_n\}$ satisfies the Lindeberg condition of order $r \geqslant 2$, that is,

$$\frac{1}{s_n^r} \sum_{j=1}^{n} \int_{|x| > \varepsilon s_n} |x|^r \, dF_j(x) \to 0 \quad \text{as} \quad n \to \infty \tag{1.4.6}$$

for all $\varepsilon > 0$ where $s_n^2 = \sigma_1^2 + \cdots + \sigma_n^2$. Then

$$\lim_{n \to \infty} E\left(\frac{S_n}{s_n}\right)^k = \frac{1}{\sqrt{2\pi}} \int_{-\infty}^{\infty} x^k e^{-x^2/2} \, dx, \qquad 1 \leqslant k \leqslant r. \tag{1.4.7}$$

It is sometimes useful to have a central limit theorem for double series of random variables $\{X_{nj}, 1 \leqslant j \leqslant k_n\}$. Suppose that, for any fixed $n \geqslant 1$, X_{nj}, $1 \leqslant j \leqslant k_n$, are independent and the sequence $\{X_{nj}\}$ is *uniformly asymptotically negligible* (u.a.n.), that is,

$$\max_{1 \leqslant k \leqslant k_n} P(|X_{nk}| > \varepsilon) \to 0 \quad \text{as} \quad n \to \infty \tag{1.4.8}$$

for all $\varepsilon > 0$. We have the following limit theorem. For proof, see, Chow and Teicher (1978, p. 434).

Proposition 1.4.5. Let $\{X_{nk}, 1 \leqslant k \leqslant k_n\}$ be u.a.n. rowwise independent random variables with mean zero and finite variance σ_{nk}^2 satisfying $\sum_{k=1}^{k_n} \sigma_{nk}^2 = 1$ for all $n \geqslant 1$. Then

$$\sum_{k=1}^{k_n} X_{nk} \xrightarrow{\mathscr{L}} N(0, 1) \tag{1.4.9}$$

if and only if

$$\sum_{k=1}^{k_n} \int_{|x| \geqslant \varepsilon} x^2 \, dF_{nk} = o(1) \quad \text{for every } \varepsilon > 0. \tag{1.4.10}$$

Another useful central limit theorem in problems of non-linear least squares is a result due to Eicker (1963).

Let $\mathscr{F}$ be a family of distribution functions and $\{X_{nk}, 1 \leqslant k \leqslant k_n\}$ be rowwise independent random variables with mean zero and finite variance σ_{nk}^2. Suppose $\{a_{nk}, 1 \leqslant k \leqslant k_n\}$ is another double sequence of real numbers such that $a_{nk_n} \neq 0$. Define

(1.4.11)
$$B_n^2 = \sum_{k=1}^{k_n} a_{nk}^2 \sigma_{nk}^2$$

and

(1.4.12)
$$S_n = \sum_{k=1}^{k_n} a_{nk} X_{nk}.$$

Let $\tau(\mathscr{F})$ be the set of all sequences of independent random variables with distribution functions belonging to $\mathscr{F}$; they need not necessarily be the same from term to term of the sequence.

Proposition 1.4.6 (Eicker, 1963). In order that

(1.4.13) (i) $\dfrac{S_n}{B_n} \xrightarrow{\mathscr{L}} N(0, 1)$,

(ii) $\{B_n^{-1} a_{nk} X_{nk}, 1 \leqslant k \leqslant k_n\}$ is u.a.n.

for every double sequence $\{X_{nk}\} \in \tau(\mathscr{F})$, it is necessary and sufficient that

(1.4.14) (i) $\displaystyle\max_{1 \leqslant k \leqslant k_n} \frac{a_{nk}^2}{\sum_{k=1}^{k_n} a_{nk}^2} \to 0$,

(ii) $\displaystyle\sup_{G \in \mathscr{F}} \int_{|x|>c} x^2 \, dG(x) \to 0$ as $c \to \infty$,

(iii) $\displaystyle\inf_{G \in \mathscr{F}} \int_{-\infty}^{\infty} x^2 \, dG(x) > 0$.

Proposition 1.4.7 (Blum et al., 1963; Prakasa Rao, 1969). Let $\{X_n, n \geqslant 1\}$ be independent random variables with $EX_n = 0$ and $EX_n^2 = 1$. Suppose that $\{N_n\}$ is a sequence of positive integer-valued random variables such that

(1.4.15)
$$\frac{N_n}{n} \xrightarrow{p} \lambda,$$

where λ is a positive random variable. If $n^{-1/2} \sum_{i=1}^{n} X_i \xrightarrow{\mathscr{L}} N(0, 1)$, then

(1.4.16)
$$N_n^{-1/2} \sum_{i=1}^{N_n} X_i \xrightarrow{\mathscr{L}} N(0, 1).$$

1.5. ASYMPTOTIC EXPANSIONS

It is important for practical purposes to obtain the rate of convergence of the distribution function of the partial sum S_n to the limiting distribution and also to obtain the asymptotic expansion for the distribution function of S_n under suitable conditions.

Proposition 1.5.1 (Berry–Esseen). If $\{X_n, n \geqslant 1\}$ are independent random variables with $EX_n = 0$, $EX_n^2 = \sigma_n^2, s_n^2 = \sum_{i=1}^n \sigma_i^2 > 0$, and $B_n^{2+\delta} = \sum_{i=1}^n E|X_i|^{2+\delta} < \infty$ for some $0 < \delta \leqslant 1$, then there exists an absolute constant c_δ such that, for every $n \geqslant 1$,

$$\sup_x \left| P\left(\frac{S_n}{s_n} < x\right) - \Phi(x) \right| \leqslant c_\delta \left(\frac{B_n}{s_n}\right)^{2+\delta}, \tag{1.5.0}$$

where $\Phi(\cdot)$ is the standard normal distribution function.

In particular, we have the following result for i.i.d. random variables.

Proposition 1.5.2. If $\{X_n, n \geqslant 1\}$ are i.i.d. random variables with mean 0 and positive variance σ^2 and if $E|X_1|^{2+\delta} = \gamma < \infty$ for some $0 < \delta \leqslant 1$, then there exists an absolute constant c_δ such that

$$\sup_x \left| P\left(\frac{S_n}{\sigma\sqrt{n}} < x\right) - \Phi(x) \right| \leqslant c_\delta \left(\frac{\gamma}{\sigma}\right)^{2+\delta} n^{-\delta/2}. \tag{1.5.1}$$

It is known that $c_\delta \leqslant 0.7975$ (Van Beeck, 1972) for $\delta = 1$.

Bounds of the Berry–Esseen type for sums of random number of random variables are given in Sreehari (1973) and Prakasa Rao (1974) among others. At times, it is useful to get nonuniform bounds, as given in the following theorem.

Proposition 1.5.3. Let $\{X_n, n \geqslant 1\}$ be i.i.d. random variables with $EX_1 = 0$, $EX_1^2 = \sigma^2 > 0$, and $E|X_1|^3 < \infty$. Then there exists an absolute constant C such that

$$\left| P\left(\frac{S_n}{\sigma\sqrt{n}} < x\right) - \Phi(x) \right| \leqslant C \frac{1}{\sqrt{n}} \frac{1}{(1+|x|)^3} \frac{E|X_1|^3}{\sigma^3} \tag{1.5.2}$$

for every $n \geqslant 1$ and for every $x \in R$.

Proposition 1.5.4. Let $X_1, \ldots, X_n$ be independent random variables with $EX_j = 0$ and $E|X_j|^{2+\delta} < \infty$ for some $0 < \delta \leqslant 1$, $1 \leqslant j \leqslant n$. Then there exists an absolute constant C such that for every x and for every $n \geqslant 1$,

$$\left| P\left(\frac{S_n}{s_n} < x \right) - \Phi(x) \right| \leqslant \frac{C}{(1 + |x|)^{2+\delta}} \left(\frac{B_n}{s_n} \right)^{2+\delta} \tag{1.5.3}$$

where $B_n^{2+\delta} = \sum_{j=1}^n E|X_j|^{2+\delta}$ and $s_n^2 = \sum_{j=1}^n EX_j^2$.

This last result is due to Bikelis (1966) [see Petrov (1975, p. 132)].

The following theorem gives a Berry–Esseen analog in the case of discounting.

Proposition 1.5.5 (Gerber, 1971). Let $X_0, X_1, X_2, \ldots$ be i.i.d. random variables with $E|X_1|^3 < \infty$. Define

$$S_v = \sum_{k=0}^{\infty} v^k X_k, \qquad 0 < v < 1,$$

and

$$Z_v = \frac{(1-v)^{1/2}}{\sigma} \left(S_v - \frac{\mu}{1-v} \right)$$

where $\mu = EX_1$ and $\sigma^2 = \operatorname{Var} X_1$. Then, as $v \to 1$,

$$Z_v \xrightarrow{\mathscr{L}} N\left(0, \frac{1}{1+v} \right)$$

and, in fact, there exists an absolute constant $C > 0$ such that

$$\sup_x |F_v(x) - \Phi_v(x)| \leqslant C \left(\frac{\rho}{\sigma} \right)^3 (1-v)^{1/2},$$

where

$$F_v(x) = P(Z_v \leqslant x),$$

$$\Phi_v(x) = \frac{(1+v)^{1/2}}{(2\pi)^{1/2}} \int_{-\infty}^{x} \exp\left(-\frac{1+v}{2} t^2 \right) dt$$

and

$$\rho = E|X_1 - \mu|^3.$$

Remark. S_v can be interpreted as the present value of periodic and identically distributed payments with discount factor v. The constant C is less than or equal to 5.4, as shown in Gerber (1971).

We now discuss some results on asymptotic expansions for distribution functions of sums of independent random variables. In order to motivate the expansions under consideration, we proceed formally as follows.

Let $F(x)$ be a distribution function with f as its characteristic function and $\{k_j\}$ its cumulants or semi invariants. Note that

$$f(t) = \exp\left(\sum_{j=1}^{\infty} k_j \frac{(it)^j}{j!} \right).$$

Let $\Psi(x)$ be a known distribution function with ψ as its characteristic function and $\{\gamma_j\}$ its cumulants. By the definition of cumulants, the characteristic functions satisfy the identity

$$f(t) = \exp\left(\sum_{j=1}^{\infty} (k_j - \gamma_j) \frac{(it)^j}{j!} \right) \psi(t) \tag{1.5.4}$$

formally. Note that $(it)^j \psi(t)$ is the *characteristic function* of the "distribution function" $(-1)^j \Psi^{(j)}(x)$, assuming that the derivatives of Ψ exist and vanish at the endpoint of the range of x. Then

$$F(x) = \exp\left(\sum_{j=1}^{\infty} (k_j - \gamma_j) \frac{(-D)^j}{j!} \right) \Psi(x) \tag{1.5.5}$$

formally, where D denotes the differential operator. Suppose $\Psi(x) = \Phi(x)$ is the standard normal distribution function and $F(x) = F_n(x)$ is the distribution function of the normalized sum of i.i.d. random variables with mean μ, variance σ^2, and higher cumulants $\{\sigma^j \lambda_j, j \geq 3\}$. Note that

$$k_1 - \gamma_1 = 0 = k_2 - \gamma_2,$$

$$k_j - \gamma_j = \frac{\lambda_j}{n^{(j/2)-1}}, \qquad j \geq 3.$$

Then

$$f_n(t) = e^{-t^2/2} \left(1 + \sum_{j=1}^{\infty} \frac{P_j(it)}{n^{j/2}} \right), \tag{1.5.6}$$

where $f_n(t)$ is the characteristic function of $F_n(x)$ and P_j is a polynomial of degree $3j$ with coefficients depending on the cumulants of orders 3 through $j+2$. If we interpret the powers of Φ as derivatives, then

$$(1.5.7)\quad F_n(x) = \Phi(x) + \sum_{j=1}^{\infty} \frac{P_j(-\Phi)(x)}{n^{j/2}}$$
$$= \Phi(x) - \frac{\lambda_3 \Phi^{(3)}(x)}{6\sqrt{n}} + \frac{1}{n}\left[\frac{\lambda_4 \Phi^{(4)}(x)}{24} + \frac{\lambda_3^2 \Phi^{(6)}(x)}{72} + \cdots\right] + \cdots$$

where $\Phi^{(k)}(x)$ denotes the kth derivative of $\Phi(\cdot)$ at x. This is known as *Edgeworth expansion.* Every term after the first term in the expansion is the product of the normal density and a polynomial in x. Another way of looking at the expansion for sums of independent random variables is as follows.

Suppose $\{X_n, n \geqslant 1\}$ are independent random variables with mean zero and finite moments of all orders. Let $s_n^2 = \sum_{j=1}^{n} EX_j^2$. Suppose $s_n > 0$ for large n. Let

$$(1.5.8)\qquad Z_n = \frac{S_n}{s_n} = \frac{X_1 + \cdots + X_n}{s_n}$$

for such n. Let $f_n(t)$ be the characteristic function of Z_n and $v_n(t)$ be that of X_n. Expanding $\log v_j(t)$ as a power series in t formally, we have

$$(1.5.9)\qquad \log v_j(t) = \sum_{k=2}^{\infty} \frac{\gamma_{kj}}{k!}(it)^k,$$

where γ_{kj}, $k \geqslant 2$, are the cumulants of the distribution of X_j. Hence

$$(1.5.10)\qquad \log f_n(t) = \sum_{j=1}^{n} \log v_j\left(\frac{t}{s_n}\right)$$
$$= \sum_{k=2}^{\infty} \frac{\lambda_{kn}}{k!\,n^{(k-2)/2}}(it)^k,$$

where

$$(1.5.11)\qquad \lambda_{kn} = \frac{n^{(k-2)/2}}{s_n^k} \sum_{j=1}^{n} \gamma_{kj}.$$

Since $\lambda_{2n} = 1$, it follows that

$$(1.5.12)\qquad f_n(t) = e^{-t^2/2} \exp\left\{\sum_{k=1}^{\infty} \frac{\lambda_{k+2,n}}{(k+2)!\,n^{k/2}}(it)^{k+2}\right\}.$$

Expand

$$\exp\left\{\sum_{k=1}^{\infty}\frac{\lambda_{k+2,n}}{(k+2)!}u^{k+2}z^{k}\right\}$$

formally as a power series

$$1+\sum_{k=1}^{\infty}P_{kn}(u)z^{k}$$

in z. Here $P_{kn}(u)$ is the coefficient of z^k in the expansion. Hence

$$f_n(u)=e^{-t^2/2}+\sum_{k=1}^{\infty}\frac{P_{kn}(it)}{n^{k/2}}e^{-t^2/2}. \tag{1.5.13}$$

Let $F_n(x)$ be the distribution function of Z_n and suppose we write

$$F_n(x)=\Phi(x)+\sum_{k=1}^{\infty}Q_{kn}(x)n^{-k/2} \tag{1.5.14}$$

so that

$$\int_{-\infty}^{\infty}e^{itx}\,dQ_{kn}(x)=P_{kn}(it)e^{-t^2/2} \tag{1.5.15}$$

from (1.5.13). The functions $P_{kn}(it)$ can be explicitly calculated by using the formula

$$\frac{d^k}{dx^k}\exp\left\{\sum_{s=1}^{\infty}a_sx^s\right\}\Bigg|_{x=0}=k!\,\Sigma^*\prod_{m=1}^{k}\frac{a_m^{j_m}}{j_m!} \tag{1.5.16}$$

where Σ^* runs over all nonnegative integers $j_1,\dots,j_k$ such that

$$j_1+2j_2+\cdots+kj_k=k. \tag{1.5.17}$$

In fact

$$P_{kn}(it)=\Sigma^*\prod_{m=1}^{k}\frac{1}{j_m!}\left\{\frac{\lambda_{m+2,n}(it)^{m+2}}{(m+2)!}\right\}^{j_m} \tag{1.5.18}$$

and it can be shown (Petrov, 1975, p. 137) that

$$Q_{kn}(x)=-\frac{1}{\sqrt{2\pi}}e^{-x^2/2}\Sigma^{**}H_{(k+2s-1)}(x)\prod_{m=1}^{k}\frac{1}{j_m!}\left(\frac{\lambda_{m+2,n}}{(m+2)!}\right)^{j_m} \tag{1.5.19}$$

where Σ^{**} is carried out over all nonnegative integers $j_1, \ldots, j_k$ such that

$$j_1 + 2j_2 + \cdots + kj_k = k,$$
$$j_1 + j_2 + \cdots + j_k = s,$$

and $\{H_k(x),\ k \geqslant 0\}$ are the Hermite polynomials given by

$$H_0(x) = 1,$$
$$H_1(x) = x,$$
$$H_2(x) = x^2 - 1,$$
$$H_3(x) = x^3 - 3x, \ldots .$$

In general,

$$H_k(x) = (-1)^k e^{-x^2/2} \frac{d^k}{dx^k} e^{-x^2/2}, \qquad k \geqslant 0.$$

If $\{X_n,\ n \geqslant 1\}$ are i.i.d. random variables with mean zero and positive variance σ^2, then λ_{kn} does not depend on n and

$$\lambda_{kn} = \gamma_k / \sigma^k,$$

where γ_k is the kth cumulant of X_1. In this case, we delete n in the definition of P and Q as given above. Proofs of the following results are given in Petrov (1975).

Proposition 1.5.6. Suppose $\{X_n, n \geqslant 1\}$ are i.i.d. random variables with $E|X_1|^r < \infty$ for some integer $r \geqslant 3$ and satisfy Cramer condition

$$\overline{\lim_{|t| \to \infty}} |v(t)| < 1, \tag{1.5.20}$$

where $v(\cdot)$ is the characteristic function of X_1. Then there exists a function $\varepsilon(u) > 0$ such that $\varepsilon(u) \to 0$ as $u \to \infty$,

$$\left| F_n(x) - \Phi(x) - \sum_{k=1}^{r-2} \frac{Q_k(x)}{n^{k/2}} \right| \leqslant \frac{\varepsilon(\sqrt{n}(1 + |x|))}{n^{(r-2)/2}(1 + |x|)^r}, \tag{1.5.21}$$

$$\sup_x \left| F_n(x) - \Phi(x) - \sum_{k=2}^{r-2} \frac{Q_k(x)}{n^{k/2}} \right| = o(n^{-(r-2)/2}), \tag{1.5.22}$$

and for any $P > 1/r$,

$$\int_{-\infty}^{\infty} \left| F_n(x) - \Phi(x) - \sum_{k=1}^{r-2} \frac{Q_k(x)}{n^{k/2}} \right|^P dx = o(n^{-(r-2)p/2}). \tag{1.5.23}$$

For the result concerning the asymptotic expansion of the distribution function of a sum of independent but not necessarily identically distributed random variables, see Theorem 7 in Petrov (1975, p. 175). We consider a special case of his result.

Proposition 1.5.7. Let $\{X_n, n \geq 1\}$ be a sequence of independent random variables with mean zero and finite variances. Let $\sigma_j^2 = EX_j^2$, $s_n^2 = \sigma_1^2 + \cdots + \sigma_n^2$, and $v_j(t) = E[e^{itX_j}]$.

(i) Suppose there exists positive constants g and G and integer $r \geq 3$ such that

$$s_n^2 \geq ng, \qquad \sum_{j=1}^{n} E|X_j|^r \leq nG \tag{1.5.24}$$

and

$$\overline{\lim_n} \frac{1}{n} \sum_{j=1}^{n} E|X_j|^{r+\delta} < \infty \tag{1.5.25}$$

for some $\delta > 0$.

(ii) Furthermore, suppose that

$$n^{(r-2)/2} \int_{|t|>\varepsilon} |t|^{-1} \prod_{j=1}^{n} |v_j(t)| \, dt \to 0 \tag{1.5.26}$$

for every $\varepsilon > 0$.

Then, for sufficiently large n,

$$F_n(x) = \Phi(x) + \sum_{k=1}^{r-2} \frac{Q_{kn}(x)}{n^{k/2}} + o(n^{-(r-2)/2}) \tag{1.5.27}$$

uniformly in x where $Q_{kn}(x)$ is as defined earlier.

For multivariate extensions and refinements of these results, see Bhattacharya (1977) or Bhattacharya and Ranga Rao (1976). We discuss these results briefly here in view of their important applications to obtaining valid expansions for distribution functions of a class of statistics that includes functions of sample moments.

Consider a sequence $\{\mathbf{X}_n, n \geq 1\}$ of i.i.d. random vectors $\mathbf{X}_n = (X_n^{(1)}, \ldots, X_n^{(k)})$ with value in R^k and with distribution function Q_1. Suppose that $E\|\mathbf{X}_1\|^2 < \infty$. Let Q_n denote the distribution function of the normalized partial sum

$$n^{-1/2}(\mathbf{X}_1 + \cdots + \mathbf{X}_n)$$

assuming without loss of generality that $E\mathbf{X}_1 = 0$ and $\operatorname{cov} \mathbf{X}_1 = I_k$, where I_k is the identity matrix of order $k \times k$. Let $\boldsymbol{\nu} = (\nu^{(1)}, \ldots, \nu^{(k)})$ denote the multiindex, that is, a k-tuple of nonnegative integers. Let

$$|\boldsymbol{\nu}| \equiv \nu^{(1)} + \cdots + \nu^{(k)}, \qquad \boldsymbol{\nu}! \equiv \nu^{(1)}!\,\nu^{(2)}! \cdots \nu^{(k)}!,$$

$$\mathbf{x} = (x^{(1)}, \ldots, x^{(k)}), \qquad \mathbf{x}^{\boldsymbol{\nu}} = (x^{(1)})^{\nu^{(1)}} \cdots (x^{(k)})^{\nu^{(k)}},$$

and

$$\|\mathbf{x}\| = \left(\sum_{i=1}^{k} x^{(i)2} \right)^{1/2}.$$

The $\boldsymbol{\nu}$th moment of $\mathbf{X}_1$ is defined to be

$$\mu_{\boldsymbol{\nu}} = E\mathbf{X}_1^{\boldsymbol{\nu}} = \int_{R^k} \mathbf{x}^{\boldsymbol{\nu}} Q_1(d\mathbf{x})$$

when it exists and for any integer s, the sth absolute moment of $\mathbf{X}_1$ is

$$\rho_s = E\|\mathbf{X}_1\|^s = \int_{R^k} \|\mathbf{x}\|^s Q_1(d\mathbf{x})$$

when it exists. Let $\hat{Q}_1(\mathbf{t})$ denote the characteristic function of $\mathbf{X}_1$. The $\boldsymbol{\nu}$th cumulant of $\mathbf{X}_1$ is, by definition,

$$\chi_{\boldsymbol{\nu}} = i^{-|\boldsymbol{\nu}|}(D^{\boldsymbol{\nu}} \log \hat{Q}_1)(0), \tag{1.5.28}$$

provided $\rho_{|\boldsymbol{\nu}|} < \infty$ where $D^{\boldsymbol{\nu}} = D_1^{\nu^{(1)}} \cdots D_k^{\nu^{(k)}}$ and D_j denotes differentiation with respect to t_j. Observe that $\chi_{\boldsymbol{\nu}} = \mu_{\boldsymbol{\nu}} = 0$ if $|\boldsymbol{\nu}| = 1$. Expand $\hat{Q}_1(\mathbf{t})$ and $\log \hat{Q}_1(\mathbf{t})$ using the Taylor expansion to obtain

$$\hat{Q}_1(\mathbf{t}) = 1 + \sum_{2 \leq |\boldsymbol{\nu}| \leq s} \frac{\mu_{\boldsymbol{\nu}}}{\boldsymbol{\nu}!}(i\mathbf{t})^{\boldsymbol{\nu}} + o(\|\mathbf{t}\|^s) \tag{1.5.29}$$

and

$$\log \hat{Q}_1(\mathbf{t}) = \sum_{2 \leqslant |\nu| \leqslant s} \frac{\chi_\nu}{\nu!}(i\mathbf{t})^\nu + o(\|\mathbf{t}\|^s) \tag{1.5.30}$$

as $\mathbf{t} \to 0$. These two lead formally to the identity

$$\sum_{2 \leqslant |\nu|} \frac{\chi_\nu}{\nu!}(i\mathbf{t})^\nu = \sum_{m=1}^{\infty} \frac{(-1)^{m+1}}{m}\left[\sum_{2 \leqslant |\nu| < \infty} \frac{\mu_\nu}{\nu!}(i\mathbf{t})^\nu\right]^m. \tag{1.5.31}$$

It is easy to check that

$$\log \hat{Q}_n(\mathbf{t}) = -\frac{\|\mathbf{t}\|^2}{2} + \sum_{3 \leqslant |\nu| \leqslant s} \frac{\chi_\nu}{\nu!}(i\mathbf{t})^\nu n^{-(|\nu|-2)/2} + o(n^{-(s-2)/2}), \tag{1.5.32}$$

provided $\rho_s < \infty$ for some $s \geqslant 2$. Note that the νth cumulant of Q_n is

$$\begin{aligned} i^{-|\nu|}(D^\nu \log \hat{Q}_n)(0) &= i^{-|\nu|}[D^\nu \log \hat{Q}_1^n(\cdot/n^{1/2})](0) \\ &= n^{-(|\nu|-2)/2}\chi_\nu, \end{aligned}$$

where χ_ν is as defined by (1.5.28). Suppose $s \geqslant 3$. Note that

$$\hat{Q}_n(\mathbf{t}) = e^{-\|\mathbf{t}\|^2/2} \exp\left\{\sum_{3 \leqslant |\nu\| \leqslant s} \frac{\chi_\nu}{\nu!}(i\mathbf{t})^\nu n^{-(|\nu|-2)/2}\right\}[1 + o(n^{-(s-2)/2})]. \tag{1.5.33}$$

Expanding the second exponential on the right-hand side of (1.5.33) and collecting powers of $n^{-1/2}$ we have

$$\exp\left\{\sum_{3 \leqslant |\nu| \leqslant s} \frac{\chi_\nu}{\nu!}(i\mathbf{t})^\nu n^{-(|\nu|-2)/2}\right\} = 1 + \sum_{r=1}^{s-2} n^{-r/2}\tilde{P}_r(i\mathbf{t}) + o(n^{-(s-2)/2}). \tag{1.5.34}$$

Therefore, formally,

$$\hat{Q}_n(\mathbf{t}) = \exp\left\{-\frac{\|\mathbf{t}\|^2}{2}\right\}\left\{1 + \sum_{r=1}^{s-2} n^{-r/2}\tilde{P}_r(i\mathbf{t})\right\} + o(n^{-(s-2)/2}). \tag{1.5.35}$$

Let

$$\chi_r(\mathbf{t}) = r! \sum_{|\nu|=r} \frac{\chi_\nu}{\nu!}\mathbf{t}^\nu. \tag{1.5.36}$$

It is easy to see that $\chi_r(\mathbf{t})$ is the rth cumulant of $\langle \mathbf{t}, \mathbf{X}_1 \rangle$, where $\langle \cdot,\cdot \rangle$ is the inner product in R^k and hence the relation (1.5.34) can be written in the form

$$\text{(1.5.37)} \quad \exp\left\{\sum_{3\leqslant r\leqslant s} \frac{\chi_r(i\mathbf{t})}{r!} n^{-(r-2)/2}\right\} = 1 + \sum_{r=1}^{s-2} \tilde{P}_r(i\mathbf{t}) n^{-r/2} + o(n^{-(s-2)/2})$$

and hence

$$\text{(1.5.38)} \quad \tilde{P}_r(i\mathbf{t}) = \sum_{m=1}^{r} \frac{1}{m!}\left\{\Sigma^* \frac{\chi_{j_1+2}(i\mathbf{t})}{(j_1+2)!}\cdots\frac{\chi_{j_m+2}(i\mathbf{t})}{(j_m+2)!}\right\},$$

where Σ^* is the sum over all $(j_1,\ldots,j_m)$ such that $j_1 + \cdots + j_m = r$. For example,

$$\text{(1.5.39)} \quad \tilde{P}_1(i\mathbf{t}) = \frac{\chi_3(i\mathbf{t})}{3!}, \qquad \tilde{P}_2(i\mathbf{t}) = \frac{\chi_4(i\mathbf{t})}{4!} + \frac{\chi_3^2(i\mathbf{t})}{2!(3!)^2}, \ldots.$$

Let $\tilde{P}_r(-D)$ be the differential operator obtained by formally replacing $(i\mathbf{t})^\nu$ by $(-1)^\nu D^\nu$ in the formula given by (1.5.31). Let

$$\text{(1.5.40)} \quad P_r(-\phi)(x) \equiv \tilde{P}_r(-D)\phi(x)$$

where ϕ is the density function $(2\pi)^{-k/2}\exp[-\|\mathbf{x}\|^2/2]$. It can be shown that $\tilde{P}_r(i\mathbf{t})\exp[-\|\mathbf{t}\|^2/2]$ is the Fourier transform of $P_r(-\phi)(x)$. Let $P_r(-\Phi)$ denote the signed measure with density $P_r(-\phi)$, where Φ denotes standard normal distribution function as earlier with density ϕ. We can now state the main theorem due to Bhattacharya (1977).

Proposition 1.5.8. Suppose the Cramer condition

$$\text{(1.5.41)} \quad \overline{\lim_{\|\mathbf{t}\|\to\infty}} |\hat{Q}_1(\mathbf{t})| < 1$$

holds. Let $\mathscr{C}$ be the class of Borel-measurable convex subsets of R^k. Furthermore suppose that $\rho_s < \infty$ for some integer $s \geqslant 3$. Then

$$\text{(1.5.42)} \quad \sup_{C\in\mathscr{C}}\left|Q_n(C) - \Phi(C) - \sum_{r=1}^{s-2} n^{-r/2} P_r(-\Phi)(C)\right| = o(n^{-(s-2)/2}).$$

A more general result is as follows.

Proposition 1.5.9. Suppose the Cramer condition (1.5.41) holds and $\rho_s < \infty$ for some integer $s \geqslant 3$. Then, for any real-valued bounded Borel-measurable function f on R^k,

$$\left| \int_{R^k} f\, d\left[Q_n - \Phi - \sum_{r=1}^{s-2} n^{-r/2} P_r(-\Phi) \right] \right| \leqslant \frac{\delta_n}{n^{(s-2)/2}} \omega_f(R^k) + \bar{\omega}_f(e^{-dn}; \Phi), \tag{1.5.43}$$

where $\delta_n \to 0$ as $n \to \infty$, $d > 0$, and δ_n and d do not depend on f. Here

$$\omega_f(R^k) \equiv \sup\{|f(\mathbf{y}) - f(\mathbf{z})| : \mathbf{y}, \mathbf{z} \in R^k\} \tag{1.5.44}$$

and

$$\bar{\omega}_f(\varepsilon; Q) \equiv \int_{R^k} \omega_f(x; \varepsilon) Q(d\mathbf{x}) \tag{1.5.45}$$

with

$$\omega_f(\mathbf{x}; \varepsilon) = \sup\{|f(\mathbf{y}) - f(\mathbf{z})| : \mathbf{y}, \mathbf{z} \in B(\mathbf{x}, \varepsilon)\} \tag{1.5.46}$$

and $B(\mathbf{x}, \varepsilon)$ is the open sphere with center $\mathbf{x}$ and radius ε.

Before we conclude this section, we state the following result, which is useful in verifying whether Cramer's condition is satisfied or not.

It is known that if a distribution function $Q_1(\cdot)$ has an absolutely continuous component, then it satisfies Cramer's condition (1.5.41) by the Riemann–Lebesgue lemma.

Proposition 1.5.10. Let $\mathbf{X}$ be a k-dimensional random vector with distribution function that has a nonzero absolutely continuous component H (relative to the Lebesgue measure on R^k). Suppose f_i, $1 \leqslant i \leqslant r$, are Borel-measurable real-valued functions on R^k. Furthermore, suppose that there exists an open sphere B of R^k in which the density of H is positive almost everywhere and in which the f_i are continuously differentiable. Then, the distribution Q_1 of $(f_1(\mathbf{X}), \ldots, f_r(\mathbf{X}))$ satisfies Cramer's condition if $1, f_1, \ldots, f_r$ are linearly independent in B.

Proof. See Bhattacharya (1977).

1.6 LARGE DEVIATIONS

Suppose $F_n(x)$ is a sequence of distribution functions converging weakly to a distribution function $F(x)$. Let $x_n \to \infty$ as $n \to \infty$. It is clear that $F_n(x_n) \to 1$ and

$F_n(-x_n) \to 0$ as $n \to \infty$. The problem of interest here is obtain the rate of convergence or in other words to study the asymptotic behavior of $1 - F_n(X_n)$ and $F_n(-x_n)$ as $n \to \infty$.

Let $\{X_i, i \geqslant 1\}$ be i.i.d. one-dimensional random variables with $EX_1 = 0$ and $\operatorname{Var} X_1 = 1$. It is known that

$$P\left[\frac{X_1 + \cdots + X_n}{\sqrt{n}} \leqslant x\right] \to \Phi(x) \quad \text{as } n \to \infty$$

for every $x \in R$. Let $F_n(x) = P(S_n \leqslant x)$, where $S_n = X_1 + \cdots + X_n$. Then

$$F_n(\sqrt{n}x) \to \Phi(x) \quad \text{as } n \to \infty$$

for every $x \in R$. In other words

$$1 - F_n(\sqrt{n}x_n) = P(S_n > x_n\sqrt{n}) \to 1 - \Phi(x) \quad \text{as } n \to \infty$$

whenever $x_n \equiv x$. Deviations x_n of this type are called *ordinary deviations.* Deviations x_n such that $x_n = o(\sqrt{n})$ and $x_n \to \infty$ are called *large deviations.* If $x_n = c\sqrt{\log n}$, then the deviation is said to be *moderate deviation.*

The first result in large deviation theory is due to Cramer (1938).

Proposition 1.6.1. Suppose $R(t) = E(e^{tX_1}) < \infty$ for some $t > 0$ and $0 < x_n = o(\sqrt{n})$. Then

$$\frac{1 - F_n(\sqrt{n}x_n)}{1 - \Phi(x_n)} = e^{x_n^2/2}\left[\rho\left(\frac{x_n}{\sqrt{n}}\right)\right]^n\left(1 + O\left(\frac{x_n + 1}{\sqrt{n}}\right)\right), \tag{1.6.1}$$

where

$$\rho(x) = \inf_t e^{-tx} R(t). \tag{1.6.2}$$

The function $\rho(\cdot)$ is called the *Chernoff function.* One of the methods for deriving large deviation results is by the method of conjugate distributions.

Given a distribution function $F(\cdot)$ and t such that $R(t) < \infty$, define the conjugate distribution function

$$\bar{F}(x) = \frac{1}{R(t)} \int_{-\infty}^{x} e^{ty}\, dF(y). \tag{1.6.3}$$

Observe that $\overline{F^{(n)}} = (\bar{F})^{(n)}$, where $F^{(n)}$ denotes the nth convolution of F. The moment-generating function of $\bar{F}$ is $R(t+s)/R(t)$, where t is fixed and s is the variable. It can be checked that

$$1 - F(x) = R(t) \int_x^\infty e^{-ty} \, d\bar{F}(y) \tag{1.6.4}$$

and hence

$$\begin{aligned} 1 - F^{(n)}(nx) &= [R(t)]^n \int_{nx}^\infty e^{-ty} \, d\bar{F}^{(n)}(y) \\ &= [e^{-tx} R(t)]^n \int_x^\infty e^{-nt(y-x)} \, d\bar{F}^{(n)}(ny). \end{aligned} \tag{1.6.5}$$

Choose t so that $e^{-tx}R(t) \simeq \rho(t)$ from (1.6.2). Note that $\bar{F}^{(n)}(\sqrt{n}\,y) \simeq \Phi(y)$. From this, one can deduce the required result. [See, for instance, Schmetterer (1974)].

For a survey of the large deviation results for sums of independent random variables, see Nagaev (1979) and Petrov (1975).

Rubin and Sethuramen (1965) proved that, if the deviation x_n is moderate, that is, $x_n = c\sqrt{\log n}$, then

$$\frac{1 - F_n(\sqrt{n}\,x_n)}{1 - \Phi(x_n)} \to 1 \quad \text{as } n \to \infty,$$

provided $E|X_1|^k < \infty$, where k depends on c.

If $x_n = x\sqrt{n}$ or $x_n \simeq c\sqrt{n}$ or $x_n/\sqrt{n} \to \infty$, then the results given above do not hold. If $x_n = x\sqrt{n}$, then we have to discuss the limit behavior of

$$P\left(\frac{S_n}{n} \geqslant x\right).$$

These are some times known as *excessive deviations* or *large deviations of the sample mean.*

The following result can be proved by the method discussed in (1.6.3)–(1.6.5). For details, see Chernoff (1952) and Bahadur and Ranga Rao (1960) [cf. Schmetterer (1974)].

Proposition 1.6.2. Let $X_1, X_2, \ldots$ be nondegenerate i.i.d. random variables with $R(t) = E[e^{tX_1}] < \infty$ for all $t \in I$ open in $[0, \infty)$. Suppose there exists $u \in I$

such that $R'(u)/R(u)=c$, where $R'(u)$ denotes the derivative of $R(u)$. Define $S_n = X_1 + \cdots + X_n$ and $\chi(t) = e^{-tc}R(t)$. Let $m = \chi(u)$. Then $m = \min_{t \in I} \chi(t)$ and

(1.6.6) $$\lim_{n \to \infty} \frac{1}{n} \log P(S_n \geqslant nc) = \log m.$$

This theorem has been extended to several general classes of statistics by Sethuraman (1964, 1965). In the multidimensional case, the Chernoff function is defined as

(1.6.7) $$\rho(\mathbf{x}) = \inf_{\mathbf{t} \in R^k} e^{-\langle \mathbf{t}, \mathbf{x} \rangle} R(\mathbf{t}), \qquad \mathbf{x} \in R^k,$$

where $R(\mathbf{t})$ is the moment-generating function of a k-dimensional random vector $\mathbf{X}_1$ and $\langle \mathbf{t}, \mathbf{x} \rangle$ denotes the inner product in R^k. We now make brief remarks regarding this function and its connection to large deviations of sample mean in the multidimensional case.

A function $f(\mathbf{x}), \mathbf{x} \in R^k$, is said to be *convex* if the set $\{(\mathbf{x}, y): y \geqslant f(\mathbf{x}), \mathbf{x} \in R^k, y \in R^1\}$ is convex, and $f(\mathbf{x})$ is said to be *closed* if this set is closed. For any closed convex function f, define

(1.6.8) $$\bar{f}(\mathbf{x}) = \sup_{\mathbf{t}} \{\langle \mathbf{t}, \mathbf{x} \rangle - f(\mathbf{t})\}.$$

Then it is known that

(1.6.9) $$\bar{\bar{f}}(\mathbf{x}) = f(\mathbf{x})$$

from Rockafeller (1970). The functions $\log R(\mathbf{t})$ and $-\log \rho(\mathbf{x})$ are convex conjugate to each other, and by the duality given above, we have the inversion formula

(1.6.10) $$R(\mathbf{t}) = \sup_{\mathbf{x} \in R^k} e^{\langle \mathbf{t}, \mathbf{x} \rangle} \rho(\mathbf{x}).$$

For any set A in R^k, define

(1.6.11) $$\rho(A) = \sup_{\mathbf{x} \in A} \rho(\mathbf{x}).$$

Bartfai (1977, 1978) proved that if A is an open set, then

(1.6.12) $$\underline{\lim}_{n} \left[P\left(\frac{S_n}{n} \in A \right) \right]^{1/n} \geqslant \rho(A),$$

and if A is a convex open set, then

$$P\left(\frac{S_n}{n}\in A\right)\leqslant[\rho(A)]^n. \tag{1.6.13}$$

In general, the result

$$\left[P\left(\frac{S_n}{n}\in A\right)\right]^{1/n}\to\rho(A) \tag{1.6.14}$$

for all Borel sets A is not true. Bartfai (1977) has given an example of a closed set A with a nonempty interior such that

$$P\left(\frac{S_n}{n}\in A\right)=0, \qquad n=1,2,\ldots,$$

and

$$\rho(A)>0.$$

For a survey of results on large deviations of the sample mean in general vector spaces, see Bahadur and Zabell (1979). The following multivariate Chernoff-type theorem has been derived by Groeneboom et al. (1979) from their general results on large deviation theorem for empirical probability measure. We state the theorem.

Proposition 1.6.3. Let $\mathbf{X}_1, \mathbf{X}_2, \ldots$ be i.i.d. k-dimensional random vectors with probability measure P absolutely continuous with respect to the Lebesgue measure on R^k. Then, for any sequence $\{\mathbf{u}_n\}$ in R^k such that $\mathbf{u}_n\to\mathbf{0}$ and for any vector $\mathbf{x}\in R^k$,

$$\lim_{n\to\infty}\frac{1}{n}\log P\left[\frac{S_n}{n}\geqslant\mathbf{x}+\mathbf{u}_n\right]=\sup_{\mathbf{t}\in R_+^k}\left\{\langle\mathbf{t},\mathbf{x}\rangle-\log\int_{R^k}e^{\langle\mathbf{t},\mathbf{x}\rangle}\,dP(\mathbf{x})\right\}. \tag{1.6.15}$$

(Here $\mathbf{x}\geqslant\mathbf{y}$ if $x^{(i)}\geqslant y^{(i)}$ for $1\leqslant i\leqslant k$ and $R_+^k=\{\mathbf{x}\in R^k:\mathbf{x}\geqslant\mathbf{0}\}$).

1.7 MARTINGALES

In this section, we discuss briefly the concept of martingales and their properties. The theory of martingales has wide applicability in the asymptotic theory of statistics, especially in the theory of statistical inference for stochastic processes [cf. Basawa and Prakasa Rao (1980) and Billingsley (1961a)].

We first study the concept of conditional expectation and its related properties.

Let $(\Omega, \mathscr{B}, P)$ be a probability space and f be a $\mathscr{B}$-measurable nonnegative P-integrable function. Let $\mathscr{B}_0$ be a sub-σ-algebra of $\mathscr{B}$. Consider

$$\nu(A) = \int_A f \, dP, \qquad A \in \mathscr{B}_0.$$

ν is a finite measure on $(\Omega, \mathscr{B}_0)$. Clearly $\nu \ll P$ on $(\Omega, \mathscr{B}_0)$. Hence, by the Radon–Nikodym theorem, there exists a function g, $\mathscr{B}_0$-measurable, unique a.e. $[P]$ such that

$$\nu(A) = \int_A g \, dP, \qquad A \in \mathscr{B}_0.$$

Hence

$$\int_A f \, dP = \int_A g \, dP, \qquad A \in \mathscr{B}_0,$$

and g is $\mathscr{B}_0$-measurable. g is called a *conditional expectation* of f given the σ-algebra $\mathscr{B}_0$, and we write

$$g = E(f \mid \mathscr{B}_0).$$

For any general P-integrable $\mathscr{B}$-measurable f, we define

$$g = E(f^+ \mid \mathscr{B}_0) - E(f^- \mid \mathscr{B}_0),$$

where $f = f^+ - f^-$ and

$$f^+ = \begin{cases} f & \text{if } f > 0 \\ 0 & \text{if } f \leqslant 0, \end{cases} \qquad f^- = \begin{cases} -f & \text{if } f < 0 \\ 0 & \text{if } f \geqslant 0. \end{cases}$$

Observe that the conditional expectation, if it exists, is unique almost everywhere.

The following results give some properties of conditional expectations. For proofs, see Chow and Teicher (1978).

Proposition 1.7.1. Let $(\Omega, \mathscr{B}, P)$ be a probability space and X be a random variable defined on $(\Omega, \mathscr{B}, P)$ such that $E|X| < \infty$. Let $\mathscr{B}_0$ be a sub-σ-algebra of $\mathscr{B}$:

(i) $E(1 \mid \mathscr{B}_0) = 1$ a.e. $[P]$.

(ii) $E(X \mid \mathscr{B}_0) \geqslant 0$ if $X \geqslant 0$ a.e. $[P]$.

(iii) $E(cX|\mathscr{B}_0) = cE(X|\mathscr{B}_0)$ a.e. $[P]$ for any $-\infty < c < \infty$.

(iv) $E(X|\mathscr{B}_0) = X$ a.e. $[P]$ if X is $\mathscr{B}_0$-measurable.

(v) Suppose $\mathscr{B}_1$ and $\mathscr{B}_2$ are sub-σ-algebras of $\mathscr{B}$ such that $\mathscr{B}_1 \subset \mathscr{B}_2 \subset \mathscr{B}$. Then

$$E\{E(X|\mathscr{B}_2)|\mathscr{B}_1\} = E\{X|\mathscr{B}_1\} \text{ a.e. } [P].$$

(vi) If $E|X+Y| < \infty$, then $E(X+Y|\mathscr{B}_0) = E(X|\mathscr{B}_0) + E(Y|\mathscr{B}_0)$ a.e. $[P]$.

(vii) Suppose Y is a $\mathscr{B}_0$-measurable random variable such that $E|XY| < \infty$. Then

$$E(XY|\mathscr{B}_0) = YE(X|\mathscr{B}_0) \text{ a.e. } [P].$$

Proposition 1.7.2. Suppose $\{X_n, n \geqslant 1\}$ and Y are random variables defined on a probability space $(\Omega, \mathscr{B}, P)$ with $E|Y| < \infty$. Let $\mathscr{B}_0$ be a sub-σ-algebra of $\mathscr{B}$.

(i) If $X_n \geqslant Y$, $n \geqslant 1$, and $X_n \uparrow X$ a.e. $[P]$, then $E(X_n|\mathscr{B}_0) \uparrow E(X|\mathscr{B}_0)$ a.e. $[P]$.

(ii) If $X_n \geqslant Y$, $n \geqslant 1$, then $E(\underline{\lim}_n X_n|\mathscr{B}_0) \leqslant \underline{\lim}_n E(X_n|\mathscr{B}_0)$ a.e. $[P]$.

(iii) If $|X_n| \leqslant |Y|$, $n \geqslant 1$, and $X_n \to X$ a.e. $[P]$, then $E(X_n|\mathscr{B}_0) \to E(X|\mathscr{B}_0)$ a.e. $[P]$.

(iv) If $X_n \to X$ a.e. and $\sup_n E\{|X_n|^r|\mathscr{B}_0\} \leqslant C < \infty$ for some $r > 1$, then $E(X_n|\mathscr{B}_0) \to E(X|\mathscr{B}_0)$ a.e. $[P]$.

In the following discussion, the index set T is $\{1, 2, \ldots\}$ or an interval in R.

Definition 1.7.1. Let $(\Omega, \mathscr{B}, P)$ be a probability space and $\{\mathscr{B}_t, t \in T\}$ be an increasing family of sub-σ-algebras contained in $\mathscr{B}$. Suppose $\{X_t, \mathscr{B}_t, t \in T\}$ is a family of random variables such that X_t is $\mathscr{B}_t$-measurable and $E|X_t| < \infty$ for every $t \in T$. The family $\{X_t, \mathscr{B}_t, t \in T\}$ is called a *martingale* if

$$E(X_t|\mathscr{B}_s) = X_s, \qquad t \geqslant s, \quad t, s \in T.$$

It is called a *submartingale* if

$$E(X_t|\mathscr{B}_s) \geqslant X_s, \qquad t \geqslant s, \quad t, s \in T,$$

and it is called a *supermartingale* if

$$E(X_t|\mathscr{B}_s) \leqslant X_s, \qquad t \geqslant s, \quad t, s \in T.$$

Observe that if $\{X_t, \mathscr{B}_t, t \in T\}$ is a submartingale, then EX_t is a nondecreasing function of t, and if $\{X_t, \mathscr{B}_t, t \in T\}$ is a martingale, then EX_t is a constant and $E|X_t|$ is non-decreasing in t.

A random variable τ defined on $(\Omega, \mathscr{B}, P)$ and taking values in T is called a *stopping time* with respect to the family of σ-algebras $\{\mathscr{B}_t, t \in T\}$ if the event $[\tau > t] \in \mathscr{B}_t$ for every $t \in T$. Let $\mathscr{B}_\tau$ denote the smallest σ-algebra containing events of the form $F_t \cap [\tau > t]$, $F_t \in \mathscr{B}_t$, $t \in T$.

We now present some examples of martingales and submartingales.

Example 1.7.1. Let $\{Z_n, n \geqslant 1\}$ be independent random variables. Define $S_n = \sum_{i=1}^n Z_i$. Suppose $EZ_i = 0$ for $i \geqslant 1$. Then $\{S_n, \mathscr{B}_n, n \geqslant 1\}$ is a martingale where $\mathscr{B}_n$ is the smallest σ-algebra generated by $Z_1, \ldots, Z_n$. If $EZ_i \geqslant 0$ for $i \geqslant 1$, then it is a submartingale.

Example 1.7.2. Let $\{Z_n, n \geqslant 1\}$ be independent random variables with $EZ_n = 0$ for $n \geqslant 1$. Let $k \geqslant 1$. Define

$$U_{kn} = \sum_{1 \leqslant i_1 < \cdots < i_k \leqslant n} Z_{i_1} Z_{i_2} \cdots Z_{i_k}, \qquad n \geqslant k.$$

Then $\{U_{kn}, n \geqslant k\}$ is a martingale with respect to the family of σ-algebras $\{\mathscr{B}_n, n \geqslant k\}$ defined above.

Example 1.7.3. A sequence of random variables $\{Z_n, n \geqslant 1\}$ is said to be *interchangeable* if the joint distribution of every finite subset of k of them depends only on k but not on the particular subset for any $k \geqslant 1$. Let $\phi(\cdot)$ be any symmetric function defined on R^m such that $E|\phi(Z_1, \ldots, Z_m)| < \infty$. Define

$$U_{mn} = \frac{1}{\binom{n}{m}} \sum_{1 \leqslant i_1 < \cdots < i_m \leqslant n} \phi(Z_{i_1}, \ldots, Z_{i_m}), \qquad n \geqslant m.$$

Let $\mathscr{F}_n$ be the σ-algebra generated by $U_{mj}, j \geqslant n$. It can be checked that

$$E(U_{mn} | \mathscr{F}_{n+1}) = U_{m,n+1} \quad \text{a.e.}$$

and hence $\{U_n^*, \mathscr{F}_n^*, n \leqslant -m\}$ is a martingale where $U_n^* = U_{m,-n}$ and $\mathscr{F}_n^* = \mathscr{F}_{-n}, n \leqslant -m$.

Example 1.7.4. Let $\{X_n, n \geqslant 1\}$ be a sequence of random variables defined on $(\Omega, \mathscr{B}, P\}$ such that the joint density of $(X_1, \ldots, X_n)$ exists and is given by $p_n(x_1, \ldots, x_n; \theta)$ where θ is a scalar parameter. Furthermore, suppose that the conditional density of X_n given $X_1, \ldots, X_{n-1}$ exists. Then, it is given by

$p_n(x_1,\ldots,x_n;\theta)/p_{n-1}(x_1,\ldots,x_{n-1};\theta)$. Note that

$$\int_{-\infty}^{\infty} p_n(x_n;\theta|x_1,\ldots,x_{n-1})\,dx_n = 1 \quad \text{a.e.}$$

Suppose we can differentiate this equation under the integral sign with respect to θ. It is clear that

$$E_\theta\left[\frac{d\log p_n(X_n;\theta|\mathbf{X}^{n-1})}{d\theta}|\mathbf{X}^{n-1}\right] = 0 \quad \text{a.e.,}$$

where $\mathbf{X}^{n-1} = (X_1,\ldots,X_{n-1})$. Let

$$S_n = \sum_{i=1}^{n} \frac{d\log p_n(X_n;\theta|\mathbf{X}^{n-1})}{d\theta}.$$

Then $\{S_n, \mathscr{B}_n, n \geqslant 1\}$ is a zero-mean martingale where $\mathscr{B}_n$ is the σ-algebra generated by X_i, $1 \leqslant i \leqslant n$.

Example 1.7.5. Let $\{Z_n, n \geqslant 0\}$ be a Galton–Watson branching process, that is, a sequence of nonnegative integer-valued random variables such that $Z_0 = 1$, $Z_1 = X_1$, $Z_{n+1} = \sum_{i=1}^{Z_n} X_i$, where X_i, $i \geqslant 1$, are i.i.d. as X_1. Suppose $EX_1 = m < \infty$ and $m > 0$. Then

$$\{W_n = Z_n/m^n, n \geqslant 1\}$$

is a martingale with respect to the sequence of σ-algebras $\mathscr{F}_n$ generated by $\{Z_1,\ldots,Z_n\}$.

Example 1.7.6. Let $\{W(t), t \geqslant 0\}$ be the standard Wiener process, that is,

- **(i)** $W(0) = 0$,
- **(ii)** $W(t) - W(s)$ is normal with mean 0 and variance $|t-s|$,
- **(iii)** $W(t_1) - W(t_2)$ and $W(t_3) - W(t_4)$ are independent if $0 \leqslant t_1 < t_2 \leqslant t_3 < t_4 < \infty$.

Define $\zeta_t = \exp[\alpha W(t) - t^2/2]$, where $\alpha > 0$. Then $\{\zeta_t, \mathscr{B}_t, t \geqslant 0\}$ is a martingale where $\mathscr{B}_t$ is the σ-algebra generated by $\{W(s), 0 \leqslant s \leqslant t\}$.

Suppose $T = \{0, 1, 2, \ldots, n\}$ and let $\{\mathscr{B}_k, 0 \leqslant k \leqslant n\}$ be a nondecreasing sequence of σ-algebras contained in $\mathscr{B}$. Suppose $\{X_k, \mathscr{B}_k, 0 \leqslant k \leqslant n\}$

is a martingale. Let $\tau_1 \leqslant \tau_2 \leqslant \cdots \leqslant \tau_s$ be a family of stopping times with respect to $\{\mathscr{B}_k, 0 \leqslant k \leqslant n\}$. Then $\{X_{\tau_k}, \mathscr{B}_{\tau_k}, 1 \leqslant k \leqslant s\}$ is a martingale.

Proposition 1.7.3.

(i) Suppose $\{S_n, \mathscr{B}_n, n \geqslant 1\}$ is a submartingale. Let $\mathscr{F}_n \subset \mathscr{B}_n$, $\mathscr{F}_n$ nondecreasing, and S_n be $\mathscr{F}_n$-measurable. Then $\{S_n, \mathscr{F}_n, n \geqslant 1\}$ is a submartingale.

(ii) Let $\{S_n = \sum_{i=1}^n X_i, \mathscr{B}_n, n \geqslant 1\}$ be a martingale with $EX_i = 0$, $i \geqslant 1$. Then

$$ES_n^2 = \sum_{i=1}^n EX_i^2.$$

(iii) Let $\{S_n, \mathscr{B}_n, n \geqslant 1\}$ be a martingale and ϕ be any real convex function such that $E|\phi(S_n)| < \infty$. Then $\{\phi(S_n), \mathscr{B}_n, n \geqslant 1\}$ is a submartingale. If $\{S_n, \mathscr{B}_n, n \geqslant 1\}$ is a submartingale, then $\{\phi(S_n), \mathscr{B}_n, n \geqslant 1\}$ is a submartingale provided $\phi(\cdot)$ is in addition nondecreasing.

(iv) Let $\{S_k, \mathscr{B}_k, 0 \leqslant k \leqslant n\}$ be a submartingale. Then, for any $\varepsilon > 0$,

$$P\left[\max_{0 \leqslant k \leqslant n} S_k \geqslant \varepsilon\right] \leqslant \frac{1}{\varepsilon} E[S_k^+].$$

Proposition 1.7.4. Let $\{S_n, \mathscr{B}_n, n \geqslant 1\}$ be a martingale. Then the following conditions are equivalent:

(i) The family $\{S_n, n \geqslant 1\}$ is uniformly integrable.

(ii) $E|S_n - S_m| \to 0$ as $m, n \to \infty$.

Furthermore if one of these conditions hold, then S_n converges to a random variable S a.s. If for some $p > 1$, $\sup_n E|S_n|^p < \infty$, then (i) and (ii) hold and hence $S = \lim S_n$ a.s. and in the sense of convergence in pth mean.

Remark. In particular, if $\{S_n, \mathscr{B}_n, n \geqslant 1\}$ is a nonnegative martingale, then S_n converges a.s. to a random variable S, and, furthermore, if $\sup_n ES_n^{1+\delta} < \infty$ for some $\delta > 0$, then $ES_n \to ES$ as $n \to \infty$.

For proofs of Propositions 1.7.3 and 1.7.4, see Chow and Teicher (1978) or Gikhman and Skorokhod (1974).

Definition 1.7.2. A *reverse martingale* or *backward martingale* $\{X_t, \mathscr{B}_t, t \in T\}$ is a family of random variables such that X_t is $\mathscr{B}_t$-measurable, $E|X_t| < \infty$ but $\mathscr{B}_t$

is a decreasing family of σ-algebras such that

$$E(X_t|\mathscr{B}_s)=X_s, \qquad s\geqslant t, \quad t,s\in T.$$

If $T=\{1,2,\ldots,n\}$, then $\{X_i,\mathscr{B}_i, 1\leqslant i\leqslant n\}$ is a reverse martingale if and only if $\{X_{n-i+1},\mathscr{B}_{n-i+1}, 1\leqslant i\leqslant n\}$ is a martingale.

Proposition 1.7.5. If $\{S_n,\mathscr{B}_n, n\geqslant 1\}$ is a reverse martingale, then S_n converges almost surely to a finite limit.

Limit theorems for reverse martingales have been studied by Loynes (1969, 1970) and Prakasa Rao (1977, 1979). An example of a reverse martingale occurs in the study of limit theorem for U-statistics [see Example 1.7.3 and Loynes (1978)].

An important application of Proposition 1.7.4 is to check whether two probability measures are absolutely continuous or not and the computation of their Radon–Nikodym derivative in the case of absolute continuity. For proofs, see Gikhman and Skorokhod (1974).

Proposition 1.7.6. Let $(\Omega,\mathscr{B})$ be a measurable space and $\{\mathscr{B}_n, n\geqslant 1\}$ be an increasing family of sub-σ-algebras of $\mathscr{B}$ such that $\vee_{n=1}^{\infty}\mathscr{B}_n=\mathscr{B}$. Let μ_1 and μ_2 be probability measures on $(\Omega,\mathscr{B})$ and $\mu_i^{(n)}$ be the restriction of μ_i to $\mathscr{B}_n$. Suppose $\mu_2^{(n)}\ll\mu_1^{(n)}$ for all $n\geqslant 1$. Let ρ_n be the Radon–Nikodým derivative of $\mu_2^{(n)}$ with respect to $\mu_1^{(n)}$ on $(\Omega,\mathscr{B}_n)$. Then $\{\rho_n, n\geqslant 1\}$ is a martingale with respect to $(\Omega,\mathscr{B},\mu_1)$. Furthermore,

$$\rho_n\to\rho \quad \text{a.e. } [\mu_1]$$

where ρ is the density (Radon–Nikodym derivative) of the absolutely continuous component of μ_2 with respect to μ_1. In order that $\mu_2\ll\mu_1$, it is necessary and sufficient that

$$\int_\Omega \rho(\mathrm{u})\mu_1(d\mathrm{u})=1$$

and in such an event $d\mu_2/d\mu_1=\rho$ a.e. $[\mu_1]$.

For further discussion on the absolute continuity and the singularity of measures, see Sections 1.3 and 1.12.

Proposition 1.7.7. Let $(\Omega,\mathscr{B},\mu)$ be a probability space and ν be another totally finite measure on $(\Omega,\mathscr{B})$ absolutely continuous with respect to μ. Let

$\{A_{nk}, k \geqslant 1\}$ be a measurable partition of Ω for every $n \geqslant 1$ and define, for every $\omega \in \Omega$,

$$\rho_n(\omega) = \begin{cases} \dfrac{\nu(A_{nk}(\omega))}{\mu(A_{nk}(\omega))} & \text{if } \mu(A_{nk}(\omega)) > 0 \\ 0 & \text{if } \mu(A_{nk}(\omega)) = 0, \end{cases}$$

where $A_{nk}(\omega)$ is that set of the sequence $\{A_{nk}, k \geqslant 1\}$ that contains ω. Let $\mathscr{B}_n$ be the σ-algebra generated by $\{A_{nk}, k \geqslant 1\}$. Then

(i) $\{\rho_n, \mathscr{B}_n, n \geqslant 1\}$ is a martingale,

(ii) $\rho_n(\cdot)$ converges to a limit (say) $\rho(\cdot)$ a.e. $[\mu]$ independent of the choice of partition, and

(iii) $(d\nu/d\mu)(\omega) = \rho(\omega)$ a.e. $[\mu]$.

This proposition provides a method for the computation of the Radon–Nikodym derivative when it exists.

We now state some results concerning WLLN and SLLN for martingales.

Proposition 1.7.8 (WLLN). Let $\{X_n, n \geqslant 1\}$ be a sequence of random variables such that $E(X_1) = 0$ and $E(X_n | X_1, \ldots, X_{n-1}) = 0, n \geqslant 2$. Let $S_n = X_1 + \cdots + X_n$. If $EX_j^2 < \infty$ for all $j \geqslant 1$ and

$$\frac{1}{n^2} \sum_{j=1}^{n} EX_j^2 \to 0,$$

then

$$\frac{S_n}{n} \xrightarrow{p} 0 \quad \text{as } n \to \infty.$$

Proposition 1.7.9 (WLLN). Define S_n as in Proposition 1.7.8. Suppose $\{b_n\}$ is a sequence of positive constants tending to ∞. Let

$$X_{ni} = \begin{cases} X_i & \text{if } |X_i| \leqslant b_n \\ 0 & \text{if } |X_i| > b_n. \end{cases}$$

Suppose

(i) $\displaystyle\sum_{i=1}^{n} P(|X_i| > b_n) \to 0,$

$$\text{(ii)}\quad b_n^{-1} \sum_{i=1}^{n} E(X_{ni}|\mathscr{F}_{i-1}) \xrightarrow{p} 0,$$

and

$$\text{(iii)}\quad b_n^{-2} \sum_{i=1}^{n} \{EX_{ni}^2 - E[E(X_{ni}|\mathscr{F}_{i-1})]^2\} \to 0$$

where $\mathscr{F}_i = \sigma(X_1, \ldots, X_i)$. Then

$$\frac{S_n}{b_n} \xrightarrow{p} 0 \quad \text{as } n \to \infty.$$

For proofs of Propositions 1.7.8 and 1.7.9, see Hall and Heyde (1980).

Proposition 1.7.10 (SLLN). Let $S_n = \sum_{i=1}^{n} X_i$, $n \geqslant 1$, be as defined in Proposition 1.7.8. Suppose that $\{b_n\}$ is a sequence of positive constants increasing to ∞ and

$$\sum_{k=1}^{\infty} \frac{E(X_k^2)}{b_k^2} < \infty.$$

Then

$$\frac{S_n}{b_n} \xrightarrow{\text{a.s.}} 0 \quad \text{as } n \to \infty.$$

For proof, see Loeve (1963). The law of iterated logarithm can also be obtained for martingales. See Stout (1970a, b) or Heyde (1977).

Observe that if $\{Z_n, n \geqslant 0\}$ is any sequence of random variables such that $E|Z_n| < \infty$, then the sequence

$$S_n = \sum_{i=1}^{n} [Z_i - E(Z_i|Z_1, \ldots, Z_{i-1})]$$

is a martingale sequence with respect to the sequence of σ-algebras $\mathscr{B}_n = \sigma(Z_0, \ldots, Z_n)$, $n \geqslant 1$.

For other variations and further discussion about the laws of large numbers in the martingale case, see Hall and Heyde (1980) and Chow and Teicher (1978).

The next result is the central limit theorem for martingales proved by Billingsley (1961b) and Ibragimov (1963).

Proposition 1.7.11 (CLT). Let $\{Z_n, n \geqslant 1\}$ be a strictly stationary ergodic process such that $E(Z_1^2) < \infty$ and $E(Z_n|Z_1, \ldots, Z_{n-1}) = 0$ for $n > 1$ and

$E(Z_1) = 0$. Then

$$n^{-1/2} \sum_{k=1}^{n} Z_k \xrightarrow{\mathscr{L}} N(0, \sigma^2),$$

where $\sigma^2 = E(Z_1^2)$.

Proposition 1.7.12 (CLT). Let $\{Z_n, n \geq 1\}$ be a sequence of random variables defined on a probability space $(\Omega, \mathscr{B}, P)$ and $\{\mathscr{B}_n, n \geq 1\}$ be a sequence of σ-algebras such that $E(Z_n | \mathscr{B}_{n-1}) = 0$ a.s. for $n > 1$ and $E(Z_1) = 0$. Suppose that there exists $\delta > 0$ such that

$$\sup_n E|Z_n|^{2+\delta} < \infty$$

and

$$\lim_{n \to \infty} n^{-1} \sum_{k=1}^{n} E(Z_k^2 | \mathscr{B}_{k-1}) = \sigma^2 > 0 \quad \text{a.s.}$$

Then $\{S_n, \mathscr{B}_n; n \geq 1\}$ is a martingale and

$$n^{-1/2} S_n \xrightarrow{\mathscr{L}} N(0, \sigma^2),$$

where $S_n = Z_1 + \cdots + Z_n$.

Proposition 1.7.12 is due to Billingsley (1961a). We now state a general central limit theorem for dependent random variables due to Helland (1982).

Proposition 1.7.13. Let $\{X_{nk}, k \geq 1, n \geq 1\}$ be an array of random variables defined on a probability space $(\Omega, \mathscr{B}, P)$ and $\{\mathscr{B}_{nk}, k \geq 0, n \geq 1\}$ be an array of σ-algebras on Ω such that X_{nk} is $\mathscr{B}_{nk}$-measurable and

$$\mathscr{B}_{n,k-1} \subset \mathscr{B}_{n,k} \subset \mathscr{B}$$

for $n \geq 1$ and $k \geq 1$. Let τ_n be a stopping time with respect to $\{\mathscr{B}_{nk}, k \geq 0\}$. Denote by $P_{k-1}[\cdot]$, $E_{k-1}[\cdot]$, and $\mathrm{Var}_{k-1}[\cdot]$ the conditional probability, the conditional expectation, and the conditional variance given $\mathscr{B}_{n,k-1}$.

Suppose that

$$\text{(i)} \quad \sum_{k=1}^{\tau_n} P_{k-1}[|X_{nk}| > \varepsilon] \xrightarrow{p} 0 \quad \text{for all } \varepsilon > 0,$$

$$\text{(ii)} \quad \sum_{k=1}^{\tau_n} E_{k-1}[X_{nk} I(|X_{nk}| \leq 1)] \xrightarrow{p} 0,$$

and

$$\text{(iii)} \quad \sum_{k=1}^{\tau_n} \operatorname{var}_{k-1}[X_{nk}I(|X_{nk}| \leqslant 1)] \xrightarrow{p} 1$$

as $n \to \infty$. Let $S_n = \sum_{k=1}^{\tau_n} X_{nk}$. Then $S_n \xrightarrow{\mathscr{L}} N(0, 1)$.

As a consequence of Proposition 1.7.13, the following central limit theorem for martingales can be obtained.

Proposition 1.7.14 (Helland, 1982). Suppose $\{X_{nk}\}$ and $\{\mathscr{B}_{nk}\}$ are as defined above and $\tau_n = k_n$ is a sequence of constants. Suppose further that $E_{k-1}(X_{nk}) = 0$ (such an array is called a *martingale difference array*). Assume that one of the following three sets of conditions [(a)–(c)] holds:

$$\text{(a)} \quad \text{(i)} \quad \sum_{k=1}^{k_n} E_{k-1}[X_{nk}^2 I(|X_{nk}| > \varepsilon)] \xrightarrow{p} 0 \quad \text{for all } \varepsilon > 0;$$

$$\text{(ii)} \quad \sum_{k=1}^{k_n} \operatorname{var}_{k-1}[X_{nk}] \xrightarrow{p} 1.$$

$$\text{(b)} \quad \text{(i)} \quad \sum_{k=1}^{k_n} X_{nk}^2 \xrightarrow{p} 1;$$

$$\text{(ii)} \quad E\left[\max_{1 \leqslant k \leqslant k_n} |X_{nk}|\right] \to 0.$$

$$\text{(c)} \quad \text{(i)} \quad \operatorname{mar}_{1 \leqslant k \leqslant k_n} |X_{nk}| \xrightarrow{p} 0;$$

$$\text{(ii)} \quad \sum_{k=1}^{k_n} X_{nk}^2 \xrightarrow{p} 1;$$

$$\text{(iii)} \quad \sum_{k=1}^{k_n} |E_{k-1}[X_{nk}I(|X_{nk}| > 1)]| \xrightarrow{p} 0.$$

Then $S_n \xrightarrow{\mathscr{L}} N(0, 1)$.

In applications to statistical inference for stochastic processes, it is some times important to obtain limit theorem with random norming.

Proposition 1.7.15 (Basawa and Scott, 1977). Let $\{Z_n, n \geqslant 1\}$ be a sequence of random variables defined on a probability space $(\Omega, \mathscr{B}, P)$ such that $E(Z_k | Z_1, \ldots, Z_{k-1}) = 0$, $k > 1$ and $E(Z_1) = 0$. Suppose that $E(Z_k^2) < \infty, k \geqslant 1$. Let

$$S_n = Z_1 + \cdots + Z_n \quad \text{and} \quad I_n = \sum_{k=1}^{n} E(Z_k^2).$$

If

$$\text{(i)} \quad I_n^{-1} \sum_{k=1}^{n} Z_k^2 \xrightarrow{p} \eta, \quad \text{where} \quad P(\eta > 0) = 1,$$

and

$$\text{(ii)} \quad I_n^{-1} E\left\{ \max_{1 \leqslant k \leqslant n} Z_k^2 \right\} \to 0 \quad \text{as} \quad n \to \infty,$$

then

$$I_n^{-1/2} S_n \xrightarrow{\mathscr{L}} \eta^{1/2} Z,$$

where Z is standard normal and independent of η.

Proposition 1.7.16 (Hall, 1977). Define $\{Z_n, n \geqslant 1\}$ as above. Let

$$\xi_n = \sum_{i=1}^{n} E(Z_i^2 | \mathscr{B}_{i-1}),$$

where $\mathscr{B}_i = \sigma(Z_j, 1 \leqslant j \leqslant i)$ and $\mathscr{B}_0 = \{\phi, \Omega\}$. If

$$\text{(i)} \quad I_n^{-1} \xi_n \xrightarrow{p} \eta \quad \text{where} \quad P(\eta > 0) = 1,$$

and

$$\text{(ii)} \quad I_n^{-1} \sum_{k=1}^{n} E_{k-1}[Z_k^2 I(|Z_k| \geqslant \varepsilon I_n^{1/2})] \to 0 \quad \text{for all } \varepsilon > 0,$$

then

$$I_n^{-1/2} S_n \xrightarrow{\mathscr{L}} \eta^{1/2} Z$$

and

$$\xi_n^{-1/2} S_n \xrightarrow{\mathscr{L}} Z,$$

where Z is standard normal and independent of η.

Before we conclude the discussion on central limit theorems for martingales, we state a random central limit theorem for martingales.

Proposition 1.7.17 (Prakasa Rao, 1969; Basawa and Prakasa Rao, 1980). Let $\{Z_n, n \geqslant 1\}$ be a martingale difference sequence as defined in Proposition

1.7.12. Let $S_n = Z_1 + \cdots + Z_n$. Furthermore, suppose that $n^{-1/2}S_n \xrightarrow{\mathscr{L}} N(0, \sigma^2)$. Suppose $\{v_n\}$ is a sequence of positive integer-valued random variables such that $n^{-1}v_n \xrightarrow{p} c$, where $c > 0$. Then

$$v_n^{-1/2} S_{v_n} \xrightarrow{\mathscr{L}} N(0, \sigma^2)$$

provided

$$\lim_{\varepsilon \to \pm 0} \overline{\lim_n} \frac{E(S^2_{[nc]}) - E(S^2_{[nc(1 \pm \varepsilon)]})}{n} = 0.$$

The following result, known as *Skorokhod embedding*, is quite useful at times to prove limit theorems or obtain the rates of convergence for sums of martingale differences.

Proposition 1.7.18 (Skorokhod, 1965). Let $S_n = \sum_{i=1}^n X_i, n \geqslant 1$, be a zero-mean, square integrable martingale (i.e., $\sup_n ES_n^2 < \infty$) defined on a probability space $(\Omega, \mathscr{B}, P)$. Then there exist a probability space $(\Omega', \mathscr{B}', P')$, a standard Wiener process $W(\cdot)$ defined on it, and a sequence of nonnegative random variables $\tau_1, \tau_2, \ldots$ with the following properties. Define

$$T_n = \sum_{i=1}^n \tau_i, \qquad S'_n = W(T_n), \qquad X'_1 = S'_1, \qquad X'_n = S'_n - S'_{n-1}$$

for $n \geqslant 2$ and $\mathscr{B}'_n$ be the σ-algebra generated by $S'_1, \ldots, S'_n$ and $W(t), 0 \leqslant t \leqslant T_n$. Then

(i) $\{S'_n, n \geqslant 1\}$ has the same distribution as $\{S_n, n \geqslant 1\}$,

(ii) T_n is $\mathscr{B}'_n$-measurable,

(iii) $E(\tau_n | \mathscr{B}'_{n-1}) = E\{(X'_n)^2 | \mathscr{B}'_{n-1}\}$ a.s., and

(iv) for each $r \geqslant 1$, there exists a constant $c_r > 0$ depending or r such that

$$E(\tau_n^r | \mathscr{B}'_{n-1}) \leqslant c_r E(|X'_n|^{2r} | \mathscr{B}'_{n-1}) \text{ a.s.}$$

See Hall and Heyde (1980, p. 269) for proof of Proposition 1.7.18. We now prove an inequality that is more general than part (iv) of this proposition. The following result and its consequences stated below are due to Sheu and Yao (1984). We introduce some notation.

Let X be a zero-mean random variable with distribution function F and

$E(X^2) < \infty$. Let $h(x) = F^{-1}(x) \equiv \inf\{t: F(t) \geqslant x\}$, $0 \leqslant x \leqslant 1$. Let $g(x)$ be a monotone solution of the equation

$$\int_x^{g(x)} h(t)\,dt = 0. \tag{1.7.0}$$

Let us construct a probability space $(\Omega, \mathscr{F}, P)$ supporting a standard Wiener process W and a random variable U independent of $\{W(t), t \geqslant 0\}$, which is uniform on $[0, 1]$. Conditioning on $U = x$, define

$$\tau(x) = \inf\{t: W(t) = h(x) \quad \text{or} \quad h(g(x))\}. \tag{1.7.1}$$

We shall show that the distribution of $W(\tau)$ is F and $E(\tau) = E(X^2)$.

Let $G(u, v)$ be the distribution that assigns probability $|u|/(|u| + |v|)$ at v and probability $|v|/(|u| + |v|)$ at u whenever $uv < 0$. It is easy to see that, given that $U = x$, $W(\tau)$ has distribution $G(h(x), h(g(x)))$ by the definition of τ and the properties of Wiener process. Hence, for any bounded measurable function ψ,

$$E\psi(W(\tau)) = \int_0^1 \frac{[\psi(h(x))|h(g(x))| + \psi(h(g(x))|h(x)|]}{(|h(g(x))| + |h(x)|)}\,dx. \tag{1.7.2}$$

Note that $g(g(x)) = x$ and $h(x) = h(g(x))g^{(1)}(x)$ a.e. Here $g^{(1)}(x)$ denotes the derivative of $g(x)$, which exists almost everywhere. Therefore

$$\int_0^1 \frac{\psi(h(g(x))|h(x)|}{(|h(x)| + |h(g(x))|)}\,dx = \int_0^1 \frac{\psi(h(x))|h(x)|}{(|h(x)| + |h(g(x)|)}\,dx \tag{1.7.3}$$

and hence

$$E\psi(W(\tau)) = \int_0^1 \psi(h(x))\,dx = \int_{-\infty}^{\infty} \psi(x)\,dF(x). \tag{1.7.4}$$

Hence $W(\tau)$ has the distribution function F. Similar reasoning shows that

$$E(\tau) = \int_0^1 |h(x)h(g(x))|\,dx = \int_0^1 h^2(x)\,dx = \int_{-\infty}^{\infty} x^2\,dF(x) = E(X^2). \tag{1.7.5}$$

The following theorem gives an inequality for expectation of a function of τ.

Theorem 1.7.19 (Sheu and Yao, 1984). Let $\phi(x)$, $x \geqslant 0$, be a nonnegative function such that $\phi(x)/x^a$ is nondecreasing and $\phi(x)/x^b$ is nonincreasing for some a, b such that $\frac{1}{2} < a < b < \infty$. Then

$$E\phi(\tau) \leqslant CE\phi(X^2), \tag{1.7.6}$$

where C is a constant depending only on a and b.

Proof. Let $T = \inf\{t: W(t) = u \text{ or } v\}$ where $-u \leqslant 0 \leqslant v$, $u + v \neq 0$. It is known from Breiman (1967) that

$$P(T \geqslant t) = \frac{4}{\pi} \sum_{n=1}^{\infty} (2n+1)^{-1} \exp\left[-\frac{\lambda_n^2 t}{2r^2}\right] \sin\left(\frac{\lambda_n u}{r}\right), \tag{1.7.7}$$

where $\lambda_n = 2n + 1$, $r = u + v$. It is easy to check that

$$\sin\left(\frac{\lambda_n u}{r}\right) \leqslant \frac{4\pi^2(2n+1)uv}{r^2},$$

using induction. Hence

$$\begin{aligned} E\phi(T) &= \int_0^\infty P(T \geqslant t)\, d\phi(t) \\ &\leqslant C_1 \frac{uv}{r^2} \int_0^\infty \left(\sum_{n=1}^{\infty} \exp\left[-\frac{\lambda_n t}{2r^2}\right]\right) d\phi(t), \end{aligned} \tag{1.7.8}$$

where C_1 is an absolute constant. Note that, for any $s > 0$,

$$\begin{aligned} \sum_{n=0}^{\infty} e^{-\lambda_n^2 s} &\leqslant e^{-8s} \sum_{n=0}^{\infty} e^{-4n^2\pi^2 s} \\ &\leqslant e^{-8s}\left(1 + \int_0^\infty e^{-\pi^2 s x^2}\, dx\right) \\ &= (1 + C_2 s^{-1/2}) e^{-8s} \end{aligned}$$

for some absolute constant C_2. Therefore

$$\begin{aligned} E\phi(T) \leqslant C_3 \Bigg[&\frac{uv}{r^2} \int_0^\infty \exp\left(-\frac{4t}{r^2}\right) d\phi(t) \\ &+ \frac{uv}{r} \int_0^\infty \exp\left(-\frac{4t}{r^2}\right) \frac{d\phi(t)}{t^{1/2}} \Bigg], \end{aligned} \tag{1.7.9}$$

where C_3 is an absolute constant. But, for $s > 0$,

$$\int_0^\infty e^{-st}\,d\phi(t) = \int_0^{1/s} e^{-st}\,d\phi(t) + \int_{1/s}^\infty e^{-st}\,d\phi(t), \tag{1.7.10}$$

where

$$\begin{aligned} \int_0^{1/s} e^{-st}\,d\phi(t) &= e^{-1}\phi\left(\frac{1}{s}\right) + s\int_0^{1/s} e^{-st}\phi(t)\,dt \\ &\leqslant e^{-1}\phi\left(\frac{1}{s}\right) + s^{a+1}\phi(s^{-1})\int_0^1 t^a e^{-st}\,dt \\ &\leqslant C_4\phi(s^{-1}), \end{aligned} \tag{1.7.11}$$

and

$$\begin{aligned} \int_{1/s}^\infty e^{-st}\,d\phi(t) &= -e^{-1}\phi(s^{-1}) + s\int_{1/s}^\infty e^{-st}\phi(t)\,dt \\ &\leqslant -e^{-1}\phi(s^{-1}) + s^{b+1}\phi(s^{-1})\int_{1/s}^\infty t^b e^{-st}\,dt \\ &\leqslant C_5\phi(s^{-1}), \end{aligned} \tag{1.7.12}$$

where C_4 and C_5 are constants depending on a and b. Furthermore,

$$\begin{aligned} \int_0^\infty t^{-1/2}e^{-st}\,d\phi(t) &= \int_0^{1/s} t^{-1/2}e^{-st}\,d\phi(t) + \int_{1/s}^\infty t^{-1/2}e^{-st}\,d\phi(t) \\ &\leqslant C_6 s^{1/2}\phi(s^{-1}), \end{aligned} \tag{1.7.13}$$

where C_6 is a constant depending on a and b by a similar argument. Combining relations (1.7.8) to (1.7.13), we have

$$E\phi(T) \leqslant C\frac{uv}{r^2}\phi\left(\frac{r^2}{4}\right) \leqslant C\frac{uv}{r^2}\phi\left(\frac{u^2+v^2}{2}\right), \tag{1.7.14}$$

where C is a constant depending on a and b. Note that

$$\phi\left(\frac{u^2+v^2}{2}\right) \leqslant \max\{\phi(u^2), \phi(v^2)\} \leqslant \phi(u^2) + \phi(v^2).$$

Hence

$$E\phi(T) \leqslant C\frac{uv}{r^2}\{\phi(u^2)+\phi(v^2)\}. \tag{1.7.15}$$

Therefore

$$E\phi(\tau) \leqslant C\int_0^1 \frac{|h(x)h(g(x))|}{(|h(x)|+|h(g(x))|)^2}|\phi(h^2(x))+\phi(h^2(g(x))|\,dx \tag{1.7.16}$$

$$= C\int_0^1 |h(x)|\,|\phi(h^2(x))|(|h(x)|+|h(g(x))|)^{-1}\,dx$$

$$\leqslant C\int_0^1 \phi(h^2(x))\,dx = CE\phi(X^2).$$

This completes the proof.

We shall now derive some consequences of Theorem 1.7.19 following Sheu and Yao (1984).

Theorem 1.7.20. Let $S_n=\sum_{j=1}^n X_j$, $n \geqslant 1$, be a zero-mean martingale and $S_n \to \infty$ a.s. Let ϕ be a function as defined in Theorem 1.7.19. Furthermore, suppose that $\phi(x+y) \leqslant \phi(x)+\phi(y)$ for all $x, y \geqslant 0$. If $0<a_n \uparrow \infty$ and

$$\sum_{n=1}^{\infty}\frac{E\phi(X_n^2)}{\phi(a_n)}<\infty, \tag{1.7.17}$$

then

$$\frac{S_n}{a_n^p} \to 0 \quad \text{a.s} \quad \text{for every } p > \tfrac{1}{2}. \tag{1.7.18}$$

Proof. Let the sequence τ_n be constructed as in Proposition 1.7.18 so that $\{W(T_n), n \geqslant 1\}$ has the same distribution as $\{S_n, n \geqslant 1\}$, where $T_n=\sum_{j=1}^n \tau_j$ and $W(\cdot)$ is the standard Wiener process. Since

$$E\phi(\tau_n) \leqslant CE\phi(X_n^2),$$

where $X_n = S_n - S_{n-1}$, it follows from Theorem 1.7.19 and the condition (1.7.17) that

$$\sum_{n=1}^{\infty}\frac{\phi(\tau_n)}{\phi(a_n)}<\infty \quad \text{a.s.}$$

This in turn implies that

$$\frac{1}{\phi(a_n)}\sum_{j=1}^{n}\phi(\tau_j)\to 0 \quad \text{a.s.}$$

by Kronecker's lemma. But

$$\phi\left(\sum_{j=1}^{n}\tau_j\right)\leqslant\sum_{j=1}^{n}\phi(\tau_j)$$

by hypothesis and hence

$$\phi\left(\sum_{j=1}^{n}\tau_j\right)\Big/\phi(a_n)\to 0 \quad \text{a.s.}$$

But

$$\phi\left(\sum_{j=1}^{n}\tau_j\right)\Big/\phi(a_n)\geqslant\min\left\{\left[\frac{1}{a_n}\sum_{j=1}^{n}\tau_j\right]^a,\left[\frac{1}{a_n}\sum_{j=1}^{n}\tau_j\right]^b\right\}$$

since $\phi(x)/x^a$ is nondecreasing and $\phi(x)/x^b$ is nonincreasing. Therefore

$$\frac{1}{a_n}\sum_{j=1}^{n}\tau_j\to 0 \quad \text{a.s.} \tag{1.7.19}$$

Since $S_n\to\infty$ a.s., it follows that $\sum_{j=1}^{n}\tau_j\to\infty$ a.s. The law of iterated logarithm for Wiener process (Levy, 1937; Csörgö and Revesz, 1981) implies that

$$\overline{\lim_{T\to\infty}}\frac{|W(T)|}{(2T\log\log T)^{1/2}}=1 \quad \text{a.s.}$$

Hence

$$\frac{W(\sum_{j=1}^{n}\tau_j)}{(\sum_{j=1}^{n}\tau_j)^p}\to 0 \quad \text{a.s.} \tag{1.7.20}$$

for $p>\frac{1}{2}$. Relations (1.7.19) and (1.7.20) prove that

$$\frac{S_n}{a_n^p}\to 0 \quad \text{a.s. for } p>\tfrac{1}{2}. \tag{1.7.21}$$

As corollaries to Theorem 1.7.20, we have the following SLLN for martingales.

Corollary 1.7.21. Let $S_n = \sum_{j=1}^{n} X_j$, $n \geqslant 1$, be a zero-mean martingale. If $E(X_n^2) < \infty$ for all $n \geqslant 1$ and $s_n^2 = \sum_{j=1}^{n} EX_j^2 \to \infty$, then

$$\frac{S_n}{s_n^p} \to 0 \quad \text{a.s. for} \quad p > 1.$$

Proof. Let $\phi(x) = x$ and $a_n = s_n^2(\log s_n^2)^{1+\varepsilon}$ for some $\varepsilon > 0$ in Theorem 1.7.20. Clearly

$$\sum_{n=1}^{\infty} \frac{E\phi(X_n^2)}{\phi(a_n)} = \sum_{n=1}^{\infty} \frac{E(X_n^2)}{s_n^2(\log s_n^2)^{1+\varepsilon}} < \infty$$

and hence

$$\frac{S_n}{s_n^{2p}(\log s_n^2)^{p+p\varepsilon}} \to 0 \quad \text{a.s. for every} \quad p > \tfrac{1}{2}$$

by Theorem 1.7.20, which implies that

$$\frac{S_n}{s_n^p} \to 0 \quad \text{a.s. for every} \quad p > 1.$$

Corollary 1.7.22. Let $S_n = \sum_{j=1}^{n} X_j$, $n \geqslant 1$, be a zero-mean martingale. If

$$\sum_{n=1}^{\infty} \frac{E|X_n|^r}{n^r} < \infty \quad \text{for some } r \text{ such that } 1 < r \leqslant 2,$$

then

$$\frac{S_n}{n^p} \to 0 \quad \text{a.s. for every } p > 1.$$

Proof. Let $\phi(x) = x^{r/2}$ and $a_n = n^2 \cdot \phi(\cdot)$ satisfies conditions in Theorem 1.7.20 and

$$\sum_{n=1}^{\infty} \frac{E\phi(X_n^2)}{\phi(a_n)} = \sum_{n=1}^{\infty} \frac{E|X_n|^r}{n^r} < \infty$$

by hypothesis. Hence

$$\frac{S_n}{n^p} \to 0 \quad \text{a.s. for every} \quad p > 1.$$

Remark. If $S_n = \sum_{j=1}^n X_j$, $n \geqslant 1$, is a zero-mean martingale and if

$$\sum_{n=1}^{\infty} \frac{EX_n^2}{n^2} < \infty,$$

then $S_n/n \to 0$ a.s. (Chow, 1967).

For further results on bounds of the Berry–Esseen type and on nonuniform bounds for martingale differences, see Hall and Heyde (1980). Helland (1982) gives an excellent discussion on central limit theorems for martingales with continuous time. No discussion of martingale theory is complete unless one includes some recent interesting developments that had an important bearing, for instance, on statistical inference for point processes. For an excellent survey of recent developments, see Shiryayev (1981). A brief discussion of the interface between martingale theory and point processes is given in Prakasa Rao (1984) following Aalen (1982). See Section 5.5.

1.8 SEMIMARTINGALES

Let $(\Omega, \mathscr{B}, P)$ be a probability space and $\{\mathscr{B}_t, t \geqslant 0\}$ be a nondecreasing family of sub-σ-algebras of $\mathscr{B}$ where $\mathscr{B}_0$ contains all P-null sets and $\{\mathscr{B}_t, t \geqslant 0\}$ is continuous from the right, that is,

$$\mathscr{B}_t = \mathscr{B}_{t+} = \bigcap_{s>t} \mathscr{B}_s.$$

Let $X = \{X_t, t \geqslant 0\}$ be a stochastic process such that X_t is $\mathscr{B}_t$-measurable for every $t \geqslant 0$. Then X is said to be *adapted* to $\{\mathscr{B}_t, t \geqslant 0\}$. We assume that $X_0 = 0$ without loss of generality. If $\mathscr{B}_t^X = \sigma(X_s, s \leqslant t)$, then $\{\mathscr{B}_t^X, t \geqslant 0\}$ is called a *flow* corresponding to the process X.

Definition 1.8.1. A *stopping time* τ is an $\overline{R^+}$-valued random variable such that $[\tau \leqslant t] \in \mathscr{B}_t$ for each $t \geqslant 0$. We denote

$$\mathscr{B}_\tau = \{B \in \mathscr{B} : B \cap [\tau \leqslant t] \in \mathscr{B}_t, t \geqslant 0\}.$$

Note that τ could take the value $+\infty$.

Definition 1.8.2. A process X is said to satisfy a property (α) *locally* if there exists a sequence of stopping times τ_n, $n \geqslant 1$, such that $\tau_n \uparrow \infty$ a.s. as $n \to \infty$ and, for each n, the stopped process $t \to I_{[\tau_n > 0]} X(t \wedge \tau_n)$ has the required property (α).

Definition 1.8.3. A *local martingale* $M = \{M_t, \mathscr{B}_t, t \geqslant 0\}$ is an adapted process for which one can find a sequence of stopping times $\{\tau_n, n \geqslant 0\}$ increasing to infinity such that the stopped process

$$M^n = \{M_{t \wedge \tau_n}, \mathscr{B}_t, t \geqslant 0\}$$

is a martingale for every $n \geqslant 1$.

Let $\mathscr{M}_{\text{loc}}$ denote the class of local martingales defined on Ω with respect to $\{\mathscr{B}_t, t \geqslant 0\}$.

Definition 1.8.4. A process $V = \{V_t, \mathscr{B}_t, t \geqslant 0\}$ is a process of *locally bounded variation* if, for arbitrary $t \geqslant 0$ and $\omega \in \Omega$, the variation

$$\int_0^t |dV_s(\omega)| < \infty.$$

Let $\mathscr{V}_{\text{loc}}$ denote the class of such processes.

Definition 1.8.5. A random process $X = \{X_t, \mathscr{B}_t, t \geqslant 0\}$ is called a *semimartingale* if there exists $V \in \mathscr{V}_{\text{loc}}$ and $M \in \mathscr{M}_{\text{loc}}$ such that

$$X = V + M.$$

Remarks. An arbitrary process X with stationary independent increments is a semimartingale and a process with independent increments is a semimartingale if the function $\phi(t) = E[e^{i\lambda X_t}]$ is a function of locally bounded variation for each $\lambda \in R$. In the case of discrete time, an arbitrary sequence of random variables $\{X_n, \mathscr{B}_n, n \geqslant 0\}$ is a semimartingale. The class of semimartingales includes many classes of point processes, Ito processes, diffusion processes, and so on. However there are processes that are not semimartingales, for instance, a function of unbounded variation.

It is known that, if $X = \{X_n, \mathscr{B}_n, n \geqslant 0\}$ is a submartingale, then X_n can be written in the form

$$X_n = A_n + M_n, \qquad n \geqslant 0,$$

where $M = \{M_n, \mathscr{B}_n, n \geqslant 0\}$ is a martingale and $A = \{A_n, \mathscr{B}_n, n \geqslant 0\}$ is an increasing process with $A_0 = 0$ and A_n is $\mathscr{B}_{n-1}$ measurable. This follows from Doob (1953). Furthermore, such a representation is unique.

Let $D \equiv D[0, \infty)$ be the space of right-continuous functions with left limits endowed with Skorokhod topology [cf. Lindvall (1973), Section 1.10], and $\mathscr{D}$ be the σ-algebra generated by open sets in D. Consider the space $\Omega \times [0, \infty)$ and functions $X(\omega, t)$ defined on it such that $X(\omega, t)$ is $\mathscr{B}_t$-measurable for any fixed t and $X(\omega, t)$ is continuous in t for fixed ω. Consider the σ-algebra generated by such functions on $\Omega \times [0, \infty)$, namely,

$$\mathscr{P} = \sigma\{(\omega, t): X(\omega, \cdot) \in A, A \in \mathscr{D}, X(\omega, \cdot) \text{ continuous}\}.$$

$\mathscr{P}$ is called the *predictable* σ-algebra over $\Omega \times [0, \infty)$. The process $X = \{X(\cdot, t), t \geqslant 0\}$ is said to be *predictable* if it is measurable with respect to $\mathscr{P}$.

Predictability of a process essentially implies that one can predict the values at any point by the values at the preceding points as in the case of processes that are left continuous. Any process with almost sure continuous sample paths is predictable. For instance, a Wiener process is predictable whereas a Poisson process is not.

If a semimartingale X admits a representation $X = A + M$, where M is a martingale and A is predictable, then it is known (Meyer, 1976) that such a representation is unique. For details, see Meyer (1976) or Jacod (1979).

In the following discussion, we consider martingales $M \equiv \{M(t), \mathscr{B}_t, t \geqslant 0\}$ that are right continuous with left limits (RCLL). Note that, if M is a square integrable martingale, that is, $\sup_{t \geqslant 0} E|M(t)|^2 < \infty$, then $\lim_{t \to \infty} M(t) = M(\infty)$ exists almost surely. Furthermore, if $\mathscr{B}_\infty$ denotes the σ-algebra generated by $\{\mathscr{B}_t, t \geqslant 0\}$, then $M = \{M(t), \mathscr{B}_t, 0 \leqslant t \leqslant \infty\}$ is a square integrable martingale.

Let M_1 and M_2 be local square integrable martingales. Then there exists a unique predictable process $\langle M_1, M_2 \rangle$ whose variation is finite and that is locally integrable such that $M_1 M_2 - \langle M_1, M_2 \rangle$ is a local martingale taking the value zero at time zero. If $M_1 = M_2$, then $\langle M_1, M_2 \rangle$ is nondecreasing. $\langle M_1, M_2 \rangle$ is called the *predictable covariation process* of M_1 and M_2. If M_1 and M_2 are square integrable martingales, then $M_1 M_2 - \langle M_1, M_2 \rangle$ is a martingale on $[0, \infty]$. $\langle M_1, M_2 \rangle$ is an RCLL process and $\langle \cdot, \cdot \rangle$ is symmetric and bilinear. For simplicity, we denote $\langle M, M \rangle$ by $\langle M \rangle$ in case M is a local square integrable martingale. $\langle M \rangle$ is called the *quadratic characteristic* of the process M.

Suppose Y on R^+ is an RCLL process and is of locally bounded variation with $Y(0-0) = 0$. Furthermore, suppose that X is Lebesgue-measurable real-valued function on R^+ such that

$$\int_0^t |X(s)||dY(s)| < \infty, \qquad t \geqslant 0,$$

that is, X is locally integrable with respect to Y. We write

$$\int_0^t X\,dY = \int_0^t X(s)\,dY(s)$$

and denote the function defined by the integral by $\int X\,dY$. In particular

$$\left(\int X\,dY\right)(0) = X(0)\,Y(0).$$

Stochastic integrals with respect to semimartingales can be defined in several ways. It is known that the class of semimartingales is the largest class of processes for which it is possible to define stochastic integral with reasonable properties such as linearity and satisfying a dominated convergence theorem [cf. Dellacherie (1980)]. One approach due to Metivier and Pellaumail (1980) is similar to the definition of Ito integral [Gikhman and Skorokhod (1972), Section 1.9]. They define a stochastic integral $\int_0^T H_t\,dM_t$ for step functions H_t in the natural way and extend to general classes for processes M with the property that there exists an increasing predictable process Q such that

$$E \sup_{t<T} \left| \int_0^t H_s\,dM_s \right|^2 \leqslant E\left\{ Q_{T-} \int_0^T H_s^2\,dQ_s \right\}.$$

It can be shown that if a process M satisfies such a condition, then M is a semimartingale.

Suppose that M_1 and M_2 are local square integrable martingales. Furthermore, suppose that H_1 and H_2 are predictable and locally bounded. In particular H_1 and H_2 satisfy these conditions if they are left continuous with right limits (LCRL) and are adapted. Then

$$\int H_i\,dM_i, \qquad i = 1, 2,$$

exist and they are local square integrable martingales with

$$\left\langle \int H_1\,dM_1, \int H_2\,dM_2 \right\rangle = \int H_1 H_2\,d\langle M_1, M_2\rangle.$$

If, further, $M_1(0) = M_2(0) = 0$ and the localizing sequence of stopping times associated with M_1, M_2, H_1, and H_2 can be taken to be sequences of constants,

then

$$E\left(\int H_i\,dM_i\right)=0, \qquad i=1,2,$$

and

$$E\left(\int H_1\,dM_1 \int H_2\,dM_2\right)=E\left(\int H_1 H_2\,d\langle M_1, M_2\rangle\right)$$

as functions.

Part of our discussion in this section is based on Gill (1980) and Shiryayev (1981).

We now state the important change-of-variable formula for semimartingales that generalizes Ito's formula (see Section 1.9). If $F=F(x)$ is a twice continuously differentiable function and X is a semimartingale, then

$$\begin{aligned} F(X_t)=F(X_0)+\int_0^t F'(X_s)\,dX_s+\frac{1}{2}\int_0^t F''(X_{s-})\,d\langle X^c\rangle_s \\ +\sum_{0<s\leqslant t}[F(X_s)-F(X_{s-})-F'(X_{s-})\Delta X_s] \end{aligned}$$

where X^c and ΔX_s are as defined below, namely, the continuous component of X, $\langle X^c\rangle$, the quadratic characteristic and $\Delta X_s=X_s-X_{s-}$ (see the discussion later in this section).

In order to state some of the functional central limit theorems for semimartingales due to Liptser and Shiryayev (1980), we now introduce some further concepts and notation.

Let $(D, \mathscr{D})$ be the measurable space of right-continuous functions $x=x(t)$, $t\geqslant 0$, having left limits with Skorokhod topology as before. Denote the jumps of a function $x(\cdot)$ by $\Delta x_t=x_t-x_{t-}$ assuming that $\Delta x_0=0$. Let $X=\{X_t, \mathscr{B}_t, t\geqslant 0\}$ be a $\mathscr{B}_t$-adapted random process with sample paths in D. For simplicity, suppose that $X_0=0$. Furthermore, suppose that $X=\{X_t, \mathscr{B}_t, t\geqslant 0\}$ is a semimartingale. Hence

$$X_t=V_t+M_t,$$

where $V_t\in\mathscr{V}_{\text{loc}}$ and $M_t\in\mathscr{M}_{\text{loc}}$. For some $a>0$, let

$$X_t^a=\sum_{0<s\leqslant t}\Delta X_s I(|\Delta X_s|>a).$$

The process $X-X^a$ with $X^a=\{X_t^a, \mathscr{B}_t, t\geqslant 0\}$ is again a semimartingale, and it admits a unique representation

$$X_t - X_t^a = A_t^a + M_t^a \tag{1.8.0}$$

with a predictable process $A^a \in \mathscr{A}_{\text{loc}}$ and $M^a \in \mathscr{M}_{\text{loc}}^2$, where $\mathscr{A}_{\text{loc}}$ is the class of $\mathscr{B}_t$-adapted locally integrable processes and $\mathscr{M}_{\text{loc}}^2$ is the class of $\mathscr{B}_t$-adapted locally square integrable martingales. Observe that if a process $A \in \mathscr{A}_{\text{loc}}$, then there exists a predictable process $\tilde{A}$ such that $A - \tilde{A} \in \mathscr{M}_{\text{loc}}$. $\tilde{A}$ is called the *compensator* of the process A. Note that in this case

$$|\Delta A_t^a| \leqslant a, \qquad |\Delta M_t^a| \leqslant 2a \quad a.s.\ [P], \quad t \geqslant 0.$$

Let μ be the integer-valued random measure of jumps of the semimartingale X, that is,

$$\mu((0,t],\Gamma) = \sum_{0<s\leqslant t} I(\Delta X_s \in \Gamma), \qquad \Gamma \in \mathscr{B}(R\backslash\{0\}) \tag{1.8.1}$$

where $\mathscr{B}(R\backslash\{0\})$ is the Borel σ-algebra of $R\backslash\{0\}$. It is known that there exists a compensating measure ν called *compensator* on $R^+ \times R\backslash\{0\}$ such that for any nonnegative function Y measurable with respect to the predictable σ-algebra

$$E\int_0^\infty \int_{R\backslash\{0\}} Y\,d\mu = E\int_0^\infty \int_{R\backslash\{0\}} Y\,d\nu. \tag{1.8.2}$$

ν is a compensator for μ in the sense that, for each $\Gamma \in \mathscr{B}(R)$, the process $(\nu((0,t], \Gamma \cap (|x| > \varepsilon)), \mathscr{B}_t)$ compensates $(\mu((0,t], \Gamma \cap (|x| > \varepsilon), \mathscr{B}_t)$ up to a local martingale for every $\varepsilon > 0$, that is,

$$(\mu(\cdot, \Gamma \cap (|x| > \varepsilon)) - \nu(\cdot, \Gamma \cap (|x| > \varepsilon)) \in \mathscr{M}_{\text{loc}}. \tag{1.8.3}$$

Let us suppose that $a = 1$ for convenience. Then

$$X_t^1 = \int_0^t \int_{|x|>1} x\mu(ds, dx), \tag{1.8.4}$$

and we have

$$X_t = A_t^1 + M_t^1 + \int_0^t \int_{|x|>1} \mu(ds, dx). \tag{1.8.5}$$

from (1.8.0). The local martingale $M^1 = (M_t^1, \mathscr{B}_t, t \geqslant 0)$ admits a unique decomposition as the sum of a continuous local martingale M^c and purely discontinuous martingale M^d. Here

$$M_t^d = \int_0^t \int_{|x| \leqslant 1} x(\mu - \nu)(ds, dx), \tag{1.8.6}$$

and hence

$$X_t = B_t + M_t^c + \int_0^t \int_{|x|>1} x\mu(ds, dx) + \int_0^t \int_{|x| \leqslant 1} x(\mu - \nu)(ds, dx), \tag{1.8.7}$$

where $B_t = A_t^1$ is the predictable process belonging to $\mathscr{V}_{\text{loc}}$. This representation is called the *canonical decomposition* of the semimartingale X.

The continuous component M^c and the compensator ν are given uniquely by the semimartingale X, and we denote M^c by X^c and call it the *continuous martingale component* of the semimartingale X. Note that

$$\Delta B_t = \int_{|x| \leqslant t} x\nu(\{t\}, dx), \tag{1.8.8}$$

where

$$\nu(\{t\}, \Gamma) = \int_{(t,t]} \int_\Gamma \nu(ds, dx), \; \nu(\{t\}, R \backslash \{0\}) \leqslant 1. \tag{1.8.9}$$

If $M \in \mathscr{M}_{\text{loc}}^2$, we know that there exists a unique predictable process $\langle M \rangle = (\langle M \rangle_t, \mathscr{B}_t)$ called the *quadratic characteristic* of M such that $M^2 - \langle M \rangle \in \mathscr{M}_{\text{loc}}$. In particular, since X^c is the continuous martingale component of X, the process $\langle X^c \rangle$ is well defined. The triple $(B, \langle X^c \rangle, \nu)$ associated with the semimartingale X is called the *triplet of local predictable characteristics* of the semimartingale X.

We shall now state two limit theorems for semimartingales due to Liptser and Shiryayev (1980).

Let $\mathscr{B}^n = \{\mathscr{B}_t^n, t \geqslant 0\}$, $n \geqslant 1$, be a family of σ-algebras satisfying the "standard" conditions for each n as described at the beginning of this section. Let $X^n = \{X_t^n, \mathscr{B}_t^n, t \geqslant 0\}$ be a semimartingale for every $n \geqslant 1$. Let $M = \{M_t, \mathscr{B}_t, t \geqslant 0\}$ be a continuous Gaussian martingale with quadratic characteristic $\langle M \rangle = \{\langle M \rangle_t, \mathscr{B}_t, t \geqslant 0\}$. Let

$$B_t^c = B_t - \sum_{0<s \leqslant t} \Delta B_s, \tag{1.8.10}$$

where B_t is as defined in (1.8.7).

Theorem 1.8.1. Let $\{X^n, n \geq 1\}$ be semimartingales with triplets $(B^n, \langle X^{nc}\rangle, \nu^n)$, and let M be a continuous Gaussian martingale with quadratic characteristic $\langle M\rangle$. Suppose that, for any $t > 0$, $\varepsilon \in (0, 1]$, the following conditions hold:

$$(1.8.11)\quad \text{(i)}\quad \text{(A)}\quad \int_0^t \int_{|x|>\varepsilon} \nu^n(ds, dx) \xrightarrow{P} 0,$$

$$\text{(B)}\quad B_t^{nc} + \sum_{0<s\leq t} \left(\int_{|x|\leq\varepsilon} x\nu^n(\{s\}, dx) \right) \xrightarrow{P} 0,$$

$$\text{(C)}\quad \langle X^{nc}\rangle_t + \int_0^t \int_{|x|\leq\varepsilon} x^2 \nu^n(ds, dx) - \sum_{0<s\leq t} \left(\int_{|x|\leq\varepsilon} x\nu^n(\{s\}, dx) \right)^2 \xrightarrow{P} \langle M\rangle_t.$$

Then $X^n \xrightarrow{\mathscr{L}^*} M$ in the sense that the finite dimensional distributions of X^n converge weakly to the corresponding finite dimensional distributions of M.

(1.8.12) (ii) If (A) and (C) hold and, further, if

$$\sup_{0\leq s\leq t} |B_t^{nc} + \sum_{0<s\leq t} \int_{|x|\leq\varepsilon} x\nu^n(\{s\}, dx)| \xrightarrow{P} 0$$

for any $t > 0$ and $0 < \varepsilon < 1$, then

$$X^n \xrightarrow{\mathscr{L}} M.$$

Corollary 1.8.2. Let $X^n \in \mathscr{M}^2_{\text{loc}}(\mathscr{B}^n, P)$, and let the Lindeberg condition be satisfied:

$$(1.8.13)\quad (\text{L}_2)\quad \int_0^t \int_{|x|>\varepsilon} \nu^n(ds, dx) \xrightarrow{P} 0, \qquad t > 0, \quad \varepsilon \in (0, 1].$$

Then (C) is equivalent to

$$(1.8.14)\quad (\text{C}')\quad \langle X^n\rangle_t \xrightarrow{P} \langle M\rangle_t, \qquad t > 0.$$

If (L_2) and (C') hold, then

$$(1.8.15)\qquad X^n \xrightarrow{\mathscr{L}} M.$$

As a consequence of Theorem 1.8.1, the following central limit theorem for martingale differences can be obtained (cf. Preposition 1.7.14).

Corollary 1.8.3. For each $n \geqslant 0$, let the sequence $\xi^n = (\xi_{nk}, \mathscr{F}^n_k)$ be a square integrable martingale difference, that is, $E(\xi_{nk} | \mathscr{F}^n_{k-1}) = 0$, $E\xi^2_{nk} < \infty$, $1 \leqslant k \leqslant n$. Suppose that $\xi_{n0} = 0$ and $\sum_{k=1}^n \xi^2_{nk} = 1$ for all $n \geqslant 1$. Let

$$X^n_t = \sum_{k=0}^{[nt]} \xi_{nk}.$$

Then the conditions

(i) $\displaystyle\sum_{k=1}^{n} E(\xi^2_{nk} I(|\xi_{nk}| > \varepsilon) | \mathscr{F}^n_{k-1}) \xrightarrow{p} 0, \qquad 0 < \varepsilon \leqslant 1, \quad t > 0,$

(ii) $\displaystyle\sum_{k=1}^{[nt]} E(\xi^2_{nk} | \mathscr{F}^n_{k-1}) \xrightarrow{p} t, \qquad t > 0,$

are sufficient (and necessary) for $X^n \xrightarrow{\mathscr{L}} W$. Where W is the standard Wiener process. The condition

(iii) $\displaystyle\sum_{k=1}^{[nt]} \xi^2_{nk} \xrightarrow{p} t$

is also sufficient (and necessary) for $X^n \xrightarrow{\mathscr{L}} W$.

In particular, for any fixed T,

$$X^n_T \xrightarrow{\mathscr{L}} N(0, \langle M \rangle_T)$$

under conditions (A), (B), and (C) of Theorem 1.8.1.

Let us now discuss an application of the theory of semimartingales to point (counting) processes.

Let $(\Omega, \mathscr{B}, P)$ be a probability space and $\{\mathscr{B}_t, t \geqslant 0\}$ be a family of sub-σ-algebras with properties stated at the begining of this section. Let $\{\tau_n, n \geqslant 0\}$ with $\tau_0 = 0$ and $\tau_n \uparrow \tau_\infty = \infty$ be a sequence of stopping times such that $\tau_0 < \tau_1 < \tau_2 < \cdots$. Clearly the trajectory of a point process is completely determined by the sequence $\{\tau_n\}$ and τ_n denote the jump times of the process. Let

$$F_n(t) = P[\tau_n \leqslant t | \tau_0, \ldots, \tau_{n-1}], \qquad n \geqslant 1, \tag{1.8.16}$$

and

$$N_t = \sum_{n \geqslant 1} I(\tau_n \leqslant t), \qquad t \geqslant 0.$$

$\{N_t, t \geqslant 0\}$ is the point (counting) process under discussion. It is clear that $\{N_t, \mathscr{B}_t, t \geqslant 0\}$ is a submartingale, and hence it is a semimartingale. It admits a unique decomposition $N = A + M$, where $A = \{A_t, \mathscr{B}_t, t \geqslant 0\}$ is a predictable and increasing process and M is a martingale. A is the *compensator* of the process N.

For example, if N is a Poisson process with parameter λ, then

$$N_t = \lambda t + (N_t - \lambda t), \qquad t \geqslant 0.$$

With $A_t = \lambda t$ being predictable trivially and $\{N_t - \lambda t, t \geqslant 0\}$ being a martingale, the uniqueness of the representation shows that $A_t = \lambda t$ is the compensator of the process N_t.

In general, it can be shown that

$$A_t = \int_{\tau_{n-1}}^{t} \frac{dF_n(s)}{1 - F_n(s-)}, \qquad \tau_{n-1} \leqslant t < \tau_n,$$

for any point process $\{N_t, \mathscr{B}_t, t \geqslant 0\}$ (Liptser and Shiryayev, 1978, Theorem 18.2), where $F_n(s)$ is defined by (1.8.16).

Suppose $N = (N_1, \ldots, N_r)$ is an r-variate counting process defined on $(\Omega, \mathscr{B}, P)$. Then, as before, we have

$$M_i = N_i - A_i, \qquad 1 \leqslant i \leqslant r,$$

where A_i is the compensator of N_i. Note that A_i is a right-continuous nondecreasing predictable process. Furthermore,

$$0 \leqslant \Delta A_i \leqslant 1,$$

where $\Delta A_i(t) = A_i(t) - A_i(t-0)$. The following proposition holds.

Proposition 1.8.4. Under the conditions stated above, the compensator A_i satisfies $0 \leqslant \Delta A_i \leqslant 1$. The M_i are local square integrable martingales with

$$\langle M_i, M_i \rangle = \int (1 - \Delta A_i)\, \Delta A_i,$$

and

$$\langle M_i, M_j \rangle = -\int \Delta A_i\, \Delta A_j, \qquad i \neq j,$$

for $1 \leqslant i, j \leqslant r$. The localizing stopping times τ_n may be taken to be a nondecreasing sequence $\{\tau_n\}$, $\tau_n \to \infty$ a.s. such that

$$E\left(\sum_{i=1}^{r} N_i(\tau_n)\right) < \infty, \qquad n = 1, 2, \ldots$$

Here $N_i(\infty) = \sup_t N_i(t)$.

For proof, see Gill (1980, p. 145).

Suppose $r = 1$. The following theorem of Murali Rao (1969) can be used to compute the compensator in an alternate way.

Proposition 1.8.5. Let N be a univariate counting process, and let $t \in (0, \infty)$ with $E(N(t)) < \infty$. Define

$$t_{n,i} = it2^{-n}, \qquad n \geqslant 1, \quad 0 \leqslant i \leqslant 2^n,$$

and

$$U_n = \sum_{i=0}^{2^n - 1} E(N(t_{n,i+1}) - N(t_{n,i}) | \mathscr{B}_{t_{n,i}}), \qquad n \geqslant 1.$$

Then there exists a subsequence $r_n \to \infty$ as $n \to \infty$ and a unique random variable U such that

$$E(HU_{rn}) \to E(HU)$$

for all bounded random variables H. Furthermore $A(t) = U$ a.s.

This section is based on Shiryayev (1981) and Gill (1980).

1.9 STOCHASTIC DIFFERENTIAL EQUATIONS

For the study of some recent applications of martingales to some problems in classic statistical theory, we develop now some results connected with stochastic integrals and stochastic differential equations. For a brief introduction to the theory of stochastic integrals and stochastic differential equations, see Basawa and Prakasa Rao (1980). Gikhman and Skorokhod (1972) deal extensively with this subject. We assume that the reader is familiar with some elementary concepts. Our discussion later in this section is based on Khmaladze (1981).

For completeness, we give a brief description of Ito stochastic integrals and stochastic differential equations.

Let $\{W(t), t \geqslant 0\}$ be a standard Wiener process, that is, a stochastic process with independent increments defined on a probability space $(\Omega, \mathscr{F}, P)$ with

$W(0)=0$ and $W(t)-W(s)$ is $N(0,|t-s|)$. Let $\{\mathscr{F}_t, t\geqslant 0\}$ be a nondecreasing family of sub-σ-algebras of $\mathscr{F}$ such that $W(t)$ is $\mathscr{F}_t$-measurable and the process $W_t(s)=W(t+s)-W(t)$, $s\geqslant 0$, is independent of $\mathscr{F}_t$ for any fixed $t\geqslant 0$. Let $H_2[0,T]$ be the class of all stochastic processes $\{f(t,\cdot), 0\leqslant t\leqslant T\}$ adapted to $\{\mathscr{F}_t, t\geqslant 0\}$ such that

$$\int_0^T f^2(t)\,dt<\infty \quad \text{a.s. } [P].$$

If f is a step function in $H_2[0,T]$, then define

$$\int_0^T f(t)\,dW(t)=\sum_{k=0}^{n-1} f(t_k)[W(t_{k+1})-W(t_k)],$$

where $0<t_1<\cdots<t_k=T$ are such that $f(t)=f(t_k)$, $t_k\leqslant t<t_{k+1}$. In general, it can be shown that, given any $f\in H_2[0,T]$, one can define

$$\int_0^T f(t)\,dW(t)$$

as a limit in probability. The following properties hold:

(i) $$\int_0^T (f_1(t)+f_2(t))\,dW(t) = \int_0^T f_1(t)\,dW(t)+\int_0^T f_2(t)\,dW(t) \quad \text{for } f_1, f_2\in H_2[0,T].$$

(ii) For any $\varepsilon>0$ and $\lambda>0$,

$$P\left\{\left|\int_0^T f(t)\,dW(t)\right|>\varepsilon\right\}\leqslant P\left\{\int_0^T f^2(t)\,dt>\lambda\right\}+\frac{\lambda}{\varepsilon^2} \quad \text{for } f\in H_2[0,T].$$

(iii) If $f\in H[0,T]$, the class of $f\in H_2[0,T]$ such that

$$\int_0^T E[f^2(t)]\,dt<\infty,$$

then

$$E\left[\int_0^T f(t)\,dW(t)\right]=0,$$

$$E\left[\int_0^T f(t)\,dW(t)\right]^2=\int_0^T E[f^2(t)]\,dt,$$

$$E\left[\int_\alpha^\beta f(t)\,dW(t)\middle|\mathscr{F}_\alpha\right]=0 \quad \text{a.s.},$$

and

$$E\left\{\left[\int_\alpha^\beta f(t)\,dW(t)\right]^2\middle|\mathscr{F}_\alpha\right\}=\int_\alpha^\beta E[f^2(t)|\mathscr{F}_\alpha]\,dt \quad \text{a.s.}$$

(iv) If $f_1, f_2 \in H[0, T]$, then

$$E\left[\int_0^T f_1(t)\,dW(t)\int_0^T f_2(t)\,dW(t)\right]=\int_0^T E[f_1(t)f_2(t)]\,dt.$$

(v) If $f \in H[0, T]$ and, for some $m \geqslant 1$,

$$\int_0^T Ef^{2m}(t)\,dt<\infty,$$

then

$$E\left[\int_0^T f(t)\,dW(t)\right]^{2m}\leqslant[m(2m-1)]^{m-1}T^{m-1}\int_0^T Ef^{2m}(t)\,dt.$$

(vi) Let f_1 and $f_2 \in H[0, T]$ and τ_1 and τ_2 be stopping times adapted to $\{\mathscr{F}_t\}$ such that $P[0\leqslant\tau_1\leqslant\tau_2\leqslant T]=1$. Then

$$E\left[\int_{\tau_1}^{\tau_2} f(t)\,dW(t)\middle|\mathscr{F}_{\tau_1}\right]=0 \quad \text{a.s.},$$

$$E\left[\left\{\int_{\tau_1}^{\tau_2} f(t)\,dW(t)\right\}^2\middle|\mathscr{F}_{\tau_1}\right]=E\left[\int_{\tau_1}^{\tau_2} f^2(t)\,dt\middle|\mathscr{F}_{\tau_1}\right] \quad \text{a.s.}$$

(vii) Let $f \in H[0, T]$ for every $T>0$ whenever f is restricted to $[0, T]$ and

$$\int_0^\infty f^2(s)\,ds=\infty \quad \text{a.s.}$$

Let

$$\tau_t=\inf\left\{u\colon \int_0^u f^2(s)\,ds\geqslant t\right\}.$$

Then

$$\zeta_t=\int_0^{\tau_t} f(s)\,dW(s), \quad t\geqslant 0,$$

is a Wiener process adapted to $\{\mathscr{F}_t\}$.

For proofs of properties (i)–(vii), see Gikhman and Skorokhod (1972) or Basawa and Prakasa Rao (1980).

The following theorem is a central limit theorem for stochastic integrals due to Kutoyants (1975) [cf. Basawa and Prakasa Rao (1980, p. 405)]. We present a univariate version for simplicity.

Proposition 1.9.0 (Kutoyants, 1975). Suppose $f \in H[0, T]$ for every $T > 0$ and

$$\frac{1}{T}\int_0^T f^2(t)\,dt \xrightarrow{p} \sigma^2 > 0 \quad \text{as} \quad T \to \infty.$$

Then

$$\frac{1}{T^{1/2}}\int_0^T f(t)\,dW(t) \xrightarrow{\mathscr{L}} N(0, \sigma^2).$$

Let $\{\zeta(t), 0 \leqslant t \leqslant T\}$ satisfy the relation

$$\zeta(t_2) - \zeta(t_1) = \int_{t_1}^{t_2} a(t)\,dt + \int_{t_1}^{t_2} b(t)\,dW(t), \qquad 0 \leqslant t_1 \leqslant t_2 \leqslant T,$$

where $|a(\cdot)|^{1/2} \in H_2[0, T]$ and $b(t) \in H_2[0, T]$. We then say that the process $\{\zeta(t), 0 \leqslant t \leqslant T\}$ has the *stochastic differential*

$$d\zeta(t) = a(t)\,dt + b(t)\,dW(t)$$

on $[0, T]$.

Consider the stochastic differential equation

$$d\eta(t) = a(t, \eta(t))\,dt + \sigma(t, \eta(t))\,dW(t), \qquad 0 \leqslant t \leqslant T, \tag{1.9.0}$$

where $a(t, x)$ and $\sigma(t, x)$ are measurable functions on $[0, T] \times R$. A random process $\{\eta(t), 0 \leqslant t \leqslant T\}$ is called a solution of the equation (1.9.0) on $[0, T]$ if it satisfies the following conditions:

(i) if $\mathscr{F}_t$ denotes the smallest σ-algebra generated by $\{W(s), \eta(s), 0 \leqslant s \leqslant t\}$, then the process $W_t(s) = W(t + s) - W(t)$, $s \geqslant 0$, is independent of $\mathscr{F}_t$;

(ii) $|a(t, \eta(t))|^{1/2} \in H_2[0, T]$, $\sigma(t, \eta(t)) \in H_2[0, T]$;

(iii) the process $\{\eta(t), 0 \leqslant t \leqslant T\}$ has the stochastic differential $d\eta(t) = \bar{a}(t)\,dt + \bar{b}(t)\,dW(t)$, where $\bar{a}(t) = a(t, \eta(t))$ a.s. and $\bar{b}(t) = \sigma(t, \eta(t))$ a.s. If $\eta(\cdot)$ is a solution of (1.9.0), then $\eta(0)$ will be independent of $W(t) - W(0)$ for $t > 0$ and

$$\eta(t) = \eta(0) + \int_0^t a(s, \eta(s))\,ds + \int_0^t \sigma(s, \eta(s))\,dW(s) \quad \text{a.s.}$$

for $0 \leqslant t \leqslant T$.

Suppose the following conditions hold:

(i) There exists a constant K such that

$$|a(t,x)-a(t,y)|+|\sigma(t,x)-\sigma(t,y)|\leqslant K|x-y|, \qquad 0\leqslant t\leqslant T, \quad x,y\in R,$$
$$|a(t,x)|^2+|\sigma(t,x)|^2\leqslant K(1+|x|^2), \quad 0\leqslant t\leqslant T, \quad x,y\in R.$$

(ii) $\eta(0)$ is independent of $W(t)$ for all $t\in[0,T]$ and $E\eta^2(0)<\infty$.

Then it can be shown that there exists a solution of (1.9.0) such that

(a) $\eta(t)$ is continuous a.s. and $\eta(t)=\eta(0)$ for $t=0$,
(b) $\sup_{0\leqslant t\leqslant T} E\eta^2(t)<\infty$.

Furthermore, if η_1 and η_2 are two solutions of (1.9.0) satisfying (a) and (b), then

$$P\left\{\sup_{0\leqslant t\leqslant T}|\eta_1(t)-\eta_2(t)|=0\right\}=1.$$

For proofs of above results, see Gikhman and Skorokhod (1972) and Basawa and Prakasa Rao (1980).

The following result known as *Ito's lemma* gives a formula for change of variables. If

$$X_t=\int_0^t f(s)\,dW(s)$$

and $F=F(\cdot)$ is twice continuously differentiable, then

$$F(X_t)=F(X_0)+\int_0^t F'(X_s)f(s)\,dW_s+\frac{1}{2}\int_0^t F''(X_s)f^2(s)\,ds.$$

Let $\{b(t),\ 0\leqslant t\leqslant 1\}$ be the standard Wiener process on the interval $[0,1]$ and $P_{[0,1]}$ be the measure induced by it on $L_2[0,1]$ with the associated Borel σ-algebra. Let $L_2^{(r)}[0,1]$ be the space of r-dimensional vector functions ψ with norm

$$\|\psi\|=\left[\int_0^1 \psi'(t)\psi(t)\,dt\right]^{1/2}. \tag{1.9.1}$$

Let $\mathbf{g}(\cdot)$ and $\mathbf{h}(\cdot)$ be vector functions such that $\nabla\mathbf{g}(t)$ and $\mathbf{h}(t)$ belong to $L_2^{(r)}[0, 1]$, where $\nabla\mathbf{g}$ denotes the r-dimensional vector of derivatives of $\mathbf{g}(\cdot)$ with respect to t. Let

$$\mathbf{B}^{\mathbf{h}}(t) = \int_0^t \nabla\mathbf{g}(s)\mathbf{h}(s)'\,ds. \tag{1.9.2}$$

Suppose $\mathbf{B}^{\mathbf{h}}(1) = \mathbf{I}_r$, where $\mathbf{I}_r$ is the identity matrix. Consider the process

$$u^{\mathbf{h}}(t) = b(t) - \mathbf{g}(t)' \int_0^1 \mathbf{h}(s)\,db(s), \tag{1.9.3}$$

where the integral on the right-hand side can be thought of as a particular case of an Ito stochastic integral. If $\mathbf{h} = \nabla\mathbf{g}$, then we write $u(t)$ for $u^{\nabla\mathbf{g}}(t)$. In other words

$$u(t) = b(t) - \mathbf{g}(t)' \int_0^1 \nabla\mathbf{g}(s)\,db(s), \tag{1.9.4}$$

with

(1.9.5) (i) $\mathbf{B}(t) = \int_0^t \nabla\mathbf{g}(s)\,\nabla\mathbf{g}(s)'\,ds$

(ii) $\mathbf{B}(1) = \mathbf{I}_r$.

We note that $u^{\mathbf{h}}$ is the projection of b along $\mathbf{g}$. This can be seen from the fact that if

$$\pi_{\mathbf{h}} b(t) = b(t) - \mathbf{g}(t)' \int_0^1 \mathbf{h}(s)\,db(s), \tag{1.9.6}$$

then $\pi_{\mathbf{h}}^2 = \pi_{\mathbf{h}}$ and $\pi_{\mathbf{h}}\mathbf{g}'\boldsymbol{\alpha} = 0$ for any vector $\boldsymbol{\alpha}$ in R^r since

$$\int_0^1 \mathbf{h}(s)\,\nabla\mathbf{g}(s)'\,ds = \mathbf{I}_r.$$

Furthermore,

$$\int_0^1 \mathbf{h}(s)\,d\pi_{\mathbf{h}} b(s) = 0.$$

Since the representation (1.9.3) involves the values of the process $b(s)$, $0 \leqslant s \leqslant 1$, the process $u^{\mathbf{h}}(t)$ is not nonanticipative with respect to

$\{\mathscr{F}_t^b, 0 \leq t \leq 1\}$ as in the case of stochastic differentials [cf. Gikhman and Skorokhod (1972) and Basawa and Prakasa Rao (1980)]. Here $\{\mathscr{F}_t^b, 0 \leq t \leq 1\}$ is the flow corresponding to process b. Let $P_{[0,T]}^{\xi}$ be the measure induced on $L_2[0,T]$ by a process ξ on $[0,T]$. Denote $P_{[0,T]}^b$ by $P_{[0,T]}$ for simplicity. Observe that the probability measures $P_{[0,1]}^{u^{\mathbf{h}}}$ and $P_{[0,1]}$ are singular with respect to each other, since from the definition of $u^{\mathbf{h}}$ as given by (1.9.3), it follows that

$$\int_0^1 \mathbf{h}(t)\, du^{\mathbf{h}}(t) = 0 \quad \text{and} \quad \int_0^1 \mathbf{h}(t)\, db(t) \neq 0 \tag{1.9.7}$$

with probability 1. Hence the process $u^{\mathbf{h}}$ cannot be represented as a diffusion-type process satisfying a stochastic differential equation of the form

$$du^{\mathbf{h}}(t) = \alpha(t)\, dt + \beta(t)\, db(t), \qquad 0 \leq t \leq 1$$

[cf. Gikhman and Skorokhod (1972)].

Proposition 1.9.1. The measures $P_{[0,T]}^u$ and $P_{[0,T]}$ are equivalent for all $T, 0 \leq T \leq 1$, if and only if the components of $\nabla \mathbf{g}(t)$ are linearly independent on $[T,1]$ for any $T < 1$.

Proof. Note that

$$\operatorname{cov}[u(t), u(s)] = t \wedge s - \mathbf{g}(t)'\mathbf{g}(s). \tag{1.9.8}$$

By Theorem 1 in Shepp (1966), the probability measures $P_{[0,T]}^u$ and $P_{[0,T]}(t)$ are equivalent if and only if the operator with the kernel $\nabla \mathbf{g}(t)' \nabla \mathbf{g}(s)$, $0 \leq t, s \leq T$, has no eigenvalue equal to 1.

Let $Q \in L_2[0,T]$ and $\phi_T(t) = \phi(t) I_{[0,T]}(t)$. Clearly $\phi_T \in L_2[0,1]$. In view of (1.9.5) (ii), the components of $\nabla \mathbf{g}$ form an orthonormal system in $L_2[0,1]$ and hence

$$\phi_T(t) = \left(\int_0^T \phi(\tau) \nabla \mathbf{g}(\tau)'\, d\tau \right) \nabla \mathbf{g}(t) + \lambda(t), \tag{1.9.9}$$

where λ is orthogonal to the components of $\nabla \mathbf{g}$.
Therefore

$$\int_0^T \phi^2(t)\, dt = \int_0^1 \phi_T^2(t)\, dt \geq \int_0^T \phi(t) \nabla \mathbf{g}(t)'\, dt \int_0^T \phi(t) \nabla \mathbf{g}(t)\, dt, \tag{1.9.10}$$

and hence the distributions $P^u_{[0,T]}$ and $P_{[0,T]}$ are equivalent if and only if strict inequality holds in (1.9.10) for all $\phi \in L_2[0, T]$, $\phi \neq 0$. This last condition holds if and only if

$$\int_0^1 \lambda^2(t)\, dt > 0,$$

which in turn holds if and only if the components of $\nabla \mathbf{g}$ are linearly independent on $[T, 1]$ for all $T < 1$.

Let us assume here after that the components of $\nabla \mathbf{g}$ are linearly independent in any neighborhood of 1. Hence the measures $P^u_{[0,T]}$ and $P_{[0,T]}$ are equivalent. Therefore there exists a Wiener process $w(t)$ with respect to the flow $\{\mathscr{F}^u_t, 0 \leqslant t \leqslant T\}$ such that $u(t)$ has a stochastic differential

$$du(t) = \left(\int_0^t L(t, \tau)\, du(\tau) \right) dt + dw(t) \tag{1.9.11}$$

on $[0, T]$, $T < 1$. This follows from Hitsuda (1968) [cf. Liptser and Shiryayev (1978, Theorem 7.11)]. We shall now obtain a more precise form of the kernel $L(t, \tau)$. Define

$$\tilde{u}(t) = \int_0^t \nabla g(\tau)\, du(\tau), \qquad 0 \leqslant t \leqslant 1. \tag{1.9.12}$$

In view of the representation (1.9.4), the matrix-valued covariance function of the process $\tilde{u}(t)$ is

$$\mathbf{B}(t \wedge s) - \mathbf{B}(t)\mathbf{B}(s), \tag{1.9.13}$$

where $\mathbf{B}$ is defined by (1.9.5).

If $g(t)$ is a scalar function, then $\tilde{u}(t)$ is a Brownian bridge with respect to the time

$$B(t) = \int_0^t \left(\frac{dg(\tau)}{d\tau} \right)^2 d\tau$$

and the representation

$$d\tilde{u}(t) = \frac{-(dg(t)/dt)^2}{1 - B(t)} \tilde{u}(t)\, dt + d\tilde{w}(t) \tag{1.9.14}$$

holds where $\tilde{w}(t)$ is a Wiener process with respect to the time $B(t)$. In the vector case,

$$d\tilde{u}(t) = -\nabla\mathbf{g}(t)\nabla\mathbf{g}(t)'[\mathbf{I}_r - \mathbf{B}(t)]^{-1}\tilde{u}(t)\,dt + d\tilde{w}(t), \tag{1.9.15}$$

where $\tilde{w}(t)$ has the covariance function $\mathbf{B}(t \wedge s)$. Note that

$$d\tilde{u}(t) = \nabla\mathbf{g}(t)\,du(t)$$

and hence

$$du(t) = -(\nabla\mathbf{g}(t)'[I_r - \mathbf{B}(t)]^{-1}\int_0^t \nabla\mathbf{g}(\tau)\,du(\tau))\,dt + dw(t), \tag{1.9.16}$$

where $w(\cdot)$ is a standard Wiener process. In order to justify the results given above, we have to prove the $\mathbf{I}_r - \mathbf{B}(t)$ is invertible for $t < 1$. Note that the equation (1.9.16) gives the Doob–Meyer decomposition for $\{u(t), 0 \leqslant t < 1\}$.

Suppose $\xi \in R^r$ and $b(\cdot)$ is a standard Wiener process. Then

$$\begin{aligned}\xi'\mathbf{B}(t)\xi &= E\left[\int_0^t \xi'\,\nabla\mathbf{g}(t)\,db(\tau)\right]^2 \\ &< E\left[\int_0^1 \xi'\nabla\mathbf{g}(\tau)\,db(\tau)\right]^2 = \xi'\xi\end{aligned}$$

since the random variables

$$\int_0^t \xi'\,\nabla\mathbf{g}(\tau)\,db(\tau) \quad \text{and} \quad \int_t^1 \xi'\,\nabla\mathbf{g}(\tau)\,db(\tau)$$

are independent and since $\xi'\,\nabla\mathbf{g}(\tau) \not\equiv 0$ on any interval $[t, 1]$ for $\xi \neq \mathbf{0}$ by the assumption that the components of $\nabla\mathbf{g}$ are linearly independent in any neighborhood of 1. Hence 0 is not an eigenvalue of $\mathbf{I}_r - \mathbf{B}(t)$ for any $t < 1$, that is, $[\mathbf{I}_r - \mathbf{B}(t)]^{-1}$ exists for $t < 1$.

Proposition 1.9.2. Suppose the components of $\nabla\mathbf{g}$ are linearly independent in a neighborhood of 1, and define the process u by the relation

$$u(t) = b(t) - \mathbf{g}(t)'\int_0^1 \nabla\mathbf{g}(s)\,db(s), \tag{1.9.17}$$

where $b(\cdot)$ is a standard Wiener process with $\nabla\mathbf{g} \in L_2^{(r)}[0, 1]$ and (1.9.5) holds.

Then the relation (1.9.17) is invertible, and if

$$w(t) = u(t) + \int_0^t \left(\nabla \mathbf{g}(s)'[\mathbf{I}_r - \mathbf{B}(s)]^{-1} \int_0^s \nabla \mathbf{g}(\tau)\, du(\tau) \right) ds, \tag{1.9.18}$$

for $0 \leqslant t \leqslant 1$, then $\{w(t), 0 \leqslant t \leqslant 1\}$ is a Wiener process with respect to the flow $\{\mathscr{F}_t^u, 0 \leqslant t \leqslant 1\}$ and

$$u(t) = w(t) + \int_0^t \left(\nabla \mathbf{g}(s)' \int_0^s [\mathbf{I}_r - \mathbf{B}(\tau)]^{-1}\, dw(\tau) \right) ds. \tag{1.9.19}$$

Proof. Let $T < 1$. Let C_1 denote the set of functions of the form $\Phi(t) = \int_t^1 \phi(s)\, ds$, where $\phi \in L_2[0, 1]$. For any $\Phi \in C_1$, define

$$\Phi_T(t) = \Phi(t) I_{[0,T]}(t).$$

$$\int_0^T \Phi(t)\, dw(t) = \int_0^1 \Phi_T(t)\, dw(t) = \int_0^1 \left[\Phi_T(t) + \left(\int_t^1 \Phi_T(s) \nabla \mathbf{g}(s)' \cdot [\mathbf{I}_r - \mathbf{B}(s)]^{-1}\, ds \right) \nabla \mathbf{g}(s) \right] du(t). \tag{1.9.20}$$

Hence the linear functional

$$\int_0^T \Phi(t)\, dw(t) \tag{1.9.21}$$

is Gaussian with mean zero by the definition of $u(\cdot)$. On the other hand,

$$E\left[\int_0^T \Phi(t)\, dw(t) \right]^2 = \int_0^1 \psi^2(t)\, dt - \int_0^1 \psi(t) \nabla \mathbf{g}(t)'\, dt \int_0^1 \psi(t) \nabla \mathbf{g}(t)\, dt$$

from (1.9.17) and (1.9.20) and hence

$$E\left[\int_0^1 \Phi_T(t)\, dw(t) \right]^2 = \int_0^1 \Phi_T^2(t)\, dt \tag{1.9.22}$$

from (1.9.5), where

$$\psi(t) = \Phi_T(t) + \left(\int_t^1 \Phi_T(s) \nabla \mathbf{g}(s)'[\mathbf{I}_r - \mathbf{B}(s)]^{-1}\, ds \right) \nabla \mathbf{g}(t). \tag{1.9.23}$$

Hence the covariance function of the process $\{w(t), 0 \leqslant t \leqslant T\}$ is $t \wedge s$ for $t, s \leqslant T$ and the process $\{w(t),\ 0 \leqslant t \leqslant T\}$ is a Wiener process. In view of relation (1.9.22), if $T \to 1$, the family $\int_0^1 \Phi_T(t)\, dw(t)$ is Cauchy in q.m. and, furthermore, the variance of the random variable

$$\int_0^1 \Phi_T(t)\, dw(t)$$

tends to $\int_0^1 \Phi_T^2(t)\, dt$. Define

$$\int_0^1 \Phi(t)\, dw(t) = \underset{T \to 1}{\text{l.i.m.}} \int_0^1 \Phi_T(t)\, dw(t), \tag{1.9.24}$$

where l.i.m. denotes the limit in quadratic mean. Then

$$\int_0^1 \Phi(t)\, dw(t)$$

is Gaussian with mean 0 and variance $\int_0^1 \Phi^2(t)\, dt$ for every $\Phi \in C_1$. Hence $P^w_{[0,1]}$ is the Wiener measure on $L_2[0, 1]$ where w is defined by relation (1.9.18). Relation (1.9.19) is an immediate consequence of (1.9.18) by taking the inverse. This completes the proof of Proposition 1.9.2.

In view of (1.9.7), we have

$$\int_0^1 \nabla \mathbf{g}(t)\, du(t) = \mathbf{0} \tag{1.9.25}$$

and hence, the representation (1.9.11) can be written in the form

$$du(t) = \left(\nabla \mathbf{g}(t)'(\mathbf{I}_r - \mathbf{B}(t))^{-1} \int_t^1 \nabla \mathbf{g}(\tau)\, du(\tau) \right) dt + dw(t). \tag{1.9.26}$$

Hence the process u defined by (1.9.4) is a solution of both (1.9.11) and (1.9.26) for any standard Wiener process b.

Proposition 1.9.3. Define

$$w^{\mathrm{h}}(t) = u^{\mathrm{h}}(t) + \int_0^t \left(\nabla \mathbf{g}(s)'[\mathbf{I}_r - \mathbf{B}(s)]^{-1} \int_0^s \nabla \mathbf{g}(\tau)\, du^{\mathrm{h}}(\tau) \right) ds, \tag{1.9.27}$$

where

$$u^{\mathbf{h}}(t) = b(t) - \mathbf{g}(t)' \int_0^1 \mathbf{h}(s)\, db(s). \tag{1.9.28}$$

Then the process $w^{\mathbf{h}}(\cdot)$ does not depend on $\mathbf{h}$ and is a standard Wiener process.

Proof. Substituting relation (1.9.28) into equation (1.9.27), due to the identity

$$\nabla\mathbf{g}(t) = \nabla\mathbf{g}(t)'(\mathbf{I}_r - \mathbf{B}(t))^{-1} \int_t^1 \nabla\mathbf{g}(\tau)\nabla\mathbf{g}(\tau)'\, d\tau, \tag{1.9.29}$$

it follows that

$$db(t) = \left(\nabla\mathbf{g}(t)'(\mathbf{I}_r - \mathbf{B}(t))^{-1} \int_0^1 \nabla\mathbf{g}(\tau)\, db(\tau) \right) dt + dw^{\mathbf{h}}(t), \tag{1.9.30}$$

and hence $w^{\mathbf{h}}(\cdot)$ is determined by $b(\cdot)$ only and does not depend on h. We have already seen that $w^{\mathbf{h}}(\cdot)$ is a standard Wiener process by choosing $\mathbf{h} = \nabla\mathbf{g}$ in Proposition 1.9.2. This completes the proof of the proposition.

Remarks. Equation (1.9.30) is the Doob–Meyer decomposition of the process $\{b(t), \mathscr{F}_t^b V \mathscr{F}^{\mathbf{g}}, 0 \leqslant t \leqslant 1\}$, where $\mathscr{F}^{\mathbf{g}}$ denotes the σ-algebra generated by the random vector $\int_0^1 \nabla\mathbf{g}(s)\, db(s)$. In particular if $g(t) = t$, then $\nabla\mathbf{g} = 1$ and

$$db(t) = \frac{b(1) - b(t)}{1 - t} + dw(t), \qquad 0 \leqslant t \leqslant 1, \tag{1.9.31}$$

and

$$dv(t) = -\frac{v(t)}{1 - t} + dw(t), \qquad 0 \leqslant t \leqslant 1, \tag{1.9.32}$$

where v is the Brownian bridge $b(t) - tb(1)$, $0 \leqslant t \leqslant 1$. The latter stochastic differential equation follows from (1.9.4) and (1.9.26).

In view of the fact that

$$\int_0^1 \nabla\mathbf{g}(\tau)\, du^{\mathbf{h}}(\tau) = 0, \tag{1.9.33}$$

relation (1.9.27) can also be written in the form

$$(1.9.34)\qquad w(t) = u^{\mathbf{h}}(t) - \int_0^t \left(\nabla \mathbf{g}(s)'[\mathbf{I}_r - \mathbf{B}(s)]^{-1} \int_t^1 \nabla \mathbf{g}(\tau)\, du^{\mathbf{h}}(\tau) \right) ds$$

using the fact that $w^{\mathbf{h}}(t)$ does not depend on $\mathbf{h}$. In particular, choosing $\mathbf{h} = 0$, we obtain the following representation of the Wiener process $w(t)$ in terms of another Wiener process $b(t)$, namely,

$$(1.9.35)\qquad dw(t) = db(t) - \left(\nabla \mathbf{g}(t)'[\mathbf{I}_r - \mathbf{B}(t)]^{-1} \int_t^1 \nabla \mathbf{g}(\tau)\, db(\tau) \right) dt.$$

Observe that Proposition 1.9.2 implies that there is a one-to-one correspondence between the process u defined by (1.9.17) and the Wiener process $w(\cdot)$ defined by (1.9.18), which is adapted to the flow $\{\mathscr{F}_t^u, 0 \leqslant t \leqslant 1\}$. Results obtained in this section will be used in Chapter 5.

1.10 PROBABILITY MEASURES ON METRIC SPACES

All along we have been assuming that one can consider sequences of random variables on a given probability space and study their properties. It is important to ask whether it is possible to construct a probability space $(\Omega, \mathscr{B}, P)$ such that one can define a sequence $\{X_n\}$ of random variables on it so that the joint distribution function of $(X_1, \ldots, X_n)$ coincides with distribution function $F_{1,\ldots,n}(x_1, \ldots, x_n)$ given a priori for every $n \geqslant 1$. This question can be answered in the affirmative subject to certain conditions.

Definition 1.10.1. A family of distribution functions $\{F_{1,\ldots,n}(x_1, \ldots, x_n), n \geqslant 1\}$ is said to be *consistent* if, for every $n \geqslant 1$,

$$F_{1,\ldots,n}(x_1, \ldots, x_n) = \lim_{x_{n+1} \to \infty} F_{1,\ldots,n+1}(x_1, \ldots, x_{n+1})$$

for every $(x_1, \ldots, x_n) \in R^n$.

Proposition 1.10.1 (Kolmogorov Consistency Theorem). If $\{F_{1,\ldots,n}, n \geqslant 1\}$ is a consistent family of distribution functions, then there exists a probability measure P on $(R^\infty, \mathscr{B}^\infty)$ such that the distribution function of the coordinate vector $(X_1, \ldots, X_n)$ under P coincides with $F_{1,\ldots,n}$ for every $n \geqslant 1$. (Here $\mathscr{B}^\infty$ is the σ-algebra on R^∞ generated by all the finite-dimensional cylinder sets).

For proof, see Chow and Teicher (1978).

Consider a probability space $(\Omega, \mathscr{B}, P)$ and $\{X_t, t \geqslant 0\}$ be a stochastic process defined on it, that is, a collection of random variables defined on $(\Omega, \mathscr{B}, P)$. In order to study probabilities of events of the type

$$\left[\sup_{0 \leqslant t \leqslant T} X_t > \varepsilon\right], \left[\int_0^T X_t^2 \, dt > \varepsilon\right],$$

it is some times convenient to treat $X_t(\omega)$ as a random element defined on $(\Omega, \mathscr{B}, P)$ taking values in a function space and study its properties. We refer the reader to Billingsley (1968) and Parthasarathy (1967) for a detailed survey of this approach. We present here some discussion that is relavant to the asymptotic theory of statistics.

Let $(\Omega, \mathscr{B}, P)$ be a probability space and $(\mathscr{X}, \mathscr{F})$ be a measurable space where $\mathscr{X}$ is a complete separable metric space with $\mathscr{F}$ the associated σ-algebra of Borel sets. $X: \Omega \to \mathscr{X}$ is said to be a *random element* taking value in $\mathscr{X}$ if $X^{-1}(F) \in \mathscr{B}$ for every $F \in \mathscr{F}$. For any $F \in \mathscr{F}$, define

$$P_X(F) = P[X^{-1}(F)].$$

$P_X(\cdot)$ is called the *probability measure induced* by X on the space $(\mathscr{X}, \mathscr{F})$. Let $\mathscr{M}$ be the class of all probability measures on $(\mathscr{X}, \mathscr{F})$. Let $\{\mu_n, n \geqslant 1\}$ and μ be members of $\mathscr{M}$. μ_n is said to *converge weakly* to μ if for every bounded continuous functional $g(\cdot)$ defined on $(\mathscr{X}, \mathscr{F})$,

$$\int_{\mathscr{X}} g \, d\mu_n \to \int_{\mathscr{X}} g \, d\mu \quad \text{as} \quad n \to \infty. \tag{1.10.1}$$

We write $\mu_n \xrightarrow{w} \mu$.

Let $C[\mathscr{Y}]$ denote the space of real-valued bounded continuous functions defined on $\mathscr{Y}$. If $\mathscr{Y} = [0, 1]$, we denote $C[\mathscr{Y}]$ by $C[0, 1]$, and if $\mathscr{Y} = (-\infty, \infty)$, we denote $C[\mathscr{Y}]$ by $C(-\infty, \infty)$. It is known that $C[0, 1]$ is a complete separable metric space under the metric

$$\|f - g\| = \sup_{0 \leqslant x \leqslant 1} |f(x) - g(x)|.$$

For statistical applications, a more general notion of uniform weak convergence is needed. We shall discuss this notion with reference to the space $C[\mathscr{X}]$ where $(\mathscr{X}, \mathscr{F})$ is as defined above. Let $\mu_\theta, \mu_{n,\theta}, n \geqslant 1$, be probability measures defined on $(\mathscr{X}, \mathscr{F})$ where $\theta \in \Theta$, a parameter space that is a subset of a metric space. $\mu_{n,\theta}$ is said to *converge weakly* to μ_θ *uniformly* in $\theta\in\Theta(\mu_{n,\theta}\overset{w}{\rightarrow}_u\mu_\theta)$ if

$$\int_{\mathscr{X}} g\,d\mu_{n,\theta}\rightarrow\int_{\mathscr{X}} g\,d\mu_\theta \tag{1.10.2}$$

for all $g\in C_u[\mathscr{X}]$ where $C_u[\mathscr{X}]$ is the space of real-valued bounded uniformly continuous functions on $\mathscr{X}$. The family $\{\mu_\theta\}$ is said to be *continuous* if $\mu_{\theta_n}\overset{w}{\rightarrow}\mu_\theta$ whenever $\theta_n\rightarrow\theta$ and a set $B\in\mathscr{F}$ is called a (μ_θ)-*continuity set* if $\mu_\theta(\partial B)=0$ for all $\theta\in\Theta$ where ∂B is the boundary of the set B. Note that, if g_n and $g\in C[\mathscr{X}]$, then $g_n(x)\rightarrow g(x)$ uniformly in $x\in\mathscr{X}$ iff $g_n(x_n)\rightarrow g(x)$ for every sequence $x_n\rightarrow x$.

Proposition 1.10.2. Suppose $\mu_{n,\theta}\overset{w}{\rightarrow}_u\mu_\theta$ and $\{\mu_\theta\}$ is continuous. Then

$$\int_{\mathscr{X}} g\,d\mu_{n,\theta}\rightarrow\int_{\mathscr{X}} g\,d\mu_\theta \tag{1.10.3}$$

for all $g\in C[\mathscr{X}]$ and $\mu_{n,\theta}(B)\rightarrow\mu_\theta(B)$ uniformly in $\theta\in\Theta$ for every (μ_θ)-continuity set B.

Proof. Let $\theta_n\rightarrow\theta$. It is clear that

$$\int_{\mathscr{X}} g\,d\mu_{n,\theta_n}\rightarrow\int_{\mathscr{X}} g\,d\mu_\theta \tag{1.10.4}$$

for all $g\in C_u[\mathscr{X}]$ by the definition and the remark made above. Hence relation (1.10.3) holds for all $g\in C[\mathscr{X}]$ [cf. Billingsley (1968, Theorem 2.1)]. This proves (1.10.3). Since $\mu_{n,\theta_n}\overset{w}{\rightarrow}\mu_\theta$ by the remark made above, it follows from Theorem 2.1 in Billingsley (1968) that $\mu_{n,\theta_n}(B)\rightarrow\mu_\theta(B)$ whenever $\mu_\theta(\partial B)=0$. Hence $\mu_{n,\theta}(B)\rightarrow\mu_\theta(B)$ uniformly in θ for every $\{\mu_\theta\}$-continuity set B.

As a consequence of Proposition 1.10.2 we have the following result on uniform convergence for random vectors. We omit the proof. See Ibragimov and Hasminskii (1981) for additional results of this type.

Proposition 1.10.3. Let $\mathbf{X}_\theta,\mathbf{X}_{n\theta},n\geqslant 1$, be k-dimensional random vectors on $(\Omega,\mathscr{B},P_\theta)$, $\theta\in\Theta\subset R^m,m\geqslant 1$. If

$$\mathbf{X}_{n\theta}\overset{p}{\rightarrow}\mathbf{X}_\theta\quad\text{uniformly in}\quad\theta\in\Theta, \tag{1.10.5}$$

then

$$\mathbf{X}_{n\theta} \xrightarrow{\mathscr{L}} \mathbf{X}_{\theta} \quad \text{uniformly in} \quad \theta \in \Theta. \tag{1.10.6}$$

Let $\{X_n\}$ be a sequence of random elements defined on $(\Omega, \mathscr{B}, P)$ taking values in a measurable space $(\mathscr{X}, \mathscr{F})$ where $(\mathscr{X}, d)$ is a complete separable metric space and $\mathscr{F}$ the associated Borel σ-algebra. $\{X_n\}$ is said to be *Renyi mixing* with limit μ if for every $D \in \mathscr{B}$ with $P(D) > 0$,

$$\lim_{n \to \infty} P[(X_n \in A) \cap D] = \mu(A)P(D),$$

for every Borel set $A \in \mathscr{F}$ that is μ-continuous.

Proposition 1.10.4 (Prakasa Rao, 1973). Let $\{\mu_n\}$ be a sequence of probability measures induced by random elements $\{X_n\}$ taking values in a complete separable metric space $(\mathscr{X}, d)$. Suppose that $\mu_n \xrightarrow{w} \mu$ and $\{N_n\}$ is a sequence of positive integer-valued random variables such that $N_n/n \xrightarrow{p} \gamma > 0$ for some constant γ as $n \to \infty$. Furthermore, suppose that for every $\eta > 0$ and $\varepsilon > 0$, there exists $0 < c < 1$ and $N_0 > 0$ such that

$$P[\sup\{d(Y_n, Y_k) : r_n \leqslant k \leqslant s_n\} > \varepsilon] < \eta, \tag{1.10.7}$$

where $r_n = [n(1-c)]$ and $s_n = [n(1+c)]$.

Then $\mu_{N_n} \xrightarrow{w} \mu$, where μ_{N_n} is the probability measure induced by X_{N_n}.

Proposition 1.10.5 (Prakasa Rao, 1973). Defined X_n, μ_n, and N_n as in Proposition 1.10.4. Suppose the sequence $\{X_n\}$ is Renyi mixing with limit μ and $\{N_n\}$ satisfies the condition

$$\frac{N_n}{n} \xrightarrow{p} M,$$

where M is a positive and discretely distributed random variable. Furthermore, suppose that $\{X_n\}$ satisfies Anscombe's condition (1.10.7). Then

$$\mu_{N_n} \xrightarrow{w} \mu.$$

Remarks. It is known that condition (1.10.7) is not necessary [cf. Richter

(1965)] for the result stated in Proposition 1.10.5. As an application of Proposition 1.10.4, one can obtain the following random central limit theorem for Hilbert-space-valued random elements.

Proposition 1.10.6. Let $\{X_n, n \geq 1\}$ be a sequence of symmetric independent random elements with values in a separables Hilbert space. Let $\{N_n\}$ be a sequence of positive integer-valued random variables such that $N_n/n \xrightarrow{p} 1$ as $n \to \infty$. Denote by μ_n the measure induced by $n^{-1/2}S_n$ where $S_n = \sum_{i=1}^{n} X_i$. If $\mu_n \xrightarrow{w} \mu$, then $\mu_{N_n} \xrightarrow{w} \mu$.

For proofs of the above results, see Prakasa Rao (1973).

A sequence of random elements $\{X_n\}$ taking values in a complete separable metric space $(\mathscr{X}, d)$ is said to be mixing with limit μ with respect to a family of events τ (or in short *τ-mixing*) if

$$\mathscr{L}(X_n | H) \xrightarrow{w} \mu$$

every $H \in \tau$ with $F(H) > 0$ where $\mathscr{L}(X|H)$ denotes the conditional distribution of the random element X given H.

The following theorem generalizes Proposition 1.10.6.

Proposition 1.10.7 (Szasz, 1973). Define X_n, μ_n, and N_n as before. Suppose the following conditions hold.

(i) $\{N_n\}$ is a sequence of positive integer-valued random variables such that

$$\frac{N_n}{n} \xrightarrow{p} M,$$

where M is a positive random variable.

(ii) The sequence $\{X_n\}$ is mixing with limit μ with respect to the σ-algebra $\mathscr{B}(M)$ generated by M.

(iii) For every $\varepsilon > 0$ and $\eta > 0$, there exists $C(\varepsilon, \eta) > 0$ such that for any $H \in \mathscr{B}(M)$ with $P(H) > 0$ there exists $n_0 = n_0(\varepsilon, \eta, H)$ such that for arbitrary $n > n_0$,

$$P\left\{ \sup_{|i-n|<cn} d(X_n, X_i) > \varepsilon | H \right\} < \eta.$$

Then the sequence $\{X_{N_n}\}$ is $\mathscr{B}(M)$-mixing with limit μ.

For related work, see Aldous (1978).

Let $D[0, 1]$ be the space of functions on $[0, 1]$ with discontinuities utmost of the first kind. For any f and g on $D[0, 1]$, define

$$\|f-g\| = \inf_{\lambda \in \Lambda}\left[\sup_{0 \leqslant x \leqslant 1} |f(x)-g(\lambda(x))| + \sup_{0 \leqslant x \leqslant 1} |x-\lambda(x)|\right],$$

where Λ is the space of all homeomorphisms λ of $[0, 1]$ onto $[0, 1]$ such that $\lambda(0) = 0$, $\lambda(1) = 1$. It is known that $D[0, 1]$ is a complete separable metric space under the topology generated by the above metric known as *Skorokhod topology* [cf. Parthasarathy (1967, p. 234)]. See Section 1.8.

The following result gives sufficient conditions for a stochastic process to have a separable version with paths in the function spaces $C[0, 1]$ or $D[0, 1]$. For definition of separability for a stochastic process, see Doob (1953).

Proposition 1.10.8 (Kolmogorov). Let $\{X(t), 0 \leqslant t \leqslant 1\}$ be a stochastic process defined on a probability space $(\Omega, \mathscr{B}, P)$. Suppose there exist constants c, α, and β all positive such that

$$E|X(t_1)-X(t_2)|^{\alpha} \leqslant c|t_1-t_2|^{1+\beta} \tag{1.10.8}$$

for all t_1, t_2 in $[0, 1]$. Then there exists a separable version $\tilde{X}$ of the process X with continuous sample paths almost surely, that is, $P\{\omega{:}\tilde{X}(t, \omega)$ is continuous in $t \in [0, 1]\} = 1$.

Proposition 1.10.9 (Chentsov). Let $\{X(t), 0 \leqslant t \leqslant 1\}$ be a stochastic process defined on a probability space $(\Omega, \mathscr{B}, P)$. Suppose there exists constants c, α_1, α_2, and β all positive such that

$$E\{|X(t_1)-X(t_2)|^{\alpha_1}|X(t_2)-X(t_3)|^{\alpha_2}\} \leqslant c|t_1-t_3|^{1+\beta} \tag{1.10.9}$$

for all $t_1 \leqslant t_2 \leqslant t_3$ in $[0, 1]$. Then there exists a separable version $\tilde{X}$ of the process with sample paths in $D[0, 1]$ almost surely.

It is easy to check that a Wiener process $\{W(t), 0 \leqslant t \leqslant 1\}$ has continuous sample paths with probability 1 and a Poisson process $\{N(t), 0 \leqslant t \leqslant 1\}$ has sample paths in $D[0, 1]$ with probability 1. For a survey of results on continuity of Gaussian process, see Jain and Marcus (1978).

Definition 1.10.2. A family $\{\mu_\alpha, \alpha \in I\}$ of probability measures in $\mathscr{M}(\mathscr{X})$ is said to be *tight* if for every $\varepsilon > 0$ there exists a compact set $K_\varepsilon \subset \mathscr{X}$ such that

$$\mu_\alpha(K_\varepsilon) > 1-\varepsilon \quad \text{for all } \alpha \in I. \tag{1.10.10}$$

We now state a result connected with the convergence of probability measures on $C[0, 1]$ and $D[0, 1]$. See Parthasarathy (1967).

Proposition 1.10.10. Let $\mathscr{X}$ be a complete separable metric space. Then the space $\mathscr{M}(\mathscr{X})$ of probability measures on $\mathscr{X}$ associated with the weak topology defined by (1.10.1) is a complete separable metric space. Let $\Gamma \subset \mathscr{M}(\mathscr{X})$. A necessary and sufficient condition that $\bar{\Gamma}$ is compact is that Γ is tight.

See Parthasarathy (1967, p. 47) for a proof of the result given above.

Proposition 1.10.11. Let X_n and X be random elements defined on a probability space $(\Omega, \mathscr{B}, P)$ taking values in a complete separable metric space. Let μ_n and μ be the corresponding induced probability measures on $(\mathscr{X}, \mathscr{F})$. Suppose $\mu_n \xrightarrow{w} \mu$. Let $g(\cdot)$ be any measurable functional defined on $\mathscr{X}$ such that the set of discontinuities of $g(\cdot)$ have zero measure with respect to μ. Then

$$g(X_n) \xrightarrow{\mathscr{L}} g(X). \tag{1.10.11}$$

Let $\{g_n(\cdot), n \geqslant 1\}$ be any sequence of measurable functionals defined on $\mathscr{X}$ such that $g_n(x_n) \to g(x)$ whenever $x_n \to x$, $x \in \mathscr{A}$, where $\mu(A) = 1$. Then

$$g_n(X_n) \xrightarrow{\mathscr{L}} g(X). \tag{1.10.12}$$

This result follows from general theorems in Billingsley (1968) or Parthasarathy (1967).

We now discuss some sufficient conditions for convergence in distribution for sequences of stochastic processes with values in $C[0, 1]$ or $D[0, 1]$.

Proposition 1.10.12. Let $X_n \equiv \{X_n(t), 0 \leqslant t \leqslant 1\}$ and $X \equiv \{X(t), 0 \leqslant t \leqslant 1\}$ be stochastic processes with trajectories in $C[0, 1]$ with probability 1. Suppose that

(i) for every $t_1, t_2, \ldots, t_k$ in $[0, 1]$ and $k \geqslant 1$,

$$[X_n(t_1), \ldots, X_n(t_k)] \xrightarrow{\mathscr{L}} [X(t_1), \ldots, X(t_k)] \tag{1.10.13}$$

and

(ii) there exists positive constants c, α, and β *independent* of n such that

$$E|X_n(t_1) - X_n(t_2)|^{\alpha} \leqslant c|t_1 - t_2|^{1+\beta}, \qquad 0 \leqslant t_1, t_2 \leqslant 1. \tag{1.10.14}$$

Then $\mu_n \xrightarrow{w} \mu$, where μ_n and μ are the measures induced by X_n and X, respectively, on $C[0,1]$. In particular, if g is any measurable functional defined on $C[0,1]$ with the set of discontinuities having μ-measure zero, then

$$g(X_n) \xrightarrow{\mathscr{L}} g(X). \tag{1.10.15}$$

This theorem as well as the following theorem are consequences of Propositions 1.10.10 and 1.10.11. Condition (ii) in Proposition 1.10.12 implies the tightness of the family $\{\mu_n\}$ of probability measures.

Proposition 1.10.13 (Chenstov, 1956). Let $X_n = \{X_n(t), 0 \leqslant t \leqslant 1\}$ and $X \equiv \{X(t), 0 \leqslant t \leqslant 1\}$ be stochastic processes with trajectories in $D[0,1]$ with probability 1. Suppose that

(i) For every $t_1, t_2, \ldots, t_k$ in $[0,1]$ and $k \geqslant 1$,

$$[X_n(t_1), \ldots, X_n(t_k)] \xrightarrow{\mathscr{L}} [X(t_1), \ldots, X(t_k)] \tag{1.10.16}$$

and

(ii) there exist positive constants c, α_1, α_2, and β *independent* of n such that

$$\begin{aligned} &E\{|X_n(t_1) - X_n(t_2)|^{\alpha_1} |X_n(t_2) - X_n(t_3)|^{\alpha_2}\} \\ &\leqslant c|t_1 - t_3|^{1+\beta}, \qquad 0 \leqslant t_1 \leqslant t_2 \leqslant t_3 \leqslant 1. \end{aligned} \tag{1.10.17}$$

Then $\mu_n \xrightarrow{w} \mu$ where μ_n and μ are the measures induced by X_n and X respectively on $D[0,1]$. In particular, if g is any measurable functional defined on $D[0,1]$ with the set of discontinuities having μ-measure zero, then

$$g(X_n) \xrightarrow{\mathscr{L}} g(X). \tag{1.10.18}$$

We remark that the space $C[0,1]$ with uniform metric topology can be taken as the subset of $D[0,1]$ with relative Skorokhod topology (Parthasarathy, 1967). Consequently, if the limiting process X of X_n with paths in $D[0,1]$ has continuous sample paths, it is enough to compute the measure of the set of discontinuities of any measurable functional g with respect to the uniform metric instead of the Skorokhod metric and show that it is zero in order to conclude that $g(X_n) \xrightarrow{\mathscr{L}} g(X)$.

The discussion given above clearly holds for $C[a,b]$ or $D[a,b]$ whenever $[a,b]$ is any finite interval in R. The results can be extended to semiinfinite intervals. For instance, define the topology on $C[0, \infty)$ by the relation that $f_n \to f$ in $C[0, \infty)$ iff

$$\sup_{0 \leqslant t \leqslant N} |f_n(t) - f(t)| \to 0 \quad \text{as } n \to \infty$$

for every $N > 0$. It can be shown that $X_n \xrightarrow{\mathscr{L}} X$ on $C[0, \infty)$ iff $X_n \xrightarrow{\mathscr{L}} X$ on $C[0, N]$ for every N. An analogous result holds for $D[0, \infty)$. See Stone (1963), Whitt (1970), and Lindvall (1973) for details. Whitt (1980) discusses some useful functions that arise in stochastic models and the corresponding limit theorems.

At times, it is important to consider tightness for the families of measures generated by random fields, that is, stochastic processes where the index is multidimensional.

Consider $C[0, 1]^k$, the space of continuous functions $f(x_1, \ldots, x_k)$, $0 \leqslant x_i \leqslant 1$, $1 \leqslant i \leqslant k$, and endow the space with supremum norm as before. Let $\{X_n(\mathbf{t}), \mathbf{t} \in [0, 1]^k\}$ and $\{X(\mathbf{t}), \mathbf{t} \in [0, 1]^k\}$ be stochastic processes such that

$$E|X_n(\mathbf{t}_1) - X_n(\mathbf{t}_2)|^\alpha \leqslant C \| \mathbf{t}_1 - \mathbf{t}_2 \|^{k+\beta} \tag{1.10.19}$$

for all $\mathbf{t}_1, \mathbf{t}_2 \in [0, 1]$. It can be shown that the family of measures $\{\mu_n\}$ generated by $\{X_n\}$ forms a tight family and one can prove theorems analogous to those stated in Proposition 1.10.12. However, the sufficient condition for tightness stated above is too strong for statistical purposes [cf. Prakasa Rao (1986)]. For general theory of random fields, see Adler (1980). Tightness for family of probability measures generated by stochastic processes on metric spaces is studied in Prakasa Rao (1975) and Pfaff (1982). Neuhaus (1971) extended Skorokhod metric to the space D_k, which generalizes the space $D[0, 1]$ to k dimensions. He studied weak convergence of the sequence of stochastic processes with sample paths in D_k. For further comments, see the end of this section.

The following criterion for tightness for a sequence of martingales with sample paths in $C[0, 1]$ or $D[0, 1]$ is due to Loynes (1976). Loynes (1976) proved that, if the finite-dimensional distributions of a sequence of martingales $X_n(t)$ taking values either in $C[0, 1]$ or $D[0, 1]$ converge weakly and if the sequence $\{X_n(t), n \geqslant 1\}$ is uniformly integrable for each t, then the weak convergence holds provided the limiting process satisfies a certain condition that, in particular, holds for the Wiener process. We now discuss some details of his results.

Suppose the limiting process X on $[0, 1]$ satisfies the following condition.

(A) For any given $\varepsilon > 0$, $\eta > 0$ there exists $0 < \delta < 1$ such that the following implication holds: For every $t = i\delta$ where i is an integer, the smallest solution $\lambda = \lambda(t)$ with the property

$$P[|X(t+\delta) - X(t)| > \lambda] \leqslant \frac{1}{\varepsilon} E[|X(t+\delta) - X(t)| I(|X(t+\delta) - X(t)| \geqslant \lambda)]$$

satisfies either

$$P[|X(t+\delta) - X(t)| \geqslant \lambda] \leqslant \eta\delta$$

or

$$E[|X(t+\delta) - X(t)| I(|X(t+\delta) - X(t)| \geqslant \lambda)] \leqslant \eta\delta.$$

Here $I(A)$ denotes the indicator function of a set A as usual and $t, t+\delta \in [0, 1]$. If the process X is separable, then condition (A) implies that X is in C with probability 1. It can be checked that the Wiener process satisfies condition (A) [cf. lemma in Loynes (1976)].

Proposition 1.10.14. In either $C[0, 1]$ or $D[0, 1]$, suppose that

- **(i)** for each n, X_n is a martingale,
- **(ii)** the finite-dimensional distributions of X_n converge weakly to the corresponding finite dimensional distributions of X,
- **(iii)** for each t, $E|X_n(t)| \to E|X(t)|$, and
- **(iv)** X satisfies the condition (A).

Then X_n converges in distribution to X.

Under conditions (i) and (iii) of Proposition 1.10.14, X is necessarily a martingale. As a consequence of the proposition, we have the following corollary.

Corollary 1.10.15. In either $C[0, 1]$ or $D[0, 1]$, suppose that

- **(i)** $EX_n(t) = \mu_n(t)$ is monotone nondecreasing for every n,
- **(ii)** $X_n - \mu_n$ is a martingale for every n,
- **(iii)** $\{X_n(t)\}$ are uniformly integrable for each t, and
- **(iv)** the finite-dimensional distributions of X_n converge weakly to the corresponding finite-dimensional distributions of the Wiener process W_{μ,σ^2} with drift μ and diffusion or scale parameter σ^2.

Then X_n converges in distribution to W_{μ,σ^2}.

For proofs, see Loynes (1976).

It is necessary to study the weak convergence for measures on Hilbert spaces, for instance, in obtaining the limiting distribution for the Cramer–von Mises statistic or in testing goodness of fit when nuisance parameters are present. We briefly discuss a few results in this area. For an extensive discussion, see Parthasarathy (1967) or Gikhman and Skorokhod (1974).

Let $\mathscr{X}$ be a real separable Hilbert space and $\langle x, y\rangle$ denote the inner product between x and y in $\mathscr{X}$. Let $\|x\|$ denote the norm in $\mathscr{X}$. For any $\mu\in\mathscr{M}(\mathscr{X})$, the space of probability measures on $\mathscr{X}$, define

$$\hat{\mu}(y) = \int_{\mathscr{X}} e^{i\langle x,y\rangle}\mu(dx).$$

$\hat{\mu}(\cdot)$ is called the *characteristic functional* of μ.

Proposition 1.10.16. If a sequence $\{\mu_n\}$ in $\mathscr{M}(\mathscr{X})$ is weakly compact and $\hat{\mu}_n(y)\to\phi(y)$ as $n\to\infty$ for every $y\in\mathscr{X}$, then there exists a $\mu\in\mathscr{M}(\mathscr{X})$ such that $\hat{\mu}(y)=\phi(y)$ for all $y\in\mathscr{X}$ and $\mu_n\xrightarrow{w}\mu$.

Let $\{e_i\}$ be an orthonormal basis in $\mathscr{X}$. The following result gives a sufficient condition for the weak compactness.

Proposition 1.10.17. A set $\Gamma\subset\mathscr{M}(\mathscr{X})$ is weakly compact if

$$\lim_{N\to\infty}\sup_{\mu\in\Gamma}\int r_N^2(x)\mu(dx)=0,$$

where

$$r_N^2(x)=\sum_{i=N}^{\infty}\langle x, e_i\rangle^2.$$

This result is a consequence of Proposition 1.10.10. For the proof of Proposition 1.10.16, see Parthasarathy (1967, p. 154).

Suppose $\mu\in\mathscr{M}(\mathscr{X})$ such that $\int_{\mathscr{X}}\|x\|^2\mu(dx)<\infty$. Define the operator S by

$$\langle Sy, y\rangle = \int_{\mathscr{X}}\langle x, y\rangle^2\mu(dx), \qquad y\in\mathscr{X}.$$

S is called the *covariance operator* of μ. Note that S is a positive definite Hermitian operator with finite trace since

$$\sum_i \langle Se_i, e_i\rangle = \int_{\mathscr{X}}\|x\|^2\mu(dx)<\infty.$$

Such an operator is also called an *S-operator*.

The following result gives necessary and sufficient conditions for a set $K \subset \mathcal{M}(\mathcal{X})$ to be weakly compact in terms of the S-operators.

Proposition 1.10.18. A set $K \subset \mathcal{M}(\mathcal{X})$ is weakly compact iff for every $\varepsilon > 0$ and for every $\mu \in K$, there exist an S-operator $S_{\mu,\varepsilon}$ such that

(i) $1 - \operatorname{Re} \hat{\mu}(y) \leqslant \langle S_{\mu,\varepsilon} y, y \rangle + \varepsilon, y \in \mathcal{X}$,

and

(ii) the set of operators $\{S_{\mu,\varepsilon}, \mu \in K\}$ is compact.

As a consequence of this proposition, we can prove the following corollary.

Corollary 1.10.19. For a sequence $\{\mu_n\}$ in $\mathcal{M}(\mathcal{X})$ to be weakly compact, it is necessary that

$$\lim_{N \to \infty} \sup_n \mu_n[r_N^2(x) > \varepsilon] = 0$$

for every $\varepsilon > 0$ where $r_N^2(x) = \sum_{i=N}^{\infty} \langle x, e_i \rangle^2$ and $\{e_i\}$ is an orthonormal basis for $\mathcal{X}$.

For proofs of the above results, see Parthasarathy (1967).

It is sometimes important to study compactness and convergence of sets of probability measure on $D[0, 1]$ when it is endowed with the supremum norm but not Skorokhod norm (see Chapter 7). In such an event, $D[0, 1]$ is a complete metric space but not separable.

What we have discussed so far may be termed as *weak invariance principle.* It is sometimes possible to represent the process $X_n = X + Y$ a.s. where X is the limiting process and Y is a process whose order of growth can be computed. This can be illustrated by the following example. Let X_i, $i \geqslant 1$, be i.i.d. random variables with mean zero and finite variance. Define $S_n = \sum_{i=1}^{n} X_i$. Strassen (1964) has proved that one can construct a probability space and a process $\{S(t), t \geqslant 0\}$ defined on it and also a Wiener process defined on the same space such that

(i) $S(n)$ has the same distribution as S_n

and

(ii) $S(t) = W(t) + o[(t \log \log t)^{1/2}]$ a.s. as $t \to \infty$.

Results of this type are discussed in Csörgö and Revesz (1981). Such results are known as *strong invariance principles.* Applications of strong invariance principles in nonparametric density estimation are given in Prakasa Rao (1983). See Heyde (1981) for a brief survey on strong invariance principles.

We now state two results due to Le Cam (1970) that we shall need in the asymptotic theory of maximum likelihood estimation in Section 3.5.

Let $\{X(t), t\in T\}$, $T\subset R$, be a separable stochastic process with $E|X(t)|^2<\infty$ for all $t\in T$. Then $\{X(t), t\in T\}$ is said to be a *second-order process.* Define

$$\|X(t)-X(s)\|^2=E|X(t)-X(s)|^2.$$

The variation L of X on T is defined by

$$L=\sup\left\{\sum_{j=1}^{n}\|X(t_{j+1})-X(t_j)\|\right\}, \tag{1.10.20}$$

where the supremum is taken over all finite sets $\{t_j\}$ such that $t_j<t_{j+1}$ and $t_j\in T$ for all j.

Proposition 1.10.20 (Le Cam, 1970). Let $\{X(t), t\in T\}$ be a separable stochastic process continuous in quadratic mean. Suppose T is closed and has infimum $t_0\in T$. Then, for any $\lambda>0$,

$$P\left\{\sup_{t\in T}|X(t)-X(t_0)|>\right\}\leqslant 24L^2\lambda^{-2},$$

where L is as defined in (1.10.20).

Proposition 1.10.21 (Le Cam, 1970). Let $\{X_0(t), t\in T\}$ be a separable stochastic process continuous in quadratic mean and T be a compact subset of R. Suppose $\{X_n(t), t\in T\}$ is a sequence of second-order separable stochastic processes such that

$$E|X_n(t)-X_n(s)|^2=E|X_0(t)-X_0(s)|^2,\qquad t,s\in T,$$

and

$$E|X_n(t_0)|^2<b<\infty\quad\text{for all } n$$

and for some $t_0\in T$. Then X_n is a.s. continuous for every n and, for any $\varepsilon>0$, there exists a bounded equicontinuous subset K of continuous functions on T such that

$$P\{X_n(\cdot)\in K\}>1-\varepsilon\quad\text{for all } n,$$

that is, the measures induced by X_n on $C(T)$ form a tight family.

We now state a fluctuation inequality for stochastic process with index belonging to a metric space.

Definition 1.10.3. Let (T, ρ) be a metric space. A subset $M \subset T$ is said to be *finite dimensional* with dimension $D > 0$ if there exists a constant $C < \infty$ such that, for every $t_0 \in M$ and every pair $0 < r < R < \infty$, there exists a finite collection of balls $B_\rho(t_i, r) = \{t \in T : \rho(t, t_i) < r\}$, $1 \leqslant i \leqslant m$, covering the set $B_\rho(t_0, R) \cap M$, where $t_i \in M$, $1 \leqslant i \leqslant m$, and

$$m \leqslant C\left(\frac{R}{r}\right)^D.$$

Proposition 1.10.22 (Pfaff, 1982). Let (T, ρ) be a metric space, and suppose $t_0 \in T$ and $\varepsilon > 0$ are such that the ball $B_\rho(t_0, \varepsilon)$ is finite dimensional with dimension $D > 0$. Assume that $\{X_t, t \in T\}$ is a separable random field on a probability space $(\Omega, \mathscr{F}, P)$ satisfying

$$E|X(t_2) - X(t_1)|^s \leqslant L\rho^k(t_1, t_2)$$

for every $t_1, t_2 \in B_\rho(t_0, \varepsilon)$ where $k > D, s > 0$, and $L < \infty$ are constants. Then there exists a constant $K > C$ such that

$$P\left\{\sup_{t_1, t_2 \in B_\rho(t_0, \varepsilon)} |X(t_2) - X(t_1)| > \delta\right\} \leqslant KL\varepsilon^{-k}\delta^{-s}$$

for every $\delta > 0$ where K depends only on k, s, D and C in the definition of the dimension D and is independent of ε, δ, and L.

This result is an improvement over an earlier result in Prakasa Rao (1975).

We now discuss an alternative criterion to that given in (1.10.19) for tightness due to Chenstov (1970) for sequences of random fields. We introduce some notation.

Let $\{X(\boldsymbol{\eta}),\ \boldsymbol{\eta} \in R^k\}$ be a random field. For any $\boldsymbol{\eta}_0$ and $\boldsymbol{\eta}_1 \in R^k$ such that $\eta_0^{(j)} \leqslant \eta_1^{(j)}$, $1 \leqslant j \leqslant k$, define a pave M in R^k and the process X is defined on M by the relation

$$X(M) = \Delta_1 \Delta_2 \cdots \Delta_k X(\boldsymbol{\eta})$$

where $\Delta_j X(\eta^{(1)}, \eta^{(2)}, \ldots, \eta^{(j)}, \ldots, \eta^{(k)})$ denotes the difference

$$X(\eta^{(1)}, \eta^{(2)}, \ldots, \eta_1^{(j)}, \ldots, \eta^{(k)}) - X(\eta^{(1)}, \eta^{(2)}, \ldots, \eta_0^{(j)}, \ldots, \eta^{(k)}).$$

In other words

$$X(M) = \sum_{\varepsilon \in \{0,1\}^k} (-1)^{r - |\varepsilon|} X(\eta_\varepsilon)$$

with $\eta_\varepsilon = (\eta^{(1)}_{\varepsilon_1}, \ldots, \eta^{(k)}_{\varepsilon_k})$ and $|\varepsilon| = \sum_{i=1}^{k} \varepsilon_1$. The process of paves is an additive process on paves having a common face. Paves of processes in multi dimensions correspond to increments of a process over intervals in the one dimension.

Proposition 1.10.23 (Chenstov, 1970); Deshayes and Picard (1984). Let X be a process with values in $C[0,1]^k$ such that there exist constants $B > 0, \gamma \geqslant 0$, and $\alpha > 1$ and a measure ν finite on $[0,1]^k$ such that for every $\varepsilon > 0$ and every pave M contained in $[0,1]^k$,

$$P(|X(M)| > \varepsilon) \leqslant \frac{B}{\varepsilon^\gamma}[\nu(M)]^\alpha.$$

Then there exists a constant C (independent of the dimension k) such that, for every $\varepsilon > 0$ and every pave M' in $[0,1]^k$,

$$P\left(\sup_{M \subset M'} |X(M)| > \varepsilon\right) \leqslant \frac{CB}{\varepsilon^\gamma}[\nu(M')]^\alpha.$$

Proof. The proof is by induction. For $k = 1$, the result can be obtained from Billingsley (1968, Theorem 12.2). Suppose the result is true for all processes with dimension less then or equal to $k-1$. Fix a pave $M' = M'_{k-1} \times [s', t']$ where M'_{k-1} is a pave in $[0,1]^{k-1}$. Clearly

$$\begin{aligned}\sup_{M \subset M'} |X(M)| &= \sup_{\substack{s,t \\ s' \leqslant s \leqslant t \leqslant t'}} \ \sup_{M_{k-1} \subset M'_{k-1}} |X(M_{k-1}, t) - X(M_{k-1}, s)| \\ &= \sup_{M_{k-1} \subset M'_{k-1}} |X(M_{k-1}, \hat{t}) - X(M_{k-1}, \hat{s})|,\end{aligned}$$

where $\hat{s}$ and $\hat{t}$ are well-defined random variables (depending on the ractangle M') [for instance, see Parthasarathy (1972)]. The process

$$X(M_{k-1}, \hat{t}) - X(M_{k-1}, \hat{s})$$

is a process of pave of dimension $k-1$ satisfying the hypothesis in the lemma with $\nu_{k-1} = \nu(\cdot \times [s', t'])$. This can be seen again from Billingsley (1968, Theorem 12.2). Hence the proof of the lemma is complete.

The following result gives tightness for a family of random fields $\{X_n, n \geqslant 1\}$ taking values in $C(R^k)$. Note that X_n is the sum of 2^k processes of paves of dimension $r \in \{0, 1, \ldots, k\}$ defined on the faces passing through a point, say the origin. A face of dimension r in R^k passing through the origin is a subspace of R^k of dimension r where $k - r$ of the coordinates are zero.

Proposition 1.10.24 (Chenstov, 1970). The family of probability measures generated by a sequence of random fields $\{X_n, n \geqslant 1\}$ taking values in $C(R^k)$ is tight if

(i) the family $\{X_n(\mathbf{0}), n \geqslant 1\}$ generates a tight family, and

(ii) for every $r \in \{1, \ldots, k\}$ and for every face of dimension r in R^k passing through the origin, there exist constants $\gamma \geqslant 0$, $\alpha > 1$, and a Radon measure v_r on R^r with diffused marginals such that for every pave K of R^r, there exists a constant $B > 0$ such that, for every $M \subset K$ and for every $\varepsilon > 0$, $n \geqslant 1$,

$$P\{|X_n(M)| > \varepsilon\} \leqslant \frac{B}{\varepsilon^\gamma}[v_r(M)]^\alpha.$$

Proof. It is sufficient to show the tightness of any process of pave on a face passing through the origin of dimension r, $1 \leqslant r \leqslant k$. By the definition of a process of pave of dimension r, it is zero on the face of dimension $r-1$ passing through the origin. It suffices therefore to show the equicontinuity on the (paved) compact K of R^k.

The additivity of the process on the paves having a common face reduces the problem to proving that for every $\varepsilon > 0$ there exists $\delta > 0$ such that

$$P\left\{\sup_{\|M\| < \delta} |X_n(M)| > \varepsilon\right\} < \varepsilon$$

where M runs through paves of K with $\|M\| \leqslant \delta$.

Consider a net in K of size δ whose points are the lower vertices of paves $M_j, j \in J$, under consideration. Then

$$\begin{aligned}
&P\left\{\sup_{\|M\| < \delta} |X_n(M)| > \varepsilon\right\} \\
&\leqslant \sum_{j \in J} P\left\{\sup_{M \subset M_j} |X_n(M)| \geqslant \frac{\varepsilon}{3}\right\} \\
&\leqslant BC\left(\frac{3}{\varepsilon}\right)^\gamma \sum_{j \in J} [v(M)]^\alpha \quad \text{(by Proposition 1.10.23)} \\
&\leqslant BC\left(\frac{3}{\varepsilon}\right)^\gamma v(K)\left[\sup_{\|M\| < \delta} v(M)\right]^{\alpha - 1}
\end{aligned}$$

and the result follows due to the fact that the measures v_r are with diffused marginals.

Remark. Suppose $\nu = \lambda$ in the result given above, where λ is the Lebesgue measure. If

$$E\left[\sup_{\theta \in K} \left| \frac{\partial^k}{\partial \theta^{(1)} \cdots \partial \theta^{(k)}} X(\theta) \right|^{\gamma} \right] < \infty$$

for some $\gamma > 1$, then the Chenstov condition holds, since, for any pave M of R^k,

$$X(M) = \int \cdots \int_M \frac{\partial^k}{\partial \theta^{(1)} \cdots \partial \theta^{(k)}} X(\theta)\, \lambda(d\theta).$$

We conclude this section with the following result useful in comparing two stochastic processes.

Proposition 1.10.25 (Lenglart, 1977). Let X and Y be adapted right-continuous nonnegative processes and suppose that Y is nondecreasing, zero at time zero, and predictable. Suppose that, for all almost surely finite stopping times T, $EX(T) \leqslant EY(T)$. Then, for any stopping time T and any $\varepsilon, \eta > 0$,

$$P\left[\sup_{s \leqslant T, s < \infty} X(s) \geqslant \varepsilon \right] \leqslant \frac{\eta}{\varepsilon} + P(Y(T) > \eta).$$

1.11 CONTIGUITY AND SEPARABILITY

The concept of contiguity plays a major role in the asymptotic theory of statistical inference.

Let $(\Omega_n, \mathscr{A}_n)$, $n \geqslant 1$, be a sequence of measurable spaces and P_n, Q_n be measures defined on $(\Omega_n, \mathscr{A}_n)$. The sequence $\{Q_n\}$ is said to be *contiguous* to $\{P_n\}$ if

(1.11.1) $$\lim_{n \to \infty} P_n(A_n) = 0, \qquad A_n \in \mathscr{A}_n \Rightarrow \lim_{n \to \infty} Q_n(A_n) = 0.$$

The sequences $\{P_n\}$ and $\{Q_n\}$ are said to be *completely separable* if there exists a subsequence $\{n_j\}$ and sets $A_{n_j} \in \mathscr{A}_{n_j}$ such that

(1.11.2) $$\lim_{j \to \infty} P_{n_j}(A_{n_j}) = 0 \quad \text{and} \quad \lim_{j \to \infty} Q_{n_j}(A_{n_j}) = 1.$$

The concepts of contiguity and complete separability are analogs of absolute continuity and singularity respectively for measures. We now discuss the concept of contiguity following Hall and Loynes (1977). The notion of contiguity was introduced by Le Cam (1960).

Let μ_n be a measure on $(\Omega_n, \mathscr{A}_n)$ dominating both P_n and Q_n. For instance, $\mu_n = P_n + Q_n$. Let

$$p_n = \frac{dP_n}{d\mu_n} \quad \text{and} \quad g_n = \frac{dQ_n}{d\mu_n}. \tag{1.11.3}$$

Define the likelihood ratio

$$L_n = \begin{cases} q_n/p_n & \text{if } p_n > 0 \\ 1 & \text{if } q_n = p_n = 0 \\ n & \text{if } q_n > p_n = 0. \end{cases} \tag{1.11.4}$$

The range space of L_n is $[0, \infty)$. Let P_n' and Q_n' be the measures induced by L_n on the Borel σ-algebra of R^+ by P_n and Q_n, respectively.

In order to study the interrelation between the concepts of contiguity and its various implications, we first state a lemma on uniform integrability.

Lemma 1.11.1. Suppose X_n is a random variable defined on $(\Omega_n, \mathscr{A}_n, P_n)$ for each $n \geqslant 1$. Then $\{X_n\}$ is uniformly integrable under $\{P_n\}$ if and only if

(1.11.5) (i) $\sup_n E|X_n| < \infty$,

(ii) $B_n \in \mathscr{A}_n, P_n(B_n) \to 0 \Rightarrow \int_{B_n} |X_n| \, dP_n \to 0.$

We omit the proof [cf. Billingsley (1968, p. 34)].

Proposition 1.11.2. $\{Q_n\}$ is contiguous to $\{P_n\}$ iff $\{L_n\}$ is uniformly integrable under $\{P_n\}$ and $Q_n(p_n = 0) \to 0$.

Proof. Note that

$$\begin{aligned} Q_n(B_n) &= \int_{B_n} dQ_n \\ &= \int_{B_n \cap [p_n = 0]} dQ_n + \int_{B_n \cap [p_n > 0]} L_n \, dP_n \\ &= \int_{B_n \cap [p_n = 0]} dQ_n + \int_{B_n} L_n \, dP_n \end{aligned}$$

and hence

$$\int_{B_n} L_n\,dP_n \leqslant Q_n(B_n) \leqslant Q_n(p_n = 0) + \int_{B_n} L_n\,dP_n.$$

Furthermore,

$$\int_{\Omega_n} L_n\,dP_n = \int_{[p_n > 0]} dQ_n \leqslant 1.$$

These relations together with Lemma 1.11.1 prove the result.

We now list nine assertions whose interrelations will be investigated in the sequel:

(A_1) $\{L_n\}$ converges in law, say to L, under $\{P_n\}$, that is, $P'_n \xrightarrow{w} P$ weakly where P'_n is the probability measure of L_n under P_n and P is the probability measure of L;

(A_2) $\{L_n\}$ converges in law under $\{Q_n\}$ with Q as the limiting probability measure;

(A_3) A_1 and A_2 hold and $dQ = L\,dP$;

(A_4) $\{Q_n\}$ is contiguous to $\{P_n\}$;

(A_5) $\{P_n\}$ is contiguous to $\{Q_n\}$;

(A_6) $\{L_n\}$ converges in law to L under $\{P_n\}$ and $E(L) = 1$;

(A_7) $\{L_n\}$ converges in law under $\{P_n\}$ and $P(\{0\}) = 0$;

(A_8) $\{L_n\}$ is uniformly integrable under $\{P_n\}$;

(A_9) $Q_n(p_n = 0) \to 0$.

In this notation, Proposition 1.11.2 can be rewritten

(1.11.6) $$A_4 \Leftrightarrow A_8 \text{ and } A_9.$$

Proposition 1.11.3. A_2 or $A_6 \Rightarrow A_8$.

Proof. Suppose A_2 holds. Clearly

$$0 \leqslant \int_{[L_n > \alpha]} L_n\,dP_n = \int_{[L_n > \alpha, p_n > 0]} L_n p_n\,d\mu_n$$
$$\leqslant \int_{[L_n > \alpha]} q_n\,d\mu_n = Q_n(L_n > \alpha)$$

and the last term tends to zero by A_2 as $\alpha \to \infty$ uniformly in n. It is clear that

$$\int_{\Omega_n} L_n \, dP_n \leqslant 1.$$

Hence, by Lemma 1.11.1, A_8 holds. On the other hand, suppose A_6 holds. Then

$$E_{P_n}(L_n) = \int_{[p_n > 0]} L_n \, dP_n = \int_{[p_n > 0]} q_n \, d\mu_n \leqslant 1.$$

But, by Fatou's lemma,

$$\varliminf_n EL_n \geqslant EL$$

[cf. Billingsley (1968, Theorem 5.3)] since $\{L_n\}$ converges in law under $\{P_n\}$. But $EL = 1$ by A_6. Hence $EL_n \to EL$. Therefore A_8 holds by Billingsley (1968, Theorem 5.4).

The main theorem on contiguity, connecting some of the relations A_1 to A_9, is as follows.

Theorem 1.11.4. (A_1 and A_4)$\Leftrightarrow A_2 \Leftrightarrow A_3 \Leftrightarrow A_6$.

Proof. We prove the theorem by the following chain of relations:

- **(i)** $A_3 \Rightarrow A_6 \Rightarrow$ (A_1 and A_4) $\Rightarrow A_3$.
- **(ii)** $A_3 \Rightarrow A_2 \Rightarrow$ (A_1 and A_4) $\Rightarrow A_3$.

(i) $A_3 \Rightarrow A_6$ since Q is a probability measure. Suppose A_6 holds. Note that A_1 holds by the definition of A_6. Furthermore, $\{L_n\}$ is uniformly integrable under $\{P_n\}$ by Proposition 1.11.3. Hence $E_{P_n}(L_n) \to E(L) = 1$ from A_6. Therefore

$$Q_n(p_n = 0) = 1 - Q_n(p_n > 0) = 1 - E_{P_n}(L_n) \to 0.$$

Since $\{L_n\}$ is uniformly integrable and $Q_n(p_n = 0) \to 0$, it follows that $\{Q_n\}$ is contiguous to $\{P_n\}$ by Proposition 1.11.2. In other words A_4 holds, that is, $A_6 \Rightarrow$ (A_1 and A_4). We now show that (A_1 and A_4) $\Rightarrow A_3$, that is, if $\{L_n\}$ converges in law under $\{P_n\}$ to L with probability measure P and if $\{Q_n\}$ is contiguous to $\{P_n\}$, then L_n converges in law to a probability measure Q such that $dQ = L\,dP$. Since $\{L_n\}$ converges in law under $\{P_n\}$ to L, the sequence of probability measures $\mathscr{L}(L_n|P_n)$, $n \geq 1$, forms a tight family. Hence, given $\varepsilon_i \downarrow 0$, there exists $b_i = b_i(\varepsilon_i) \uparrow \infty$ such that

$$P_n(L_n > b_i) \leq \varepsilon_i, \qquad i = 1, 2, \ldots$$

for all $n \geq 1$. Suppose $\mathscr{L}(L_n|Q_n)$ is not tight. Then there exists $\varepsilon > 0$ and $\{n_i\}$ such that $n_i \uparrow \infty$ and

$$Q_{n_i}(L_{n_i} > b_i) > \varepsilon, \qquad i = 1, 2, \ldots.$$

Let $E_{n_i} = [L_{n_i} > b_i]$. Then $P_{n_i}(E_{n_i}) \to 0$ and $Q_{n_i}(E_{n_i}) \not\to 0$, contradicting the fact that $\{Q_n\}$ is contiguous to $\{P_n\}$. Hence $\{\mathscr{L}(L_n|Q_n)\}$ is tight. In particular, there exists a subsequence $\{n_j\}$ such that $\{L_{n_j}\}$ converges in law under $\{Q_{n_j}\}$ to a limiting measure, say Q. Since $\{Q_n\}$ is contiguous to $\{P_n\}$, Proposition 1.11.2 implies that $\{L_n\}$ is uniformly integrable under $\{P_n\}$ and $Q_n(p_n = 0) \to 0$. Hence $1 = E_{P_n}(L_n) \to E_P(L) = 1$ or, in other words, $\int L\,dP = 1$. But $dQ = L\,dP$. Hence Q is a probability measure. Since this is true for any subsequence and all the subsequences converge to the same limit, it follows that $\mathscr{L}(L_n|Q_n)$ converges to a probability measure Q with $dQ = L\,dP$. This proves that

$$(\mathrm{A}_1 \quad \text{and} \quad \mathrm{A}_4) \Rightarrow \mathrm{A}_3.$$

(ii) It is trivial to see that $\mathrm{A}_3 \Rightarrow \mathrm{A}_2$, and we have already proved that (A_1 and A_4)$\Rightarrow \mathrm{A}_3$. It is sufficient to show that $\mathrm{A}_2 \Rightarrow$(A_1 and A_4). Note that $[p_n = 0] \subset [L_n = n] \cup [q_n = 0]$ and hence $0 \leq Q_n[p_n = 0] \leq Q_n(L_n = n)$. Therefore $Q_n[p_n = 0] \to 0$ since $\{L_n\}$ converges in law under $\{Q_n\}$ from A_2. In other words A_9 holds. But $\mathrm{A}_2 \Rightarrow \mathrm{A}_8$ by Proposition 1.11.3. Hence A_8 and A_9 hold, which implies A_4 by Proposition 1.11.2. This proves that $\mathrm{A}_2 \Rightarrow \mathrm{A}_4$. We now prove that $\mathrm{A}_2 \Rightarrow \mathrm{A}_1$. Note that $\{L_n\}$ is tight under $\{P_n\}$ since, for any $\alpha > 0$,

$$\begin{aligned} P_n(L_n > \alpha) &\leq \int_{[q_n > \alpha p_n]} p_n\, d\mu_n \\ &\leq \frac{1}{\alpha} \int_{[q_n > \alpha p_n]} q_n\, d\mu_n \leq \frac{1}{\alpha} \to 0 \quad \text{as} \quad \alpha \to \infty. \end{aligned}$$

Hence A_1 holds provided every convergent subsequence of L_n under $\{P_n\}$ has the same limit. Suppose that, for some subsequence, $P'_{n_j} \to P$ (say). Clearly $dQ = L\,dP$ or $dP = L^{-1}\,dQ$ on $(0, \infty)$. Hence the limit is *uniquely* defined and hence the whole sequence $\{L_n\}$ converges in law under $\{P_n\}$.

In particular Theorem 1.11.4 shows that the contiguity of $\{Q_n\}$ to $\{P_n\}$ and

P_n convergence of $\{L_n\}$ are together equivalent to Q_n convergence of $\{L_n\}$. In other words, the only way L_n can converge under the alternative $\{Q_n\}$ is for it to converge under $\{P_n\}$ *and* for contiguity of $\{Q_n\}$ to $\{P_n\}$ to hold.

As a consequence of Theorem 1.11.4, we have the following important corollary useful in applications.

Corollary 1.11.5. (i) Suppose the sequence $\{\log L_n\}$ converges in law under $\{P_n\}$ to $N(\mu, \sigma^2)$. Then $\mu \leqslant -\frac{1}{2}\sigma^2 \leqslant 0$. Furthermore, $\{\log L_n\}$ converges in law to Z (say) under $\{Q_n\}$ if and only if $\mu = -\frac{1}{2}\sigma^2$. In such a case Z is $N(\frac{1}{2}\sigma^2, \sigma^2)$.

(ii) Suppose $\{\log L_n\}$ converges in law under $\{Q_n\}$ to $N(\mu, \sigma^2)$. Then $\mu \geqslant \frac{1}{2}\sigma^2 \geqslant 0$ and $\{L_n\}$ converges in law under $\{P_n\}$ to a mixture of e^Z and 0 with probabilities $p = e^{-\mu+\sigma^2/2}$ and $1 - p$, respectively, where Z is $N(\mu - \sigma^2, \sigma^2)$.

Another important fact is the following proposition, which relates contiguity and tightness.

Proposition 1.11.6. $\{Q_n\}$ is contiguous to $\{P_n\}$ if and only if $\{\log L_n\}$ or equivalently $\{L_n\}$ is tight under $\{Q_n\}$. (Here we define $\log L_n = -\infty$ if $L_n = 0$).

Proof. This follows from the fact $\{L_n\}$ is always tight under $\{P_n\}$ and Theorem 1.11.4.

The usefulness and the power of the contiguity theory in applications is that the weak convergence properties of a statistic S_n under $\{Q_n\}$ can be derived by studying the weak convergence of $(S_n, \log L_n)$ under $\{P_n\}$.

Let S_n denote a statistic defined on $(\Omega_n, \mathscr{A}_n)$ taking values in a metric space M. Let $\mathscr{Y} = M \times R^+$ be the range space of (S_n, L_n). Let P_n'' and Q_n'' be the probability measures of (S_n, L_n) on $\mathscr{Y}$ under P_n and Q_n, respectively. Consider now the following assertions.

(B_1) $\{(S_n, L_n)\}$ converges in law under $\{P_n\}$ to a probability measure P on $\mathscr{Y}$.

(B_2) $\{(S_n, L_n)\}$ converges in law under $\{Q_n\}$ to a probability measure Q on $\mathscr{Y}$.

(B_3) B_1 and B_2 hold and $dQ(u, v) = v\,dP(u, v)$.

Theorem 1.11.7.

(i) (B_1 and A_4) $\Rightarrow B_3$.
(ii) (B_2 and A_3) $\Rightarrow$ (B_3 and A_4).

Proof. (i) Given that $\{(S_n, L_n)\}$ converges in law under $\{P_n\}$ to P on $\mathscr{Y}$ and that $\{Q_n\}$ is contiguous to $\{P_n\}$, we have to prove that $\{(S_n, L_n)\}$ converges in law under $\{Q_n\}$ to Q on $\mathscr{Y}$ where $dQ(u, v) = v\,dP(u, v)$. Since $\{(S_n, L_n)\}$ converges in law under $\{P_n\}$ to P on $\mathscr{Y}$, $\{L_n\}$ convergence in law under $\{P_n\}$ to a random variable with probability measure P_L on R^+ where P_L denotes the marginal of P. By the contiguity of $\{Q_n\}$ to $\{P_n\}$ it follows that $\{L_n\}$ converges in law under $\{Q_n\}$ to L with a probability measure, say Q_L on R^+. Furthermore $dQ_L = L\,dP_L$. Define

$$Q(A) = \int_A v\,dP(u, v).$$

Clearly

$$Q(M \times R^+) = \int_{M \times R^+} v\,dP(u, v) = \int_{R^+} L\,dP_L = 1.$$

Hence Q is a probability measure on $M \times R^+$. In order to prove that $\{(S_n, L_n)\}$ converges in law under $\{Q_n\}$ to Q, it is sufficient to prove that for all bounded continuous functions f on $M \times R^+$

$$\int f(u, v)\,dQ_n''(u, v) \to \int f(u, v)\,dQ(u, v).$$

Since $Q(\cdot)$ is a probability measure on $M \times R^+$, it is sufficient to prove the relation given above for f vanishing outside any compact set $K \subset M \times R^+$ [cf. Chung (1974, Sections 4.3 and 4.4)]. Let f be such a function. Now

$$\begin{aligned}\int f(u, v)\,dQ_n'' &= \int_{[p_n q_n > 0]} f(u, v)v\,dP_n'' + \int_{[p_n = 0, q_n > 0]} f(u, v)\,dQ_n'' \\ &= \int f(u, v)v\,dP_n'' - \int_{[p_n > 0, q_n = 0]} f(u, v)v\,dP_n'' \\ &\quad + \int_{[p_n = 0, q_n > 0]} f(u, v)\,dQ_n'' \\ &= \int f(u, v)v\,dP_n'' + \int_{[p_n = 0, q_n > 0]} f(u, v)\,dQ_n''.\end{aligned}$$

Since f is bounded and $Q_n''(p_n = 0) \to 0$ by contiguity of $\{Q_n\}$ to $\{P_n\}$, it follows that

$$\begin{aligned}\int f(u, v)\,dQ_n'' &= \int f(u, v)v\,dP_n'' + o(1) \\ &\to \int f(u, v)v\,dP\end{aligned}$$

as $P''_n \xrightarrow{w} P$ and $f(u, v)v$ is bounded and continuous. Hence

$$(B_1 \quad \text{and} \quad A_4) \Rightarrow B_3.$$

(ii) It is clear that $B_2 \Rightarrow A_2$ and $A_2 \Rightarrow A_4$ by Theorem 1.11.4. We have to show that B_2 and A_3 imply B_3. In other words, suppose $\{(S_n, L_n)\}$ converges in law under $\{Q_n\}$ to Q and (A_3) holds. We have to prove that $\{(S_n, L_n)\}$ converges in law under $\{P_n\}$. Let

$$L_n^* = \begin{cases} L_n & \text{if } L_n > 0 \\ 1/n & \text{if } L_n = 0 \end{cases}$$

and M_n be the likelihood ratio of P_n to Q_n. Define M_n^* as before. Then $M_n^* = (L_n^*)^{-1}$. Moreover,

$$\begin{aligned} 0 \leqslant Q_n(L_n \neq L_n^*) &\leqslant Q_n(L_n = 0) \\ &\leqslant Q_n(q_n = 0) = 0 \end{aligned}$$

and

$$0 \leqslant Q_n(M_n \neq M_n^*) \leqslant Q_n(L_n = n) \to 0 \quad \text{as } n \to \infty.$$

Since $\{(S_n, L_n, L_n)\}$ converges in law under $\{Q_n\}$, it follows that $\{(S_n, L_n, L_n^*)\}$, $\{(S_n, L_n, M_n^*)\}$, and $\{(S_n, L_n, M_n)\}$ converge in law under $\{Q_n\}$. Let $T_n = (S_n, L_n)$. Interchanging $\{Q_n\}$ and $\{P_n\}$ in part (i) of this theorem, in view of A_5, it follows that $\{(S_n, L_n, M_n)\}$ converges in law under $\{P_n\}$ by part (i) and hence $\{(S_n, L_n)\}$ converges in law under $\{P_n\}$.

As a consequence of this theorem, we have the following important result for applications.

Theorem 1.11.8. If $\{(S_n, \log L_n)\}$ converges in law under $\{P_n\}$ and $\{Q_n\}$ is contiguous to $\{P_n\}$ or if $\{(S_n, \log L_n)\}$ converges in law under $\{Q_n\}$ and $\{P_n\}$ is contiguous to $\{Q_n\}$, then $\{(S_n, \log L_n)\}$ converges in law both under $\{Q_n\}$ and $\{P_n\}$ to, say, Q and P, respectively. Furthermore $dQ(u, v) = e^v\, dP(u, v)$.

As consequences or corollaries to the theorems studied above, we have the following immediate results.

Corollary 1.11.9 (Le Cam's First Lemma). If $\{L_n\}$ converges in law to L with probability measure P under $\{P_n\}$ and $E_P(L) = 1$, then $\{Q_n\}$ is contiguous to $\{P_n\}$.

Corollary 1.11.10. If $L_n \xrightarrow{\mathscr{L}}$ log-normal $(-\frac{1}{2}\sigma^2, \sigma^2)$, then $\{Q_n\}$ contiguous to $\{P_n\}$.

Corollary 1.11.11 (Le Cam's Third Lemma). Suppose $\{(S_n, \log L_n)\}$ converges in law under $\{P_n\}$ to a bivariate normal distribution with parameter $(\mu_1, \mu_2, \sigma_1^2, \sigma_2^2, \sigma_{12})$ with $\mu_2 = -\frac{1}{2}\sigma_2^2$. Then S_n converges in law to $N(\mu_1 + \sigma_{12}, \sigma_1^2)$ under $\{Q_n\}$.

Remarks. We should remark that, if $\{Q_n\}$ is contiguous to $\{P_n\}$, it does *not* follow that Q_n is absolutely continuous with respect to P_n even for large n. It only implies that

$$Q_n(p_n = 0) \to 0 \quad \text{as } n \to \infty.$$

Contiguity is a type of asymptotic absolute continuity. However, the following result is true.

Theorem 1.11.12. If $\| P_n - Q_n \| \to 0$, then $\{P_n\}$ and $\{Q_n\}$ are contiguous with respect to each other. Conversely, if $\{P_n\}$ and $\{Q_n\}$ are contiguous, then there exist sequences $\{P'_n\}$ and $\{Q'_n\}$ of probability measures such that

(i) P'_n and Q'_n are equivalent,
(ii) $\{P'_n\}$ and $\{Q'_n\}$ are contiguous, and
(iii) $\| P_n - P'_n \| + \| Q_n - Q'_n \| \to 0 \quad \text{as } n \to \infty$.

Here

$$\| P - Q \| = \sup_A |P(A) - Q(A)| = \frac{1}{2}\int |f - g| \, d\mu,$$

where

$$f = \frac{dP}{d\mu} \quad \text{and} \quad g = \frac{dQ}{d\mu}.$$

For proof, see Roussas (1972, Lemma 2.1, Chapter 1) and Basawa and Scott (1983, p. 158).

1.12 ABSOLUTE CONTINUITY AND SINGULARITY

In problems connected with the asymptotic theory of statistical inference, it is crucial to know whether the probabilities measures defined on a measure space are absolutely continuous or not. An earlier discussion on this aspect is given in Sections 1.3 and 1.7. We now present a brief survey of some important results. Chatterjee and Mandrekar (1978) give an excellent review regarding the equivalence (mutual absolute continuity) or singularity (orthogonality) of Gaussian measures.

Let $(\Omega, \mathscr{F})$ be a measurable space and P_i, $i = 1, 2$, be two probability measures, on $\mathscr{F}$. By the Lebesgue decomposition of P_2 with respect to P_1,

$$(1.12.0) \qquad P_2(A) = \int_{A \cap [p_1 > 0]} \frac{p_2}{p_1} dP_1 + P_2(A \cap [p_1 = 0]), \qquad A \in \mathscr{F},$$

where p_i is the Radon–Nikodým derivative of P_i with respect to $P_1 + P_2$. In particular, we have

$$(1.12.1) \quad \text{(i)} \quad P_1 \perp P_2 \quad \text{if and only if } P_1[(p_2/p_1) = 0] = 1,$$
$$\text{(ii)} \quad P_1 \ll P_2 \quad \text{if and only if } P_1[(p_2/p_1) > 0] = 1.$$

The last relation can be obtained by the following argument. Suppose $P_1[(p_2/p_1) > 0] = 1$. Then $P_1(A) > 0$ implies that $P_2(A) > 0$, which in turn shows that $P_1 \ll P_2$. Conversely, suppose $P_1 \ll P_2$. Note that $p_1 + p_2 = 1$ a.e. $[P_1 + P_2]$. Hence

$$\begin{aligned} P_2(p_2 = 0) = 0 &\Rightarrow P_2(p_1 = 1) = 0 \\ &\Rightarrow P_1(p_1 = 1) = 0 \\ &\Rightarrow P_1(0 < p_1 < 1) = 1 \\ &\Rightarrow P_1\left[\left(\frac{p_2}{p_1}\right) > 0\right] = 1. \end{aligned}$$

The following theorem due to Kakutani (1948) gives necessary and sufficient conditions for the equivalence or orthogonality of infinite product measures.

Theorem 1.12.1 (Kakutani). Let P_i and Q_i be equivalent probability measures on $(\Omega_i, \mathscr{F}_i)$ and P_1^n and Q_1^n be the product measures on the product space $(\Omega_1^n, \mathscr{F}_1^n) = \prod_{i=1}^n (\Omega_i, \mathscr{F}_i)$. Then the measures P_1^∞ and Q_1^∞ on $(\Omega_1^\infty, \mathscr{F}_1^\infty)$ are either orthogonal or equivalent according as the infinite product

$$(1.12.2) \qquad \prod_{i=1}^{\infty} \int_{\Omega_i} (p_i(x) q_i(x))^{1/2} \mu(dx)$$

tends to zero or not where $p_i = dP_i/d\mu$, $q_i = dQ_i/d\mu$, and μ is a σ-finite measure dominating $\{P_i\}$ and $\{Q_i\}$.

We shall prove a more general theorem and deduce the above result as a special case.

Let P and Q be two probability measures on $(\Omega, \mathscr{F})$. Define

$$\rho_\alpha(P, Q) = \int_\Omega \left(\frac{dP}{d\mu}\right)^\alpha \left(\frac{dQ}{d\mu}\right)^{1-\alpha} d\mu, \qquad 0 \leqslant \alpha \leqslant 1, \tag{1.12.3}$$

where μ is a σ-finite measure dominating P and Q. For $\alpha = \frac{1}{2}$, see Section 1.3 (Proposition 1.3.18). The following lemma gives properties of ρ_α [cf. Nemetz (1974)]. We omit the proof.

Lemma 1.12.2. (i) $\rho_\alpha(P, Q)$ does not depend on the choice of μ.

(ii) $\rho_\alpha(P, Q) \geqslant 0$ equality occurring if and only if $P \perp Q$.

(iii) $\rho_\alpha(P, Q) \leqslant 1$ equality occurring if and only if $P = Q$.

(iv) $\rho_\alpha(P, Q) = \rho_{1-\alpha}(Q, P), \qquad 0 < \alpha < 1.$

(v) $\rho_\alpha(P, Q)$ is a continuous convex function of $\alpha, 0 < \alpha < 1$, for fixed P and Q.

(vi) $\rho_\alpha(P_1^n, Q_1^n) = \prod_{i=1}^{n} \rho_\alpha(P_i, Q_i), \qquad 0 < \alpha < 1.$

(vii) $[\rho_{1/2}(P, Q)]^2 \leqslant \rho_\alpha(P, Q)$. In fact

$$\rho_{1/2}(P, Q) \leqslant \rho_\alpha^{1/2}(P, Q)\rho_{1-\alpha}^{1/2}(P, Q), \qquad 0 < \alpha < 1.$$

Lemma 1.12.3. The following relations hold:

$$\text{(i)} \quad \lim_{\alpha \to 0} \rho_\alpha(P, Q) = Q[p > 0] \tag{1.12.4}$$

and

$$\text{(ii)} \quad \lim_{\alpha \to 1} \rho_\alpha(P, Q) = P[q > 0], \tag{1.12.5}$$

where $p = dP/d\mu$ and $q = dQ/d\mu$.

Proof. Let us choose $\mu = P + Q$ without loss of generality. Then $0 \leqslant p, q \leqslant 1$. Let $A = \{x : p(x) > 0\}$ and $A_\varepsilon = \{x : p(x) > \varepsilon\}$ where $\varepsilon > 0$. Given $\delta > 0$ and $\varepsilon > 0$, there exists $\alpha(\delta, \varepsilon)$ such that

$$1 \geqslant p^\alpha(x) \geqslant 1 - \delta$$

whenever $0 \leqslant \alpha \leqslant \alpha(\delta, \varepsilon)$ and $x \in A_\varepsilon$. Hence,

$$\begin{aligned}\rho_\alpha(P, Q) &\geqslant \int_{A_\varepsilon} p^\alpha(x) q^{1-\alpha}(x)\, d\mu(x)\\ &\geqslant (1-\delta) \int_{A_\varepsilon} q^{1-\alpha}(x)\, d\mu(x)\end{aligned}$$

for $0 \leqslant \alpha \leqslant \alpha(\delta, \varepsilon)$. Letting $\alpha \to 0$, we have

$$\begin{aligned}\varliminf_{\alpha\to 0} \rho_\alpha(P, Q) &\geqslant (1-\delta) \varliminf_{\alpha \to 0} \int_{A_\varepsilon} q^{1-\alpha}(x)\, d\mu(x)\\ &\geqslant (1-\delta) Q(A_\varepsilon)\end{aligned}$$

[by Fatou's lemma (Proposition 1.2.8)] for all $\varepsilon > 0$. Therefore

$$\varliminf_{\alpha\to 0} \rho_\alpha(P, Q) \geqslant (1-\delta) Q(A)$$

by letting $\varepsilon \to 0$. Let $\delta \to 0$. Then it follows that

$$\varliminf_{\alpha\to 0} \rho_\alpha(P, Q) \geqslant Q(A). \tag{1.12.6}$$

On the other hand,

$$\rho_\alpha(P, Q) \leqslant \int_A q^{1-\alpha}(x)\, d\mu(x)$$

since $0 \leqslant p \leqslant 1$, and hence

$$\varlimsup_{\alpha\to 0} \rho_\alpha(P, Q) \leqslant \varlimsup_{\alpha\to 0} \int_A q^{1-\alpha}(x)\, d\mu(x) = Q(A) \tag{1.12.7}$$

by the dominated convergence theorem. Relations (1.12.6) and (1.12.7) prove (1.12.4). Relation (1.12.5) follows from (1.12.4) by using (iv) of Lemma 1.12.2.

Remarks. We define $\rho_0(P, Q)$ and $\rho_1(P, Q)$ as the limits of $\rho_\alpha(P, Q)$ as $\alpha \to 0+$ and $\alpha \to 1-$, respectively. Properties (ii), (iv), (v), and (vi) of Lemma 1.12.2 hold now for $0 \leqslant \alpha \leqslant 1$. As a consequence of (1.12.4) of Lemma 1.12.3, we have the following result.

Proposition 1.12.4. $\rho_0(P, Q) \leqslant 1$ with equality occurring if and only if Q is absolutely continuous with respect to P.

Lemma 1.12.5. (i) Let ζ be a sub-σ-algebra of $\mathscr{F}$ and $\tilde{P}, \tilde{Q}$ denote the restrictions of P, Q to ζ. Then

(1.12.8) $$\rho_\alpha(\tilde{P}, \tilde{Q}) \geqslant \rho_\alpha(P, Q).$$

(ii) Let $\{\zeta_n\}$ be a sequence of sub-σ-algebras of $\mathscr{F}$ increasing to $\mathscr{F}$ and P_n, Q_n denote the restrictions of P, Q to ζ_n. Then

(1.12.9) $$\lim_{n\to\infty} \rho_\alpha(P_n, Q_n) = \rho_\alpha(P, Q).$$

Proof of this lemma is left as an exercise for the reader. It follows, for instance, from Csiszàr (1967). We now state and prove a generalization of Theorem 1.12.1.

Theorem 1.12.6 (Nemetz, 1974). Let P and Q be probability measures on the infinite-dimensional product space $(\Omega_1^\infty, \mathscr{F}_1^\infty)$ and denote the restrictions of P and Q to $(\Omega_1^n, \mathscr{F}_1^n)$ by P_1^n and Q_1^n, respectively. Then

(1.12.10) $$P \perp Q \quad \text{iff} \quad \lim_{n\to\infty} \rho_\alpha(P_1^n, Q_1^n) \to 0 \quad \text{for} \quad 0 < \alpha < 1$$

and

(1.12.11) $Q \ll P$ if and only if for every $\varepsilon > 0$ there exists $\alpha_0 = \alpha(\varepsilon) > 0$

such that

$$\rho_\alpha(P_1^n, Q_1^n) \geqslant 1 - \varepsilon \quad \text{when} \quad 0 < \alpha \leqslant \alpha_0$$

for every $n \geqslant 1$.

Proof. $P \perp Q$ on $(\Omega_1^\infty, \mathscr{F}_1^\infty)$ if and only if $\rho_\alpha(P, Q) = 0$ for all $0 < \alpha < 1$ by (ii) of Lemma 1.12.2. An application of (ii) of Lemma 1.12.5 proves (1.12.10).

In order to prove (1.12.11), it is sufficient to show that $\rho_0(P, Q) = 1$ (by Proposition 1.12.4) iff for every $\varepsilon > 0$ there exists α_0 such that

(1.12.12) $$\rho_\alpha(P_1^n, Q_1^n) \geqslant 1 - \varepsilon \quad \text{when} \quad 0 < \alpha \leqslant \alpha_0.$$

Since $\rho_\alpha(P, Q)$ is continuous in α over $[0, 1]$, given $\varepsilon > 0$, there exists $\alpha_0 = \alpha(\varepsilon)$

such that

$$\rho_\alpha(P,Q) \geqslant 1-\varepsilon \quad \text{if} \quad 0<\alpha\leqslant\alpha(\varepsilon)$$

holds if and only if $\rho_0(P,Q)=1$. But

$$\rho_\alpha(P_1^n, Q_1^n) \geqslant \rho_\alpha(P,Q) \geqslant 1-\varepsilon$$

for every n. Conversely, it is easy to see that, if

$$\rho_\alpha(P_1^n, Q_1^n) \geqslant 1-\varepsilon \quad \text{when} \quad 0<\alpha\leqslant\alpha_0$$

for every $n \geqslant 1$, then, by (ii) of Lemma 1.12.5, it follows that

$$\rho_\alpha(P,Q) \geqslant 1-\varepsilon \quad \text{for} \quad 0<\alpha\leqslant\alpha_0.$$

Hence, by the continuity of $\rho_\alpha(P,Q)$, it follows that $\rho_0(P,Q)=1$.

Remarks. We leave it to the reader to derive Kakutani's theorem (Theorem 1.12.1) from Theorem 1.12.6. We should caution that the equivalence of P_1^n and Q_1^n for every n need not imply that P and Q are either equivalent or orthogonal, as the following example indicates.

Example 1.12.1. Let P and Q be orthogonal on $(\Omega_1^\infty, \mathscr{F}_1^\infty)$ with equivalent finite-dimensional restrictions P_1^n and Q_1^n on $(\Omega_1^n, \mathscr{F}_1^n)$. Let $\bar{Q}=(P+Q)/2$. Then P_1^n and $\bar{Q}_1^n$ are equivalent for all n, but P and $\bar{Q}$ are neither orthogonal nor equivalent.

Chatterjee and Mandrekar (1978) extended Theorem 1.12.6. Their result is as follows. We omit the proof.

Theorem 1.12.7. Let P and Q be probability measures on $(\Omega, \mathscr{F})$ and $\{\mathscr{F}_n, n\geqslant 1\}$ be a nondecreasing family of sub-σ-algebras of $\mathscr{F}$ generating $\mathscr{F}$. Let ρ_n be the Radon–Nikodým derivative of the absolutely continuous part of Q with respect to P on $\mathscr{F}_n$. Then

(1.12.13) (i) $P \perp Q$ if and only if $\lim_n \int \rho_n^\alpha \, dP = 0$ for $0<\alpha<1$,

(ii) $P \ll Q$ if and only if for every $\varepsilon>0$ there exists $\alpha_0 = \alpha(\varepsilon) \in (0,1)$

such that

$$\int \rho_n^\alpha \, dP > 1 - \varepsilon \quad \text{when} \quad 0 < \alpha \leqslant \alpha_0 \tag{1.12.14}$$

for every $n \geqslant 1$.

We now discuss a theorem due to Kabanov et al. (1977) that generalizes Theorem 1.12.1 on the dichotomy of equivalence and orthogonality for product measures. Our approach is that of Eagleson and Gundy (1983) via martingales.

Let P and Q be two measures on a measurable space $(\Omega, \mathscr{F})$, and let $\{\mathscr{F}_n, n \geqslant 0\}$ be an increasing sequence of sub-σ-algebras of $\mathscr{F}$ generating $\mathscr{F}$. Let P_n and Q_n be the restrictions of P and Q, respectively, to $\mathscr{F}_n$. Suppose $Q_n \ll P_n$ for all n. Let $\rho_n = dQ_n/dP_n$. Define

$$\rho_n^* = \begin{cases} \rho_n^{-1} & \text{if} \quad \rho_n \neq 0 \\ 0 & \text{if} \quad \rho_n = 0 \end{cases} \tag{1.12.15}$$

and

$$\alpha_n = \rho_n \rho_{n-1}^*. \tag{1.12.16}$$

α_n is the density of Q_n with respect to P_n conditional on $\mathscr{F}_{n-1}$. Observe that $[\rho_{n-1} = 0] \subset [\rho_n = 0]$ by the property that $(\rho_n, \mathscr{F}_n, n \geqslant 1)$ is a martingale under P. Furthermore, it is a nonnegative martingale. Hence ρ_n converges a.s. $[P]$ to, say, ρ_∞. Therefore $\lim \alpha_n$ exists a.s. $[P]$ and the limit is either 0 or 1 a.s. $[P]$.

Theorem 1.12.8 (Kabanov et al., 1977). The measure $Q \ll P$ if and only if

$$Q\left[\sum_{j=1}^{\infty} E_P(1 - \alpha_j^{1/2} | \mathscr{F}_{j-1}) < \infty\right] = 1, \tag{1.12.17}$$

and $Q \perp P$ if and only if

$$Q\left[\sum_{j=1}^{\infty} E_P(1 - \alpha_j^{1/2} | \mathscr{F}_{j-1}) = \infty\right] = 1. \tag{1.12.18}$$

Remarks. Theorem 1.12.8 and the comments made earlier imply that

$$[\rho_\infty = 0] = \left[\sum_{j=1}^{\infty} E_P(1 - \alpha_j^{1/2} | \mathscr{F}_{j-1}) = \infty\right] \quad \text{a.s.} \quad [P] \tag{1.12.19}$$

and

$$(1.12.20)\quad [\rho_\infty > 0] = \left[\sum_{j=1}^{\infty} E_P(1-\alpha_j^{1/2}|\mathscr{F}_{j-1}) < \infty\right] \quad \text{a.s.} \quad [P].$$

(For two sets A and B in $\mathscr{F}$, $A = B$ a.s. $[P]$ if the sets A and B differ by a set of P measure zero.) We now state and prove a generalized version of Theorem 1.12.8 due to Eagleson and Gundy (1983).

Let $(\Omega_n, \mathscr{A}_n)$ be a sequences of measurable spaces and $P^{(n)}$ and $Q^{(n)}$ be probability measures defined on $(\Omega_n, \mathscr{A}_n)$. Let $\rho_n = dQ^{(n)}/dP^{(n)}$ when it exists and $\rho_n = \infty$ otherwise. $\rho^{(n)}$ is the Radon–Nikodým derivative of the absolutely continuous component of $Q^{(n)}$ with respect to $P^{(n)}$. Suppose $\{\mathscr{F}_{nj}, 1 \leqslant j \leqslant n\}$ is a nondecreasing sequence of sub-σ-algebras of $\mathscr{A}_n$ with $\mathscr{F}_{nn} = \mathscr{A}_n$. Let P_{nj} and Q_{nj} be the restrictions of $P^{(n)}$ and $Q^{(n)}$ to $\mathscr{F}_{nj}$. Define $\rho_{nj} = dQ_{nj}/dP_{nj}$, which is defined P_{n_j} – a.s. Let

$$\alpha_{nj} = \rho_{nj}\rho^*_{n,j-1},$$

where $\rho^*_{n,j-1} = \rho^{-1}_{n,j-1}$ if $\rho_{n,j-1} \neq 0$ and 0 otherwise. Then $\{\rho_{nj}, \mathscr{F}_{nj}, 1 \leqslant j \leqslant n\}$ is a supermartingale on $(\Omega_n, \mathscr{A}_n, P^{(n)})$. Furthermore,

$$[\rho_{n,j-1} = 0] \subset [\rho_{n,j} = 0] \quad \text{a.s.} \quad [P^{(n)}]$$

and hence $\alpha_{n,k} = 0$ for $k \geqslant j$ whenever $\alpha_{nj} = 0$, $P^{(n)}$ – a.s. It is clear that

$$E_{P^{(n)}}(\alpha_{nj}|\mathscr{F}_{n,j-1}) \leqslant 1$$

for all n and j.

Definition 1.12.1. A sequence $\{X_n\}$ of random variables defined on $(\Omega_n, \mathscr{A}_n, P^{(n)})$ is said to be *tight* if $\lim_{N\to\infty} P^{(n)}(|X_n| > N) = 0$ uniformly in $n \geqslant 1$. In such an event, we write $X_n, n \geqslant 1$, is tight $(P^{(n)})$.

Remark. This is the same as the Definition 1.10.2 expressed in terms of random variables.

Theorem 1.12.9 (Eagleson and Gundy, 1983). Let $B_n \in \mathscr{A}_n$ and $P^{(n)}_{B_n}(\cdot)$ denote the conditional probability under $P^{(n)}$ given B_n. If, for all $k > 0$,

$$(1.12.21)\quad \text{(a)}\quad \lim_{n\to\infty} P^{(n)}_{B_n}\left\{\sum_{j=1}^{n} E(1-\alpha_{nj}^{1/2}|\mathscr{F}_{n,j-1}) > k\right\} = 1,$$

then, for every $\varepsilon > 0$,

$$(1.12.22)\quad \text{(b)}\quad \lim_{n\to\infty} P_{B_n}^{(n)}(\rho_n > \varepsilon) = 0.$$

On the other hand, if the sequences

$$(1.12.23)\quad \text{(i)}\quad \sum_{j=1}^{n} E(1-\alpha_{nj}^{1/2}|\mathscr{F}_{n,j-1}),$$

$$\text{(ii)}\quad \max_{1\leqslant j\leqslant n} \alpha_{nj}^{-1/2}$$

are tight ($P_{B_n}^{(n)}$), then

$$(1.12.24)\qquad \log\rho_n \quad \text{is tight} \quad (P_{B_n}^{(n)}).$$

Before we give a proof of this theorem, we derive Theorem 1.12.8 as a consequence of Theorem 1.12.9 in the light of the remarks made following the statement of Theorem 1.12.8.

Suppose $\Omega_n = \Omega$, $P^{(n)} = P$, and $E(\rho_n) = 1$ for all n in Theorem 1.12.9. The fact that $E(\rho_n) = 1$ implies that $Q_n \ll P_n$ for all $n \geqslant 1$. Let

$$B_n = B = \left[\sum_{j=1}^{\infty} E_P(1-\alpha_j^{1/2}|\mathscr{F}_{j-1}) = \infty\right].$$

We shall prove that $B \subset [\rho_\infty = 0]$ a.s. $[P]$. If $\alpha_j = 0$ for some j, then $\alpha_k = 0$ for $k \geqslant j$ and hence $\rho_k = 0$ for $k \geqslant j$, which implies that $\rho_\infty = 0$. Let us suppose that $\alpha_j > 0$ for $j = 1, 2, \ldots$. Hence

$$E_P[\alpha_j^{1/2}|\mathscr{F}_{j-1}] > 0, \qquad j = 1, 2, \ldots.$$

Observe that

$$\begin{aligned} 0 \leqslant \rho_n^{1/2} &= \prod_{j=1}^{n} \alpha_j^{1/2} \\ &= \prod_{j=1}^{n} [\{\alpha_j^{1/2}/E_P(\alpha_j^{1/2}|\mathscr{F}_{j-1})\} E_P(\alpha_j^{1/2}|\mathscr{F}_{j-1})] \\ &= M_n D_n, \end{aligned}$$

where

$$M_n = \sum_{j=1}^{n} \{\alpha_j^{1/2}/E_P(\alpha_j^{1/2}|\mathscr{F}_{j-1})\}$$

and

$$D_n = \prod_{j=1}^{n} E_P(\alpha_j^{1/2} | \mathscr{F}_{j-1}).$$

$\{M_n, n \geqslant 1\}$ is a nonnegative martingale and hence converges a.s. $[P]$. But

$$D_n \leqslant \exp\left(- \sum_{j=1}^{n} E_P(1 - \alpha_j^{1/2} | \mathscr{F}_{j-1}) \right)$$

since

$$0 \leqslant E_P[\alpha_j^{1/2} | \mathscr{F}_{j-1}] \leqslant 1$$

and therefore $\rho_\infty = 0$ from the definition of B. This proves that

$$B \subset [\rho_\infty = 0] \quad \text{a.s.} \quad [P].$$

Let

$$B_n = C = \left[\sum_{j=1}^{\infty} E_P(1 - \alpha_j^{1/2} | \mathscr{F}_{j-1}) < \infty \right].$$

We shall show that $C \subset [\rho_\infty > 0]$ a.s. $[P]$. Note that

$$0 \leqslant E_P(1 - \alpha_j^{1/2} | \mathscr{F}_{j-1}) \leqslant 1$$

and

$$E_P\{(1 - \alpha_j^{1/2})^2 | \mathscr{F}_{j-1}\} \leqslant 2E_P(1 - \alpha_j^{1/2} | \mathscr{F}_{j-1}).$$

Hence

$$\sum_{j=1}^{\infty} (1 - \alpha_j^{1/2})^2 < \infty \quad \text{a.s.} \quad [P] \quad \text{on} \quad C$$

and

$$\sum_{j=1}^{\infty} (1 - \alpha_j^{1/2}) < \infty \quad \text{a.s.} \quad [P] \quad \text{on} \quad C.$$

It is easy to check that there exists $c > 0$ and $d > 0$ such that

$$-c \sum_{j=1}^{\infty} (1 - \alpha_j^{1/2})^2 \leqslant \log \rho_n + \sum_{j=1}^{n} (1 - \alpha_j^{1/2}) \leqslant d \sum_{j=1}^{n} (1 - \alpha_j^{1/2})^2,$$

which in turn proves that

$$C \subset [\rho_\infty > 0] \quad \text{a.s.} \quad [P].$$

This completes the proof of Theorem 1.12.8 as a consequence of Theorem 1.12.9.

We first prove a lemma before giving the proof of Theorem 1.12.9.

Lemma 1.12.10. In the notation of Theorem 1.12.9, suppose that

$$(1.12.25)\qquad \sum_{j=1}^{n} E_{P^{(n)}}\{1-\alpha_{nj}^{1/2}|\mathscr{F}_{n,j-1}\},\qquad n\geqslant 1 \quad\text{is tight}\quad (P_{B_n}^{(n)}).$$

Then

$$(1.12.26)\qquad \text{(i)}\quad \sum_{j=1}^{n} E_{P^{(n)}}\{(1-\alpha_{nj}^{1/2})^2|\mathscr{F}_{n,j-1}\},\qquad n\geqslant 1 \quad\text{is tight}\quad (P_{B_n}^{(n)}),$$

$$\text{(ii)}\quad \sum_{j=1}^{n} (1-\alpha_{nj}^{1/2})^2,\qquad n\geqslant 1 \quad\text{is tight}\ (P_{B_n}^{(n)}),$$

$$\text{(iii)}\quad \sum_{j=1}^{n} (1-\alpha_{nj}^{1/2}),\qquad n\geqslant 1 \quad\text{is tight}\quad (P_{B_n}^{(n)}).$$

Proof. Relation (1.12.26) (i) holds by the inequality

$$E_{P^{(n)}}[(1-\alpha_{nj}^{1/2})^2|\mathscr{F}_{n,j-1}]\leqslant 2E_{P^{(n)}}(1-\alpha_{nj}^{1/2}|\mathscr{F}_{n,j-1}),$$

and in view of (1.12.25). For every $n\geqslant 1$ and $c>0$, define

$$\nu_c^{(n)}=\inf\left\{k\colon \sum_{j=1}^{k+1} E_{P^{(n)}}[(1-\alpha_{nj}^{1/2})^2|\mathscr{F}_{n,j-1}]>c\right\}$$

if such a k exists and $\nu_c^{(n)}=n$ otherwise. Then $\nu_c^{(n)}$ is a predictable stopping time (see Section 1.8) [cf. Neveu (1974)] with respect to the σ-algebras $\{\mathscr{F}_{nj}, j=1,2,\ldots,n\}$ and

$$E_{P^{(n)}}\left[\sum_{j=1}^{\nu_c^{(n)}}(1-\alpha_{nj}^{1/2})^2\right]\leqslant c.$$

Hence the sequence

$$\sum_{j=1}^{\nu_c^{(n)}}(1-\alpha_{nj}^{1/2})^2,\qquad n\geqslant 1,$$

is tight $(P_{B_n}^{(n)})$. But

$$\begin{aligned}
P_{B_n}^{(n)}\left[\sum_{j=1}^{n}(1-\alpha_{nj}^{1/2})^2\neq\sum_{j=1}^{\nu_c^{(n)}}(1-\alpha_{nj}^{1/2})^2\right]
&=P_{B_n}^{(n)}[\nu_c^{(n)}\neq n]\\
&=P_{B_n}^{(n)}\left[\sum_{j=1}^{n}E_{P^{(n)}}\{(1-\alpha_{n}^{1/2})^2|\mathscr{F}_{n,j-1}\}>c\right]\\
&\to 0
\end{aligned}$$

uniformly in n as $c \to \infty$. Hence $\sum_{j=1}^{n}(1-\alpha_{nj}^{1/2})^2$, $n = 1, 2, \ldots$, is tight. We now prove part (iii) of the lemma. Consider

$$\begin{aligned}
&E_{P^{(n)}}\left|\sum_{j=1}^{\nu_c^{(n)}}\{1-\alpha_{nj}^{1/2}-E(1-\alpha_{nj}^{1/2}|\mathscr{F}_{n,j-1})\}\right|^2 \\
&= E_{P^{(n)}}\left\{\sum_{j=1}^{\nu_c^{(n)}}[1-\alpha_{nj}^{1/2}-E(1-\alpha_{nj}^{1/2}|\mathscr{F}_{n,j-1})]^2\right\} \\
&\leqslant 2E_{P^{(n)}}\left\{\sum_{j=1}^{\nu_c^{(n)}}E[(1-\alpha_{nj}^{1/2})^2|\mathscr{F}_{n,j-1}]\right\} \\
&\leqslant 2c.
\end{aligned}$$

Hence the sequence

$$\sum_{j=1}^{\nu_c^{(n)}}\{1-\alpha_{nj}^{1/2}-E(1-\alpha_{nj}^{1/2}|\mathscr{F}_{n,j-1})\}, \qquad n \geqslant 1,$$

is tight $(P_{B_n}^{(n)})$. But

$$\sum_{j=1}^{\nu_c^{(n)}}E(1-\alpha_{nj}^{1/2}|\mathscr{F}_{n,j-1}), \qquad n \geqslant 1,$$

is tight $(P_{B_n}^{(n)})$ by hypothesis. Hence

$$\sum_{j=1}^{\nu_c^{(n)}}(1-\alpha_{nj}^{1/2}), \qquad n \geqslant 1,$$

is tight $(P_{B_n}^{(n)})$. This in turn implies that

$$\sum_{j=1}^{n}(1-\alpha_{nj}^{1/2}), \qquad n \geqslant 1,$$

is tight $(P_{B_n}^{(n)})$. This completes the proof of the Lemma.

Proof of Theorem 1.12.9. Suppose (1.12.21) holds. Then there exists a sequence $k_n \to \infty$ such that

$$\lim_{n\to\infty} P_{B_n}^{(n)}\left\{\sum_{j=1}^{n}E(1-\alpha_{nj}^{1/2}\big|\mathscr{F}_{n,j-1}) > k_n\right\} = 1. \tag{1.12.27}$$

Let $B_n = B_{n1} \cup B_{n2}$, where

$$B_{n1} = B_n \cap [\alpha_{n,j+k} = 0 \quad \text{for some} \quad j < n, k = 1, 2, \ldots, n-1]$$

and

$$B_{n2} = B_n \cap [\alpha_{nj} > 0 \quad \text{for} \quad j = 1, 2, \ldots, n].$$

Note that $\{\rho_{nj}, \mathscr{F}_{nj}, P^{(n)}, 1 \leqslant j \leqslant n\}$ is a supermartingale and $\rho_{n,j+k} = 0$ on $[\rho_{nj} = 0]$ for $k = 1, 2, \ldots,$ a.s. $[P^{(n)}]$. Furthermore, $\rho_n = 0$ on the set B_{n1}. Hence it is sufficient to consider the set B_{n2} in order to prove (1.12.22). As was noted earlier

$$0 < E(\alpha_{nj}^{1/2} | \mathscr{F}_{n,j-1}) \leqslant 1$$

on B_{n2} and

$$\begin{aligned} \rho_n^{1/2} &= \prod_{j=1}^{n} \{\alpha_{nj}^{1/2} / E(\alpha_{nj}^{1/2} | \mathscr{F}_{n,j-1})\} \left\{ \prod_{j=1}^{n} E(\alpha_{nj}^{1/2} | \mathscr{F}_{n,j-1}) \right\} \\ &= M_n D_n \quad \text{(say)}, \end{aligned}$$

where M_n are positive random variables and $E_{P^{(n)}}(M_n) = 1$ for all n. Hence $\{M_n, n \geqslant 1\}$ is tight $(P^{(n)})$ and therefore $I_{B_{n2}} M_n$ is tight, where I_A denotes the indicator function of the set A. On the other hand,

$$D_n \leqslant \exp\left\{ -\sum_{j=1}^{n} E(1 - \alpha_{nj}^{1/2} | \mathscr{F}_{n,j-1}) \right\}$$

and (1.12.27) implies that there exist $\varepsilon_n \downarrow 0$ such that

$$\lim_{n \to \infty} P^{(n)}(I_{B_{n2}} D_n > \varepsilon_n) = 0.$$

Hence

$$\lim_{n \to \infty} P^{(n)}(I_{B_{n2}} \rho_n > \varepsilon) = 0,$$

which implies the assertion (1.12.22).

Let us now assume that (1.12.23) holds. Lemma 1.12.10 implies that

$$\sum_{j=1}^{n} (1 - \alpha_{nj}^{1/2})^2, \qquad n \geqslant 1,$$

is tight $(P_{B_n}^{(n)})$, and therefore

$$\max_{1 \leqslant j \leqslant n} \alpha_{nj}^{1/2}, \qquad n \geqslant 1,$$

is tight $(P_{B_n}^{(n)})$. Furthermore,

$$\sum_{j=1}^{n} (1-\alpha_{nj}^{1/2}), \qquad n \geqslant 1,$$

is tight $(P_{B_n}^{(n)})$ by Lemma 1.12.10. This, together with tightness of the sequence (ii) of (1.12.23) imply that for every $\varepsilon > 0$, there exist $c > 0$ and $d > 0$ not depending on n such that

$$P_{B_n}^{(n)}\left\{ -c \sum_{j=1}^{n} (1-\alpha_{nj}^{1/2})^2 \leqslant \log \rho_n + \sum_{j=1}^{n} (1-\alpha_{nj}^{1/2}) \leqslant d \sum_{j=1}^{n} (1-\alpha_{nj}^{1/2})^2 \right\} > 1-\varepsilon,$$

which in turn proves (1.12.24). This completes the proof of Theorem 1.12.9.

As a consequence of this theorem, we have the following corollary giving sufficient conditions for the contiguity of product measures due to Oosterhoff and Van Zwet (1979).

Corollary 1.12.11. Let P_{nj}, Q_{nj} be probability measures defined on measurable space $(\Omega_{nj}, \mathscr{A}_{nj})$, $1 \leqslant j \leqslant n$. Let $\mathscr{A}_{nj}$ be any σ-finite measure dominating $P_{nj} + Q_{nj}$. Define

$$\alpha_{nj} = \frac{dQ_{nj}/d\mu_{n_j}}{dP_{nj}/d\mu_{nj}}.$$

Let $(\Omega_n, \mathscr{A}_n)$ be the product space of $(\Omega_{nj}, \mathscr{A}_{nj})$, $1 \leqslant j \leqslant n$, and $Q_n = \prod_{j=1}^{n} Q_{nj}$ and $P_n = \prod_{j=1}^{n} P_{nj}$ be the product measures. Then $\{Q_n\}$ is contiguous to $\{P_n\}$ if and only if

$$\overline{\lim_{n\to\infty}} \sum_{j=1}^{n} E_{P_{nj}}(1-\alpha_{nj}^{1/2})^2 < \infty \tag{1.12.27}$$

and

$$\lim_{n\to\infty} \sum_{j=1}^{n} Q_{nj}(\alpha_{nj} \geqslant c_n) = 0 \tag{1.12.28}$$

whenever $c_n \to \infty$.

Proof. Since we are dealing with product measures,

$$E_{P_{nj}}(1-\alpha_{nj}^{1/2} | \mathscr{F}_{n,j-1}) = E_{P_{nj}}(1-\alpha_{nj}^{1/2})$$

and

$$E_{P_{nj}}[(1-\alpha_{nj}^{1/2})^2 | \mathscr{F}_{n,j-1}] = E_{P_{nj}}(1-\alpha_{nj}^{1/2}).$$

Note that $E_{P_{nj}}(\alpha_{nj}) = E_{P_{nj}}(\alpha_{nj} I[\alpha_{nj} < \infty]) = Q_{nj}[\alpha_{nj} < \infty]$ and hence

$$(1.12.29) \quad 2\sum_{j=1}^{n} E_{P_{nj}}(1-\alpha_{nj}^{1/2}) - \sum_{j=1}^{n} E_{P_{nj}}(1-\alpha_{nj}^{1/2})^2 = \sum_{j=1}^{n} Q_{nj}(\alpha_{nj} = \infty).$$

Suppose relations (1.12.27) and (1.12.28) hold. Then the right-hand side of relation (1.12.29) tends to zero as $n \to \infty$ by (1.12.28). This in turn implies that

$$(1.12.30) \qquad \overline{\lim_{n}} \sum_{j=1}^{n} E_{P_{nj}}(1-\alpha_{nj}^{1/2}) < \infty$$

by (1.12.27). Let $B_n = \Omega_n$ in Theorem 1.12.9. Interchanging the role of $P^{(n)}$ and $Q^{(n)}$ in the theorem (α_{nj} must be replaced by $1/\alpha_{nj}$) and observing that

$$E_{P_{nj}}(1-\alpha_{nj}^{1/2}) = E_{Q_{nj}}(1-\alpha_{nj}^{-1/2}),$$

we note that (1.12.29) and (1.12.27) imply that relation (1.12.23) of Theorem 1.12.9 holds (use Lemma 1.12.10). Hence $\{\log \rho_n\}$ is tight ($Q^{(n)}$). This implies in turn that $\{Q^{(n)}\}$ is contiguous to $\{P^{(n)}\}$ by Proposition 1.11.6 (Hall and Loynes, 1977).

On the other hand, suppose that

$$(1.12.31) \qquad \overline{\lim_{n}} \sum_{j=1}^{n} E_{P_{nj}}(1-\alpha_{nj}^{1/2})^2 = \infty.$$

Then $\{Q^{(n)}\}$ and $\{P^{(n)}\}$ are completely separable by the first part of the Theorem 1.12.9. For, if possible, suppose $\{Q^{(n)}\}$ is contiguous to $\{P^{(n)}\}$. Then

$$(1.12.32) \qquad \overline{\lim_{n}} \sum_{j=1}^{n} E_{P_{nj}}(1-\alpha_{nj}^{1/2}) < \infty$$

by (1.12.29) and hence

$$(1.12.33) \qquad \overline{\lim_{n}} E_{P_{nj}}(1-\alpha_{nj}^{1/2})^2 < \infty,$$

contradicting relation (1.12.31).

Suppose now $\{Q^{(n)}\}$ is contiguous to $\{P^{(n)}\}$. Relation (1.12.28) follows from the relation

$$\lim_{n\to\infty}\sum_{j=1}^{n} P_{nj}\{(1-\alpha_{nj}^{1/2})^2 > c_n\} \leqslant \lim_{n\to\infty} c_n^{-1}\sum_{j=1}^{n} E_{P_{nj}}(1-\alpha_{nj}^{1/2})^2 = 0 \tag{1.12.34}$$

and the fact that $\{Q^{(n)}\}$ is contiguous to $\{P^{(n)}\}$, which in turn implies that

$$\lim_{n\to\infty}\sum_{j=1}^{n} P_{nj}(A_{nj}) = 0 \Rightarrow \lim_{n\to\infty}\sum_{n=1}^{\infty} Q_{nj}(A_{nj}) = 0. \tag{1.12.35}$$

This completes the proof the corollary.

Remarks. The results of this section indicate that the contiguity–separability and absolute continuity–singularity properties can be studied at the same time in a martingale approach. We discuss some connections in this direction in the next section.

1.13 CONTIGUITY–SEPARABILITY AND ABSOLUTE CONTINUITY–SINGULARITY

Liptser et al. (1982) obtained necessary and sufficient conditions for contiguity and complete separability for probability measures by methods analogous to those discussed in Section 1.12. We now present their main results. The proofs are similar to those in Section 1.12. We omit the proofs.

Let P_n and Q_n be probability measures defined on $(\Omega_n, \mathscr{F}_n)$, $n \geqslant 1$. We denote by $(Q_n) \lhd (P_n)$ if $\{Q_n\}$ is contiguous to $\{P_n\}$ and by $(Q_n) \triangle (P_n)$ if $\{Q_n\}$ and $\{P_n\}$ are completely separable. Let $\{\mathscr{F}_{nk}, k \geqslant 0\}$ be a nondecreasing sequence of sub-σ-algebras of $\mathscr{F}_n$ generating $\mathscr{F}_n$, and let P_{nk} be the restriction of P_n to $\mathscr{F}_{nk}$. Here $\mathscr{F}_{n0} = \{\phi, \Omega_n\}$. Set

$$R_n = \tfrac{1}{2}(P_n + Q_n), \qquad R_{nk} = \tfrac{1}{2}(P_{nk} + Q_{nk}). \tag{1.13.0}$$

R_{nk} is the restriction of R_n to $\mathscr{F}_{nk}$. For any $0 \leqslant a \leqslant \infty$, define

$$a^* = \begin{cases} 0 & \text{if } a = 0 \\ a^{-1} & \text{if } 0 < a < \infty \\ \infty & \text{if } a = \infty. \end{cases} \tag{1.13.1}$$

Let

$$\rho_{nk} = \frac{dP_{nk}}{dR_{nk}} \quad \text{and} \quad \tilde{\rho}_{nk} = \frac{dQ_{nk}}{dR_{nk}} \tag{1.13.2}$$

be the Radon–Nikodym derivatives of the measures P_{nk} and Q_{nk}, respectively, with respect to R_{nk}. Set

$$\beta_{nk} = \rho_{nk}\rho^*_{n,k-1} \quad \text{and} \quad \tilde{\beta}_{nk} = \tilde{\rho}_{nk}\tilde{\rho}^*_{n,k-1} \tag{1.13.3}$$

with $\rho_{n0} = \tilde{\rho}_{n0} = 1$. Define

$$z_{nk} = \frac{\tilde{\rho}_{nk}}{\rho_{nk}} \tag{1.13.4}$$

and

$$\alpha_{nk} = z_{nk}\tilde{z}^*_{n,k-1}, \qquad k \geq 1, \tag{1.13.5}$$

The following theorem is due to Liptser et al. (1982).

Theorem 1.13.1. $(Q_n) \lhd (P_n)$ if and only if

(A) $$\lim_{N\to\infty} \overline{\lim_{n\to\infty}}\, Q_n\left(\sup_k \alpha_{nk} \geq N\right) = 0$$

and

(B) $$\lim_{N\to\infty} \overline{\lim_{n\to\infty}}\, Q_n\left(\sum_{k=1}^{\infty} E_{R_n}[(\tilde{\beta}^{1/2}_{nk} - \beta^{1/2}_{nk})^2 | \mathscr{F}_{n,k-1}] \geq N\right) = 0.$$

Remarks. This theorem generalizes Theorem 1.12.9 and Corollary 1.12.11 of Section 1.12. Suppose that P_n and Q_n are product measures given by

$$P_n = \mu_1^{(n)} \times \mu_2^{(n)} \times \cdots$$

and

$$Q_n = \tilde{\mu}_1^{(n)} \times \tilde{\mu}_2^{(n)} \times \cdots,$$

where

$$\Omega_n = \Omega_{n1} \times \Omega_{n2} \times \cdots$$

and

$$\mathscr{F}_n = \mathscr{F}_{n1} \times \mathscr{F}_{n2} \times \cdots.$$

Then

$$\begin{aligned} E_{R_n}[(\tilde{\beta}^{1/2}_{nk} - \beta^{1/2}_{nk})^2 | \mathscr{F}_{n,k-1}] &= E_{R_n}[(\tilde{\beta}^{1/2}_{nk} - \beta^{1/2}_{nk})^2] \\ &= \int_{\Omega_{nk}} (\tilde{\beta}^{1/2}_{nk} - \beta^{1/2}_{nk})^2 \, d\left(\frac{\mu_k^{(n)} + \tilde{\mu}_k^{(n)}}{2}\right), \end{aligned}$$

which is the square of the Hellinger distance $H(\tilde{\mu}_k^{(n)}, \mu_k^{(n)})$ (cf. Section 1.3) between the measures $\tilde{\mu}_k^{(n)}$ and $\mu_k^{(n)}$. As consequences of Theorem 1.13.1, the following corollaries hold.

Corollary 1.13.2. Let $X_{n1}, \ldots, X_{nn}$ be i.i.d. random variables with densities $p_n(x)$ or $q_n(x)$. Let P_n and Q_n be the corresponding joint distributions of $(X_{n1}, \ldots, X_{nn})$ over $(R^n, \mathscr{B}_n)$, $n \geqslant 1$. Then $(Q_n) \lhd (P_n)$ if and only

$$\text{(i)}\quad \lim_{N \to \infty} \overline{\lim_{n}}\, nQ_n\left[\frac{q_n(X_{n1})}{p_n(X_{n1})} \geqslant N\right] = 0,$$

$$\text{(ii)}\quad \overline{\lim_{n \to \infty}}\, n \int_{-\infty}^{\infty} (q_n^{1/2}(x) - p_n^{1/2}(x))^2\, dx < \infty.$$

Corollary 1.13.3. Suppose P and Q are probability measures on $(\Omega, \mathscr{F})$ and $\{\mathscr{F}_n, n \geqslant 0\}$ is a nondecreasing sequence of sub-σ-algebras of $\mathscr{F}$ generating $\mathscr{F}$ with $\mathscr{F}_0 = \{\phi, \Omega\}$. Let P_n and Q_n denote the restrictions of P and Q, respectively, to $\mathscr{F}_n$. Suppose further $Q \overset{\text{loc}}{\ll} P$ (Q is *locally absolutely continuous* with respect to P), that is, $Q_n \ll P_n$ for all $n \geqslant 1$. Then $Q \ll P$ iff

$$Q\left(\sum_{k=1}^{\infty} E(1 - \alpha_k^{1/2} | \mathscr{F}_{k-1}) < \infty\right) = 1,$$

where

$$\alpha_k = z_k z_{k-1}^* \quad \text{and} \quad z_k = \frac{dQ_k}{dP_k}.$$

Furthermore, if $Q \overset{\text{loc}}{\ll} P$, then $Q(\sup_k \alpha_k < \infty) = 1$.

Lemma 1.13.4. If condition (A) of Theorem 1.13.1 holds, then condition (B) is equivalent to

$$\text{(B*)}\quad \lim_{N \to \infty} \overline{\lim_{n \to \infty}}\, Q_n\left(\sum_{k=1}^{\infty} I_{[\alpha_{n,k-1} < \infty]} E_{P_n}(1 - \alpha_{nk}^{1/2} | \mathscr{F}_{n,k-1}) \geqslant N\right) = 0$$

or

$$\text{(B**)}\quad \lim_{N \to \infty} \overline{\lim_{n \to \infty}}\, Q_n\left(\sum_{k=1}^{\infty} E_{P_n}[(1 - \alpha_{nk}^{1/2})^2 | \mathscr{F}_{n,k-1}] \geqslant N\right) = 0.$$

The next few results give necessary and sufficient conditions for complete separability.

Theorem 1.13.5. Suppose that

$$\text{(C)}\quad \overline{\lim_n}\, Q_n\left(\sum_{k=1}^{\infty} I_{[\alpha_{n,k-1}<\infty)} E_{P_n}(1-\alpha_{nk}^{1/2}\,|\,\mathscr{F}_{n,k-1}) \geq N\right) = 1 \text{ for all } N>0.$$

Then $(Q_n) \triangle (P_n)$.

Theorem 1.13.6. Suppose that $Q_n \overset{\text{loc}}{\ll} P_n, n \geq 1$, and that

$$\text{(A)}\quad \lim_{N\to\infty} \overline{\lim_{n\to\infty}}\, Q_n\left(\sup_k \alpha_{nk} \geq N\right) = 0,$$

$$\text{(D)}\quad \lim_{N\to\infty} \overline{\lim_{n\to\infty}}\, Q_n\left(\inf_k \alpha_{nk} \leq 1/N\right) = 0.$$

Then

$$\text{(E)}\quad \overline{\lim_n}\, Q_n\left(\sum_{k=1}^{\infty} E_{P_n}(1-\alpha_{nk}^{1/2}\,|\,\mathscr{F}_{n,k-1}) \geq N\right) = 1 \quad \text{for all} \quad N>0$$

is necessary and sufficient for $(Q_n) \triangle (P_n)$ to hold.

As a corollary to Theorem 1.13.1, Lemma 1.13.4, and Theorem 1.13.6, the following result can be obtained.

Theorem 1.13.7. Suppose $Q_n \overset{\text{loc}}{\ll} P_n, n \geq 1$, and conditions (A) and (D) hold. Then

$$(1.13.6)\quad (Q_n) \lhd (P_n) \quad \text{iff} \quad \lim_{N\to\infty} \overline{\lim_n}\, Q_n\left(\sum_{k=1}^{\infty} E_{P_n}(1-\alpha_{nk}^{1/2}\,|\,\mathscr{F}_{n,k-1}) \geq N\right) = 0$$

and

$$(1.13.7)\quad (Q_n) \triangle (P_n) \quad \text{iff} \quad \overline{\lim_n}\, Q_n\left(\sum_{k=1}^{\infty} E_{P_n}(1-\alpha_{nk}^{1/2}\,|\,\mathscr{F}_{n,k-1}) \geq N\right) = 1$$

for all $N>0$.

Theorem 1.12.8 of Section 1.12, due to Kabanov et al. (1977), follows as a consequence of Theorem 1.13.7 since conditions (A) and (D) take the form

$$Q\left(\sup_k \alpha_k < \infty\right) = 1, \qquad Q\left(\inf_k \alpha_k = 0\right) = 0,$$

which follow from (B^{**}) and (E). For comparison, we restate Theorem 1.12.8.

Theorem 1.13.8. Suppose $Q \overset{\text{loc}}{\ll} P$ and Q and P are probability measures on $(\Omega, \mathscr{F})$ as in Corollary 1.13.3. Then

$$Q \ll P \Leftrightarrow Q\left(\sum_{k=1}^{\infty} E_P(1 - \alpha_k^{1/2} | \mathscr{F}_{k-1}) < \infty \right) = 1 \tag{1.13.8}$$

and

$$Q \perp P \Leftrightarrow Q\left(\sum_{k=1}^{\infty} E_P(1 - \alpha_k^{1/2} | \mathscr{F}_{k-1}) = \infty \right) = 1. \tag{1.13.9}$$

Theorem 1.13.9. Suppose P_n and Q_n are product measures as discussed in the remarks made after Theorem 1.13.1. Furthermore, suppose that $Q_n \overset{\text{loc}}{\ll} P_n$ for all $n \geqslant 1$,

$$\text{(A}') \quad \lim_N \overline{\lim_{n\to\infty}}\, Q_n\left(\sup_k \frac{\tilde{\beta}_{nk}}{\beta_{nk}} \geqslant N \right) = 0$$

and

$$\text{(D}') \quad \lim_N \overline{\lim_n}\, Q_n\left(\inf_k \frac{\tilde{\beta}_{nk}}{\beta_{nk}} \leqslant 1/N \right) = 0.$$

Then

$$(Q_n) \lhd (P_n) \quad \text{iff} \quad \sum_{k=1}^{\infty} H^2(\tilde{\mu}_k^{(n)}, \mu_k^{(n)}) < \infty \tag{1.13.10}$$

and

$$(Q_n) \triangle (P_n) \quad \text{iff} \quad \sum_{k=1}^{\infty} H^2(\tilde{\mu}_k^{(n)}, \mu_k^{(n)}) = \infty, \tag{1.13.11}$$

where $H(\cdot,\cdot)$ is the Hellinger distance.

Remarks. Under assumptions (A′) and (D′), this result is an analog of Kakutani's theorem (Theorem 1.12.1) and either $(Q_n) \lhd (P_n)$ or $(Q_n) \triangle (P_n)$, that is, a dichotomy between contiguity and complete separability holds for product measures.

Definition 1.13.1. A family of random variables $\{\xi_n, n \geqslant 1\}$ defined on $(\Omega_n, \mathscr{F}_n, P_n)$ is said to be *asymptotically tight* if

$$\lim_{N\to\infty} \overline{\lim_n}\, P_n(|\xi_n| \geqslant N) = 0. \tag{1.13.12}$$

In such an event we write (ξ_n, P_n)-a.s. tight.

Proofs of Theorems 1.13.1, 1.13.5, and 1.13.6 consist in showing that

$$\text{(i)}\quad (Q_n) \lhd (P_n) \Leftrightarrow \left(\sup_k z_{nk}, Q_n\right)\text{-a.s. tight}$$

$$\Leftrightarrow \text{(A) and (B) hold}$$

and

$$\text{(ii)}\quad (Q_n) \triangle (P_n) \Leftrightarrow \overline{\lim_n}\, Q_n\left(\sup_k z_{nk} \geqslant N\right) = 1 \quad \text{for all} \quad N > 0$$

and

$$\text{(D)} \Rightarrow \overline{\lim_n}\, Q_n\left(\sup_k z_{nk} \geqslant N\right) = 1 \quad \text{for all} \quad N > 0.$$

For proofs, see Liptser et al. (1982). Absolute continuity and singularity of locally absolute continuous probability distributions are extensively investigated in Kabanov et al. (1978, 1979).

For diffusion processes, such results were obtained in Gikhman and Skorokhod (1972) [cf. Basawa and Prakasa Rao (1980)]. Absolute continuity or otherwise of the measures corresponding to semimartingales is discussed in Kabanov et al. (1978, 1979). Kabonov et al. (1980) studied limit theorems for point processes. Shiryayev (1980) gives a discussion on the absolute continuity and singularity of probability measures in functional spaces. Pukelsheim (1986) obtained a criterion for absolute continuity and singularity of probability measures that dispenses with the assumption on local absolute continuity and allows an arbitrary root instead of the square root of the likelihood ratio as in Theorem 1.13.8. Besides, this criterion is in terms of probabilities and expectations under Q as opposed to the criterion in terms of Q and P as given in (1.13.8) and (1.13.9).

1.14 SOME INEQUALITIES

We now list some of the important inequalities that are necessary to prove several results stated in the earlier sections and are important and useful in applications to be discussed in later chapters.

Proposition 1.14.1 (Chebyshev Inequality). For any random variable X with mean μ and variance σ^2,

$$P(|X - \mu| \geqslant \varepsilon) \leqslant \frac{\sigma^2}{\varepsilon^2}$$

for every $\varepsilon > 0$. In general, for any random variable X with $E|X| < \infty$.

$$P(|X| \geqslant \varepsilon) \leqslant \frac{1}{\varepsilon} E|X|$$

for every $\varepsilon > 0$.

Proposition 1.14.2 (Cauchy–Schwarz Inequality). Let X and Y be random variables such that $E(X^2) < \infty$ and $E(Y^2) < \infty$. Then

$$E|XY| \leqslant (EX^2)^{1/2}(EY^2)^{1/2},$$

equality occurring if and only if X and Y are linearly related a.s.

Proposition 1.14.3 (Holder Inequality). Let X and Y be random variables defined on a probability space $(\Omega, \mathscr{F}, P)$ such that $E|X|^p < \infty$ and $E|Y|^q < \infty$, where $1 < p < \infty$ and $1/p + 1/q = 1$. Then

$$E|XY| \leqslant (E|X|^p)^{1/p}(E|Y|^q)^{1/q}.$$

Let ζ be a sub-σ-algebra of $\mathscr{F}$. Then

$$E\{|XY|\,|\zeta\} \leqslant [E\{|X|^p|\zeta\}]^{1/p}[E\{|Y|^q|\zeta\}]^{1/q}.$$

Proposition 1.14.4 (Minkowski Inequality). Let X_1 and X_2 be random variables such that $E|X_1|^p < \infty$ and $E|X_2|^p$ for some $p \geqslant 1$. Then

$$\{E|X_1 + X_2|^p\}^{1/p} \leqslant (E|X_1|^p)^{1/p} + (E|X_2|^p)^{1/p}.$$

Proposition 1.14.5 (Marcinkiewicz–Zygmund Inequality). Let $\{X_i, 1 \leqslant i \leqslant n\}$ be independent random variables such that $EX_i = 0$ for $1 \leqslant i \leqslant n$ and $E|X_i|^p < \infty$ for some $p \geqslant 1$. Then there exist positive constants A_p and B_p depending on p such that

$$A_p \left\| \left(\sum_{j=1}^{n} X_j^2 \right)^{1/2} \right\|_p \leqslant \left\| \sum_{j=1}^{n} X_j \right\|_p \leqslant B_p \left\| \left(\sum_{j=1}^{n} X_j^2 \right)^{1/2} \right\|_p,$$

where $\|X\|_p = (E|X|^p)^{1/p}$.

Proposition 1.14.6 (Levy Inequality). For any random variable $X, m(X)$ is called a *median* of X if $P(X \leqslant m(X)) \geqslant \frac{1}{2}$ and $P(X \geqslant m(X)) \geqslant \frac{1}{2}$. Let $\{X_j, 1 \leqslant j \leqslant n\}$ be independent random variables and $S_j = \sum_{i=1}^{j} X_i$. Then, for every $\varepsilon > 0$,

$$P\left\{\max_{1 \leqslant j \leqslant n} [S_j - m(S_j - S_n)] \geqslant \varepsilon\right\} \leqslant 2P(S_n \geqslant \varepsilon)$$

and

$$P\left\{\max_{1 \leqslant j \leqslant n} |S_j - m(S_j - S_n)| \geqslant \varepsilon\right\} \leqslant 2P(|S_n| \geqslant \varepsilon).$$

In particular, if $\{X_j, 1 \leqslant j \leqslant n\}$ are independent random variables symmetric about zero, then

$$P\left\{\max_{1 \leqslant j \leqslant n} S_j \geqslant \varepsilon\right\} \leqslant 2P(S_n \geqslant \varepsilon)$$

and

$$P\left\{\max_{1 \leqslant j \leqslant n} |S_j| \geqslant \varepsilon\right\} \leqslant 2P(|S_n| \geqslant \varepsilon).$$

Proposition 1.14.7 (Hajek–Renyi Inequality). Let $\{S_n = \sum_{i=1}^n X_i, \mathscr{F}_n, n \geqslant 1\}$ be a martingale with $E(S_n^2) < \infty$, $n \geqslant 1$, and let $\{b_n, n \geqslant 1\}$ be a positive nondecreasing sequence. Then, for any $\lambda > 0$,

$$P\left\{\max_{1 \leqslant j \leqslant n} \left|\frac{S_j}{b_j}\right| \geqslant \lambda\right\} \leqslant \frac{1}{\lambda^2} \sum_{j=1}^{n} \frac{EX_j^2}{b_j^2}.$$

In particular, if $b_j \equiv 1$ in Proposition 1.14.7, then

$$P\left\{\max_{1 \leqslant j \leqslant n} |S_j| \geqslant \lambda\right\} \leqslant \frac{1}{\lambda^2} ES_n^2,$$

which is the *Kolmogorov inequality* for martingales. As a further special case we have the following result.

Proposition 1.14.8 (Kolmogorov Inequality). If $\{X_j, 1 \leqslant j \leqslant n\}$ are independent random variables with $EX_j = 0$, $EX_j^2 < \infty$, then

$$P\left\{\max_{1 \leqslant j \leqslant n} |X_1 + \cdots + X_j| \geqslant \varepsilon\right\} \leqslant \frac{1}{\varepsilon^2} \sum_{j=1}^{n} EX_j^2$$

for every $\varepsilon = 0$.

Proposition 1.14.9 (Doob Inequality). Let $\{S_n = \sum_{j=1}^n X_j, \mathscr{F}_n, n \geqslant 1\}$ be nonnegative submartingale. Then

$$P\left\{\max_{1\leqslant j\leqslant n} S_j \geqslant \lambda\right\} \leqslant \frac{1}{\lambda}\int_{\substack{[\max S_j \geqslant \lambda]\\ 1\leqslant j\leqslant n}} S_n$$

for every $\lambda > 0$.

Proposition 1.14.10 (Burkholder Inequality). Let $\{S_n, \mathscr{F}_n, n \geqslant 1\}$ be a martingale and $p > 1$. Define

$$V_n = \left[\sum_{i=1}^{n} (S_i - S_{i-1})^2\right]^{1/2}$$

and

$$V_\infty = \left[\sum_{i=1}^{\infty} (S_i - S_{i-1})^2\right]^{1/2}.$$

Then there exist positive constants A_p and B_p such that

$$A_p \| V_n \|_p \leqslant \| S_n \|_p \leqslant B_p \| V_n \|_p$$

and

$$A_p \| V_\infty \|_p \leqslant \sup_{n\geqslant 1} \| S_n \|_p \leqslant B_p \| V_\infty \|_p,$$

where $\| X \|_p = (E|X|^p)^{1/p}$, $A_p = \{18p^{3/2}/(p-1)\}^{-1}$ and $B_p = 18p^{3/2}/(p-1)^{1/2}$.

Proposition 1.14.11 (Jensen Inequality). Let X and Y be random variables with $EX < \infty$ and $EY < \infty$ defined on $(\Omega, \mathscr{F}, P)$. Let ζ be a sub-σ-algebra of $\mathscr{F}$. Suppose g is convex on R. If either (i) $X = E(Y|\zeta)$ or (ii) $X \leqslant E(Y|\zeta)$ and $g\uparrow$, then

$$g(X) \leqslant E\{g(Y)|\zeta\} \quad \text{a.s.}$$

In particular, if $EX < \infty$ and g is convex, then

$$Eg(X) \geqslant g(EX).$$

Proposition 1.14.12 (c_r Inequality). For any two random variables X and Y with $E|X|^r < \infty$ and $E|Y|^r < \infty$ for some $r > 0$,

$$E|X+Y|^r \leqslant c_r(E|X|^r + E|Y|^r),$$

where $c_r = 2^{r-1}$ if $r \geqslant 1$ and $c_r = 1$ if $0 < r \leqslant 1$.

Proposition 1.14.13 (Whittle Inequality). Let $X_j, 1 \leqslant j \leqslant n$, be independent random variables with $EX_j = 0$ and $E|X_j|^m < \infty$ for some $m \geqslant 2$ for $1 \leqslant j \leqslant n$. Then, for any real sequence $b_j, 1 \leqslant j \leqslant n$,

$$E\left|\sum_{i=1}^{n} b_i X_i\right|^m \leqslant c_m \left[\sum_{i=1}^{n} b_i^2 \{E(|X_i|^m\}^{2/m}\right]^{m/2},$$

where c_m is an absolute constant depending on m.

Proposition 1.14.14 (Dharmadhikari–Jogdeo Inequality). Let $X_j, 1 \leqslant j \leqslant n$, be independent random variables with $EX_j = 0$ and $E|X_j|^m < \infty$ for some $m \geqslant 2$. Let $S_n = \sum_{j=1}^{n} X_j$. Then there exists an absolute constant c_m depending on m such that

$$E|S_n|^m \leqslant c_m n^{m/2-1} \sum_{i=1}^{n} E|X_i|^m.$$

Proofs of Propositions 1.14.1–1.14.11 may be found in Chow and Teicher (1978). Proposition 1.14.12 follows from the elementary inequality

$$|a+b|^r \leqslant c_r(|a|^r + |b|^r),$$

where $c_r = 2^{r-1}$ if $r \geqslant 1$ and $c_r = 1$ if $0 < r \leqslant 1$. Proof of Proposition 1.14.13 is given in Whittle (1960). Proposition 1.14.14 is proved in Dharmadhikari and Jogdeo (1969).

Proposition 1.14.15 (Hoeffding Inequality). Let $X_j, 1 \leqslant j \leqslant n$, be independent random variables such that $a \leqslant X_j \leqslant b$, $1 \leqslant j \leqslant n$ for some $-\infty < a < b < \infty$. Then, for every $\lambda > 0$.

$$P\left(\sum_{i=1}^{n} [X_i - E(X_i)] \geqslant n\lambda\right) \leqslant \exp\left[\frac{-2n\lambda^2}{(b-a)^2}\right].$$

For proof, see Hoeffding (1963).

The following result gives the rate of convergence in the weak law of large numbers.

Proposition 1.14.16. Let $X_i, i \geqslant 1$, be i.i.d. random variables with $EX_1 = 0$ and $E|X_1|^{r+1} < \infty$ for some $r > 0$. Let $S_n = \sum_{j=1}^{n} X_j$. Then, for any $\varepsilon > 0$,

$$P\left(\left|\frac{S_n}{n}\right| \geqslant \varepsilon\right) = o(n^{-r}).$$

In fact, for any $\varepsilon > 1$,

$$P\left(\left|\frac{S_n}{n}\right| \geqslant \varepsilon\right) = \varepsilon^{-r-1}\eta_n,$$

where $\eta_n = o(n^{-r})$ and does not depend on ε.

Proof of this result follows from Petrov (1975, Chapter 4, Theorems 28 and 29). The following result is useful in studying the asymptotic efficiency of estimators in statistical inference.

Observe that if X is a random variable with a unimodal density and symmetric about zero, then, for any $c > 0$ and $-\infty < a < \infty$,

$$P(-c < X + a < c) \leqslant P(-c < X < c),$$

and if Y is a random variable independent of X, then

$$\begin{aligned} P(-c < X + Y < c) &= \int_{-\infty}^{\infty} P(-c < X + y < c)\, dF_Y(y) \\ &\leqslant P(-c < X < c), \end{aligned}$$

where F_Y is the distribution function of the random variable Y. The following result is due to Anderson (1955).

Proposition 1.14.17 (Anderson, 1955). Suppose $\mathbf{X}$ is a random vector with density f such that $f(\mathbf{x}) = f(-\mathbf{x})$ and $\{\mathbf{x}: f(\mathbf{x}) \geqslant u\}$ is convex for every $u > 0$. Then, for any random vector $\mathbf{Y}$ independent of $\mathbf{X}$ and any convex set C symmetric about the origin,

$$P(\mathbf{X} + \mathbf{Y} \in C) \leqslant P(\mathbf{X} \in C).$$

(C is said to be *symmetric* if $\mathbf{x} \in C \Rightarrow -\mathbf{x} \in C$).

1.15 STOCHASTIC ORDER RELATIONS

We now give some notation and define some concepts that are useful in handling statements concerning limiting properties for sequences of random variables. The discussion including examples is based completely on Chernoff (1965). For earlier work in this area, see Mann and Wald (1943) and Pratt (1959).

Definition 1.15.1. Let $\{r_n\}$ be a sequence of positive real numbers and $\{\mathbf{y}_n\}$ with $\mathbf{y}'_n = (y_{n1}, \ldots, y_{nk})$ be a sequence of vectors in extended R^k with $-\infty \leqslant y_{ni} \leqslant \infty$. Then $\mathbf{y}_n = o(r_n)$ if, for every $\eta > 0$, there exists an integer N such that

$$(\|\mathbf{y}_n\|/r_n) \leqslant \eta \quad \text{for } n \geqslant N$$

and $\mathbf{y}_n = O(r_n)$ if there exists $\eta > 0$ and an integer N such that

$$(\|\mathbf{y}_n\|/r_n) \leqslant \eta \quad \text{for all } n \geqslant N.$$

In analogy with the above definition, let us introduce the following notions o_p and O_p for a sequence of random vectors taking values in the extended Euclidean space R^k.

Definition 1.15.2a. Let $\{\mathbf{Y}_n\}$ be a sequence of random vectors with $\mathbf{Y}_n$ defined on a probability space $(\Omega_n, \mathscr{B}_n, P_n)$ and taking values in the extended R^k. Let $\{r_n\}$ be a sequence of positive real numbers. Then $\mathbf{Y}_n = o_p(r_n)$ if, for every $\varepsilon > 0$ and $\eta > 0$, there exists an integer N such that

$$P_n\{(\|\mathbf{Y}_n\|/r_n) \leqslant \eta\} \geqslant 1 - \varepsilon \quad \text{for } n > N$$

and $\mathbf{Y}_n = O_p(r_n)$ if, for every $\varepsilon > 0$, there exists $\eta > 0$ and an integer N such that

$$P_n\{(\|\mathbf{Y}_n\|/r_n) \leqslant \eta\} \geqslant 1 - \varepsilon \quad \text{for } n > N.$$

Remarks. Note that $\mathbf{Y}_n = o_p(r_n)$ iff for every $\eta > 0$

$$P_n\{(\|\mathbf{Y}_n\|/r_n) \leqslant \eta\} \to 1 \quad \text{as } n \to \infty.$$

Hereafter we use the symbol ⓞ to represent that the statement holds if either o is used throughout or O is used throughout. We now present alternate equivalent definition for Definition 1.15.2a.

Definition 1.15.2b. $\mathbf{Y}_n =$ ⓞ$_p(r_n)$ if for every $\varepsilon > 0$ there is a sequence $c_n =$ ⓞ(r_n) such that

$$P_n(\|\mathbf{Y}_n\| \leqslant c_n) \geqslant 1 - \varepsilon \quad \text{for all } n.$$

Definition 1.15.2c. $\mathbf{Y}_n =$ ⓞ$_p(r_n)$ if for every $\varepsilon > 0$ there is a sequence T_n of Borel sets such that

(i) $P_n(\mathbf{Y}_n \in T_n) \geqslant 1 - \varepsilon$,
(ii) $\mathbf{y}_n \in T_n$, $n \geqslant 1 \Rightarrow y_n =$ ⓞ(r_n).

Definition 1.15.2d. $\mathbf{Y}_n = \circledcirc_p(r_n)$ if for every $\varepsilon > 0$ there is a sequence $\{S_n\}$ of measurable sets such that

(i) $P_n\{X_n \in S_n\} \geqslant 1 - \varepsilon$,
(ii) $x_n \in S_n$ for all n implies that $\mathbf{y}_n = f_n(x_n) = \circledcirc(r_n)$.

It can be shown that Definitions 1.15.2a–1.15.2d are equivalent. Note that $Y_n \xrightarrow{p} c$ iff $Y_n - c = o_p(1)$.

Let $(\mathscr{X}_n, \mathscr{B}_n, P_n)$ be a sequence of probability spaces. We write $P_n(X_n \in S_n)$ as $P_n(S_n)$ for any measurable set $S_n \in \mathscr{B}_n$. Let S denote a set of sequences $\mathbf{x} = (x_1, x_2, \ldots)$ with $x_n \in \mathscr{X}_n$. S need not be measurable.

Definition 1.15.3. A set S *occurs in probability* (written $\mathscr{P}(S)$) if for every $\varepsilon > 0$ there is a sequence S_n such that

(i) $P_n(X_n \in S_n) \geqslant 1 - \varepsilon$ for all n

(ii) $x_n \in S_n$ for all n implies $\mathbf{x} \in S$.

In terms of the notation $\mathscr{P}$, we can rewrite Definition 1.15.2d in the following way.

Definition 1.15.2d′. $Y_n = \circledcirc_p(r_n)$ if $\mathscr{P}(S)$ where

$$S = \{\mathbf{x} : y_n = f_n(x_n) = \circledcirc(r_n)\}.$$

The following results are easy consequences of Definition 1.15.3.

Proposition 1.15.1. If $S \subset T$, then $\mathscr{P}(S) \Rightarrow \mathscr{P}(T)$.

Proposition 1.15.2. $\mathscr{P}(S^{\alpha_i})$ for $i \geqslant 1 \Leftrightarrow \mathscr{P}(\bigcap_{i=1}^{\infty} S^{\alpha_i})$

Proposition 1.15.3. $\mathscr{P}(S^{\alpha})$ for some $\alpha \in I \Rightarrow \mathscr{P}(\bigcup_{\alpha \in I} S^{\alpha})$.

As a consequence of Propositions 1.15.1 and 1.15.2, we have the following theorem, which is useful in asymptotic theory.

Theorem 1.15.4. If

$$f_n^{(j)}(\mathbf{X}_n) = O_p(r_n^j), \qquad 1 \leqslant j \leqslant J,$$
$$g_n^{(k)}(\mathbf{X}_n) = o_p(s_n^k), \qquad 1 \leqslant k \leqslant K,$$

and

$$h_n(\mathbf{x}_n) = \circledcirc(t_n)$$

whenever

$$f_n^{(j)}(\mathbf{x}_n) = O(r_n^j), \qquad 1 \leqslant j \leqslant J$$

and

$$g_n^{(k)}(\mathbf{x}_n) = o(s_n^k), \qquad 1 \leqslant k \leqslant K,$$

then

$$h_n(\mathbf{X}_n) = \circledcirc_p(t_n).$$

The importance of this result is that it separates the nonstochastic or deterministic limit problems and the stochastic or random limit problems involved in a particular context.

Example 1.15.1. It is easy to check that

(a) $o_p(1)O_p(1) = o_p(1)$,

(b) $\dfrac{1}{1 + o_p(1)} = 1 + o_p(1)$

from Theorem 1.15.4. For instance, if $Y_{n1} = g_n(\mathbf{X}_n) = o_p(1)$ and $Y_{n2} = f_n(\mathbf{X}_n) = O_p(1)$, then $Y_{n1}Y_{n2} = h_n(\mathbf{X}_n) = o_p(1)$. Similarly, if $Y_{n1} = 1/(1 + Y_{n2})$ where $Y_{n2} = g_n(\mathbf{X}_n) = o_p(1)$, then $Y_{n1} = 1 + Y_{n3}$ where $Y_{n3} = Y_{n1} - 1 = h_n(\mathbf{X}_n) = o_p(1)$.

Example 1.15.2. Suppose $Y_n = f_n(X_n) \xrightarrow{p} c$ and $g(\cdot)$ is continuous at c. Then $g(Y_n) \xrightarrow{p} g(c)$. This can be seen as follows. Let

$$S_1 = \{x : y_n = f_n(x_n) \to c\}$$

and

$$S_2 = \{x : g(y_n) = g(f_n(x_n)) \to g(c)\}.$$

By continuity, it follows that $S_1 \subset S_2$. The result is now a consequence of Definition 1.15.2d and Proposition 1.15.1.

Example 1.15.3. If $Y_{nj} = f_{nj}(X_n) \xrightarrow{p} c_j$, $1 \leqslant j \leqslant k$, then $(Y_{n1}, \ldots, Y_{nk}) \xrightarrow{p} (c_1, \ldots, c_k)$, and conversely. To see this, let $S^{(i)} = \{\mathbf{x} : y_{ni} \to c_i\}$, $1 \leqslant i \leqslant k$, and $S = \{\mathbf{x} : y_{ni} \to c_i, 1 \leqslant i \leqslant k\}$. The result follows from Definition 1.15.2d and Proposition 1.15.2.

Example 1.15.4. If $Y_{nj} = f_{nj}(X_n) \xrightarrow{p} c_j$, $1 \leqslant j \leqslant k$, and $g(y_1, \ldots, y_k)$ is continuous at $c=(c_1,\ldots,c_k)$, then $g(Y_{n1},\ldots,Y_{nk})\xrightarrow{p} g(c_1,\ldots,c_k)$. This follows from Examples 1.15.2 and 1.15.3.

Remark. Slutsky's lemma is a consequence of this example.

Example 1.15.5. Suppose $\mathbf{Y}_n-\mathbf{Y}'_n=o_p(1)$, $\mathbf{Y}'_n=O_p(1)$ and g is a continuous function. Then $g(\mathbf{Y}_n)-g(\mathbf{Y}'_n)=o_p(1)$. To prove this, let

$$S^{(1)}=\{\mathbf{x}:\mathbf{y}_n-\mathbf{y}'_n=o(1)\},$$

$$S^{(2)}=\{\mathbf{x}:\mathbf{y}'_n=O(1)\},$$

and

$$S=\{\mathbf{x}:g(\mathbf{y}_n)-g(\mathbf{y}'_n)=o(1)\}.$$

Since the continuity of g implies uniform continuity on bounded sets, it follows that $S^{(1)}\cap S^{(2)}\subset S$, and the result follows from Proposition 1.15.1 and 1.15.2 or equivalently from Theorem 1.15.4.

Example 1.15.6. A sequence $\{\mathbf{y}_n\}$ is said to be *restrained from D* if there is an open set $U\supset D$ such that $\mathbf{y}_n\in U^c$ for n sufficiently large. A sequence of random vectors $\{\mathbf{Y}_n\}$ is said to be *restrained from D in probability* $\mathscr{P}(S)$ where

$$S=\{\mathbf{x}:\mathbf{y}_n=f_n(\mathbf{x})\text{ is restrained from }D\}.$$

Note that if $\mathbf{Y}_n-\mathbf{Y}'_n=o_p(1)$, $\mathbf{Y}'_n=O_p(1)$ and $\mathbf{Y}'_n$ is restrained from D_g in probability, where D_g is the set of discontinuities of g, then $g(\mathbf{Y}_n)-g(\mathbf{Y}'_n)=o_p(1)$. (Here U^c is the complement of the set U).

Example 1.15.7 (Pratt, 1959) Theorem 1.15.4 should be applied with care. For instance, while $y_n/n\to 0$ and y_n finite for all n implies that $\max(y_1,\ldots,y_n)/n\to 0$, it is not true that $Y_n/n\xrightarrow{p}0$ implies that $\max(Y_1,\ldots,Y_n)/n\xrightarrow{p}0$. The reason that Theorem 1.15.4 is not applicable is that the nth element of the sequence depends not on $(\mathscr{X}_n,\mathscr{B}_n,P_n)$ alone but on all the probability spaces $(\mathscr{X}_j,\mathscr{B}_j,P_j)$, $1\leqslant j\leqslant n$.

Example 1.15.8 (Pratt, 1959). If $Y_n\xrightarrow{p}Y$ where $P(Y<\infty)=1$ and if $P(Y\in D_g)=0$ where D_g is the set of discontinuities of $g(\cdot)$, then $g(Y_n)\xrightarrow{p}g(Y)$. This is a consequence of Example 1.15.6 by the following argument. Let (Y_{n1},Y_{n2}) be a bivariate random vector defined on $(\mathscr{X}_n,\mathscr{B}_n,P_n)$ such that (Y_{n1},Y_{n2}) has the same distribution as (Y_n,Y). Note that $Y_{n2}=O_p(1)$ and, given $\varepsilon>0$, there exists an open set $D\supset D_g$ such that $P(Y\in D)<\varepsilon$.

Thus Y_{n2} is restrained from D_g in the sense of Example 1.15.6. Hence $g(Y_{n1}) - g(Y_{n2}) = o_p(1)$. In other words, $g(Y_n) \xrightarrow{p} g(Y)$.

Example 1.15.9 (Pratt, 1959). If $Y_n \xrightarrow{p} Y$ and $P(Y < \infty) = 1$ and G_n is a random function such that

$$\sup_{|y| \leqslant M} |G_n(y) - g(y)| \xrightarrow{p} 0 \quad \text{for each } M > 0$$

and $P(Y \in D_g) = 0$ where D_g is the set of discontinuities of g, then

$$G_n(Y_n) \xrightarrow{p} g(Y).$$

This example gives a sufficient condition for the random function of a random variable to converge in probability to a specific function of a limiting random variable. In fact $G_n(Y_n) \xrightarrow{\mathscr{L}} g(Y)$, as we have seen in Section 1.10.

Example 1.15.10. If $Y_n - c = O_p(r_n)$ and $r_n = o(1)$ and $g(y) = T_l(y, c) + o(|y - c|^l)$ where $y = c + o(1)$ and $T_k(y, c)$ is the kth-order Taylor expansion of g about c, then

$$g(Y_n) = T_l(Y_n, c) + o_p(r_n^l).$$

This follows from Theorem 1.15.4.

Example 1.15.11. If $Y_n - Y_n' = O_p(r_n)$, $r_n = o(1)$, $Y_n' = O_p(1)$ is restrained from D in probability and g has a continuous lth-order derivative on D^c, then $g(Y_n) = T_l(Y_n, Y_n') + o_p(r_n^l)$. This result again is a consequence of Theorem 1.15.4.

We shall now prove the following theorem.

Theorem 1.15.5. Definitions 1.15.2a–1.15.2d are equivalent.

Proof. We shall prove that the Definitions 1.15.2a–1.15.2d imply the following chains of relations, which prove the theorem:

(i) Definition 1.15.2b ⇒ Definition 1.15.2c ⇒ Definition 1.15.2b ⇒ Definition 1.15.2a,

(ii) Definition 1.15.2c ⇔ Definition 1.15.2d, and

(iii) Definition 1.15.2a ⇒ Definitions 1.15.2b and 1.15.2c.

To prove (i). It is clear that Definition 1.15.2b implies Definition 1.15.2c. Suppose that Definition 1.15.2c holds. Let $c_n = \sup_{y_n \in T_n} \| y_n \|$. It is clear that condition (ii) of Definition 1.15.2c implies that $c_n = \circledcirc(r_n)$ and Definition 1.15.2b holds. It is easy to see that Definition 1.15.2b implies Definition 1.15.2a.

To prove (ii) Suppose that Definition 1.15.2c holds. Let $S_n = \{x_n : y_n = f_n(x_n) \in T_n\}$. Then S_n is measurable since f_n is measurable and Definition 1.15.2d holds. Conversely suppose that Definition 1.15.2d holds. Note that $T_n = \overline{f_n(S_n)}$ is measurable and $P_n(Y_n \in T_n) \geqslant P(X_n \in S_n) \geqslant 1 - \varepsilon$. Furthermore, given a sequence $y_n \in T_n$, there is a sequence $x_n \in S_n$ with $|y_n - x_n| \leqslant r_n/n$. Hence $|y_n| \leqslant |x_n| + r_n/n = \circledcirc(r_n)$ and therefore condition (ii) of Definition 1.15.2c holds. (Here $\bar{A}$ denotes the closure of a set A.)

To prove (iii). Suppose that Definition 1.15.2a holds for o_p. Let $\varepsilon > 0$ and $0 < \eta_i \to 0$. Then there an increasing sequence $N_i = N(\varepsilon, \eta_i)$ such that

$$P_n\{ \| Y_n \| \leqslant \eta_i r_n\} \geqslant 1 - \varepsilon \quad \text{for } n > N_i$$

and hence for $N_i < n \leqslant N_{i+1}$. Let $c_n = \infty$ and $T_n = [-\infty, \infty]$ for $n \leqslant N_1$, and let $c_n = \eta_i r_n$ and $T_n = [-c_n, c_n]$ for $N_i < n \leqslant N_{i+1}$. Then Definitions 1.15.2b and 1.15.2c hold.

Suppose that Definition 1.15.2a holds for O_p. Then there is an $\eta > 0$ and N_1 such that

$$P_n\{ \| Y_n \| \leqslant \eta r_n\} \geqslant 1 - \varepsilon \quad \text{for } n > N_1.$$

Let $c_n = \infty$ and $T_n = [-\infty, \infty]$ for $n \leqslant N_1$, and let $c_n = \eta r_n$ and $T_n = [-c_n, c_n]$ for $n > N_1$. Then Definitions 1.15.2b and 1.15.2c hold. This completes the proof of the theorem.

Remarks. Definitions 1.15.2a–1.15.2d are not in general equivalent when Y_n are random elements taking values in a function space. For an extension of these definitions to random elements taking values in general topological spaces and their implications, see Chernoff (1965).

REFERENCES

Aalen, O. (1982). Practical applications of the nonparametric statistical theory for counting processes. Research report, Institute of Mathematics, University of Oslo.

Adler, R. J. (1980). *The Geometry of Random Fields*, Wiley, New York.

Aldous, D. J. (1978). Weak convergence of randomly indexed sequences of random variables. *Math. Proc. Camb. Phil. Soc.* **83**, 117–126.

Anderson, T. W. (1955). The integral of a symmetric unimodal function. *Proc. Amer. Math. Soc.* **6**, 170–176.

Bahadur, R. R. and Ranga Rao, R. (1960). On deviations of the sample mean. *Ann. Math. Statist.* **31**, 1015–1027.

Bahadur, R. R. and Zabell, S. L. (1979). Large deviations of the sample mean in general vector spaces. *Ann. Probability* **7**, 587–621.

Bartfai, P. (1977). On the multivariate Chernoff theorem. Preprint, Mathematical Institute, Hungarian Academy of Sciences.

Bartfai, P. (1978). Large deviations of the sample mean in Euclidean spaces. Technical Report No. 78–13, Purdue University.

Basawa, I. V. and Prakasa Rao, B. L. S. (1980). *Statistical Inference for Stochastic Processes.* Academic Press, New York.

Basawa, I. V. and Scott, D. (1977). Efficient tests for stochastic processes. *Sankhya Ser. A* **39**, 21–31.

Basawa, I. V. and Scott, D. (1983). *Asymptotic Optimal Inference for Non-ergodic Models.* Lecture Notes in Statistics No. 17, Springer-Verlag, New York.

Bhattacharya, R. N. (1977). Refinements of the multidimensional central limit theorem and applications. *Ann. Probability* **5**, 1–27.

Bhattacharya, R. N. and Ranga Rao, R. (1976). *Normal Approximation and Asymptotic Expansions*, Wiley, New York.

Bikelis, A. (1966). Estimates of the remainder term in the central limit theorem. *Litovskii Math. sb.* **6**(3), 323–346. (In Russian).

Billingsley, P. (1961a). *Statistical Inference for Markov Processes.* University of Chicago Press, Chicago.

Billingsley, P. (1961b). The Lindeberg–Levy theorem for martingales. *Proc. Amer. Math. Soc.* **12**, 788–792.

Billingsley, P. (1968). *Convergence of Probability Measures.* Wiley, New York.

Blum, J., Hanson, D., and Rosenblatt, J. (1963). On the CLT for sums of a random number of random variables. *Z. Wahr. verw. Geb.* **1**, 389–393.

Breiman, L. (1967). On the tail behaviour of sums of independent random variables. *Z. Wahr. verw. Geb.* **9**, 20–25.

Chatterjee, S. D. and Mandrekar, V. (1978). Equivalence and singularity of Gaussian measures and applications. *Probabilistic Analysis and Related Topics*, Vol. 1 (Ed. A. T Bharucha-Reid), Academic Press, New York.

Chenstov, N. N. (1956). Weak convergence of stochastic processes whose trajectories have no discontinuities of the second kind and the "Heuristic approach to the Kolmogorov–Smirnov tests" *Theory Prob. Appl.* **1**, 140–144.

Chenstov, N. N. (1970). Limit theorems for some classes of random functions. *Selected Transl. Math. Statist. and Prob. Amer. Math. Soc.* **9**, 37–42.

Chernoff, H. (1952). A measure of asymptotic efficiency for tests of a hypothesis based on the sum of observations. *Ann. Math. Statist.* **23**, 493–507.

Chernoff, H. (1965). In probability. Unpublished Lecture Notes. Stanford University.

Chow, Y. S. (1967). On a strong law of large numbers for martingales. *Ann. Math. Statist.* **38**, 610.

Chow, Y. S. and Teicher, H. (1978). *Probability Theory*, Springer-Verlag, New York.

Chung, K. L. (1974). *A Course in Probability Theory*, 2nd ed., Academic Press, New York.

Cramer, H. (1938). *Random Variables and Probability Distributions*, Cambridge University Press, Cambridge, England.

Csiszàr, I. (1967). Information-type measures of divergence of probability distributions and indirect observations. *Studia Sci. Math. Hungar.* **2**, 299–318.

Csörgö, M. and Revesz, P. (1981). *Strong Approximations in Probability and Statistics*, Academiai Kiado, Budapest.

Dellacherie, C. (1980). Un survol de la theorie de l' integrale stochastique. *Proc. Int. Congress of Math. Helsinki* **2**, 733–739.

Deshayes, J. and Picard, D. (1984). Principe d'invariance sur le processus de vraisemblance. *Annales de l'Inst. Henri Poincare* **20**, 1–20.

Dharmadhikari, S. W. and Jogdeo, K. (1969). Bounds on moments of certain random variables. *Ann. Math. Statist.* **40**, 1506–1508.

Doob, J. L. (1953). *Stochastic processes*, Wiley, New York.

Eagleson, G. K. and Gundy, R. (1983). On a theorem of Kabanov, Liptser and Sirjaev. *Probability, Statistics and Analysis* (Ed. J. F. C. Kingman and G. E. H. Reuter), London Math. Soc. Lecture Notes Series 79, Cambridge University Press, Cambridge, England.

Eicker, F. (1963). Central limit theorems for families of sequences of random variables. *Ann. Math. Statist.* **34**, 439–446.

Gerber, H. U. (1971). The discounted central limit theorem and its Berry–Esseen analogue. *Ann. Math. Statist.* **42**, 389–392.

Gikhman, I. and Skorokhod, A. V. (1972). *Stochastic Differential Equations*, Springer-Verlag, New York.

Gikhman, I. and Skorokhod, A. V. (1974). *The Theory of Stochastic Processes*, Vol. 1, Springer-Verlag, Berlin.

Gill, R. D. (1980). *Censoring and Stochastic Integrals*, Math. Centrum Tract. 124, Amsterdam.

Groeneboom, P., Oosterhoff, J., and Ruymgaart, F. H. (1979). Large deviations for empirical probability measures, *Ann. Probability* **7**, 553–586.

Hall, P. (1977). Martingale invariance principles, *Ann. Probability* **5**, 875–887.

Hall, P. and Heyde, C. C. (1980). *Martingale Limit Theory and Its Application*, Academic Press, New York.

Hall, W. J. and Loynes, R. M. (1977). On the concept of contiguity. *Ann. Probab.* **5**, 278–282.

Halmos, P. R. (1950). *Measure Theory*, Von Nostrand, Princeton.

Helland, I. S. (1982). Central limit theorems for martingales with discrete or continuous time. *Scandinavian J. of Statist.* **9**, 79–94.

Heyde, C. C. (1977). On central limit theorem and iterated logarithm supplements to the martingale convergence theorem. *J. Appl. Probab.* **14**, 758–775.

Heyde, C. C. (1981). Invariance principles in statistics. *International Statistical Review* **49**, 143–152.

Hitsuda, M. (1968). Representations of Gaussian processes equivalent to Wiener process. *Osaka J. Math.* **5**, 299–312.

Hoeffding, W. (1963). Probability inequalities for sums of bounded random variables. *J. Amer. Statist. Assoc.* **58**, 13–30.

Ibragimov, I. A. (1963). A central limit theorem for a class of dependent random variables. *Theory Probab. Appl.* **8**, 83–89.

Ibragimov, I. V. and Hasminskii, R. Z. (1981). *Statistical Estimation: Asymptotic Theory*, Springer-Verlag, Berlin.

Jacod, J. (1979). *Calcul Stochastique et Problemes de Martingales*, Lecture Notes in Mathematics, No. 714, Springer-Verlag, Berlin.

Jain, N. C. and Marcus, M. B. (1978). Continuity of subgaussian processes. *Probability on Banach Spaces* (Ed. J. Kuelbs), Marcel Dekker, New York.

Kabanov, Yu. M., Liptser, R., and Shiryayev, A. N. (1977). On the question of the absolute continuity and singularity of probability measures. *Math. USSR Sb.* **33**, 203–221.

Kabanov, Yu. M., Liptser, R., and Shiryayev, A. N. (1978). Absolute continuity and singularity of locally absolute continuous probability distributions I, *Math. USSR Sb.* **35**, 631–680.

Kabanov, Yu. M., Liptser, R., and Shirayayev, A. N. (1979). Absolute continuity and singularity of locally absolute continuous probability distributions II. *Math. USSR Sb.* **36**, 31–58.

Kabanov, Yu. M., Liptser, R., and Shiryayev, A. N. (1980). Some limit theorems for simple point processes (martingale approach). *Stochastics* **3**, 203–216.

Kakutani, S. (1948). On equivalence of infinite product measures. *Ann. of Math.* **49**, 214–226.

Khmaladze, E. V. (1981). The martingale approach to the theory of non-parametric tests of fit. *Theory Probab. Appl.* **26**, 240–257.

Kutoyants, Y. (1975). On a hypothesis testing problem and asymptotic normality of stochastic integrals. *Theory Probab. Appl.* **20**, 376–384.

LeCam, L. (1960). Locally asymptotically normal families of distributions, *Univ. of Calif. Publ. Statist.* **3**, 37–88.

LeCam, L. (1970). On the assumptions used to prove asymptotic normality of maximum likelihood estimates. *Ann. Math. Statist.* **41**, 802–828.

Lenglart, E. (1977). Relation de domination entre deux processus. *Ann. Inst. Henri Poincaré* **13**, 171–179.

Levy, P. (1937). *Theorie de l'addition des variables aleatoires*, Gauthier-Villars, Paris.

Lindvall, T. (1973). Weak convergence of probability measures and random functions in the function space $D[0, \infty)$. *J. Appl. Probability* **10**, 109–121.

Liptser, R. and Shiryayev, A. N. (1978). *Statistics of Random Processes II*, Springer-Verlag, Berlin.

Liptser, R. and Shiryayev, A. N. (1980). A functional central limit theorem for semi-martingales. *Theory Probab. Appl.* **25**, 667–688.

Liptser, R., Pukelshein, F., and Shiryayev, A. N. (1982). Necessary and sufficient conditions for contiguity and entire asymptotic separation of probability measures. *Russian Math. Surveys* **37**(6), 107–136.

Loeve, M. (1963). *Probability Theory*, Van Nostrand, Princeton.

Loynes, R. M. (1969). The central limit theorem for backwards martingales. *Z. Wahr. ver. Geb.* **13**, 1–8.

Loynes, R. M. (1970). An invariance principle for reversed martingales. *Proc. Amer. Math. Soc.* **25**, 56–64.

Loynes, R. M. (1976). A criterion for tightness for a sequence of martingales. *Ann. Probability* **4**, 859–862.

Loynes, R. M. (1978). On the weak convergence of U-statistics and of the empirical process. *Math. Proc. Camb. Phil. Soc.* **83**, 269–272.

Lukacs, E. (1975). *Stochastic Convergence*, 2nd ed., Academic Press, New York.

Mann, H. B. and Wald, A. (1943). On stochastic limit and order relationships. *Ann. Math. Statist.* **14**, 217–226.

McKean, H. P. (1969). *Stochastic Integrals*, Academic Press, New York.

Metivier, M. and Pellaumail, J. (1980). *Stochastic Integration*, Academic Press, New York.

Meyer, P. (1976). *Un cours ser les integrales stochastiques*, Lecture Notes in Math, No. 511, Springer-verlag, Berlin.

Muralirao, K. (1969). On decomposition theorem of Mayer. *Math. Scand.* **24**, 66–78.

Nagaev, S. V. (1979). Large deviations of sums of independent random variables. *Ann. Probability* **7**, 745–789.

Nemetz, T. (1974). Equivalence and singularity dichotomes of probability measures. *Limit Theorems of Probability Theory* (Ed. P. Revesz), North-Holland, Amsterdam.

Neuhaus, G. (1971). On weak convergence of stochastic processes with multi-dimensional time parameter. *Ann. Math. Statist.* **42**, 1285–1294.

Neveu, J. (1974). *Discrete Parameter Martingales*, North-Holland, Amsterdam.

Oosterhoff, J. and Van Zwet, W. R. (1979). A note on contiguity and Hellinger distance. *Contributions to Statistics*, (Ed. J. Jureckova), Academia, Prague.

Parthasarathy, K. R. (1967). *Probability Measures on Metric Spaces*, Academic Press, London.

Parthasarathy, T. (1972). *Selection Theorems and Their Applications*, Lecture Notes in Mathematics No. 263, Springer-Verlag, Berlin.

Petrov, V. (1975). *Sums of Independent Random Variables*, Springer-Verlag, Berlin.

Pfaff, T. (1982). Quick consistency of quasimaximum likelihood estimators. *Ann. Statistics* **10**, 990–1005.

Prakasa Rao, B. L. S. (1969). Random central limit theorems for martingales. *Acta Math. Acad. Sci. Hunger.* **20**, 217–222.

Prakasa Rao, B. L. S. (1973). Limit theorems for random number of random elements on complete separable metric spaces. *Acta Math. Acad. Sci. Hungar.* **24**, 1–4.

Prakasa Rao, B. L. S. (1974). On the rate of convergence in the random central limit theorem for martingales. *Bull. Acad. Polon. Sci. Ser. Sci. Math. Astronom. Phys.* **22**, 1255–1260.

Prakasa Rao, B. L. S. (1975). Tightness of probability measures generated by stochastic processes on metric spaces. *Bull. Inst. Math. Acad. Sinica* **3**, 353–367.

Prakasa Rao, B. L. S. (1977). On central limit theorems for backwards martingales arrays. *Proc. 41st Session Bull. Int. Statist. Inst.* **37**, 697–700.

Prakasa Rao, B. L. S. (1979). On central limit theorems, invariance principles and rates of convergence for backwards martingales arrays. *Lithuanian Math. J.* **19**, 538–546.

Prakasa Rao, B. L. S. (1983). *Nonparametric Functional Estimation*, Academic Press, New York.

Prakasa Rao, B. L. S. (1984). On some applications of the nonparametric statistical analysis for counting processes. *Gujarat Statistical Review* 21–34.

Prakasa Rao, B. L. S. (1986). Weak convergence of least squares random field in the smooth case. *Stat. Decisions* **4**, (to appear).

Pratt, J. W. (1959). On a general concept of "In probability" *Ann. Math. Statist.* **30**, 549–558.

Pukelsheim, F. (1986). Predictable criteria for absolute continuity and singularity of two probability measures. (Preprint).

Richter, W. (1965). Limit theorems for sequences of random variables with sequences of random indices. *Theory Probab. Appl.* **10**, 74–84.

Rockafeller, R. T. (1970). *Convex Analysis*. Princeton University Press, Princeton, New Jersey.

Roussas, G. G. (1972). *Contiguity of Probability Measures: Some Applications in Statistics*, Cambridge University Press, London.

Rubin, H. and Sethuraman, J. (1965). Probabilities of moderate deviations. *Sankhya Ser A.* **27**, 325–346.

Schmetterer, L. (1974). *Introduction to Mathematical Statistics*, Springer-Verlag, Berlin.

Serfling, R. J. (1980). *Approximation Theorems of Mathematical Statistics*, Wiley, New York.

Sethuraman, J. (1964). On the probability of large deviation of families of sample means. *Ann. Math. Statist.* **35**, 1304–1316.

Sethuraman, J. (1965). On the probability of large deviations of the mean for random variables in $D[0,1]$. *Ann. Math. Statist.* **36**, 280–285.

Shepp, L. A. (1966). Radon–Nikodým derivatives of Gaussian measures. *Ann. Math. Statist.* **37**, 321–354.

Sheu, S. S. and Yao, Y. S. (1984). A strong law of large numbers for martingales. *Proc. Amer. Mer. Math. Soc.* **92**, 283–287.

Shiryayev, A. (1980). Absolute continuity and singularity of probability measures in functional spaces. *Proc. Int. Congress. Math. Helsinki*, Vol. I, 209–225.

Shiryayev, A. N. (1981). Martingales: Recent developments, results and applications. *International Statistical Review* **49**, 199–233.

Skorokhod, A. V. (1965). *Studies in the Theory of Random Processes*, Addison-Wesley, Reading, Massachusetts.

Slutsky, E. (1925). Uber stochastische asymptoten and grenzwerte. *Metron* **5**, 1–90.

Sreehari, M. (1973). An estimate of the residual in the central limit theorem for sums of a random number of random variables. Preprint. M. S. University of Baroda.

Stone, C. (1963). Weak convergence of stochastic processes defined on semi-infinite time intervals. *Proc. Amer. Math. Soc.* **14**, 694–696.

Stout, W. F. (1970a). A martingale analogue of Kolmogorov's law of the iterated logarithm. *Z. Wahr. verw. Geb.* **15**, 279–290.

Stout, W. F. (1970b). The Hartman–Winter law of the iterated logarithm for martingales. *Ann. Math. Statist.* **41**, 2158–2160.

Strassen, V. (1964). An invariance principle for the law of iterated logarithm. *Z. Wahr. verw. Geb.* **3**, 211–226.

Szasz, D. (1973). Limit theorems for randomly indiced sequences of random variables, Preprint 58, Mathematics Institute, Hungarian Academy of Sciences.

Szynal, D. (1972). On almost complete convergence for the sums of a random number of independent random variables. *Bull. Acad. Polon. Sci. Ser. Math. Astronom. Phys.* **20**, 571–574.

Van Beeck, P. (1972). An application of Fourier methods to the problem of sharpening the Berry–Esseen inequality. *Z. Wahr. verw. Geb.* **23**, 187–196.

Whitt, W. (1970). Weak convergence of probability measures on the function space $C[0,\infty]$. *Ann. Math. Statist.* **41**, 939–944.

Whitt, W. (1980). Some useful functions for functional limit theorems. *Math. of Operations Res.* **5**, 67–85.

Whittle, P. (1960). Bounds for the moments of linear and quadratic forms in independent random variables. *Theory. Probab. Appl.* **5**, 302–305.

CHAPTER 2

Limit Theorems for Some Statistics

2.1 ASYMPTOTIC EXPANSIONS OF FUNCTIONS OF STATISTICS

Let X_i, $1 \leqslant i \leqslant n$ be i.i.d. random variables with distribution function $F(\cdot)$. Assuming that the population consists of sample observations $X_1, \ldots, X_n$ and assigning probability $1/n$ to each of the observations, one can define analogs of the distribution function and the moments. The function

$$F_n(x) = \frac{\text{Number of observations} \leqslant x}{n}, \qquad -\infty < x < \infty, \tag{2.1.0}$$

is the *empirical distribution function* of the sample $X_i, 1 \leqslant i \leqslant n$.

Let

$$m_k = \frac{1}{n} \sum_{i=1}^{n} X_i^k, \qquad k \geqslant 1 \tag{2.1.1}$$

and

$$s_k^2 = \frac{1}{n} \sum_{i=1}^{n} (X_i - m_1)^k, \qquad k \geqslant 1. \tag{2.1.2}$$

The $m_k, k \geqslant 1$, are the *sample raw moments* and s_k^2 are the *sample central moments.* We denote m_1 by $\bar{X}$ and s_1^2 by s^2 for convenience. One can define sample moments when $\mathbf{X}_i$ are multidimensional vectors in a similar fashion.

For instance, if $\mathbf{X}_i = (X_{i1}, X_{i2})$ is a bivariate random vector, then define

(2.1.3) $$\bar{X}_1 = \frac{1}{n}\sum_{i=1}^{n} X_{i1}, \qquad \bar{X}_2 = \frac{1}{n}\sum_{i=1}^{n} X_{i2}$$

and

(2.1.4) $$m_{ij} = \frac{1}{n}\sum_{k=1}^{n} (X_{k1} - \bar{X}_1)^i (X_{k2} - \bar{X}_2)^j, \qquad 0 \leqslant i, j \leqslant 2.$$

The quantity

(2.1.5) $$r = \frac{m_{11}}{(m_{20} m_{02})^{1/2}}$$

is called the *sample correlation coefficient*. It is easy to see that, in the one-dimensional case,

$$E(\bar{X}) = \mu,$$

where μ is the mean, and $\operatorname{Var}(\bar{X}) = \sigma^2/n$, where σ^2 is the variance provided μ and σ^2 are finite. It can be seen either by the WLLN or by the SLLN for i.i.d. random variables that $\bar{X} \xrightarrow{p} \mu$ and $\bar{X} \to \mu$ a.s.

Let $\mu_k(\bar{X})$ denote the kth central moment of $\bar{X}$ and μ_k be the kth central moment of X_1 when they exist. Observing that

$$\bar{X} - \mu = \frac{1}{n}\sum_{i=1}^{n} (X_i - \mu)$$

is the average of i.i.d. random variables with mean zero, it is clear that $\mu_2(\bar{X}) = n^{-1}\mu_2$,

(2.1.6) $$\mu_3(\bar{X}) = \frac{\mu_3}{n^2},$$

(2.1.7) $$\mu_4(\bar{X}) = \frac{\mu_4}{n^3} + \frac{3(n-1)}{n^3}\mu_2^2,$$

and in general

(2.1.8) $$\mu_{2k-1}(\bar{X}) = O(n^{-k}), \qquad \mu_{2k}(\bar{X}) = O(n^{-k}).$$

Cramer (1946) proved that

$$(2.1.9) \qquad E(m_j - \mu_j)^{2k-1} = O(n^{-k}) \quad \text{and} \quad E(m_j - \mu_j)^{2k} = O(n^{-k})$$

under some conditions on the existence of moments.

It is important to obtain in practice the mean and the variance of functions of some sample moments, for instance, the expectation of the sample standard deviation s. Cramer (1946) derived the asymptotic expansion for the mean and variance of a function of two sample moments m_k and m_l. Lomnicki and Zaremba (1957) studied the behavior of the first two moments of functions of multidimensional vector statistics. We now discuss a result due to Hurt (1976, 1979). We first discuss the case in which $X_i, 1 \leqslant i \leqslant n$, is a one-dimensional i.i.d. sample. Let T_n be a statistic based on the sample $X_i, 1 \leqslant i \leqslant n$.

Theorem 2.1.1. Let $g = g(t, n)$ be a function defined on $R \times \mathcal{N}$. Suppose that, for all n and for some $q \geqslant 1, g$ admits a continuous $(q+1)$th derivative for $t \in [\theta - \delta, \theta + \delta]$ for some $\delta > 0$ independent of n. Suppose $g^{(i)}(t, n)$, $q \leqslant i \leqslant q+1$, are bounded on $[\theta - \delta, \theta + \delta] \times \mathcal{N}$, where $g^{(0)} = g$ and $g^{(i)}$ denotes ith partial derivative of g with respect to t. Suppose that $\{T_n\}$ is a sequence of statistics with finite moments of order $2(q+1)$ such that

$$(2.1.10) \qquad \text{(i)} \quad E|T_n - \theta|^{2(q+1)} \to 0 \quad \text{as } n \to \infty,$$

$$\text{(ii)} \quad E|T_n - \theta|^{2(q+1)} = O([E|T_n - \theta|^{q+1}]^2).$$

Then

$$(2.1.11) \quad E[g(T_n, n) - g(\theta, n)] = \sum_{j=1}^{q} \frac{1}{j!} g^{(j)}(\theta, n) E(T_n - \theta)^j + O(E|T_n - \theta|^{q+1})$$

and

$$(2.1.12) \quad \operatorname{Var}[g(T_n, n)] = \sum_{\substack{j=1 \\ \langle j+k \leqslant q+1 \rangle}}^{q} \sum_{k=1}^{g} \frac{1}{j!} \frac{1}{k!} g^{(j)}(\theta, n) g^{(k)}(\theta, n) \operatorname{cov}[(T_n - \theta)^j,$$
$$\cdot (T_n - \theta)^k] + O(E|T_n - \theta|^{q+2}).$$

Proof. We prove (2.1.11). Proof of (2.1.12) is left as an exercise for the reader (Hurt, 1979). Observe that

$$E[g(T_n, n) - g(\theta, n)] = \int_{-\infty}^{\infty} [g(t, n) - g(\theta, n)] \, dH_n(t),$$

where H_n is the distribution function of T_n. Let $0 < \varepsilon < \delta$ and $M = (\theta - \varepsilon, \theta + \varepsilon)$. Let

$$I_1 = \int_M [g(t,n) - g(\theta,n)]\, dH_n(t)$$

and

$$I_2 = \int_{M^c} [g(t,n) - g(\theta,n)]\, dH_n(t).$$

From the boundedness of g and the Chebyshev inequality, we have

$$I_2 = O(E|T_n - \theta|^{q+1}).$$

For any $t \in M$,

$$g(t,n) - g(\theta,n) = \sum_{j=1}^{g} \frac{1}{j!} g^{(j)}(\theta,n)(t-\theta)^j + \frac{1}{(q+1)!} g^{(q+1)}(x,n)(t-\theta)^{q+1},$$

where $x = \theta + \xi(t-\theta)$ with $\xi = \xi(\theta,t,n) \in (0,1)$. Let $b_j = g^{(j)}(\theta,n)$, $1 \leqslant j \leqslant q$, and $b_{q+1} = g^{(q+1)}(x,n)$. Observe that b_j, $1 \leqslant j \leqslant q+1$, are bounded by hypothesis. Furthermore,

$$I_1 = \sum_{j=1}^{q} b_j I_{1j} + \int_M b_{q+1}(t-\theta)^{q+1}\, dH_n(t), \tag{2.1.13}$$

where

$$I_{1j} = \int_M (t-\theta)^j\, dH_n(t), \qquad 1 \leqslant j \leqslant q.$$

Since b_{q+1} is bounded, the integral on the right-hand side of (2.1.13) is $O(E|T_n - \theta|^{q+1})$. On the other hand,

$$I_{1j} = E(T_n - \theta)^j - \int_{M^c} (t-\theta)^j\, dH_n(t)$$

and

$$\begin{aligned} \left| \int_{M^c} (t-\theta)^j\, dH_n(t) \right|^2 &\leqslant E|T_n - \theta|^{2j} P(|T_n - \theta| > \varepsilon) \\ &\leqslant \varepsilon^{-2q} E|T_n - \theta|^{2q} E|T_n - \theta|^{2j} \\ &\leqslant \varepsilon^{-2q} (E|T_n - \theta|^{2q})^{1+(j/q)} \end{aligned} \tag{2.1.14}$$

from the inequality $(E|X|^r)^{1/r} \leqslant (E|X|^s)^{1/s}$ if $r \leqslant s$. Since $E|T_n - \theta|^{2(q+1)} \to 0$ as $n \to \infty$, it follows that $E|T_n - \theta|^{2q} \to 0$ and hence, for large $n > n_0$, $E|T_n - \theta|^{2q} \leqslant 1$. Therefore, for $n > n_0$,

$$(E|T_n - \theta|^{2q})^{1+(j/q)} \leqslant (E|T_n - \theta|^{2q})^{1+(1/q)} \leqslant E|T_n - \theta|^{2(q+1)},$$

which implies that

$$\left|\int_{M^c} (t-\theta)^j \, dH_n(t)\right|^2 = O([E|T_n - \theta|^{q+1}]^2)$$

from (ii) of (2.1.10). Hence

$$E(T_n - \theta)^j = I_{1j} + O(E|T_n - \theta|^{q+1})$$

for $1 \leqslant j \leqslant q$. Since the b_j are bounded, it follows from the earlier arguments that

$$E[g(T_n, n) - g(\theta, n)] = \sum_{j=1}^{q} \frac{1}{j!} g^{(j)}(\theta, n) E(T_n - \theta)^j + O(E|T_n - \theta|^j)$$

This completes the proof of (2.1.11).

Remarks. If, in particular, $E|T_n - \theta|^{2(q+1)} = O(n^{-(q+1)})$, then the term $O(E|T_n - \theta|^{q+1})$ in (2.1.11) can be replaced by $O(n^{-(q+1)/2})$ and $O(E|T_n - \theta|^{q+2})$ in (2.1.12) can be replaced by $O(n^{-(q+2)/2})$ by analogous arguments.

Example 2.1.1. Let X_i, $1 \leqslant i \leqslant n$, be i.i.d. uniform on $(0, \theta)$, θ being the unknown parameter. It is well known that $X_{(n)} = \max_{1 \leqslant i \leqslant n} X_i$ is a sufficient statistic and

$$\hat{\theta}_n = \frac{n+1}{n} X_{(n)}$$

is a minimum variance unbiased estimator of θ. Suppose the function

$$\gamma(\theta) = \exp(-2y/\theta),$$

where y is a fixed number has to be estimated. An estimator of $\gamma(\theta)$ is $\gamma(\hat{\theta}_n) = \exp\{-2yn/(n+1)X_{(n)}\}$. It can be checked that $EX_{(n)}^j = O(n^{-j})$ for all

j. In fact

$$E(X_{(n)} - \theta) = -\frac{\theta}{n+1}$$

and

$$E(X_{(n)} - \theta)^2 = \frac{2\theta^2}{(n+1)(n+2)}.$$

Let $g(t,n) = \exp\{-2yn/(n+1)t\}$. Applying Proposition 2.1.1, it follows that

$$E[\gamma(\hat{\theta}_n)] = \exp\left\{-\frac{2y}{\theta}\right\}\left[1 + \frac{1}{n^2}\frac{y}{\theta}\left(\frac{2y}{\theta} - 2\right)\right] + O(n^{-3})$$

by choosing $q = 2$ and

$$\text{var}[\gamma(\hat{\theta}_n)] = \frac{1}{n^2}\exp\left\{-\frac{4y}{\theta}\right\}\frac{4y^2}{\theta^2} + O(n^{-3}).$$

We now state a result due to Cramer (1946) that again gives asymptotic expansions for the mean and the variance of functions of sample moments under stronger conditions.

Proposition 2.1.2. Let m_k denote the kth sample raw moment based on an i.i.d. sample of size n and $H(u,v)$ be a continuous function with continuous first- and second-order partial derivatives with respect to u and v. Furthermore, suppose that $|H(m_k, m_l)| < cn^\alpha$ for some positive constants c and α. Then

$$E(H(m_k, m_l)) = H(\mu_k, \mu_l) + O(n^{-1}) \tag{2.1.15}$$

and

$$\begin{aligned} \text{var}(H(m_k, m_l)) &= \text{var}(m_k)\left[\left\{\frac{\partial H}{\partial u}\right\}\bigg|_{(\mu_k,\mu_l)}\right]^2 \\ &\quad + \text{var}(m_l)\left[\left\{\frac{\partial H}{\partial v}\right\}\bigg|_{(\mu_k,\mu_l)}\right]^2 + 2\,\text{cov}(m_k, m_l) \\ &\quad \cdot\left\{\frac{\partial H}{\partial u}\frac{\partial H}{\partial v}\right\}\bigg|_{(\mu_k,\mu_l)} + O(n^{-3/2}). \end{aligned} \tag{2.1.16}$$

Hence, in particular, there exists a constant $d > 0$ such that

$$\operatorname{var}(H(m_k, m_l)) = \frac{d}{n} + O(n^{-3/2}) \tag{2.1.17}$$

since there exists c_1 and c_2 positive such that

$$\operatorname{var}(m_k) = \frac{c_1}{n} + O(n^{-2})$$

and

$$\operatorname{cov}(m_k, m_l) = \frac{c_2}{n} + O(n^{-2}).$$

For proof, see Cramer (1946, p. 354).

At times, it is important to obtain the limiting distribution of statistics since the exact distribution might not be computable in a closed form. It is also necessary to obtain valid asymptotic expansions for purposes of tabulation of the distribution functions.

We discuss some elementary properties regarding asymptotic distributions of sampling statistics.

Proposition 2.1.3. Suppose that

$$\sqrt{n}(T_n - \mu) \xrightarrow{\mathscr{L}} N(0, \sigma^2). \tag{2.1.18}$$

Let $g(\cdot)$ be a measurable function such that the derivative $g'(x)$ exists in a neighborhood of μ. Then

$$\sqrt{n}(g'(T_n) - g(\mu)) \xrightarrow{\mathscr{L}} N(0, (g'(\mu)\sigma)^2). \tag{2.1.19}$$

Proof. It is clear that (2.1.18) shows that

$$\sqrt{n}(T_n - \mu) = O_p(1)$$

and hence $(T_n - \mu) = o_p(1)$. Since $g(\cdot)$ is differentiable at μ,

$$\frac{g(x) - g(\mu)}{x - \mu} - g'(\mu) \to 0 \quad \text{as } x \to \mu.$$

In other words

$$g(x)-g(\mu)=(x-\mu)[g'(\mu)+\varepsilon(x)],$$

where $\varepsilon(x)\to 0$ as $x\to\mu$. Since $T_n\xrightarrow{p}\mu$, for any $\varepsilon>0$ and $\eta>0$, there exists N such that

$$P(|T_n-\mu|\leqslant\varepsilon)\geqslant 1-\eta \quad \text{for } n\geqslant N.$$

In particular, for sufficiently small $\varepsilon>0$, the relation

$$g(T_n)-g(\mu)=(T_n-\mu)[g'(\mu)+\varepsilon(T_n)]$$

holds with probability greater than or equal to $1-\eta$ for $n\geqslant N$. Therefore

$$\sqrt{n}[g(T_n)-g(\mu)]=\sqrt{n}(T_n-\mu)[g'(\mu)+\varepsilon(T_n)].$$

Since $\varepsilon(T_n)\xrightarrow{p}0$ and $\sqrt{n}(T_n-\mu)=O_p(1)$, it follows that

$$\sqrt{n}(T_n-\mu)\varepsilon(T_n)=o_p(1).$$

Therefore

$$\sqrt{n}[g(T_n)-g(\mu)]-\sqrt{n}(T_n-\mu)g'(\mu)\xrightarrow{p}0,$$

which in turn proves the result (2.1.19).

Remarks. If, in addition to the hypothesis of Proposition 2.1.3, $g'(\cdot)$ is continuous and $g'(\mu)\neq 0$, then it is easy to check that

$$\frac{\sqrt{n}[g(T_n)-g(\mu)]}{g'(T_n)}\xrightarrow{\mathscr{L}}N(0,\sigma^2)$$

from the Slutsky lemma.

A more general from of Proposition 2.1.3 is as follows.

Proposition 2.1.4. Let $\{\mathbf{T}_n\}$ be a sequence of k-dimensional statistics such that

$$\sqrt{n}(\mathbf{T}_n-\boldsymbol{\theta})\xrightarrow{\mathscr{L}}N_k(\mathbf{0},\boldsymbol{\Sigma}).$$

Let $\mathbf{g}=(g_1,\ldots,g_q)$, where each component g_i is a function of k variables, and suppose g_i is totally differentiable for $1 \leqslant i \leqslant q$. Then

$$\sqrt{n}[\mathbf{g}(\mathbf{T}_n)-\mathbf{g}(\boldsymbol{\theta})] \xrightarrow{\mathscr{L}} N_q(\mathbf{0}, \mathbf{G}\boldsymbol{\Sigma}\mathbf{G}'),$$

where $\mathbf{G}=((\partial g_i/\partial\theta_j))_{q\times k}$.

We omit the proof [see Rao (1974, p. 388).]. The next result deals with the limiting distribution when g is a function of T_n and n as in Proposition 2.1.1.

Proposition 2.1.5. Let $\{\mathbf{T}_n\}$ be a sequence of k-dimensional statistics such that

$$\sqrt{n}(\mathbf{T}_n-\boldsymbol{\theta}) \xrightarrow{\mathscr{L}} N_k(\mathbf{0},\boldsymbol{\Sigma}).$$

Let $g=g(\mathbf{T}_n,n)$ be such that $g(\mathbf{t},n)$ is partially differentiable with respect to t_i, $1 \leqslant i \leqslant k$, and

$$\frac{\partial g}{\partial t_i}(t_1,\ldots,t_k) \to G_i(\theta_1,\ldots,\theta_k), \qquad 1 \leqslant i \leqslant k,$$

as $\mathbf{t}\to\boldsymbol{\theta}$. Then

$$v_n^{-1}\sqrt{n}[g(\mathbf{T}_n,n)-g(\boldsymbol{\theta},n)] \xrightarrow{\mathscr{L}} N(0,1),$$

where

$$v_n=\left[\sum_i\sum_j \frac{\partial g}{\partial t_i}\frac{\partial g}{\partial t_j}\sigma_{i,j}\right]_{(T_{1n},\ldots,T_{kn})}$$

provided $v_n \neq 0$ when T_{in} are replaced by θ_i, $1 \leqslant i \leqslant k$.

Proof of this result is left as an exercise for the reader [see Rao (1974, p. 441)].

An interesting special result is the following theorem due to Cramer (1946, p. 366).

Proposition 2.1.6. Let X_i, $1 \leqslant i \leqslant n$, be i.i.d. random variables with μ_k and μ_l finite. Suppose there exists a neighborhood W of (μ_k,μ_l) such that $H(u,v)$ is continuous in V and has continuous partial derivatives of the first and second order with respect to u and v in W. Then $H(m_k,m_l)$ is asymptotically normal with asymptotic mean and asymptotic variance given by the leading terms of (2.1.15) and (2.1.16), respectively.

Proposition 2.1.6 can be easily extended to function of several moments.

Let us now consider a sequence of i.i.d. m-dimensional random vectors $\mathbf{X}_i$, $i \geqslant 1$. Let f_i, $1 \leqslant i \leqslant k$, be measurable functions on R^m. Consider the statistic

$$W_n = n^{1/2}(H(\bar{\mathbf{Z}}) - H(\boldsymbol{\mu})), \tag{2.1.20}$$

where H is a real-valued measurable function on R^k,

$$\mathbf{Z}_n = (f_1(\mathbf{X}_n), \ldots, f_k(\mathbf{X}_n)) = (Z_n^{(1)}, \ldots, Z_n^{(k)}), \tag{2.1.21}$$

$$\bar{\mathbf{Z}} = \frac{1}{n}\sum_{i=1}^{n} \mathbf{Z}_i = (\bar{Z}^{(1)}, \ldots, \bar{Z}^{(k)}),$$

and

$$\boldsymbol{\mu} = E\mathbf{Z}_n = (\mu^{(1)}, \ldots, \mu^{(k)}).$$

Observe that the functions of sample moments are of the from $H(\bar{Z})$ for a suitable choice of f_i. It now follows from Proposition 2.1.4 that, if $\mathbf{Z}_1$ has a finite covariance matrix and H is continuously differentiable in a neighborhood of $\boldsymbol{\mu}$, then W_n is asymptotically normal with mean zero and variance

$$\sigma^2 = \sum_{i,j=1}^{k} v_{ij}\gamma_i\gamma_j, \tag{2.1.22}$$

where $((v_{ij}))$ is the covariance matrix of $\mathbf{Z}_1$ and

$$\gamma_i = \left.\frac{\partial H}{\partial z_i}\right|_{\mathbf{z}=\boldsymbol{\mu}}, \qquad 1 \leqslant i \leqslant k. \tag{2.1.23}$$

Note that $H(\boldsymbol{\mu})$ and σ^2/n are the mean and variance of the asymptotic distribution of $H(\bar{\mathbf{Z}})$ and are *not* the mean and variance of $H(\bar{\mathbf{Z}})$. In practice, the "approximate moments" of W_n are calculated by first expanding $H(\bar{\mathbf{Z}})$ around μ and keeping a certain number of terms, raising the truncated expansion to the appropriate power, and then taking expectations term by term. This method is called the *delta method* (δ method). For an application of this method, see, for instance, David (1949). These approximate moments are then used to obtain a formal asymptotic expansion of the distribution function of W_n. We illustrate this approach by a simple example.

Let $X_1, \ldots, X_n$ be i.i.d. random variables and $H(\cdot)$ be an arbitrary function not depending on n. Suppose $H'(\mu) > 0$, where $\mu = E(X_i)$, and $\sigma^2 = \operatorname{Var}(X_i)$. Let $\bar{X} = n^{-1}\sum_{i=1}^{n} X_i$,

$$W_n = n^{1/2}(H(\bar{X}) - H(\mu)),$$

and

$$V_n = W_n/\sigma H'(\mu).$$

Suppose that

$$\begin{aligned} F_n(x) &\equiv P(V_n \leqslant x) \\ &= P\left(n^{1/2}\frac{(\bar{X}-\mu)}{\sigma} \leqslant \frac{n^{1/2}\{J(H(\mu)+xn^{-1/2})-\mu\}}{\sigma} \right) + O(n^{-p}) \end{aligned}$$

for $p > 0$, where J is the inverse function of $H(\cdot)$ in a neighborhood of μ. Therefore the asymptotic behavior of V_n is similar to that of a normalized sum of independent and identically distributed random variables. If the distribution of X_1 satisfies Cramer's condition [see (1.5.20)] and $E|X_1|^r < \infty$ for some $r \geqslant 3$, then

$$P\left(\frac{n^{1/2}(\bar{X}-\mu)}{\sigma} \leqslant u\right) = \Phi(u) + \sum_{j=1}^{r-2} \frac{Q_j(u)}{n^{j/2}} + o(n^{-(r-2)/2}) \tag{2.1.24}$$

from Proposition 1.5.6 uniformly in u. This will in turn give an *Edgeworth expansion* for the distribution function of W_n. Alternatively, one can formally compute the moments and from then the cumulants of V_n by the δ method described earlier. The formal cumulant expressions of V_n will be of the form

$$\begin{aligned} k_1(V_n) &= 0 + O(n^{-1/2}), \\ k_2(V_n) &= 1 + O(n^{-1}), \\ k_r(V_n) &= O(n^{-(r/2)+1}), \qquad r > 2, \end{aligned}$$

so that the leading terms behave exactly as for the normalized sums of i.i.d. random variables. These formal cumulant expansions may be substituted in the identity [see (1.5.5)]

$$\tilde{F}_n(x) = \exp\left(\sum_{j=1}^{\infty} (k_j - \gamma_j)\frac{(-D)^j}{j!} \right)\Phi(x),$$

where $\gamma_1 = k_1, \gamma_2 = k_2$, and $\gamma_j = k_j + O(n^{-(j/2)+1})$ for $j \geqslant 3$, and $\tilde{F}_n(x)$ is the distribution function of $n^{1/2}(\bar{X}-\mu)/\sigma$. To obtain the Edgeworth expansion, the exponential operator may be expanded formally and the terms collected according to powers of $n^{1/2}$. This will formally give the same expression as in (2.1.24) if we truncate the expansion at $r-2$. The advantage of the δ method is that it is easier to compute. Whether the expansions obtained by the δ method are valid or not was settled by Bhattacharya and Ghosh (1978). We present their results. The δ-method moment expansions are known to be valid in some cases. For instance, if a function of sample moments is uniformly bounded by a power of the sample size as in Proposition 2.1.2, then the expansion can be proved to be valid by extending Proposition 2.1.2 [cf. Wallace (1958)]. The general result is as follows.

Define W_n, $\mathbf{Z}_n$, and so on as in (2.1.20)–(2.1.23). Let

$$d_{i_1,\ldots,i_p} = (D_{i_1} D_{i_2} \cdots D_{i_p} H)(\boldsymbol{\mu}), \qquad 1 \leq i_1, \ldots, i_p \leq k, \tag{2.1.25}$$

where D_i denotes differentiation with respect to the ith component. Expanding W_n by the Taylor expansion, we have

$$\begin{aligned} W'_n = n^{1/2} \Bigg[& \sum_{i=1}^{k} d_i(\bar{Z}^{(i)} - \mu^{(i)}) \\ & + \frac{1}{2} \sum_{i,j} d_{i,j} (\bar{Z}^{(i)} - \mu^{(i)})(\bar{Z}^{(j)} - \mu^{(j)}) + \cdots \\ & + \frac{1}{(s-1)!} \sum_{i_1,\ldots,i_{s-1}} d_{i_1,\ldots,i_{s-1}} (\bar{Z}^{(i_1)} - \mu^{(i_1)}) \cdots (\bar{Z}^{(i_{s-1})} - \mu^{(i_{s-1})}) \Bigg], \end{aligned} \tag{2.1.26}$$

where $W_n - W'_n = o_p(n^{-(s-2)/2})$. Let $k_{j,n}$ be the jth cumulant of W'_n. It can be checked that

$$k_{j,n} = \bar{k}_{j,n} + o(n^{-(s-2)/2}), \qquad j \geq 1, \tag{2.1.27}$$

where

$$\bar{k}_{j,n} = \begin{cases} \displaystyle\sum_{i=1}^{s-2} n^{-i/2} b_{ji} & \text{if } j \neq 2 \\ \sigma^2 + \displaystyle\sum_{i=1}^{s-2} n^{-i/2} b_{2i} & \text{if } j = 2 \end{cases} \tag{2.1.28}$$

and the b_{ji} depend only on the moments of components of $\mathbf{Z}_1$ and on the partial derivatives of H at $\boldsymbol{\mu}$ of order $s-1$ and less. The $\bar{k}_{j,n}$ are called the *approximate cumulants* of W_n. The expression

$$\exp\left\{ it\bar{k}_{1,n} + \frac{(it)^2}{2}(\bar{k}_{2,n} - \sigma^2) + \sum_{j=3}^{s} \frac{(it)^j}{j!} \bar{k}_{j,n} \right\} \exp\left\{ -\frac{\sigma^2 t^2}{2} \right\} \tag{2.1.29}$$

is an approximate for the characteristic function of W_n. Expanding the first exponential factor, expression (2.1.29) reduces to

$$\exp\left\{-\frac{\sigma^2 t^2}{2}\right\}\left[1+\sum_{r=1}^{s-2} n^{-r/2} P_r(it)\right]+o(n^{-(s-2)/2}) \tag{2.1.30}$$

$$=\hat{\psi}_{s,n}(t)+o(n^{-(s-2)/2}),$$

where $P_r(\cdot)$ are polynomials with coefficients not depending on n. The *formal Edgeworth expansion* of the distribution function of W_n is defined using

$$\psi_{s,n}(v)=\left[1+\sum_{r=1}^{s-2} n^{-r/2} P_r\left(-\frac{d}{dv}\right)\right]\phi_\sigma(v) \tag{2.1.31}$$

and

$$\Psi_{s,n}(u)=\int_{-\infty}^{u} \psi_{s,n}(v)\,dv.$$

Here $\phi_\sigma(v)$ is the density of $N(0,\sigma^2)$ and the characteristic function of $\psi_{s,n}$ is $\hat{\psi}_{s,n}(t)$. Let $\|\cdot\|$ denote the ordinary Euclidean norm.

Theorem 2.1.7. Suppose that, for some integer $s \geqslant 3$, all the partial derivatives of H of order s and less are continuous in a neighborhood of $\boldsymbol{\mu}=E\mathbf{Z}_1$ and $E\|\mathbf{Z}_1\|^s<\infty$. Furthermore, suppose that (i) the distribution of $\mathbf{X}_1$ has a nonzero absolutely continuous component (with respect to the Lebesgue measure of R^m) and (ii) the density of this component is strictly positive on some nonempty open set U on which $f_i, 1 \leqslant i \leqslant k$, are continuously differentiable and $1, f_1, \ldots, f_k$ are linearly independent as elements of the space of continuous functions on U. Then

$$\sup_B \left| P(W_n \in B)-\int_B \psi_{s,n}(v)\,dv\right| = o(n^{-(s-2)/2}), \tag{2.1.32}$$

where B runs over all Borel sets.

For proof, see Bhattacharya and Ghosh (1978). The proof is based on Propositions 1.5.4 and 1.5.5. They have also obtained a bound of the Berry–Esseen type for the distribution function of W_n.

Theorem 2.1.8. Suppose $E\|\mathbf{Z}_1\|^3<\infty$ and all the third-order partial derivatives of H are continuous in a neighborhood of $\boldsymbol{\mu}=E\mathbf{Z}_1$. Then

$$\sup_{B\in\mathscr{B}} \left| P(W_n \in B)-\int_B \phi_\sigma(v)\,dv\right| = O(n^{-1/2}), \tag{2.1.33}$$

where $\mathscr{B}$ is the class of Borel sets satisfying

$$\sup_{B\in\mathscr{B}} \int_{(\partial B)^{\varepsilon}} \phi_{\sigma}(v)\,dv = O(\varepsilon) \quad \text{as } \varepsilon \to 0. \tag{2.1.34}$$

Here ∂B is the boundary of B and $(\partial B)^{\varepsilon}$ is an ε-neighborhood of B.

Applications of Theorem 2.1.8 for obtaining the asymptotic expansions of distribution functions of maximum likelihood estimators and minimum contrast estimators are given in Bhattacharya and Ghosh (1978). We shall not deal with them here.

2.2 SAMPLE QUANTILES AND THEIR PROPERTIES

Let F be a distribution function. For any $0 < p < 1$, let

$$\xi_p = \inf\{x : F(x) \geqslant p\} \equiv F^{-1}(p).$$

ξ_p is called pth *quantile* (or pth *fractile*) of F. It is clear that

$$F(\xi_p - 0) \leqslant p \leqslant F(\xi_p).$$

Given a sample $X_1, \ldots, X_n$ of i.i.d. random variables, a pth *sample quantile* $\hat{\xi}_{p,n}$ is defined to be a pth quantile of the empirical distribution function F_n corresponding to X_i, $1 \leqslant i \leqslant n$.

Theorem 2.2.1. Suppose $0 < p < 1$ and the pth quantile is unique. Then

(i) $\hat{\xi}_{p,n} \to \xi_p$ a.s.

and

(ii) for every $\varepsilon > 0$,

$$P(|\hat{\xi}_{p,n} - \xi_p| > \varepsilon) \leqslant 2e^{-n\delta_{\varepsilon}^2}, \qquad n \geqslant 1, \tag{2.2.1}$$

where $\delta_{\varepsilon} = \min[F(\xi_p + \varepsilon) - p, p - F(\xi_p - \varepsilon)]$.

For proof of Theorem 2.2.1, see Serfling (1980). The probability inequality in (ii) follows from Hoeffding inequality stated in Proposition 1.14.15 and the observation that

$$P(\hat{\xi}_{p,n} > x) = P(p > nF_n(x)),$$

where F_n is the empirical distribution function. Note that $F_n(x)$ is a binomial variable defined by

$$P\left(F_n(x) = \frac{m}{n}\right) = \binom{n}{m}[F(x)]^m[1 - F(x)]^{n-m}, \qquad 0 \leqslant m \leqslant n.$$

Observe that, as an approximation,

$$F_n(x) = F_n(\xi_p) + F(x) - F(\xi_p)$$

and hence

$$\begin{aligned} p = F_n(\hat{\xi}_{p,n}) &\doteq F_n(\xi_p) + F(\hat{\xi}_{p,n}) - F(\xi_p) \\ &\doteq F_n(\xi_p) + (\hat{\xi}_{p,n} - \xi_p) f(\xi_p), \end{aligned}$$

where f is the density of F (assuming that it exists). Therefore

$$\hat{\xi}_{p,n} - \xi_p \doteq \frac{p - F_n(\xi_p)}{f(\xi_p)}$$

or

$$\hat{\xi}_{p,n} \doteq \xi_p + \frac{Z_n - nq}{nf(\xi_p)},$$

where Z_n is the number of observations in the sample X_i, $1 \leqslant i \leqslant n$, such that $X_i > \xi_p$ and $q = 1 - p$. Kiefar (1967) proved the following theorem justifying the approximation given above, improving over an earlier result of Bahadur (1966).

Theorem 2.2.2. Suppose $0 < p < 1$ and $F(\xi_p) = p$. Assume that F is differentiable twice in a neighborhood of ξ_p, $F^{(2)}$ is bounded in the neighborhood, and $F'(\xi_p) = f(\xi_p) > 0$. Let $\sigma_p = [p(1-p)]^{1/2}$. Define

$$\begin{aligned} R_n &= \hat{\xi}_{p,n} - \xi_p + \frac{Z_n - nq}{nf(\xi_p)} \\ &= \hat{\xi}_{p,n} - \xi_p - \frac{F_n(\xi_p) - p}{f(\xi_p)}. \end{aligned}$$

Then

$$\overline{\lim_{n\to\infty}} \pm f(\xi_p) R_n [2^{5/4} 3^{-3/4} \sigma_p^{1/2} n^{-3/4} (\log\log n)^{3/4}]^{-1} = 1 \quad \text{a.s.} \tag{2.2.2}$$

In particular, it follows that

$$\hat{\xi}_{p,n} = \xi_p + \frac{Z_n - nq}{nf(\xi_p)} + O(n^{-3/4}(\log\log n)^{3/4}) \quad \text{a.s.} \tag{2.2.3}$$

We omit the proof. We shall now discuss a weaker version of this representation due to Ghosh (1971) that is sufficient for applications in statistical inference.

Theorem 2.2.3. Suppose $0 < p < 1$ and $F(\xi_p) = p$. Furthermore, suppose that F is differentiable at ξ_p with $F'(\xi_p) = f(\xi_p) > 0$. Let $p_n - p = o(n^{-1/2})$. Define

$$R_n = \hat{\xi}_{p_n,n} - \xi_{p_n} + \frac{G_n(\xi_p) - q}{f(\xi_p)}, \tag{2.2.4}$$

where $nG_n(x)$ denotes the number of observations in the sample $X_1, \ldots, X_n$ that are greater than x and

$$\xi_{p_n} = \xi_p + (p_n - p)/f(\xi_p). \tag{2.2.5}$$

Then

$$n^{1/2} R_n \xrightarrow{\text{p}} 0. \tag{2.2.6}$$

We first prove a lemma.

Lemma 2.2.4. Let $\{V_n\}$ and $\{W_n\}$ be two sequences of random variables such that

(i) for any $\delta > 0$, there exists λ depending on δ such that $P(|W_n| > \lambda) < \delta$ for large $n > n(\delta)$,

(ii) for all k and for all $\varepsilon > 0$

$$\lim_{n\to\infty} P(V_n \leqslant k, W_n \geqslant k + \varepsilon) = 0,$$

and

$$\lim_{n\to\infty} P(V_n \geqslant k + \varepsilon, W_n \leqslant k) = 0.$$

Then $V_n - W_n \xrightarrow{\text{p}} 0$.

Proof. Let $\varepsilon > 0$. Note that, for $n > n(\delta)$,

$$P(|V_n - W_n| > 2\varepsilon) \leqslant \delta + P(|V_n - W_n| > 2\varepsilon, |W_n| \leqslant \lambda).$$

Divide the interval $[-\lambda, \lambda]$ into m subintervals $[a_j, b_j]$ such that $b_j - a_j < \varepsilon$. Then, for $n > n(\delta)$,

$$P(|V_n - W_n| > 2\varepsilon) \leqslant \delta + \sum_{j=1}^{m} P(a_j \leqslant W_n \leqslant b_j, |V_n - W_n| > 2\varepsilon)$$

and

$$\begin{aligned} P(a_j \leqslant W_n \leqslant b_j, |V_n - W_n| > 2\varepsilon) & \\ \leqslant P(a_j \leqslant W_n \leqslant b_j, V_n \geqslant b_j + \varepsilon) & \\ + P(a_j \leqslant W_n \leqslant b_j, V_n \leqslant b_j - \varepsilon). & \end{aligned}$$

The last two terms in the latter inequality tend to zero by (ii). Hence the lemma is proved.

Proof of Theorem 2.2.3. Note that the event

$$[n^{1/2}(\hat{\xi}_{p_n,n} - \xi_{p_n}) \leqslant t] \tag{2.2.7}$$

is equivalent to the event

$$[p_n \leqslant F_n(\xi_{p_n} + tn^{-1/2})],$$

which in turn is equivalent to the event

$$[n^{1/2}(G_n(\xi_{p_n} + tn^{-1/2}) - G(\xi_{p_n} + tn^{-1/2}))[f(\xi_p)]^{-1} \leqslant t_n], \tag{2.2.8}$$

where $t_n = n^{1/2}(F(\xi_{p_n} + tn^{-1/2}) - p_n)[f(\xi_p)]^{-1}$ and $G(x) = 1 - F(x)$. Note that $G_n(x) = 1 - F_n(x)$. But $t_n \to t$ as $n \to \infty$ since

$$t_n = n^{1/2}[F(\xi_p) + \{tn^{-1/2} + (p_n - p)(f(\xi_p))^{-1}\}\{f(\xi_p) + o(1)\} - p_n][f(\xi_p)]^{-1}.$$

Let

$$Z_{t,n} = n^{1/2}\{G_n\{\xi_{p_n} + tn^{-1/2}) - G(\xi_{p_n} + tn^{-1/2})\}[f(\xi_p)]^{-1}$$

and

$$W_n = n^{1/2}\{G_n(\xi_p) - G(\xi_p)\}[f(\xi_p)]^{-1}.$$

It is easy to check that

$$E(Z_{t,n} - W_n)^2 = r_n(1 - r_n)[f(\xi_p)]^{-2},$$

where

$$r_n = |F(\xi_p) - F(\xi_{p_n} + tn^{-1/2})| \to 0.$$

Hence

$$Z_{t,n} - W_n \overset{p}{\to} 0.$$

Furthermore, W_n is asymptotically normal by the central limit theorem. Let

$$V_n = n^{1/2}(\hat{\xi}_{p_n,n} - \xi_{p_n}).$$

It is easy to see that V_n and W_n satisfy the conditions of Lemma 2.2.4. Hence

$$V_n - W_n \overset{p}{\to} 0,$$

completing the proof of the theorem.

Remarks. In particular, it follows that

$$\text{(2.2.9)} \qquad n^{1/2}(\hat{\xi}_{p_n,n} - \xi_{p_n}) \overset{\mathscr{L}}{\to} N\left(0, \frac{p(1-p)}{[f(\xi_p)]^2}\right).$$

In general, one can prove the following theorem concerning the asymptotic normality of quantiles.

Theorem 2.2.5. Let $0 < p_1 < \cdots < p_k < 1$. Suppose F is differentiable in a neighborhood of ξ_{p_i}, $1 \leqslant i \leqslant k$, and $f(\xi_{p_i}) > 0$. Assume that f is continuous at ξ_{p_i}, $1 \leqslant i \leqslant k$. Then

$$\{\sqrt{n}(\hat{\xi}_{p_i,n} - \xi_{p_i}), 1 \leqslant i \leqslant k\} \overset{\mathscr{L}}{\to} N_k(\mathbf{0}, \Sigma),$$

where $\Sigma = ((\sigma_{ij}))$ and $\sigma_{ij} = p_i(1 - p_j)/f(\xi_{p_i})f(\xi_{p_j}), i \leqslant j, \sigma_{ij} = \sigma_{ji}$ for $i > j$.

For proof, see Serfling (1980) or Cramer (1946). Reiss (1974) obtained the accuracy of the normal approximation and he derived the Edgeworth-type asymptotic expansions for the distribution function of the sample quantile in Reiss (1976).

Theorem 2.2.6 (Reiss, 1974). Let $0 < p < 1$. Suppose F has a bounded second derivative and $f(\xi_p) = F'(\xi_p) > 0$. Let $\| f' \| = \sup_x |f'(x)|$ and $\sigma_p = (p(1-p))^{1/2}$.

Then

$$\sup_t \left| P\left\{ \frac{n^{1/2} f(\xi_p)}{\sigma_p} (\hat{\xi}_{p,n} - \xi_p) < t \right\} - \Phi(t) \right|$$
$$\leqslant n^{-1/2} \left[\frac{\sigma \| f' \| \sigma_p}{10 f^2(\xi_p)} + \frac{\| f' \|^2}{f^4(\xi_p) n^{1/2}} + R_{p,n} \right],$$

where

$$R_{p,n} \equiv C \frac{1 - 2\sigma_p^2 q_n^2}{\sigma_p q_n} + \frac{3(|1-2p| + ((\log n)/n)^{1/2})}{10 \sigma_p q_n^2}$$

and

$$q_n \equiv [1 - \sigma_p^{-2} \{ |1-2p| ((\log n)/n)^{1/2} + (\log n)/n \}]^{1/2}$$

and C is an absolute constant.

Remarks. The proof of this theorem is obtained using the Berry–Esseen theorem for independent binomial random variables. An upper bound for the constant C is 0.7975, which is the one given by Van Beeck (1972). We omit the proof.

As a special case of the above result, we have the following theorem when F is the uniform distribution on $[0, 1]$.

Theorem 2.2.7. (Reiss, 1974). Let F be uniform on $[0, 1]$. Then, for any $0 < p < 1$,

$$\sup_t \left| P\left\{ \frac{n^{1/2}}{\sigma_p} (\hat{\xi}_{p,n} - p) < t \right\} - \Phi(t) \right| \leqslant n^{-1/2} R_{p,n},$$

where $R_{p,n}$ is as defined above.

Remarks. As far as the author is aware, there are no nonuniform bounds obtained in this case. However, it should be possible to derive the same from the method used in Reiss (1974) using nonuniform bounds for sums of independent random variables given in Chapter 1 [cf. Petrov (1975)].

We now discuss briefly some results connected with the quantile processes. For an extensive and illuminating discussion, see Csörgö and Revesz (1981).

Let U_i, $i \geqslant 1$, be i.i.d. random variables uniform on $(0, 1)$, and let $0 = U_0^{(n)} \leqslant U_1^{(n)} \leqslant \cdots \leqslant U_{n+1}^{(n)} = 1$ be the corresponding order statistics based on the random sample $U_1, \ldots, U_n$ of size n. Define the *uniform quantile function*

$$U_n(y) = \begin{cases} U_k^{(n)} & \text{if } (k-1)/n < y \leqslant k/n, \quad 1 \leqslant k \leqslant n \\ 0 & \text{if } y = 0 \end{cases}$$

and the *uniform sample quantile process*

$$u_n(y) = n^{1/2}(U_n(y) - y), \qquad 0 \leqslant y \leqslant 1.$$

Let $X_i, i \geqslant 1$, be i.i.d. random variables with continuous distribution function F, and let $X_1^{(n)} \leqslant \cdots \leqslant X_n^{(n)}$ be the corresponding order statistics based on a sample $X_1, \ldots, X_n$ of size n. Define the *quantile function*

$$Q_n(y) = X_k^{(n)} \quad \text{if } (k-1)/n < y \leqslant k/n, \quad k = 1, 2, \ldots, n$$

and the *sample quantile process*

$$q_n(y) = n^{1/2}(Q_n(y) - F^{-1}(y)), \qquad 0 < y \leqslant 1.$$

Recall that a stochastic process $\{B(t), 0 \leqslant t \leqslant 1\}$ is called a *Brownian bridge* if (i) $B(0) = B(1) = 0$; (ii) the process is a Gaussian process with mean zero and covariance function $R(s, t) = \min(s, t) - st, 0 \leqslant s, t \leqslant 1$; and (iii) the sample paths of the process $B(t; \omega)$ are continuous in t with probability 1.

Note that, if $\{W(t), t \geqslant 0\}$ is a Wiener process, then $B(t) = W(t) - tW(1)$, $0 \leqslant t \leqslant 1$, is a Brownian bridge. On the other hand, if $\{B(t), 0 \leqslant t \leqslant 1\}$ is a Brownian bridge, then

$$W(t) = (t+1)B\left(\frac{t}{t+1}\right), \qquad t \geqslant 0,$$

is a Wiener process. For further relations between a Brownian bridge and Wiener process, see Section 1.9.

Given a random field $\{X(\mathbf{z}), \mathbf{z} = (x, y) \in R_+^2\}$, define the increment of the process over a rectangle $M = [x_1, x_2) \times [y_1, y_2) \subset R_+^2$ by

$$X(M) = X(x_2, y_2) - X(x_1, y_2) - X(x_2, y_1) + X(x_1, y_1).$$

A stochastic process $\{W(\mathbf{z}), \mathbf{z} \in R_+^2\}$ is called a *Wiener random field* (or *Wiener sheet*) if

(i) $W(M)$ is $N(0, (x_2 - x_1)(y_2 - y_1))$ when $M = [x, x_2) \times [y_1, y_2)$,
(ii) $W(0, y) = W(x, 0) = 0$ for $0 \leqslant x, y < \infty$,
(iii) $W(R_1), W(R_2), \ldots, W(R_n)$ are independent random variables if $R_1, R_2, \ldots, R_n$ are nonoverlapping ractangles in R_+^2, and
(iv) $W(\mathbf{z})$ is continuous in $\mathbf{z}$ a.s.

Note that

$E[W(\mathbf{z}_1)W(\mathbf{z}_2)] = \min(x_1, x_2)\min(y_1, y_2)$ when $\mathbf{z}_1 = (x_1, y_1)$ and $\mathbf{z}_2 = (x_2, y_2)$. For some properties of Wiener sheet, see Csörgö and Revesz (1981).

A *Kiefer process* $\{K(x, y); 0 \leqslant x \leqslant 1, 0 \leqslant y < \infty\}$ is defined by

$$K(x, y) = W(x, y) - xW(1, y),$$

where $W(\mathbf{z}), \mathbf{z} \in R_+^2$ is a Wiener sheet.

Theorem 2.2.8 (Csörgö and Revesz, 1975). Given independent random variables uniform on $[0, 1]$, there exists a sequence of Brownian bridges $\{B_n(y): 0 \leqslant y \leqslant 1\}$ such that

$$\sup_{0 \leqslant y \leqslant 1} |u_n(y) - B_n(y)| = O(n^{-1/2} \log n) \quad \text{a.s.}$$

and a Kiefer process $\{K(y, t);\ 0 \leqslant y \leqslant 1,\ t \geqslant 0\}$ such that

$$\sup_{0 \leqslant y \leqslant 1} |u_n(y) - n^{-1/2} K(y, n)| = O(n^{-1/4} (\log\log n)^{1/4} (\log n)^{1/2}) \quad \text{a.s.}$$

For proof, see Csörgö and Revesz (1981). The following theorem gives strong approximation for a general quantile process.

Theorem 2.2.9 (Csörgö and Revesz, 1978). Suppose $X_i, i \geqslant 1$, are i.i.d. random variables with a distribution function F that is twice differentiable on (a, b) where $a = \sup\{x: F(x) = 0\}$ and $b = \inf\{x: F(x) = 1\}$. Suppose $F' = f \neq 0$ on (a, b) and

$$\sup_{a < x < b} F(x)(1 - F(x)) \left| \frac{f'(x)}{f^2(x)} \right| \leqslant \gamma$$

for some $\gamma > 0$. Let $\delta_n = 25 n^{-1} \log\log n$. Then there exists a sequence of Brownian bridges $\{B_n(y), 0 \leqslant y \leqslant 1\}$ and a Kiefer process $\{K(y, t): 0 \leqslant y \leqslant 1, t \geqslant 0\}$ such that

$$\sup_{\delta_n \leqslant y \leqslant 1 - \delta_n} |f(F^{-1}(y)) q_n(y) - B_n(y)| = O(n^{-1/2} \log n) \quad \text{a.s.}$$

and

$$\sup_{\delta_n \leqslant y \leqslant 1 - \delta_n} |n^{1/2} f(F^{-1}(y)) q_n(y) - K(y, n)| = O((n \log\log n)^{1/4} (\log n)^{1/2}) \quad \text{a.s.}$$

If, in addition $\gamma < 2$ and f is nondecreasing on an interval to the right of a

and nonincreasing on an interval to the left of b, then the results given above hold when y ranges over $(0, 1)$.

Weak convergence of the quantile process was studied by Shorack (1972) and Parthasarathy (1967). As a consequence of Theorem 2.2.9, one can obtain the following law of iterated logarithm for the quantile process.

Theorem 2.2.10. Suppose the conditions stated in Theorem 2.2.9 hold. Then

$$\overline{\lim_{n\to\infty}} (\log\log n)^{-1/2} \sup_{0<y<1} f(F^{-1}(y))|q_n(y)| = 2^{-1/2} \quad \text{a.s.}$$

2.3 EMPIRICAL DISTRIBUTION FUNCTION AND ITS PROPERTIES

Let $X_1, X_2, \ldots$ be i.i.d. random variables with distribution function F, and let $F_n(x)$ denote the empirical distribution function based on the sample $X_1, X_2, \ldots, X_n$. It is clear that

$$F_n(x) \to F(x) \quad \text{a.s.}$$

for every x, and it is known that

$$\sup_x |F_n(x) - F(x)| \to 0 \quad \text{a.s.}$$

by the Glivenko–Cantelli theorem stated in Section 1.3. Furthermore, for any given $x, nF_n(x)$ is a binomial variable with parameters n and $F(x)$ and hence

$$EF_n(x) = F(x), \qquad \operatorname{Var} F_n(x) = \frac{F(x)(1 - F(x))}{n}.$$

It follows by the central limit theorem that

$$n^{1/2}(F_n(x) - F(x)) \xrightarrow{\mathscr{L}} N(0, F(x)(1 - F(x)).$$

Let

$$D_n(x) = n^{1/2}(F_n(x) - F(x)), \qquad -\infty < x < \infty.$$

The process $\{D_n(x), -\infty < x < \infty\}$ is called the *empirical process*. It is easily seen that the process has sample paths in $D(-\infty, \infty)$. The following theorem is due to Kolmogorov and Smirnov. This gives the limiting distribution of the statistic

$$\sup_x |D_n(x)|$$

where F is continuous. Note that the limiting distribution does not depend on F as long as F is continuous.

Theorem 2.3.1 (Kolmogorov, 1933; Smirnov, 1939). If F is continuous, then, for any $r > 0$,

$$P\left\{ \sup_{-\infty < x < \infty} |D_n(x)| \leqslant r \right\} \to \sum_{j=-\infty}^{\infty} (-1)^j e^{-2j^2 r^2} \quad \text{as } n \to \infty.$$

Theorem 2.3.2 (Dvoretzky et al., 1956). Let F be any distribution function on R. There exists an absolute constant C such that, for all $n \geqslant 1$ and $r > 0$,

$$P\left(\sup_{-\infty < x < \infty} |D_n(x)| \geqslant r \right) \leqslant Ce^{-2r^2}.$$

Kiefer (1961) extended this result to the multidimensional case.

Theorem 2.3.3 (Kiefer, 1961). Let F be a distribution function on R^k. Then, for every $\varepsilon > 0$, there exists a constant C depending on ε and k but *not* on F such that

$$P\left(\sup_{\mathbf{x} \in R^k} |D_n(\mathbf{x})| > d \right) \leqslant Ce^{-(2-\varepsilon)d^2}$$

for any $d > 0$.

We omit the proofs of Theorems 2.3.2 and 2.3.3.

Suppose that $X_i, i \geqslant 1$, are i.i.d. random variables with a continuous distribution function F. Clearly $U_i \equiv F(X_i)$, $i \geqslant 1$, are i.i.d. with uniform distribution on $[0, 1]$. Let

$$\alpha_n(y) = n^{1/2}(F_n(F^{-1}(y)) - y) = \sqrt{n}(E_n(y) - y), \qquad 0 \leqslant y \leqslant 1,$$

be the empirical process based on $U_1, \ldots, U_n$. Note that $E\alpha_n(y) = 0$ and $\operatorname{cov}(\alpha_n(y_1), \alpha_n(y_2)) = \min(y_1, y_2) - y_1 y_2$ and $\alpha_n(y)$ is asymptotically normal. It is easy to see that the trajectories of the process α_n belong to $D[0, 1]$. It can be shown that the process α_n converges weakly to a Brownian bridge

$B \equiv \{B(y),\ 0 \leqslant y \leqslant 1\}$ using the theorems developed in Section 1.10. For instance, see Chenstov (1956) or Parthasarathy (1967). Hence, if g is any measurable functional on $D[0,1]$, it follows that

$$g(\alpha_n) \xrightarrow{\mathscr{L}} g(B)$$

provided the set of discontinuities has zero measure with respect to the measure generated by process B on $D[0,1]$. Let $g(f) = \sup_{0 \leqslant y \leqslant 1} |f(y)|$ for any $f \in D[0,1]$. It is easy to see that the set of discontinuities of $g(\cdot)$ has zero measure with respect to the measure μ generated by the Brownian bridge on $D[0,1]$. This follows since the Brownian bridge has continuous sample paths on $[0,1]$ with probability 1 and the supnorm topology on $C[0,1]$ and the relative Skorokhod topology on $C[0,1]$ as subset of $D[0,1]$ coincide. Hence

$$n^{1/2} \sup_x |D_n(x)| = n^{1/2} \sup_y |\alpha_n(y)| \xrightarrow{\mathscr{L}} \sup_{0 \leqslant y \leqslant 1} |B(y)|.$$

It can be shown, by using the reflection principle, that

$$P\left(\sup_{0 \leqslant y \leqslant 1} |B(y)| \leqslant r \right) = \sum_{j=-\infty}^{\infty} (-1)^j e^{-2j^2 r^2}.$$

This argument in particular proves Theorem 2.3.1.

Komlos et al. (1975) proved the following theorem for empirical process. We omit the proofs as they are too long. See Csörgo and Revesz (1981, p. 133).

Theorem 2.3.4 (Komlos et al. 1975). Let $U_1, U_2, \ldots$ be independent random variables uniform on $[0,1]$. Then there exists a sequence of Brownian bridges $\{B_n(y),\ 0 \leqslant y \leqslant 1\}$ such that, for all n and x,

$$P\left\{ \sup_{0 \leqslant y \leqslant 1} |\alpha_n(y) - B_n(y)| > n^{-1/2}(C \log n + x) \right\} \leqslant L e^{-\lambda x}, \tag{2.3.0}$$

where C, L, λ are positive absolute constants. In particular

$$\sup_{0 \leqslant y \leqslant 1} |\alpha_n(y) - B_n(y)| = O(n^{-1/2} \log n) \quad \text{a.s.}$$

Furthermore, there exists a Kiefer process $\{K(y,t): 0 \leqslant y \leqslant 1, 0 \leqslant t < \infty\}$ such that

$$P\left\{\sup_{1\leqslant k\leqslant n}\sup_{0\leqslant y\leqslant 1}|k^{1/2}\alpha_k(y)-K(y,k)|>(C\log n+x)\log n\right\}\leqslant Le^{-\lambda x}$$

for all x and n and hence, in particular,

$$\sup_{0\leqslant y\leqslant 1}|n^{1/2}\alpha_n(y)-K(y,n)|=O(\log^2 n)\quad \text{a.s.}$$

Remarks. The result in Theorem 2.3.4 clearly holds for empirical processes when $X_1, X_2, \ldots X_n$ are i.i.d. as F where F is continuous since $U_i = F(X_i)$, $1 \leqslant i \leqslant n$, are independent uniformly distributed on $[0,1]$. However, the result holds even if F is an arbitrary distribution function. This can be seen as follows.

Let U_i, $1 \leqslant i \leqslant n$, be i.i.d. uniform on $[0,1]$ for which Theorem 2.3.4 holds and $E_n(y)$, $0 \leqslant y \leqslant 1$, be the empirical distribution function of this sample. Let F be an arbitrary distribution function. Define $X_k = F^{-1}(U_k)$ where $F^{-1}(y) = \inf\{t: F(t) \geqslant y\}$ is the right-continuous inverse to F. Let $F_n(x)$ be the empirical distribution function based on these random variables. Note that $P(X_k \leqslant x) = F(x)$, $1 \leqslant k \leqslant n$, and hence $F_n(x) = E_n(F(x))$ for every point $\omega \in \Omega$ of the basic probability space $(\Omega, \mathscr{B}, P)$. Let

$$D_n(x) = n^{1/2}(F_n(x) - F(x)), \qquad -\infty < x < \infty,$$

as before. Then

$$\begin{aligned}\sup_{-\infty<x<\infty}|D_n(x)-B_n(F(x))| &= \sup_{-\infty<x<\infty}|n^{1/2}(E_n(F(x))-F(x))-B_n(F(x))|\\ &\leqslant \sup_{0\leqslant y\leqslant 1}|n^{1/2}(E_n(y)-y)-B_n(y)|\\ &= \sup_{0\leqslant y\leqslant 1}|\alpha_n(y)-B_n(y)|,\end{aligned}$$

where $\{B_n\}$ is the sequence of Brownian bridges specified in Theorem 2.3.4. This inequality proves that Theorem 2.3.4 holds for arbitrary F.

It is easy to check that

$$\text{(2.3.1)}\qquad \overline{\lim_{n\to\infty}}\left\{\sup_{-\infty<x<\infty}|F_n(x)-F(x)|\right\}\left(\frac{n}{\log\log n}\right)^{1/2}<\infty\quad \text{a.s.}$$

for an arbitrary distribution function F. However, if F is continuous, then

$$\overline{\lim_{n\to\infty}}\left\{\sup_{-\infty<x<\infty}|F_n(x)-F(x)|\right\}\left(\frac{n}{\log\log n}\right)^{1/2}=2^{-1/2}\quad \text{a.s.}$$

from Smirnov (1944) and Chung (1949) and

$$\underline{\lim_{n\to\infty}}\left\{\sup_{-\infty<x<\infty}|F_n(x)-F(x)|\right\}(n\log\log n)^{1/2}=8^{-1/2}\pi\quad \text{a.s.}$$

by Mogulskii (1977).

REFERENCES

Bahadur, R. R. (1966). A note on quantiles in large samples. *Ann. Math. Statist.* **37**, 577–580.

Bhattacharya, R. N. and Ghosh, J. K. (1978). On the validity of the formal Edgeworth expansion. *Ann. Statist.* **6**, 434–451.

Chenstov, N. N. (1956). Weak convergence of stochastic processes whose trajectories have no discontinuities of the second kind and the "Heuristic approach to the Kolmogorov–Smirnov tests", *Theory prob. Appl.* **1**, 140–144.

Chung, K. L. (1949). An estimate concerning the Kolmogorov limit distribution. *Trans. Amer. Math. Soc.* **67**, 26–50.

Cramer, H. (1946). *Mathematical Methods of Statistics*, Princeton University Press, Princeton, New Jersey.

Csörgö, M. and Revesz, P. (1975). Some notes on the empirical distribution function and the quantile process. *Colloq. Math. Soc. J. Bolyai 11 Limit Theorems of probability Theory* (P. Revesz, Ed.), North-Holland, Amsterdam.

Csörgö, M. and Revesz, P. (1978). Strong approximations of the quantile process. *Ann. Statistics* **6**, 882–894.

Csörgö, M. and Revesz, P. (1981). *Strong Approximations in Probability and Statistics*, Akademia Kiado, Budapest.

David, F. W. (1949). The moments of the Z and F distributions. *Biometrika* **36**, 394–403.

Dvorotzky, A., Kiefer, J., and Wolfowitz, J. (1956). Asymptotic minimax character of the sample distribution function and of the classical multinomial estimator. *Ann. Math. Statist.* **27**, 642–669.

Ghosh, J. K. (1971). A new proof of the Bahadur representation of quantiles and an application. *Ann. Math. Statist.* **42**, 1957–1961.

Hurt, J. (1976). Asymptotic expansions of functions of statistics, *Aplikace Matematiky* **21**, 444–456.

Hurt, J. (1979). Asymptotic expansions for moments of functions of statistics. *Proc. 2nd Prague Conf. Asym. Stat.* 233–238.

Kiefer, J. (1961). On large deviations of the empiric D.F. of vector chance variables and a law of iterated logarithm. *Pacific J. Math.* **11**, 649–660.

Kiefer, J. (1967). On Bahadur's representation of sample quantiles. *Ann. Math. Statist.* **38**, 1323–1342.

Kolmogorov, A. N. (1933). Sulla determinaziono empirica di una loggo di distributsione. *Giono Inst. Ital. Attuari* **4**, 83–91.

Komlos, J., Major, P., and Tusnady, G. (1975). An approximation for partial sums of independent R. V.'s and the sample D. F. I, *Z. Wahr. verw. Geb.* **32**, 111–131.

Lomnicki, Z. A., and Zaremba, S. K. (1957). On the estimation of autocorrelation in time series. *Ann. Math. Statist.* **28**, 140–158.

Mogulskii, A. A. (1977). Laws of iterated logarithm on function spaces (Abstracts) *Second Vilnius Conf. on Prob. Theory and Math. Statist.* **2**, 44–47. (In Russian.)

Parthasarthy, K. R. (1967). *Probability Measures on Metric Spaces*, Academic Press, London.

Petrov, V. (1975). *Sums of Independent Random Variables*, Springer-Verlag, Berlin.

Rao, C. R. (1974). *Linear Statistical Inference and its Applications*, Wiley, New York.

Reiss, R. (1974). On the accuracy of the normal approximation for quantiles. *Ann. Probability* **2**, 741–744.

Reiss, R. (1976). Asymptotic expansions for sample quantiles. *Ann. Probability* **4**, 249–258.

Serfling, R. J. (1980). *Approximation Theorems of Mathematical Statistics*, Wiley, New York.

Shorack, G. R. (1972). Convergence of quantile and spacings processes with applications. *Ann. Math. Statist.* **43**, 1400–1411.

Smirnov, N. V. (1939). On the estimation of the discrepancy between empirical curves of distribution for two independent samples. *Bull. Math. de l' Universite de Moscow* 2 (Fasc. 2).

Smirnov, N. V. (1944). Approximate laws of distribution of random variables from empirical data. *Uspekhi Mat. Nauk* **10**, 179–206. (In Russian.)

Van Beeck, P. (1972). An application of Fourier methods to the problem of sharpening the Berry-Esseen inequality. *Z. Wahr. verw. Geb.* **23**, 187–196.

Wallace, D. L. (1958). Asymptotic approximations to distributions. *Ann. Math. Statist.* **29**, 635–654.

CHAPTER 3

Asymptotic Theory of Estimation

3.1 EXISTENCE AND CONSISTENCY OF BAYES AND MINIMUM CONTRAST ESTIMATORS

Among the several methods of estimation, the method of maximum likelihood and the method of least squares are the methods most often employed for inference purposes in all branches of science, particularly in economics and agricultural statistics. In this chapter we study some theoretical aspects of the method of maximum likelihood and related methods of estimation. We expect the reader to be familiar with basic properties for such estimators.

In this section, we shall discuss the general concepts of Bayes estimators and minimum contrast estimators. We follow Strasser (1973). Minimum contrast estimators generalize the notion of maximum likelihood estimators (Pfanzagl, 1969a, 1971).

Let $(\mathscr{X}, \mathscr{A})$ be a measurable space and $\mathscr{P}$ be a family of probability measures on $(\mathscr{X}, \mathscr{A})$. Let $\mathscr{C}$ be a σ-algebra over $\mathscr{P}$ such that the mapping $P \to P(A)$ from $\mathscr{P}$ to $[0,1]$ is $\mathscr{C}$-measurable for any fixed $A \in \mathscr{A}$. Let λ be a probability measure over $(\mathscr{P}, \mathscr{C})$. λ is called a *prior* distribution. For any $A \in \mathscr{A}$ and $B \in \mathscr{C}$, define

$$Q(A \times B) = \int_B P(A)\lambda(dP) \tag{3.1.0}$$

and extend Q to the product σ-algebra $\mathscr{A} \times \mathscr{C}$. Any version of the conditional probability $Q(B|A)$, $B \in \mathscr{C}$, is called an *a posteriori* distribution.

Let (T, τ) be a topological space of decisions. Assume that (T, τ) is separable. A function $L: T \times \mathscr{P} \to R^+$ is called a *loss function* if $L(t, P)$ is

$\mathscr{C}$-measurable for every $t \in T$ and $L(t, P)$ is continuous in t for every $P \in \mathscr{P}$.

Since (T, τ) is separable, for any $P \in \mathscr{P}$ and for any $A \in \tau$, the function $\sup_{s \in A} L(s, P)$ is $\mathscr{C}$-measurable. *We assume throughout the discussion that given* $t \in T$, *there exists an open set* $U_t \in \tau$ *such that*

$$\sup_{s \in U_t} L(s, P)$$

is λ*-integrable.* Let $\mathscr{B}$ be the σ-algebra of Borel subsets of T.

Definition 3.1.1. Let H be the set of all measurable functions $h(\cdot)$ from $(\mathscr{X}, \mathscr{A})$ into $(T, \mathscr{B})$ such that $L(h(x), P)$ is Q-integrable. A measurable function $h_0 \in H$ is called a *Bayes estimator* with respect to L and λ if

$$\int L(h_0(x), P) Q(dx \times dP) = \inf_{h \in H} \int L(h(x), P) Q(dx \times dP). \tag{3.1.1}$$

Let $t \in T$. By assumption

$$\sup_{s \in U_t} L(s, P)$$

is λ-integrable. Hence, by Fubini's theorem,

$$\sup_{s \in U_t} L(s, P) \tag{3.1.2}$$

is $Q(\cdot | \mathscr{A})(x)$-integrable for Q almost all $x \in \mathscr{X}$. Since T is separable, the exceptional set of measure zero can be chosen *independent* of $t \in T$. Hence we can and do assume that for each $t \in T$, there exists an open set U_t such that the function defined in (3.1.2) is λ-integrable and $Q(\cdot | \mathscr{A})$ (x)-integrable for all $x \in \mathscr{X}$.

Let $t_0 \in T$ and $t_n \in T$ such that $t_n \to t_0$. Let U_{t_0} be an open neighborhood of t_0 such that

$$\sup_{s \in U_{t_0}} L(s, P)$$

is $Q(\cdot | \mathscr{A})$ (x)-integrable for all $x \in \mathscr{X}$. Note that $t_n \in U_{t_0}$ for large n and hence $L(t_n, P)$ is $Q(\cdot | \mathscr{A})$ (x)-integrable for all $x \in \mathscr{X}$. An application of the Lebesgue dominated convergence theorem proves that

$$\lim_{n \to \infty} E(L(t_n, \cdot) | \mathscr{A})(x) = E(L(t_0, \cdot) | \mathscr{A})(x)$$

for every $x \in \mathscr{X}$. Therefore the function

$$E(L(t,\cdot)|\mathscr{A})(x) \tag{3.1.3}$$

is continuous in t for all $x \in \mathscr{X}$. Let D be a countable dense subset of T. Clearly

$$\inf_{t \in T} E(L(t,\cdot)|\mathscr{A})(x) = \inf_{t \in D} E(L(t,\cdot)|\mathscr{A})(x).$$

It is easy to see that, if $h_0 \in H$ is a Bayes estimator, then

$$\begin{aligned} E(L(h_0(x),\cdot)|\mathscr{A}](x) &= \inf_{t \in T} E(L(t,\cdot)|\mathscr{A})(x) \quad Q-\text{a.e.} \\ &= \inf_{t \in D} E(L(t,\cdot)|\mathscr{A})(x) \quad Q-\text{a.e.} \end{aligned} \tag{3.1.4}$$

We first state and prove some measure-theoretic lemmas needed in the sequel.

Lemma 3.1.1. Let $(\mathscr{X}, \mathscr{A})$ be a measurable space and (T, τ) be a separable metric space. Suppose $f(t,x)$ is a real-valued function such that $f(t,x)$ is continuous in t for any $x \in \mathscr{X}$ and is $\mathscr{A}$-measurable for any $t \in T$. Then

$$\inf_{t \in A} f(t,x)$$

is $\mathscr{A}$-measurable for every subset $A \subset T$.

Proof. Obvious.

Lemma 3.1.2. Let $(\mathscr{X}, \mathscr{A})$ be a measurable space and (T, τ) be a locally compact space with a countable base. Let f be a nonnegative function on $T \times \mathscr{X}$ such that

(i) $f(t,x)$ is continuous in t for all $x \in \mathscr{X}$,
(ii) $f(t,x)$ is $\mathscr{A}$-measurable for all $t \in T$, and
(iii) for all $x \in \mathscr{X}$ and for all $\rho > 0$, there exists a compact set $C_{x,\rho} \subset T$ such that

$$\inf\{f(t,x) | t \notin C_{x,\rho}\} > \rho.$$

Then there exists a *measurable* map $h: \mathscr{X} \to T$ such that

$$f(h(x),x) = \inf_{s \in T} f(s,x).$$

For Proof, we refer to Landers (1968) and Strasser (1973).

Theorem 3.1.3 (Existence). Suppose (T, τ) is a compact metric space. Then there *exists* a Bayes estimator under the conditions stated earlier.

Proof. Define

$$f(t,x) = E(L(t,\cdot)|\mathscr{A})(x).$$

The result follows from Lemma 3.1.2 in view of equation (3.1.4).

Remarks. If (T, τ) is locally compact, then we need an additional condition on the loss function in order to prove the existence of a Bayes estimate.

Definition 3.1.2. A loss function $L: T \times \mathscr{P} \rightarrow R^+$ is said to *tend to infinity* on $T \times \mathscr{P}$ (with respect to a topology τ) if there exists $\Sigma_n \in \mathscr{C}$ such that $\Sigma_n \uparrow \mathscr{P}$ and for each Σ_n and $\rho > 0$ there exists a compact set $C(\Sigma_n, \rho) \subset T$ such that

$$\inf\{L(t,P): P \in \Sigma_n, t \notin C(\Sigma_n, \rho)\} > \rho.$$

For instance, if $L(t,P) = \|t - \tau(P)\|$ when $T = R^k$ and τ is a measurable mapping from $\mathscr{P}$ to R^k, then $L(t,P)$ tends to infinity. In fact $\Sigma_n = \{P \in \mathscr{P}: \|\tau(P)\| \leqslant n\}$ and $C(\Sigma_n, \rho) = \{t \in T | \; \|t\| \leqslant n + \rho\}$ will serve the purpose.

Theorem 3.1.4 (Existence). Let (T, τ) be locally compact with a countable base, and let L tend to infinity on $T \times \mathscr{P}$. Then there *exists* a Bayes estimator.

Proof. Define

$$f(t,x) = E(L(t,\cdot)|\mathscr{A})(x), \qquad x \in \mathscr{X}, \quad t \in T.$$

Choose Σ_n and $C(\Sigma_n, \rho)$ as given in the definition since L tends to infinity. Since $\Sigma_n \uparrow \mathscr{P}$, for each $x \in \mathscr{X}$, there exists n such that

$$Q(\Sigma_n|\mathscr{A})(x) > 0.$$

Let $\alpha = Q(\Sigma_n|\mathscr{A})(x)$ and $\rho > 0$. Then, for any $t \notin C(\Sigma_n, \rho/\alpha)$,

$$\begin{aligned} f(t,x) &= \int L(t,P)Q(dP|\mathscr{A})(x) \\ &\geqslant \int_{\Sigma_n} L(t,P)Q(dP|\mathscr{A})(x) \\ &\geqslant \rho. \end{aligned}$$

The theorem now follows from Lemma 3.1.2 and equation (3.1.4).

Let $(\mathscr{X}^\infty, \mathscr{A}^\infty)$ be the countable infinite product space corresponding to $(\mathscr{X}, \mathscr{A})$, and let $\mathscr{A}^n$ denote the product σ-algebra on the n-fold product $\mathscr{X}^n$ of $\mathscr{X}$.

A measurable mapping γ from $(\mathscr{P}, C)$ into $(T, \mathscr{B})$ is called a *parameter*. Let $h_n: \mathscr{X}^n \to T$ be $(\mathscr{A}^n, \mathscr{B})$-measurable and γ be a parameter. h_n is said to be (*weakly*) *consistent* for γ if for all neighborhoods U of $\gamma(P)$

$$\lim_{n\to\infty} P^\infty\{h_n \in U\} = 1$$

for all $P \in \mathscr{P}$. It is said to be *strongly consistent* if

$$P^\infty\left\{\lim_{n\to\infty} h_n = \gamma(P)\right\} = 1$$

for all $P \in \mathscr{P}$.

For every $t \in T$ and $n \geqslant 1$, let $f_n(t)$ be an $\mathscr{A}^n$-measurable real-valued function defined on $\mathscr{X}^n$. The family $\{f_n, n \geqslant 1\}$ is said to be a family of *contrast functions* for γ in $P \in \mathscr{P}$ if

$$\overline{\lim_{n\to\infty}} \inf_{t\in T} f_n(t) < \underline{\lim_{n\to\infty}} \inf_{t\in T\setminus U} f_n(t) \quad P^\infty - \text{a.s.} \tag{3.1.5}$$

for each neighborhood U of $\gamma(P)$. The contrast is said to be *strong* if the exceptional set in (3.1.5) *does not depend* on U.

A sequence of $(\mathscr{A}^n, \mathscr{B})$-measurable mappings $h_n: \mathscr{X}^\infty \to T$ is called a sequence of *minimum contrast estimators* (MCE) with respect to the family of contrast functions $\{f_n, n \geqslant 1\}$ if

$$f_n(h_n(x))(x) = \inf_{t\in T} f_n(t)(x), \qquad n \geqslant 1, \quad x \in \mathscr{X}^\infty. \tag{3.1.6}$$

Theorem 3.1.5 (Pfanzagl–Wald) (Consistency). Let $\{f_n, n \geqslant 1\}$ be a family of contrast functions for the parameter γ in $P \in \mathscr{P}$. Then any sequence $\{h_n\}$ of minimum contrast estimators is consistent. If the contrast is strong, then strong consistency of $\{h_n\}$ holds.

Proof. Let U be a neighborhood of $\gamma(P)$ and

$$M_U = \bigcap_{N=1}^{\infty} \bigcup_{n>N} [h_n \notin U]. \tag{3.1.7}$$

Let $x_0 \in M_U$ and $N_0 = \{n: h_n(x_0) \notin U\}$. Then

$$\text{(3.1.8)} \qquad \varliminf_{n\to\infty} \inf_{t\in T\setminus U} f_n(t)(x_0) \leqslant \varliminf_{n\in N_0} f_n(h_n(x_0))(x_0) \quad \text{(by (3.1.6))}$$

$$\leqslant \varlimsup_{n\in N_0} f_n(h_n(x_0))(x_0) \leqslant \varlimsup_{n\to\infty} f_n(h_n(x_0))(x_0)$$

$$= \varlimsup_{n\to\infty} \inf_{t\in T} f_n(t)(x_0) \quad \text{(by (3.1.6))}.$$

Since $\{f_n\}$ is a family of contrast functions, it follows that

$$\text{(3.1.9)} \qquad \varlimsup_{n\to\infty} \inf_{t\in T} f_n(t) < \varliminf_{n\to\infty} \inf_{t\in T\setminus U} f_n(t) \quad P^\infty - \text{a.e.}$$

Therefore $P^\infty(M_U) = 0$. In other words

$$P^\infty\left\{ \bigcup_{N=1}^{\infty} \bigcap_{n\geqslant N} [h_n \in U] \right\} = 1,$$

that is,

$$\lim_{n\to\infty} P^N\left\{ \bigcap_{n\geqslant N} [h_n \in U] \right\} = 1.$$

This implies consistency. It can be easily checked that strong consistency holds if the contrast is strong.

Remarks. For an equivalent way of describing MCE and for conditions on the existence of a measurable MCE, see Pfanzagl (1969a). Under some conditions, it is easy to see that $f_n = -\sum_{i=1}^{n} \log p_t(x_i)$ where $p_t(x)$ is the density of X with a parameter t forms a family of contrast functions and the corresponding MCE is an MLE [see Pfanzagl (1971)].

Definition 3.1.3. A loss function $L(t, P)$ is said to *contrast* the parameter γ (with respect to the topology τ on T) if

$$L(\gamma(P), P) = \inf_{t\in T} L(t, P), \qquad P \in \mathscr{P},$$

and

$$L(\gamma(P), P) < \inf_{t\in T\setminus U} L(t, P)$$

for every neighborhood U of $\gamma(P)$, $P \in \mathscr{P}$.

Proposition 3.1.6. Suppose (T, τ) is a compact metric space and the loss function L attains $\inf\{L(t,P): t\in T\}$ for each P in at most one point. Then there exists a parameter γ such that the loss function L contrasts γ.

This is a consequence of Lemma 3.1.2. The following result is again a consequence of Lemma 3.1.2.

Proposition 3.1.7. Let (T, τ) be a locally compact space with countable base. Suppose the loss function L tends to infinity and, for each $P\in\mathscr{P}$, the loss function L attains $\inf\{L(t,P): t\in T\}$ in at most one point of T. Then there exists a parameter γ that is contrasted by the loss function L.

Definition 3.1.4. A $(\mathscr{A}^\infty, \mathscr{C})$-measurable map $f: \mathscr{X}^\infty \to \mathscr{P}$ is called a *λ-almost exact estimator* if

$$P^\infty\{f = P\} = 1 \quad \lambda - \text{a.e.}$$

The following result gives the relation between Bayes estimators and minimum contrast estimators.

Theorem 3.1.8. Suppose (T, τ) is a separable metric space and L contrasts the parameter γ. Assume that there exists a λ-almost exact estimator $f: \mathscr{X}^\infty \to \mathscr{P}$. Then the family

$$f_n(t) = E(L(t,\cdot)|\mathscr{A}^n), \qquad n \geq 1, \quad t\in T,$$

strongly contrasts the parameter γ in λ-almost all points $P\in\mathscr{P}$.

Proof. Let $\{U_k, k \geq 1\}$ be a countable base for the topology τ on T and D be a countable dense subset of T. Under the conditions stated above, for every $n \geq 1$, there exists a version of the *a posteriori* distribution such that the functions

$$E(L(t,f)|\mathscr{A}^n)(x) \quad \text{and} \quad E(L(t,\cdot)|\mathscr{A}^n)(x)$$

are continuous in t for all $x\in\mathscr{X}^\infty$ [see (3.1.3)]. Let

$$M_1 = \bigcap_{t\in T} \bigcap_{n=1}^{\infty} \{(x,P)\in\mathscr{X}^\infty \times \mathscr{P}: E(L(t,\cdot)|\mathscr{A}^n)(x) = E(L(t,f)|\mathscr{A}^n)(x)\},$$

which reduces to

$$\bigcap_{t\in D} \bigcap_{n=1}^{\infty} \{(x,P)\in\mathscr{X}^\infty \times \mathscr{P}: E(L(t,\cdot)|\mathscr{A}^n)(x) = E(L(t,f)|\mathscr{A}^n)(x)\}$$

by the remark made above. Since f is λ-exact, it follows that $P^\infty(M_1)=1$. Let

$$M_2=\bigcap_{k=1}^{\infty}\left\{(x,P)\in\mathscr{X}^\infty\times\mathscr{P}:\lim_{n\to\infty}E\left(\inf_{t\in T\setminus U_k}L(t,f)|\mathscr{A}^n\right)(x)\right.$$
$$\left.=\inf_{t\in T\setminus U_k}L(t,f(x))\right\},$$

$$M_3=\bigcap_{t\in D}\{(x,P):\lim_{n\to\infty}E(L(t,f)|\mathscr{A}^n)(x)=L(t,f(x))\},$$

and

$$M_4=\{(x,P):f(x)=P\}.$$

By the martingale convergence theorem, $Q^\infty(M_2)=Q^\infty(M_3)=1$ and $Q^\infty\{M_4\}=1$ since f is λ-exact where Q is the measure on $\mathscr{X}\times\mathscr{P}$ as defined by (3.1.0). Hence

$$Q^\infty\left\{\bigcap_{i=1}^{4}M_i\right\}=1.$$

Now for any $(x,P)\in\bigcap_{i=1}^{4}M_i$ and for any $k\geqslant 1$ with $\gamma(P)\in U_k$,

$$\begin{aligned}
\overline{\lim_{n\to\infty}}\inf_{t\in T}E(L(t,\cdot)|\mathscr{A}^n)(x)&=\overline{\lim_{n\to\infty}}\inf_{t\in D}E(L(t,\cdot)|\mathscr{A}^n)(x)\\
&\leqslant\inf_{t\in D}\overline{\lim_{n\to\infty}}E(L(t,\cdot)|\mathscr{A}^n)(x)\\
&=\inf_{t\in D}\overline{\lim_{n\to\infty}}E(L(t,f)|\mathscr{A}^n)(x)\\
&=\inf_{t\in D}L(t,f(x))=L(\gamma(P),P)\\
&<\inf_{t\in T\setminus U_k}L(t,P)\\
&=\inf_{t\in T\setminus U_k}L(t,f(x))\\
&=\lim_{n\to\infty}E\left(\inf_{t\in T\setminus U_k}L(t,f)|\mathscr{A}^n\right)(x)\\
&\leqslant\underline{\lim}_{n\to\infty}\inf_{t\in T\setminus U_k}E\left(L(t,f)|\mathscr{A}^n\right)(x)\\
&=\underline{\lim}_{n\to\infty}\inf_{t\in T\setminus U_k}E(L(t,\cdot)|\mathscr{A}^n)(x).
\end{aligned}$$

Since $Q^{\infty}(\bigcap_{i=1}^{4} M_i) = 1$, it follows that

$$Q^{\infty}\left[\bigcap_{k>1;\gamma(P)\in U_k}\left\{(x,P)\in x^{\infty}\times\mathscr{P}: \overline{\lim_{n\to\infty}}\inf_{t\in T} f_n(t)(x) < \underline{\lim}_{n\to\infty}\inf_{t\in T\setminus U_k} f_n(t)(x)\right\}\right] = 1$$

and therefore, by Fubini's theorem,

$$P^{\infty}\left[\bigcap_{k\geq 1;\gamma(P)\in U_k}\left\{(x,P)\in x^{\infty}\times\mathscr{P}: \overline{\lim_{n\to\infty}}\inf_{t\in T} f_n(t)(x) < \underline{\lim}_{n\to\infty}\inf_{t\in T\setminus U_k} f_n(t)(x)\right\}\right] = 1$$

for λ-almost all $P\in\mathscr{P}$. Hence $f_n(t)$ strongly contrasts the parameter γ.

In view of Theorem 3.1.3, Proposition 3.1.6, Theorem 3.1.5, and Proposition 3.1.7, we can obtain the following result on the strong consistency of a Bayes estimator. We omit the details.

Theorem 3.1.9. Let (T,τ) be a compact metric space. Suppose the loss function is such that $\inf\{L(t,P): t\in T\}$ is attained for each $P\in\mathscr{P}$ in utmost one point $t\in T$. Furthermore, assume that there exists a λ-almost exact estimate $f: x^{\infty}\to\mathscr{P}$. Then there exists a sequence $\{h_n, n\geq 1\}$ of Bayes estimates that is strongly consistent in λ-almost all points $P\in\mathscr{P}$ for the parameter γ defined by

$$L(\gamma(P),P) = \inf_{t\in T} L(t,P).$$

Remarks. A similar result can be proved if T is locally compact with a countable base provided the loss function tends to infinity. For details, see Strasser (1973).

3.2 EXPONENTIAL RATE FOR CONSISTENCY FOR PITMAN ESTIMATORS

In this section, we introduce the concept of Pitman estimators and study the consistency properties of such estimators. These estimators include Bayes estimators for some special classes of loss functions and maximum probability estimators. Our approach is again that of Strasser (1981b, c).

Let $(\mathscr{X},\mathscr{A})$ be a measurable space and $\mathscr{P}$ be a family of probability measures defined on $(\mathscr{X},\mathscr{A})$. Let $(\mathscr{X}^n,\mathscr{A}^n)$ denote the n-fold product of $(\mathscr{X},\mathscr{A})$ as before and P^n denote the product measure on $(\mathscr{X}^n,\mathscr{A}^n)$ for any $P\in\mathscr{P}$. Let C be a σ-algebra over $\mathscr{P}$ such that the mappings $P\to P^n(A_n)$, $n\geq 1$, $A_n\in\mathscr{A}^n$ are measurable. The topology on $\mathscr{P}$ is induced by the family of test functions ϕ_n defined on $(\mathscr{X}^n, \mathscr{A}^n)$, $n \geqslant 1$. Let $\gamma(P)$ be a C-measurable function defined on $\mathscr{P}$. We are interested in estimation of $\gamma(P)$ based on i.i.d. random variables X taking values in $\mathscr{X}$. An *estimator* T_n is an $\mathscr{A}^n$-measurable function from $\mathscr{X}^n$ to R.

Let μ be a measure defined on C not necessarily finite. Define

$$(3.2.1) \qquad P_n(\mu)(A_n) = \int_{\mathscr{P}} Q^n(A_n)\mu(dQ), \qquad A_n \in \mathscr{A}^n,$$

and in general

$$(3.2.2) \qquad P_n(\gamma\mu)(A_n) = \int_{\mathscr{P}} Q^n(A_n)\gamma(Q)\mu(dQ), \qquad A_n \in \mathscr{A}^n,$$

whenever $\gamma \geqslant 0$. We assume that the measures $P_n(\mu)$ and $P_n(\gamma\mu)$ are σ-finite. It is clear that $P_n(\gamma\mu) \ll P_n(\mu)$. The *Pitman estimator* of γ with respect to μ is defined by

$$(3.2.3) \qquad T_n\gamma = \frac{dP_n(\gamma\mu)}{dP_n(\mu)}.$$

In general, for any C-measureable function $\gamma(P)$, the *Pitman estimator* of γ is defined by

$$(3.2.4) \qquad T_n\gamma = \frac{dP_n(\gamma^+\mu)}{dP_n(\mu)} - \frac{dP_n(\gamma^-\mu)}{dP_n(\mu)},$$

where γ^+ and γ^- are the positive part and negative part of γ provided $P_n(\mu)$ and $P_n(|\gamma|\mu)$ are σ-finite.

For any $P, Q \in \mathscr{P}$, define $H(P, Q)$ to be the Hellinger distance (see Section 1.3) between P and Q, and for any $P \in \mathscr{P}$ and $a \geqslant 0$, let $B(P, a) = \{Q \in \mathscr{P} : H(P, Q) < a\}$.

Definition 3.2.1. A measure μ on $(\mathscr{P}, C)$ is said to be *diametrically regular on* $\mathscr{P}$ if there exists $\beta \geqslant 1$ such that for every $Q \in \mathscr{P}$ and every $a \geqslant 0$,

$$(3.2.5) \qquad \mu(B(Q, 2a)) < \beta\mu(B(Q, a)).$$

The measure μ on $(\mathscr{P}, C)$ is said to be *diametrically regular at* $P \in \mathscr{P}$ if there exists a neighborhood U_P of P in $\mathscr{P}$ such that μ is diametrically regular on U_P.

Remark. The diametric regularity of μ at $P\in\mathscr{P}$ implies that there exists $\beta \geqslant 1$ such that

$$\mu(U_P \cap B(Q, 2a)) < \beta\mu(U_P \cap B(Q, a))$$

for every $a > 0$ and $Q \in U_P$.

We now give some sufficient conditions for a measure to be diametrically regular.

Suppose $\mathscr{P} = \{P_\theta, \theta\in\Theta\}$ where $\theta\in\Theta\subset R^k$ is open. Suppose that $\theta_1 \neq \theta_2$ if and only if $P_{\theta_1} \neq P_{\theta_2}$. Consider $(\Theta, \mathscr{B})$, where $\mathscr{B}$ is the σ-algebra of Borel subsets of Θ. Let π be the mapping from Θ to $\mathscr{P}$ defined by $\pi(\theta) = P_\theta$. Suppose the following conditions hold:

(R_1) π is a homeomorphism;

(R_2) for every $\theta\in\Theta$, there exist constants $a > 0, b < \infty, d > 0$, and $\alpha > 0$ such that

$$a|\sigma - \tau|^\alpha \leqslant H(P_\sigma, P_\tau) \leqslant b|\sigma - \tau|^\alpha, \tag{3.2.6}$$

whenever $|\sigma - \theta| < d, |\tau - \theta| < d$;

(R_3) $\mu \ll \lambda$ where λ is the Lebesgue measure on R^k and the density of μ is positive and continuous on Θ.

Strasser (1981b) has shown that the regularity conditions R_1 to R_3 imply that the measure $\mu\circ\pi^{-1}$ is diametrically regular at every $P_\theta\in\mathscr{P}$. We omit the proof.

Theorem 3.2.1. Suppose that, for some $n \geqslant 1$, $P_n(\mu)$ and $P_n(\gamma\mu)$ are σ-finite. Let $P\in\mathscr{P}$. If μ is diametrically regular at P and $\gamma(\cdot)$ is nonnegative and continuous at P, then, for every $\varepsilon > 0$, there exists a neighborhood V_p of P such that

$$\sup_{Q\in V_p} Q\left\{\bigcup_{m>n} [|T_m\gamma - \gamma(Q)| > \varepsilon]\right\} \leqslant Ce^{-cn}, \qquad n \geqslant 1, \tag{3.2.7}$$

for some positive constants C and c.

Theorem 3.2.1 gives an exponential rate of convergence of Pitman estimator to the true parameter. We now sketch a proof of this theorem. We have stated the results for the case $\gamma \geqslant 0$. The result can be easily extended to the general case.

Proposition 3.2.2. Let $\gamma(\cdot)$ be a nonnegative C-measurable function such that $P_n(\mu)$ and $P_n(\gamma\mu)$ are σ-finite. Then, for every $P\in\mathscr{P}$, $0 < \varepsilon \leqslant 1$, and $M\in C$ such that $P \in M$,

$$P\{T_n\gamma > \varepsilon\} \leqslant (2n)^{1/2} \sup_{Q \in M} H(P, Q) + \frac{1}{\varepsilon\mu(M)} \pi_n(I_M, \gamma), \tag{3.2.8}$$

where

$$\pi_n(f, g) = \inf_{\phi_n \in \Phi_n} \left[\int E_Q[\phi_n] f(Q)\mu(dQ) + \int E_Q[1 - \phi_n] g(Q)\mu(dQ) \right]. \tag{3.2.9}$$

(Here $E_Q(\phi_n)$ denotes the expectation of ϕ_n under Q^n on $(\mathscr{X}^n, \mathscr{A}^n)$, $\Phi_n = \{\phi_n : \phi_n$ is an $\mathscr{A}_n$-measurable test function$\}$ and I_M denotes the indicator function of the set M.)

Proof. Let $M \in C$ such that $P \in M$ and $\varepsilon > 0$. Observe that

$$\frac{dP_n(\gamma\mu)}{dP_n(\mu)} \leqslant \frac{dP_n(\gamma\mu)}{dP_n(I_M\mu)} \quad \text{a.e. } P_n(I_M\mu).$$

Hence

$$\begin{aligned}
P_n(I_M\mu)\{[T_n\gamma > \varepsilon]\} &\leqslant P_n(I_M\mu)\left\{\left[\frac{dP_n(\gamma\mu)}{dP_n(I_M\mu)} > \varepsilon\right]\right\} \\
&= \int_{J_n} \phi_n \, dP_n(I_M\mu) + \int_{J_n} (1 - \phi_n) \, dP_n(I_M\mu) \\
&\quad \cdot \left(\text{here } J_n = \left[\frac{dP_n(\gamma\mu)}{dP_n(I_M\mu)} > \varepsilon\right]\right) \\
&\leqslant \frac{1}{\varepsilon} \int_M E_Q[\phi_n]\mu(dQ) + \frac{1}{\varepsilon} \int E_Q[1 - \phi_n]\gamma(Q)\mu(dQ)
\end{aligned}$$

by (3.2.2) and Chebyshev's inequality. Since $\phi_n \in \Phi_n$ is arbitrary, we have

$$\frac{1}{\mu(M)} P_n(I_M\mu)\{[T_n\gamma > \varepsilon]\} \leqslant \frac{1}{\varepsilon\mu(M)} \pi_n(I_M, \gamma). \tag{3.2.10}$$

But, for any $A \in \mathscr{A}^n$,

$$\begin{aligned}
\left| P^n(A) - \frac{1}{\mu(M)} P_n(I_M\mu)(A) \right| &\leqslant \frac{1}{\mu(M)} \int_M |Q^n(A) - P^n(A)| \, \mu(dQ) \\
&\qquad \text{(by (3.2.2))} \\
&\leqslant (2n)^{1/2} \sup_{Q \in M} H(P, Q).
\end{aligned} \tag{3.2.11}$$

Relations (3.2.10) and (3.2.11) prove the proposition.

Lemma 3.2.3. Suppose μ is regular at $P \in \mathscr{P}$. Then there is a neighborhood U_P of P such that, for some $a > 0$ and $0 \leqslant \alpha < \infty$,

$$\inf_{Q \in U_P} \mu(B(Q, \varepsilon)) \geqslant a\varepsilon^{\alpha} \tag{3.2.12}$$

for every $\varepsilon > 0$.

Proof. Since μ is diametrically regular at P, there exists a neighborhood U_P of P and $\beta \geqslant 1$ such that, for every $Q \in U_P$,

$$\mu(U_P \cap B(Q, 2^k r)) < \beta^k \mu(U_P \cap B(Q, r))$$

for $k = 0, 1, 2, \ldots$. Let $\rho > 0$ and $2^k \leqslant \rho \leqslant 2^{k+1}$. Then $k \leqslant (\log \rho / \log 2)$ and

$$\beta^k = \exp(k \log \beta) \leqslant \exp\left(\frac{\log \rho}{\log 2} \log \beta\right) = \rho^{\alpha},$$

where $\alpha = \log \beta / \log 2$. Therefore

$$\begin{aligned} \mu(U_P \cap B(Q, \rho r)) &\leqslant \beta^{k+1} \mu(U_P \cap B(Q, r)) \\ &\leqslant \beta \rho^{\alpha} \mu(U_P \cap B(Q, r)) \end{aligned}$$

and hence

$$\begin{aligned} 0 < \mu(U_P) &= \mu(U_P \cap B(Q, 1)) \\ &= \mu\left(U_P \cap B\left(Q, \varepsilon \cdot \frac{1}{\varepsilon}\right)\right) \\ &\leqslant \beta \left(\frac{1}{\varepsilon}\right)^{\alpha} \mu(U_P \cap B(Q, \varepsilon)), \end{aligned}$$

which proves the lemma.

Lemma 3.2.4. Let ϕ be a test function defined on $(\mathscr{X}, \mathscr{A})$ and $0 < r < 1$. Define

$$\psi_n(\mathbf{X}) = \frac{1}{n} \sum_{i=1}^{n} \phi(X_i),$$

where X_i are i.i.d. random variables with probability measure P. Then

$$\text{(i) if } E\phi(X_1) \leqslant r, \text{ then } P[\psi_n(\mathbf{X}) \geqslant r] \leqslant \left\{ \left(\frac{1 - E\phi}{1 - r}\right)^{1-r} \left(\frac{E\phi}{r}\right)^{r} \right\}^{n} \tag{3.2.13}$$

and

$$\text{(3.2.14)} \quad \text{(ii)} \quad \text{if } E\phi(X_1) \geqslant r, \text{ then } P[\psi_n(\mathbf{X}) \leqslant r] \leqslant \left\{\left(\frac{1-E\phi}{1-r}\right)^{1-r}\left(\frac{E\phi}{r}\right)^r\right\}^n.$$

Proof. The lemma clearly holds if $r = E\phi(X_1)$. Let $Y_j = \phi(X_j)$. Then $0 \leqslant Y_j \leqslant 1$ and suppose $EY_j < r < 1$. It is easy to see that, for any $t \geqslant 0$ and $0 \leqslant x \leqslant 1$,

$$e^{tx} \leqslant 1 - x + xe^t$$

and hence

$$E[e^{tY_j}] \leqslant 1 - E(Y_j) + E(Y_j)e^t.$$

Therefore

$$g(t) = E\left[\exp\left(t \sum_{j=1}^n Y_j\right)\right] \leqslant [1 - E(Y_1) + E(Y_1)e^t]^n.$$

Note that

$$\begin{aligned} P[\psi_n \geqslant r] &= P\left[\sum_{j=1}^n Y_j \geqslant nr\right] \\ &= P\left[t \sum_{j=1}^n Y_j \geqslant nrt\right] \\ &\leqslant e^{-nrt} E\left[\exp\left(t \sum_{j=1}^n Y_j\right)\right] \\ &\leqslant e^{-nrt}[1 - E(Y_1) + E(Y_1)e^t]^n \\ &= [\{1 - E(Y_1) + E(Y_1)e^t\}e^{-rt}]^n. \end{aligned}$$

Let $\gamma = E(Y_1)$. The function

$$g^*(t) = (1 - \gamma + \gamma e^t)e^{-rt}$$

attains its minimum at t_0 given by

$$e^{t_0} = \frac{(1-\gamma)r}{\gamma(1-r)}$$

and it can be checked that

$$g^*(t_0) = \left(\frac{\gamma}{r}\right)^r \left(\frac{1-\gamma}{1-r}\right)^{1-r}.$$

This proves the first part of the lemma. The other relation follows by choosing $Y_j = 1 - \phi(X_j)$.

Given any two disjoint subsets H and K of $\mathscr{P}$, denote

(3.2.15) $$\pi_n(I_H, I_K) = \pi_n(H, K),$$

where $\pi_n(f, g)$ is as defined in (3.2.9). It is easy to see that

(3.2.16) $$\pi_n(H, K) = \inf_{\phi_n \in \Phi_n} \left\{ \sup_{P \in H} E_P \phi_n + \sup_{Q \in K} E_Q(1 - \phi_n) \right\},$$

where Φ_n is the class of all $\mathscr{A}^n$-measurable test functions. This is the minimax risk for testing H versus K.

Lemma 3.2.5 (Le Cam). For any two subsets H and K of $\mathscr{P}$,

(3.2.17) $$\pi_{2kn}(H, K) \leqslant \{\pi_n(H, K)(2 - \pi_n(H, K))\}^k, \qquad k \geqslant 1, \quad n \geqslant 1.$$

Furthermore, if there exists n such that $\pi_n(H, K) < 1$, then there exist $0 < c$, $C < \infty$ such that

(3.2.18) $$\pi_n(H, K) \leqslant Ce^{-cn}$$

This lemma is due to Le Cam (1973a). We sketch a proof. Let $\rho(P, Q)$ denote the affinity (Section 1.3) between two probability measures P and Q. Given two sets A and B of probability measures over $(\mathscr{X}, \mathscr{A})$, let

(3.2.19) $$\rho_n = \sup\{\rho(P^n, Q^n): P \in A, Q \in B\}.$$

It can be checked easily that

(3.2.20) $$\rho_{m+n} \leqslant \rho_m \rho_n.$$

Suppose there is a σ-finite measure that dominates the elements of A and B. Let $\pi_n(A, \mathrm{B}) = s$. Then, for any $P \in A$ and $Q \in B$, $\| P^n \wedge Q^n \| \leqslant s$, where $\| P^n \wedge Q^n \|$ is the L_1-norm of the minimum of P^n and Q^n. Furthermore (see Section 1.3),

(3.2.21) $$\begin{aligned} \pi_n^2(P, Q) \leqslant \rho^2(P^n, Q^n) &\leqslant 1 - (1 - \pi_n(P, Q))^2 \\ &= \pi_n(P, Q)(2 - \pi_n(P, Q)) \end{aligned}$$

(here $\pi_n(P, Q) = \pi_n(\{P\}, \{Q\})$). By repeated application of this inequality, it is easy to see that

$$\|P^n \wedge Q^n\| \leqslant \rho(P^n, Q^n) \leqslant [s(2-s)]^k, \qquad n \geqslant 1, \quad k \geqslant 1, \tag{3.2.22}$$

and hence

$$\pi_{2kn}(A, B) \leqslant (s[2-s))^k \quad \text{where } s = \pi_n(A, B). \tag{3.2.23}$$

Suppose $A \cup B$ is not dominated and $\pi_n(A, B) = s > 0$. Then for any $\varepsilon > 0$ there exists a test function ϕ_n such that $\pi_n(A, B; \phi_n) < (1 + \varepsilon)s$, where

$$\pi_n(A, B; \phi_n) = \sup\left\{\int (1 - \phi_n)\, dP^n + \int \phi_n\, dQ^n; P \in A, Q \in B\right\} \tag{3.2.24}$$

and hence there exists a test function ψ_n taking only a finite number of values and such that

$$\pi_n(A, B; \psi_n) < (1 + \varepsilon)s.$$

The argument goes through on the σ-algebra generated by ψ_n. Since ε is arbitrary, the inequality follows. The second part is an easy consequence of the first part.

Definition 3.2.2. A probability measure $P \in \mathscr{P}$ is said to be *μ-distinguished* if there exist $n \geqslant 1$ and a test function $\phi_n \in \Phi_n$ such that

$$E_P \phi_n < 1 \quad \text{and} \quad \int E_Q[1 - \phi_n] \mu(dQ) < \infty. \tag{3.2.25}$$

Proposition 3.2.6. Suppose $P \in \mathscr{P}$ is μ-distinguished. Then, for every neighborhood U_P of P in $\mathscr{P}$, there is another neighborhood V_P of P in $\mathscr{P}$ such that $V_P \subset U_P$ and

$$\pi_n(V_P, U_P) \leqslant Ce^{-cn}, \qquad n \geqslant 1, \tag{3.2.26}$$

for some $0 < c$, $C < \infty$. (Here U_P^c is the complement of U_P in $\mathscr{P}$.)

Proof. Since $P \in \mathscr{P}$ is μ-distinguished, there exists an integer $N \geqslant 1$ and a test function $\psi_N \in \Phi_N$ such that

$$E_P \psi_N < 1 \quad \text{and} \quad \int E_Q(1 - \Psi_N] \mu(dQ) < \infty. \tag{3.2.27}$$

Let $E_P\psi_N < r_0 < 1$ and $r_0 > 0$. Define

$$\bar{\psi}_{k,N}(\mathbf{x}) = \frac{1}{k}\sum_{i=1}^{k} \psi_N(x_{(i-1)N+1}, \ldots, X_{iN})$$

when $\mathbf{x} \in \mathscr{X}^{kN}$. Lemma 3.2.4 implies that

$$Q[\bar{\psi}_{k,N} \geqslant r_0] \leqslant \left\{ \left(\frac{1 - E_Q\psi_N}{1 - r_0} \right)^{1 - r_0} \left(\frac{E_Q\psi_N}{r_0} \right)^{r_0} \right\}^k \tag{3.2.28}$$

if $E_Q\psi_N < r_0$. Let

$$V_P^{(0)} = \{Q \in \mathscr{P} : E_Q\psi_N < \tfrac{1}{2}(E_P\psi_N + r_0)\}.$$

Then $V_P^{(0)}$ is a neighborhood of P. Let $r_1 = E_Q\psi_N$. Note that $E_Q\psi_N < r_0$ and

$$Q \in V_P^{(0)} \Rightarrow Q[\bar{\psi}_{k,N} \geqslant r_0] \leqslant \left\{ \left(\frac{1 - r_1}{1 - r_0} \right)^{1 - r_0} \left(\frac{r_1}{r_0} \right) \right\}^k \tag{3.2.29}$$

from (3.2.28). Define

$$\bar{\psi}_m = \begin{cases} 1 & \text{if } \bar{\psi}_{k,N} \geqslant r_0 \\ 0 & \text{if } \bar{\psi}_{k,N} < r_0 \end{cases}$$

whenever $kN \leqslant m < (k+1)N, m \geqslant 1, k \geqslant 1$. It can be checked that

$$\sup_{Q \in V_P^{(0)}} E_Q[\bar{\psi}_m] < Ce^{-cm}, \qquad m \geqslant 1, \tag{3.2.30}$$

and

$$\overline{\lim_m} \int E_Q(1 - \bar{\psi}_m)\mu(dQ) < \infty \tag{3.2.31}$$

for some $0 < c, C < \infty$ using (3.2.29) and the definition of $\psi_{k,N}$.

Let U_P be a neighborhood of P. In view of Lemma 3.2.5, there exists a neighborhood $V_P^{(1)}$ of P, $V_P \subset U_P$, such that

$$\pi_n(V_P^{(1)}, U_P^c) \leqslant Ce^{-cn}, \qquad n \geqslant 1, \tag{3.2.32}$$

for some $0 < c, C < \infty$. Note that $V_P^{(1)}$ and U_P^c can be separated by open subsets of $\mathscr{P}$ and $\pi_n(V_P^{(1)}, U_P^c) < 1$ for some $n \geqslant 1$. Hence there exists $n_1 \geqslant 1$ and a test function $\eta_{n_1} \in \Phi_{n_1}$ and a neighborhood $V_P^{(1)}$ of P, $V_P^{(1)} \subset U_P$, such that

$$\sup_{Q \in V_P^{(1)}} E_Q[\eta_{n_1}] < \tfrac{1}{4} < \tfrac{1}{2} < \inf_{Q \in U_P^c} E_Q[\eta_{n_1}]. \tag{3.2.33}$$

In view of (3.2.30) and (3.2.31), there exists $n_2 \geqslant 1$ such that

$$\sup_{Q \in V_P^{(0)}} E_Q[\bar{\psi}_{n_2}] < \tfrac{1}{4}, \qquad \int E_Q(1 - \bar{\psi}_{n_2})\mu(dQ) < \infty. \tag{3.2.34a}$$

In particular, choosing $V_P = V_P^{(0)} \cap V_P^{(1)}$, $M = \max(n_1, n_2)$ and $\bar{\phi}_M = \max(\eta_{n_1}, \bar{\psi}_{n_2})$, it follows that there exists an integer $M \geqslant 1$ such that

$$\text{(3.2.34b)} \quad \text{(i)} \quad \sup_{Q \in V_P} E_Q[\bar{\phi}_M] < \inf_{Q \in U_P^c} E_Q[\bar{\phi}_M]$$

$$\text{(ii)} \quad \int_{\mathscr{P}} E_Q[1 - \bar{\phi}_M)\mu(dQ) < \infty.$$

By arguments analogous to those given earlier, it can be shown (using Lemma 3.2.4) that there is a sequences of tests ϕ_n based on $\bar{\phi}_M$ such that

$$\text{(3.2.35)} \quad \text{(i)} \quad \sup_n \int E_Q(1 - \phi_n)\mu(dQ) < \infty,$$

$$\text{(ii)} \quad \sup_{Q \in V_P} E_Q[\phi_n] \leqslant Ce^{-cn}, \qquad n \geqslant 1,$$

$$\text{(iii)} \quad \int_{U_P^c} \{1 - E_Q[\phi_n]\}\mu(dQ) \leqslant Ce^{-cn}, \qquad n \geqslant 1.$$

Proposition 3.2.6 now follows from relations (3.2.35) provided V_P is such that $\mu(V_P) < \infty$. It can be shown that there is a neighborhood U_P of P such that $\mu(U_P) < \infty$ since P is μ-distinguished. Since $V_P \subset U_P$, it follows that $\mu(V_P) < \infty$.

As a consequence of the Proposition 3.2.6, we have the following result.

Proposition 3.2.7. Let γ be greater than or equal to 0 and be C-measurable. Suppose that, for some $n \geqslant 1$, the measures $P_n(\mu)$ and $P_n(\gamma\mu)$ are σ-finite. Then, for every neighborhood U_P of P, there is another neighborhood V_P of P, $V_P \subset U_P$, such that

$$\pi_n(I_{V_P}, \gamma I_{U_P^c}) \leqslant Ce^{-cn}, \qquad n \geqslant 1, \tag{3.2.36}$$

for some $0 < c, C < \infty$.

This follows from Proposition 3.2.6 and the fact that every $P \in \mathscr{P}$ is $\max(\gamma, 1)\mu$-distinguished whenever $P_n(\mu)$ and $P_n(\gamma\mu)$ are σ-finite for some $n \geqslant 1$.

Proposition 3.2.8. Let $\gamma \geqslant 0$ be C-measurable. Suppose that, for some $n \geqslant 1$, the measures $P_n(\mu)$ and $P_n(\gamma\mu)$ are σ-finite. If μ is diametrically regular at $P \in \mathscr{P}$, then, for every neighborhood U_P of P, there is another neighborhood V_P of P, $V_P \subset U_P$, such that

$$\sup_{Q \in V_P} Q[T_n(\gamma I_{U_P^c}) > e^{-cn}] \leqslant Ce^{-cn}, \qquad n \geqslant 1, \tag{3.2.37}$$

for some $0 < c, C < \infty$.

Proof. In view of Proposition 3.2.7, there is a neighborhood W_P of P, $W_P \subset U_P$, such that

$$\pi_n(W_P, U_P^c) \leqslant De^{-dn}, \qquad n \geqslant 1,$$

for some $0 < d, D < \infty$. Let $\delta > 0$ and V_P be a neighborhood of P such that $V_P \subset W_P \subset U_P$ and such that $Q \in V_P \Rightarrow B(Q, \delta) \subset W_P$. Choose $a > 0$ and $\alpha < \infty$ as in Lemma 3.2.3. Let

$$c = \frac{d}{4 \max(\alpha, 1)}$$

Then, for any $Q \in V_P$ and n large, by Proposition 3.2.2, we have

$$Q[T_n(\gamma I_{U_P^c}) > e^{-cn}] \leqslant \sqrt{2}\, e^{-cn} + \frac{1}{e^{-cn}\mu(M)} \pi_n(I_M, \gamma I_{U_P^c}),$$

where $M = B(Q, \delta)$, $\delta = e^{-cn} n^{-1/2}$. Therefore

$$\begin{aligned} Q[T_n(\gamma I_{U_P^c}) > e^{-cn}] &\leqslant \sqrt{2}\, e^{-cn} + \frac{De^{-Dn} n^{\alpha/2}}{ae^{-cn} e^{-\alpha cn}} \\ &\leqslant e^{-cn} + De^{-cn} \leqslant Ce^{-cn} \end{aligned}$$

for every $n \geqslant 1$ for some $0 < c, C < \infty$.

We now prove Theorem 3.2.1, giving the exponential rate of convergence of Pitman estimators.

Proof of Theorem 3.2.1. Since γ is continuous at P, there exists a neighborhood U_P of P such that

(3.2.38) $|\gamma(Q_1)-\gamma(Q_2)|<\varepsilon/2$ whenever $Q_1 \in U_P$ and $Q_2 \in U_P$.

For any $Q \in U_P$,

$$|T_n\gamma-\gamma(Q)| \leqslant \frac{\varepsilon}{2} + T_n(|\gamma-\gamma(Q)| I_{U_P^c}). \tag{3.2.39}$$

Since $\gamma(\cdot)$ is bounded on U_P, it follows from Proposition 3.2.8 that there exists a neighborhood $V_P \subset U_P$ of P such that

$$\sup_{Q \in V_P} Q[|T_n\gamma-\gamma(Q)|>\varepsilon] \leqslant C_0 e^{-cn}, \qquad n \geqslant 1, \tag{3.2.40}$$

for some $0<c$, $C_0<\infty$. The theorem follows from the inequality

$$\sum_{m=n}^{\infty} e^{-cm} \leqslant Ce^{-cn}, \qquad n \geqslant 1,$$

for some $0<c$, $C<\infty$.

Suppose $\hat{P}_n: \mathscr{X}^n \to \mathscr{P}$ is a consistent sequence of estimators of P in the sense that $H(\hat{P}_n, P)$ converges to zero in probability as $n \to \infty$. Here $H(P,Q)$ is the Hellinger distance between the probability measures P and Q as before. The consistency discussed above does not describe the asymptotic behavior of $H(\hat{P}_n^n, P^n)$ as $n \to \infty$. The sequence $\{\hat{P}_n\}$ is said to be *quickly consistent* at $P \in \mathscr{P}$ if for every $\varepsilon>0$, there is a $\delta(\varepsilon)>0$ such that

$$\overline{\lim_n}\, P\{H(\hat{P}_n^n, P^n) > 1-\delta(\varepsilon)\} < \varepsilon. \tag{3.2.41}$$

In other words, the Hellinger distance between P^n and $\hat{P}_n^n$ on $(\mathscr{X}^n, \mathscr{A}^n)$ stays bounded away from 1 as $n \to \infty$. Strasser (1981c) proved that a sequence $\{\hat{P}_n\}$ is quickly consistent at P if and only if for every $\varepsilon>0$, there is a $0<C(\varepsilon)<\infty$ such that

$$\sup_n P\{n^{1/2} H(\hat{P}_n, P) > C(\varepsilon)\} < \varepsilon. \tag{3.2.42}$$

This follows as an application of the inequality

$$1-e^{-nH^2(P,Q)} \leqslant [H(P^n, Q^n)]^2, \qquad n \geqslant 1. \tag{3.2.43}$$

He further proved that, for any sequence of estimators $\hat{P}_n: \mathscr{X}^n \to \mathscr{P}$, the following statements are equivalent.

(3.2.44) (i) There exists $a \geqslant 1$ and $0 < \varepsilon < \frac{1}{2}$ such that

$$\sup_n P\left\{n^{1/2} H(\hat{P}_n, P) > a\left(\log\frac{1}{\varepsilon}\right)^{1/2}\right\} < \varepsilon.$$

(ii) There exists $0 < C < \infty$, $h > 0$, such that for every $\varepsilon > 0$,

$$\overline{\lim_n}\, P\{H(\hat{P}_n^n, P^n) > 1 - \varepsilon\} < C\varepsilon^h.$$

Let us now recall the Definition 1.10.3 concerning finite dimensionality of a set.

Definition 3.2.3. A family $\mathscr{P}$ of probability measures on $(\mathscr{X}, \mathscr{A})$ is said to be of *finite dimension* if there exist constants $0 < C, D < \infty$ such that for every $P \in \mathscr{P}$ and $0 < r < R < 1$, the ball $B(P, R)$ can be covered by $C(R/r)^D$ balls $B(Q, r)$, $Q \in \mathscr{P}$.

Suppose $\mathscr{P} = \{P_\theta, \theta \in \Theta \subset R^k\}$ satisfies the following condition: for every $\theta \in \Theta$, there exist constants $0 < a$, $b < \infty$, $d > 0$, and $\alpha > 0$ such that

$$a|\sigma - \tau|^\alpha \leqslant H(P_\sigma, P_\tau) \leqslant b|\sigma - \tau|^\alpha.$$

Then the set $\{P_\sigma: |\sigma - \theta| < d\}$ is of finite dimension. Strasser (1981c) proved the following theorem concerning quick consistency of Pitman estimators. We omit the proof.

Theorem 3.2.9. Suppose $(\mathscr{P}, H)$ is a separable metric space. Let $\gamma(P)$ be a C-measurable function such that $P_n(\gamma\mu)$ is σ-finite for some $n \geqslant 1$. Suppose $P_n(\mu)$ is σ-finite for some $n \geqslant 1$ and the measure μ on $(\mathscr{P}, C)$ is diametrically regular at P and there is a finite-dimensional neighborhood of P. Furthermore, suppose that there is a neighborhood U_P of P and constants $0 < \alpha$, $K < \infty$ such that

$$|\gamma(Q_1) - \gamma(Q_2)| \leqslant K[H(Q_1, Q_2)]^\alpha \tag{3.2.45}$$

for all Q_1, Q_2 in U_P. Then there is a neighborhood V_P of P, $V_P \subset U_P$, and $0 < \varepsilon_0$, $C < \infty$ such that, for every $0 < \varepsilon < \varepsilon_0$, and $n \geqslant 1$,

$$\sup_{Q \in V_P} Q\left\{|T_n\gamma - \gamma(Q)|_1 > C\left(\log\frac{1}{\varepsilon}\right)^{\alpha/2} n^{-\alpha/2}\right\} < \varepsilon, \tag{3.2.46}$$

where $|a|_1 = \min\{1, |a|\}$ for any $a \in R$.

3.3 QUICK CONSISTENCY FOR QUASI-MAXIMUM LIKELIHOOD ESTIMATORS

We have discussed quick consistency of Pitman estimators, in particular Bayes estimators, in Section 3.2. The relation between the consistency of MLE and consistency of Bayes estimators will be discussed in Section 3.4. We now briefly study the concept of a quasi-maximum likelihood estimator (QMLE) and discuss quick consistency properties of such estimators. The results here are due to Pfaff (1982). Proofs are sketched.

Let $(\mathscr{X}, \mathscr{A})$ be a measurable space and $\mathscr{P}$ be a family of probability measures on $(\mathscr{X}, \mathscr{A})$. Suppose $\mathscr{P} \ll \mu$, μ σ-finite. Let

$$h_Q = \frac{dQ}{d\mu} \tag{3.3.0}$$

for $Q \in \mathscr{P}$. As before, let $H(P, Q)$ denote the Hellinger distance between P and Q and $\rho(P, Q)$ the affinity between P and Q. Recall that

$$H^2(P, Q) = 2[1 - \rho(P, Q)] = \int (h_Q^{1/2} - h_P^{1/2})^2 \, d\mu. \tag{3.3.1}$$

Assume that $(\mathscr{P}, \mathscr{H})$ is a separable metric space and $\mathscr{B}$ be the Borel σ-algebra over $\mathscr{P}$. Define

$$h_{n,Q}(\mathbf{x}) = \prod_{i=1}^{n} h_Q(x_i), \mathbf{x} \in \mathscr{X}^n. \tag{3.3.2}$$

$h_{n,Q}(\mathbf{x})$ is the likelihood function under Q. It can be shown [cf. Strasser (1981)] that, given any countable dense set S in $\mathscr{P}$, there exists a separable random function *equivalent* to $h_{n,Q}$ for any fixed $n \geq 1$. In other words, there exist functions $f_{n,Q}: \mathscr{X}^n \to [0, \infty)$ and $Q \in \mathscr{P}$ such that $f_{n,Q}(\mathbf{x})$ is $\mathscr{A}^n$-measurable for every n and Q,

$$f_{n,Q} = h_{n,Q} \quad \text{a.e.} \quad [\mu^n]$$

and for every $n \geq 1$, $Q \in \mathscr{P}$, and $\mathbf{x} \in \mathscr{X}^n$, there exists $Q_n \in S$ such that

$$\lim_n H(Q_n, Q) = 0 \quad \text{and} \quad \lim_n f_{n,Q_n}(\mathbf{x}) = f_{n,Q}(\mathbf{x}).$$

Hereafter we assume that $h_{n,Q}$, defined by (3.5.0), satisfies such conditions.

Definition 3.3.1. Let $0<\gamma\leqslant 1$ and $P\in\mathscr{P}$. A sequence of $(\mathscr{A}^n,\mathscr{B})$-measurable functions $\hat{P}_n:\mathscr{X}^n\to\mathscr{P}$, $n\geqslant 1$, is called a sequence of *quasi-maximum likelihood estimators* (QMLE) for P if

$$h_{n,\hat{P}_n}(\mathbf{x})\geqslant\gamma\sup_{Q\in\mathscr{P}}h_{n,Q}(\mathbf{x}) \tag{3.3.3}$$

or

$$\sup_{Q\in\mathscr{P}}h_{n,Q}(\mathbf{x})=\infty. \tag{3.3.4}$$

Remark. If $\gamma=1$, then these reduce to the maximum likelihood estimators.

Recall that a subset $M\subset T$ of a metric space (T,Δ) is called *finite dimensional* with dimension $D>0$ if there exists a constant $0<C<\infty$ such that for every $t_0\in M$ and $0<r<R<\infty$, there is a finite set of balls

$$B_\Delta(t_i,r)=\{t\in T:\Delta(t,t_i)<r\},$$

$1\leqslant i\leqslant m$ covering $B_\Delta(t_0,R)\cap M, t_i\in M$, and

$$m\leqslant C\left(\frac{R}{r}\right)^D. \tag{3.3.5}$$

Proposition 3.3.1. Let $P\in\mathscr{P}$ and suppose the family $\mathscr{P}$ satisfies the following conditions.

(B$_1$) There exists a compact subset $K\subset\mathscr{P}$ with $P\in K$ and $n_1\geqslant 1$ with $p>1$, $q>1$ with $1/p+1/q=1$ such that

$$\int\left\{\sup_{Q\in K^c}h_{n_1,Q}^{1/p}\right\}h_{n_1,P}^{1/q}\,d\mu^{n_1}<1. \tag{3.3.6}$$

(B$_2$) For the compact subset K given by (B_1), for every $Q_0\in K$, there exists a ball $B(Q_0,\delta)$ and $n_1\geqslant 1$ with $p>1$, $q>1$ with $1/p+1/q=1$ such that

$$\int\left\{\sup_{Q\in B(Q_0,\delta)}h_{n_1,Q}^{1/p}\right\}h_{n_1,P}^{1/q}\,d\mu^{n_1}\begin{cases}<1 & \text{if } Q_0\neq P\\ <\infty & \text{if } Q_0=P.\end{cases} \tag{3.3.7}$$

Let $\{\hat{P}_n\}$ be any sequence of QMLE. Then, for arbitrary $\varepsilon>0$, there exists $n_0\geqslant 1$ and $c>0$ such that, for every $n\geqslant n_0$,

(3.3.8) (i) $P\{H(\hat{P}_n, P) > \varepsilon\} \leqslant e^{-cn}$,

(ii) $P\left\{\sup_{Q \in \mathscr{P}} h_{n,Q} = \infty\right\} = 0.$

Proof. Let $\varepsilon > 0$. Let K be the compact set given by (B_1). It is sufficient to show that there exists $c > 0$ such that

(3.3.9) (i) $P\left\{\sup_{Q \in K^c} \frac{h_{n,Q}}{h_{n,P}} > \gamma\right\} \leqslant e^{-cn}$,

(ii) $P\left\{\sup_{Q \in K - B(Q_0, \delta)} \frac{h_{n,Q}}{h_{n,P}} > \gamma\right\} \leqslant e^{-cn}$,

for all large n and for every $Q_0 \in K - B(P, \varepsilon)$ and for some $\delta > 0$, which may depend on Q_0 [cf. Wald (1949)]. Let

$$g_k = \sup_{Q \in K^c} h_{n_1+k,Q}, \qquad k = 0, 1, \ldots, n_1 - 1, \tag{3.3.10}$$

where n_1 is as given by the condition (B_1). For $n \geqslant n_1$, write $n = mn_1 + k$, $0 \leqslant k < n_1$. Then

(3.3.11)

$$\sup_{Q \in K^c} h_{n,Q}(\mathbf{x}) \leqslant \left\{\prod_{r=0}^{m-2} g_0(x_{rn_1+1}, \ldots, x_{(r+1)n_1})\right\} g_k(x_{(m-1)n_1+1}, \ldots, x_n)$$

a.e. P^n. Furthermore,

$$\{g_{k+1}(x_1, \ldots, x_{n_1+k+1})\}^{n_1+k} \leqslant \prod_{j=1}^{n_1+k+1} \left[\sup_{Q \in K^c} \left\{\prod_{i=1}^{n_1+k+1} \frac{h_{1,Q}(x_i)}{h_{1,Q}(x_j)}\right\}\right] \tag{3.3.12}$$

a.e. P^{n_1+k+1} and hence

$$\int \left\{\frac{g_{k+1}}{h_{n_1+k+1,P}}\right\}^{(n_1+k)/p(n_1+k+1)} dP^{n_1+k+1} \tag{3.3.13}$$

$$\leqslant \int \left\{\frac{g_k}{h_{n_1+k,P}}\right\}^{1/p} dP^{n_1+k}, \qquad 0 \leqslant k \leqslant n_1 - 2.$$

But the condition (B_1) implies that

$$\alpha(k) = \int g_k^{1/p} h_{n_1+k,P}^{1/q}\, d\mu^{n_1+k} < 1, \qquad 0 \leqslant k \leqslant n_1 - 1, \tag{3.3.14}$$

for some $p > 1$, $q > 1$, $1/p + 1/q = 1$. Hence

$$\begin{aligned} P\left\{\sup_{Q \in K^c} h_{n,Q} > \gamma h_{n,P}\right\} &\leqslant P\left(\left[\left\{\prod_{r=0}^{m-2} g_0(X_{rn_1+1}, \ldots, X_{(r+1)n_1})\right\}\right.\right. \\ &\qquad \left.\left. \cdot g_k(X_{(m-1)n_1+1}, \ldots, X_n)\right] h_{n,P}(\mathbf{X})^{-1} > \gamma\right) \\ &\leqslant \gamma^{-1/p} \alpha(0)^{m-1} \alpha(k) \\ &\leqslant \gamma^{-1/p} \exp\left\{\frac{1}{2n} \log \alpha(0) n\right\} \end{aligned} \tag{3.3.15}$$

for sufficiently large n. This inequality holds with K^c replaced by $B(Q, \delta)$ by the condition (B_2) for every $Q \in K - B(P, \varepsilon)$. Again, conditions (B_1) and (B_2) and the inequality (3.3.12) imply that

$$P\left\{\sup_{Q \in \mathscr{P}} h_{n,Q} = \infty\right\} = 0$$

for all large n.

Theorem 3.3.2. Let $P \in \mathscr{P}$. Suppose the family $\mathscr{P}$ satisfies the following conditions (B_3) and (B_4) in addition to (B_1) and (B_2) stated above.

(B_3) There exists $\delta' > 0$ such that the ball $B(P, \delta')$ is finite dimensional with dimension $D > 0$.

(B_4) There are numbers $\delta'' > 0$, $s \geqslant 2$, $C < \infty$ such that $s > D$ [D as given by (B_3)] such that

$$\int |\log h_Q|\, dP \leqslant C \quad \text{for every} \quad Q \in B(P, \delta'') \tag{3.3.16}$$

and

$$\int |L_{Q_1} - L_{Q_2}|^s\, dP \leqslant CH(Q_1, Q_2)^s \tag{3.3.17}$$

for all $Q_1, Q_2 \in B(P, \delta'')$, where

$$L_Q = \log h_Q - \int \log h_Q \, dP, \qquad Q \in \mathscr{P}. \tag{3.3.18}$$

Let $\hat{P}_n$ be any sequence of QMLE for P. Then

(i) there exist constants $\alpha > 0$ and $\varepsilon_0 > 0$ such that

$$P\left\{ n^{1/2} H(\hat{P}_n, P) > a\left(\log\left(\frac{1}{\varepsilon}\right)\right)^{1/2} \right\} \leqslant \varepsilon \tag{3.3.19}$$

for every $n \geqslant 1$ and $0 < \varepsilon \leqslant \varepsilon_0$, and

(ii) there exist constants $h_0 > 0$ and $c > 0$ such that

$$P\{n^{1/2} H(\hat{P}_n, P) > h\} \leqslant e^{-ch^2} \tag{3.3.20}$$

for every $n \geqslant 1$ and $h \geqslant h_0$.

Proof. Without loss of generality, assume that $\delta = \delta' = \delta''$ in the conditions (B_2) to (B_4). Let $n \geqslant 1$ and $2 \leqslant h \leqslant n^{1/2}\delta - 1$. Then

$$\begin{aligned} P\{n^{1/2} H(\hat{P}_n, P) > h\} &\leqslant P\{H(\hat{P}_n, P) \geqslant \delta - n^{-1/2}\} \\ &+ \sum_{j=0}^{[n^{1/2}\delta - 1 - h]} P\{h + j \leqslant n^{1/2} H(\hat{P}_n, P) \leqslant h + j + 1\}. \end{aligned} \tag{3.3.21}$$

The first term on the right-hand side of the inequality in (3.3.21) can be estimated using Proposition 3.3.1. We now estimate the second term. Since $\hat{P}_n$ is a QMLE, it is sufficient to prove that there exist $c > 0$, $C > 0$ such that

$$P\left\{ \sup_{Q \in G(\lambda, n)} \frac{h_{n,Q}}{h_{n,P}} > \tfrac{1}{2}\gamma \right\} \leqslant C e^{-c\lambda^2} \tag{3.3.22}$$

for every $n \geqslant 1$ and $2 \leqslant \lambda \leqslant n^{1/2}\delta - 1$ where

$$G(\lambda, n) = \{Q \in \mathscr{P} : \lambda \leqslant n^{1/2} H(P, Q) < \lambda + 1\}. \tag{3.3.23}$$

Choose β with $0 < \beta < D^{-1}$ and let $\varepsilon = n^{-1/2} e^{-\beta\lambda^2}$. Note that $G(\lambda, n) \subset B(P, \delta) \cap B(P, n^{-1/2}(\lambda + 1))$. Since $B(P, \delta)$ is of dimension D by (B_3), it follows that $G(\lambda, n)$ can be covered by Balls $B(Q_i, \varepsilon/2)$, $1 \leqslant i \leqslant m$, with $Q_i \in B(P, \delta)$ and

$$m \leqslant Cn^{-D/2}\lambda^D \varepsilon^{-D}. \tag{3.3.24}$$

Without loss of generality, we assume that

$$B(Q_i, \varepsilon/2) \cap G(\lambda, n) \neq \phi, \qquad 1 \leqslant i \leqslant m.$$

Choose $P_i \in B(Q_i, \varepsilon/2) \cap G(\lambda, n)$ such that

$$\sup_{Q \in B(Q_i, \varepsilon/2) \cap G(\lambda, n)} \int \log h_Q \, dP < \int \log h_{P_i} \, dP + \left(\frac{1}{2n}\right) \log 2 \tag{3.3.25}$$

for $1 \leqslant i \leqslant m$. This is possible by (B_4) since supremum is finite. Observe that the sets $B_i = B(P_i, \varepsilon) \cap B(P, \delta)$, $1 \leqslant i \leqslant m$, cover $G(\lambda, n)$. Therefore

$$\begin{aligned} &P\left\{ \sup_{Q \in G(\lambda, n)} \frac{h_{n,Q}}{h_{n,P}} > \tfrac{1}{2}\gamma \right\} \\ &< \sum_{i=1}^{m} P\left\{ \frac{h_{n,P_i}^{1/2}}{h_{n,P}^{1/2}} \geqslant (\tfrac{1}{4}\gamma)^{1/2} \right\} + \sum_{i=1}^{m} P\left\{ \sup_{Q \in B_i} [L_{n,Q} - L_{n,P_i}] \geqslant \tfrac{1}{2} \log 2 \right\}, \end{aligned} \tag{3.3.26}$$

where $L_{n,Q}(\mathbf{x}) = \log h_{n,Q}(\mathbf{x}) - n \int \log h_Q \, dP$. It is clear that

$$\begin{aligned} P\left\{ \frac{h_{n,P_i}^{1/2}}{h_{n,P}^{1/2}} > \left(\frac{\gamma}{4}\right)^{1/2} \right\} &\leqslant \left(\frac{4}{\gamma}\right)^{1/2} \rho(P_i^n, P^n) \\ &\leqslant \left(\frac{4}{\gamma}\right)^{1/2} e^{-(n/2)H^2(P_i, P)} \\ &\leqslant \left(\frac{4}{\gamma}\right)^{1/2} e^{-\lambda^2/2}, \end{aligned} \tag{3.3.27}$$

where $\rho(P, Q)$ is the affinity between P and Q and $H(P, Q)$ is the Hellinger distance between P and Q. The inequalities given above follow from the following relations

(3.3.28) (i) $H^2(P, Q) = 2[1 - \rho(P, Q)]$,

(ii) $\rho(P, Q) = \int \frac{h_Q^{1/2}}{h_P^{1/2}} \, dP$,

(iii) $\rho(P^n, Q^n) = \rho^n(P, Q)$, and

(iv) $2[1 - H^2(P^n, Q^n)] \leqslant \exp((-n/2)H^2(P, Q))$.

Let us now consider the random field $\{L_{n,Q}; Q \in B(P,\delta)\}$. For a fixed n, $L_{n,Q}$ is uniformly P^n-continuous by (B_4). Hence we can and will assume that it is separable. Applying Proposition 1.10.22 with $(T,\rho)=(B(P,\delta),H)$, $X_Q = L_{n,Q}$, and the ball B_i, we have

$$E|L_{n,Q_1} - L_{n,Q_2}|^s \leqslant C n^{s/2} H^s(Q_1, Q_2)$$

for every $Q_1, Q_2 \in B(P,\delta)$. This follows from (B_4) by the inequality for absolute moments of orders for sums of i.i.d. random variables by Proposition 1.14.14 [see Dharmadhikari and Jogdeo (1969)]. Therefore

$$P\left\{\sup_{Q_1,Q_2 \in B_i} |L_{n,Q_2} - L_{n,Q_1}| > \tfrac{1}{2}\log 2\right\} \leqslant C n^{s/2} \varepsilon^s, \tag{3.3.29}$$

Combining inequalities (3.3.27) and (3.3.29) along with (3.3.24), we obtain inequality (3.3.22). Relations (3.3.21) and (3.3.22) and Proposition 3.3.1 prove that

$$\begin{aligned} P\{n^{1/2} H(\hat{P}_n, P) > h\} &\leqslant \sum_{j=0}^{\infty} C_1 e^{-c_1(h+j)^2} + e^{-c_2 n} \\ &\leqslant C_2 e^{-c_1 h^2} + e^{-c_2 n} \end{aligned}$$

for all $n \geqslant n_0$ (say) and $h \geqslant 2$. This proves (ii). (i) is clear from (ii).

Remarks. Pfaff (1982) discusses the rates of convergence for parametric families as a consequence of the result given above. For further discussion about conditions (B_1)–(B_4), see Pfaff (1982).

3.4 RELATION BETWEEN CONSISTENCY OF BAYES AND MAXIMUM LIKELIHOOD ESTIMATORS

In Section 3.1–3.3, we studied strong consistency of MCE and Bayes estimators and obtained exponential bounds on the rates of convergence of the Pitman estimators and quasi-maximum likelihood estimators.

We now show that, in general, consistency of MLE implies the consistency of the corresponding Bayes estimators for a large class of prior distributions. However, we should caution that there are situations in which MLE are not consistent but Bayes estimates are [cf. Schwarz (1965)].

Let $(\mathscr{X}, \mathscr{A})$ be a measurable space and $\mathscr{P} = \{P_\theta, \theta \in \Theta\}$ be a family of probability measures on $(\mathscr{X}, \mathscr{A})$. Suppose $\mathscr{P} \ll \mu$, μ σ-finite. Let $h_\theta = dP_\theta/d\mu$.

Let $\Delta(\sigma,\tau)=\|P_\sigma-P_\tau\|=\sup_{A\in\mathscr{A}}|P_\sigma(A)-P_\tau(A)|$ be the variational distance between P_σ and P_τ. Assume that the metric space (Θ,Δ) is separable. Let $\mathscr{B}$ denote the Borel σ-algebra of (Θ,Δ) and

$$f_\theta(x)\equiv f(x,\theta)=-\log h_\theta(x),\qquad x\in\mathscr{X},\quad \theta\in\Theta.$$

Define

$$f_n(\mathbf{x},\theta)=\frac{1}{n}\sum_{i=1}^{n} f(x_i,\theta),\qquad \mathbf{x}\in\mathscr{X}^n,\quad n\geqslant 1,\quad \theta\in\Theta.$$

Assume that for any fixed n, $f_n(\mathbf{x},\theta)$ is separable as a stochastic process in $\theta\in\Theta$ and is $\mathscr{A}^n\times\mathscr{B}$-measurable. Furthermore, we assume that

(A_1) $\{h_\theta,\theta\in\Theta\}$ are lower semicontinuous, that is,

$$\overline{\lim_{n\to\infty}}\, h_{\theta_n}\leqslant h_\theta \quad \text{a.e. } [\mu] \text{ if } \Delta(\theta_n,\theta)\to 0$$

and

(A_2) for every $\theta\in\Theta$ and $\sigma\in\Theta$, there is an open neighborhood $U_{\theta,\sigma}$ of σ such that

$$E_\theta\left[\inf_{\psi\in U_{\theta,\sigma}} f_n(\cdot,\psi)\right]>-\infty \tag{3.4.0}$$

for some $n\geqslant 1$. Let $\phi(x)=x/1+|x|$. Note that ϕ maps $\bar{R}=[-\infty,+\infty]$ onto $[-1,+1]$.

Definition 3.4.1. A sequence of $(\mathscr{A}^n,\mathscr{B})$-measurable estimators $T_n:\mathscr{X}^n\to\theta$ is called a sequence of *approximate maximum likelihood estimators* (AMLE) at $\theta\in\Theta$ if

$$\overline{\lim_{n\to\infty}}\left\{\phi\circ f_n(\mathbf{x},T_n)-\inf_{\sigma\in\theta}\phi\circ f_n(\mathbf{x},\sigma)\right\}\leqslant 0,\quad P_\theta^\infty-\text{a.e.} \tag{3.4.1}$$

Remark. Observe that AMLE always exists since Θ is separable.

Proposition 3.4.1. Suppose $\theta\in M\subset\Theta$ and M is closed. Then the following statements are equivalent:

(i) every sequence of AMLE $\{T_n\}$ at θ satisfies

$$P_\theta^\infty\left\{\underline{\lim_{n\to\infty}}\,[T_n\in M]\right\}=1; \tag{3.4.2}$$

(ii) there exists on open set $U \supset M$ such that

$$\lim_{n\to\infty}\left\{\inf_{\sigma\in M^c}\phi\circ f_n(\cdot,\sigma)-\inf_{\sigma\in U}\phi\circ f_n(\cdot,\sigma)\right\}>0,\quad P_\theta^\infty-\text{a.e.} \tag{3.4.3}$$

(iii) there exists a sequence of open sets $U_i \supset M$, $i \geqslant 1$, such that

$$\lim_{i}\lim_{n\to\infty}\left\{\inf_{\sigma\in M^c}\phi\circ f_n(\cdot,\sigma)-\inf_{\sigma\in U_i}\phi\circ f_n(\cdot,\sigma)\right\}>0,\quad P_\theta^\infty-\text{a.e.} \tag{3.4.4}$$

Proof. It is obvious that (iii)$\Rightarrow$(ii). (ii)$\Rightarrow$(i) from the definition of $\{T_n\}$ [cf. Wald (1949)]. We now prove that (i)$\Rightarrow$(iii).

Suppose (iii) does not hold for a sequence $\{U_i\}$ of open sets with $U_i \supset M$. Let

$$A=\left[\inf_{i}\lim_{n\to\infty}\left\{\inf_{\sigma\in M^c}\phi\circ f_n(\cdot,\sigma)-\inf_{\sigma\in U_i}\phi\circ f_n(\cdot,\sigma)\right\}\leqslant 0\right]$$

and

$$A_{n,k}=\bigcup_{i=1}^{\infty}\left[\left\{\inf_{\sigma\in M^c}\phi\circ f_n(\cdot,\sigma)-\inf_{\sigma\in U_i}\phi\circ f_n(\cdot,\sigma)\right\}<\frac{1}{k}\right]$$

for $n \geqslant 1$ and $k \geqslant 1$. Observe that $P_\theta^\infty(A)=\alpha>0$ and

$$A\subset\bigcap_{m=1}^{\infty}\bigcup_{n=m}^{\infty}A_{nk}$$

for every $k \geqslant 1$. Hence there exists a sequence $n_k \uparrow \infty$ such that

$$P_\theta^\infty\left[\bigcup_{n=n_k}^{n_{k+1}-1}(A_{nk}\cap A)\right]\geqslant\alpha\left(1-\frac{1}{2^{k+1}}\right),\qquad k\geqslant 1,$$

and therefore

$$P_\theta^\infty\left[\bigcap_{k=1}^{\infty}\bigcup_{n=n_k}^{n_{k+1}-1}(A_{nk}\cap A)\right]\geqslant\frac{\alpha}{2}>0. \tag{3.4.5}$$

For any $n \geqslant 1$, choose $k_n \geqslant 1$ such that $n_{k_n} \leqslant n < n_{k_n+1}$. Clearly $A_{n,k_n} \in \mathscr{A}^n$ and, for any $\mathbf{x} \in A_{n,k_n}$, choose $T_n(\mathbf{x}) \in M^c$ such that

$$\phi\circ f_n(\mathbf{x},T_n(\mathbf{x}))-\inf_{\sigma\in M^c}\phi\circ f_n(\mathbf{x},\sigma)\leqslant\frac{1}{k_n}$$

and, for $\mathbf{x} \notin A_{n,k_n}$, choose $T_n(\mathbf{x}) \in \Theta$ such that

$$\phi \circ f_n(\mathbf{x}, T_n(\mathbf{x})) - \inf_{\sigma \in \Theta} \phi \circ f_n(\mathbf{x}, \sigma) \leqslant \frac{1}{n}.$$

Since (Θ, Δ) is separable, this can be done so that T_n is $(\mathscr{A}^n, \mathscr{B})$-measurable. Furthermore, T_n is an AML estimator at θ, since for $\mathbf{x} \in A_{n,k_n}$,

$$\begin{aligned}
\phi \circ f_n(\mathbf{x}, T_n(\mathbf{x})) &\leqslant \inf_{\sigma \in M^c} \phi \circ f_n(\mathbf{x}, \sigma) + \frac{1}{k_n} \\
&\leqslant \min\left[\inf_{\sigma \in M^c} \phi \circ f(\mathbf{x}, \sigma), \sup_i \inf_{\sigma \in U_i} \phi \circ f_n(\mathbf{x}, \sigma)\right] + 2/k_n \\
&\leqslant \inf_{\sigma \in \Theta} \phi \circ f(\mathbf{x}, \sigma) + \frac{2}{k_n}
\end{aligned}$$

and clearly this relation holds for $\mathbf{x} \notin A_{n,k_n}$ by the definition of $T_n(\mathbf{x})$. Hence $\{T_n\}$ is a sequence of AMLE. However

$$A_{n,k_n} \subset [T_n \notin M] \tag{3.4.6}$$

for every $n \geqslant 1$ and

$$\begin{aligned}
\bigcap_{m=1}^{\infty} \bigcup_{n=m}^{\infty} A_{n,k_n} &= \bigcap_{k=1}^{\infty} \bigcup_{n=n_k}^{\infty} A_{n,k_n} \\
&\supset \bigcap_{k=1}^{\infty} \bigcup_{n=n_k}^{n_{k+1}} A_{n,k_n} \\
&= \bigcap_{k=1}^{\infty} \bigcup_{n=n_k}^{n_{k+1}-1} A_{n,k}.
\end{aligned} \tag{3.4.7}$$

But

$$P_\theta^\infty\left(\bigcap_{k=1}^{\infty} \bigcup_{n=n_k}^{n_{k+1}-1} A_{n,k}\right) \geqslant \frac{\alpha}{2} > 0 \tag{3.4.8}$$

from (3.4.5). Relations (3.4.6)–(3.4.8) prove that (3.4.2) does not hold for the AMLE $\{T_n\}$. This is a contradiction. Hence (i) $\Rightarrow$ (iii).

Proposition 3.4.2. Suppose conditions (A_1) and (A_2) hold. Let $\theta \in M \subset \Theta$ where M is compact. Then the following statements are equivalent:

(i) Every sequence of AMLE $\{T_n\}$ at θ satisfies

$$P_\theta^\infty\left[\varliminf_{n\to\infty}[T_n\in M]\right]=1; \tag{3.4.9}$$

$$\text{(ii)}\quad \liminf_{n\to\infty}\ \inf_{\sigma\in M^c} f_n(\cdot,\sigma) > E_\theta f_\theta, \quad P_\theta^\infty - \text{a.e.} \tag{3.4.10}$$

Proof. Let $U \subset \Theta$ be open such that, for some $n_0 \geqslant 1$,

$$E_\theta\left(\inf_{\sigma\in U} f_{n_0}(\cdot,\sigma)\right) > -\infty.$$

Let $\mathscr{F}_n$ denote the sub-σ-algebra of $\mathscr{A}^n$ generated by symmetric sets. The sequence $\{Z_n, \mathscr{F}_n\}$, where $Z_n = \inf_{\sigma\in U} f_n(\cdot,\sigma)$, is a reversed supermartingale in the sense that

$$E[Z_n|\mathscr{F}_m] \geqslant Z_m, \quad \text{a.e.} \tag{3.4.11}$$

for all $m \geqslant n$. By the martingale convergence theorem and the Hewitt–Savage zero–one law (Hall and Heyde, 1980; Loeve, 1963) it follows that

$$\liminf_{n\to\infty}\ \inf_{\sigma\in U} f_n(\cdot,\sigma) = \sup_n E_\theta\left(\inf_{\sigma\in U} f_n(\cdot,\sigma)\right), \quad P_\theta^\infty - \text{a.e.} \tag{3.4.12}$$

Now, let us choose and fix $\tau\in\Theta$. Let $\{U_k\}$ be a sequence of open neighborhoods of τ such that $U_k \downarrow \{\tau\}$, $U_k \subset U$, U open in Θ, and $k \geqslant 1$. Since $f_n(\cdot,\sigma)$ is lower semicontinuous by condition (A_1), for any fixed n, it follows that

$$\lim_{k\to\infty}\ \inf_{\sigma\in U_k} f_n(\cdot,\sigma) = f_n(\cdot,\tau).$$

Hence, by the monotone convergence theorem,

$$\lim_{k\to\infty} E_\theta\left(\inf_{\sigma\in U_k} f_n(\cdot,\sigma)\right) = E_\theta(f_n(\cdot,\tau)) = E_\theta(f(\cdot,\tau)) \tag{3.4.13}$$

[cf. Pfanzagl (1969a) and Michel and Pfanzagl (1971)]. Therefore, for every compact $M \subset \Theta$ and for every $\varepsilon > 0$, there is an open set $U_\varepsilon \subset M$ such that

$$\begin{aligned} -\infty < \sup_n E_\theta\left(\inf_{\sigma\in U_\varepsilon} f_n(\cdot,\sigma)\right) &\leqslant \inf_{\sigma\in M} E_\theta(f(\cdot,\sigma)) \\ &\leqslant \sup_n E_\theta\left(\inf_{\sigma\in U_\varepsilon} f_n(\cdot,\sigma)\right) + \varepsilon. \end{aligned}$$

The result now follows from Proposition 3.4.1.

We shall now show that statement (ii) in Proposition 3.4.2 implies the consistency of Bayes estimates. Hence, if AMLE is consistent, it implies that Bayes estimators are also consistent.

Let $\lambda(\cdot)$ be a probability measure on $(\Theta, \mathscr{B})$. Suppose the following conditions are satisfied by $\lambda(\cdot)$.

(A_3) $\lambda\{\sigma: E_\theta(f_\sigma) < E_\theta(f_\theta) + \varepsilon\} > 0$ for all $\theta \in \Theta$ and $\varepsilon > 0$.

(A_4) For every $\theta \in \Theta$, there exists $n_\theta \geqslant 1$ such that

$$P_\theta^n\left\{\int_\Theta \prod_{i=1}^n h_\sigma(X_i)\lambda(d\sigma) < \infty\right\} = I \quad \text{for } n \geqslant n_\theta.$$

Define

$$(3.4.14) \qquad F_{n,\mathbf{x}}(B) = \begin{cases} \dfrac{\int_B \prod_{i=1}^n h_\sigma(x_i)\lambda(d\sigma)}{\int_\Theta \prod_{i=1}^n h_\sigma(x_i)\lambda(d\sigma)} & \text{if } \int_\Theta \prod_{i=1}^n h_\sigma(x_i)\lambda(d\sigma) > 0 \\ 0 & \text{otherwise} \end{cases}$$

for $n \geqslant n_\theta$, $\mathbf{x} \in \mathscr{X}^n$, and $B \in \mathscr{B}$. $F_{n,\mathbf{x}}(\cdot)$ is the *posteriori distribution* of θ given $\mathbf{x}$.

Theorem 3.4.3. Suppose that regularity conditions, (A_1)–(A_4) hold. Let $\theta \in M \subset \Theta$, where M is compact. If every sequence of AMLE $\{T_n\}$ satisfies

$$(3.4.15) \qquad P_\theta^\infty\left\{\varliminf_{n\to\infty} [T_n \in M]\right\} = 1,$$

then

$$(3.4.16) \qquad P_\theta^\infty\left\{\varliminf_{n\to\infty} F_{n,x}(M) = 1\right\} = 1.$$

Proof. For any set $A \in \mathscr{A}$ with $P(A) > 0$ and an $\mathscr{A}$-measurable function g,

$$\int I_A g\, dP \leqslant \left\{\int g^n\, dP\right\}^{1/n} \left\{\int I_A\, dP\right\}^{1-1/n}$$

by the Holder inequality, and hence

$$\frac{1}{n}\log \int g^n\, dP \geqslant \log \frac{1}{P(A)}\int_A g\, dP + \frac{1}{n}\log P(A).$$

Let $\varepsilon > 0$ and $M_\varepsilon = \{\sigma \in \Theta : E_\theta f_\sigma < E_\theta f_\theta + \varepsilon\}$. Let $g = \exp[-f_n(\mathbf{x}, \sigma)]$ and $P(\cdot) = \lambda(\cdot)$. Applying the inequality given above, we have

$$(3.4.17)\quad \underline{\lim}_{n\to\infty} \frac{1}{n} \log \int_\Theta \exp(-nf_n(\mathbf{x},\sigma))\,\lambda(d\sigma)$$
$$\geqslant \underline{\lim}_{n\to\infty} \log \frac{1}{\lambda(M_\varepsilon)} \int_{M_\varepsilon} \exp(-nf_n(\mathbf{x},\sigma))\,\lambda(d\sigma) + \lim_{n\to\infty} \frac{1}{n} \log \lambda(M_\varepsilon)$$
$$\geqslant \log \frac{1}{\lambda(M_\varepsilon)} \int_{M_\varepsilon} \exp(-E_\theta(f_\sigma))\,\lambda(d\sigma) \quad \text{(by Fatou's lemma)}$$
$$\geqslant -E_\theta(f_\theta) - \varepsilon, \quad P_\theta^\infty - \text{a.e.}$$

Hence

$$(3.4.18)\qquad \int_\Theta \exp(-nf_n(\mathbf{x},\sigma))\,\lambda(d\sigma) > 0$$

for large $n \geqslant n_0$, P_θ^∞ – a.e. Therefore

$$(3.4.19)\quad P_\theta^\infty \left\{ \overline{\lim}_{n\to\infty} F_{n,x}(M^c) \neq \lim_{n\to\infty} \frac{\int_{M^c} \exp(-nf_n(\mathbf{x},\sigma))\lambda(d\sigma)}{\int_\Theta \exp(-nf_n(\mathbf{x},\sigma))\lambda(d\sigma)} \right\} = 0.$$

Now

$$(3.4.20)\quad \overline{\lim}_{n\to\infty} \frac{1}{n} \log \int_{M^c} \exp(-nf_n(\mathbf{x},\sigma))\,\lambda(d\sigma)$$
$$= \overline{\lim}_{n\to\infty} \frac{1}{n} \log \int_{M^c} \exp\left(-\sum_{i=n_0+1}^{n} f(x_i,\sigma)\right) \exp\left(-\sum_{i=1}^{n_0} f(x_i,\sigma)\right) \lambda(d\sigma)$$
$$\leqslant \overline{\lim}_{n\to\infty} \left(-\frac{1}{n} \inf_{\sigma\in M^c} \sum_{i=n_0+1}^{n} f(x_i,\sigma)\right) + \overline{\lim}_{n\to\infty} \frac{1}{n} \log \int \prod_{i=1}^{n_0} h(x_i,\sigma)\,\lambda(d\sigma)$$
$$= -\overline{\lim}_{n\to\infty} \inf_{\sigma\in M^c} f_n(\mathbf{x}_{n_0},\sigma), \quad P_\theta^\infty - \text{a.e.},$$

where

$$\mathbf{x}_{n_0} = (x_{n_0+1}, x_{n_0+2}, \ldots, x_n).$$

Therefore

$$(3.4.21)\quad \overline{\lim}_{n\to\infty} \log F_{n,x}(\mathbf{M}^c) \leqslant -\overline{\lim}_{n\to\infty} \inf_{\sigma\in M^c} f_n(\mathbf{x}_{n_0},\sigma) + E_\theta(f_\theta) + \varepsilon, \quad P_\theta^\infty - \text{a.e.}$$

Let

(3.4.22) $$A_r = \left[\overline{\lim_{n\to\infty}} \inf_{\sigma \in M^c} f_n(\mathbf{x}_{n_0}, \sigma) \geqslant E_\theta(f_\theta) + \frac{1}{r} \right]$$

for $r \geqslant 1$. Since (3.4.15) holds for every AMLE sequence $\{T_n\}$ by hypothesis, it follows that

(3.4.23) $$P_\theta^\infty \left(\bigcup_{r=1}^{\infty} A_r \right) = 1$$

by Proposition 3.4.2. Note that

(3.4.24) $$P_\theta^\infty \left\{ \bigcup_{r=1}^{\infty} \left[\overline{\lim_n} \frac{1}{n} \log F_{n,\mathbf{x}}(M^c) \leqslant -\frac{1}{2r} \right] \right\}$$
$$= P_\theta^\infty \left\{ \bigcup_{r=1}^{\infty} \left[\overline{\lim_n} \frac{1}{n} \log F_{n,\mathbf{x}}(M^c) \leqslant -\frac{1}{r} + \frac{1}{2r} \right\} \right.$$
$$\geqslant P_\theta^\infty \left\{ \bigcup_{r=1}^{\infty} \left[-\overline{\lim_n} \inf_{\sigma \in M^c} f_n(\mathbf{x}_{n_0}, \sigma) + E_\theta(f_\theta) + \frac{1}{2r} \leqslant -\frac{1}{r} + \frac{1}{2r} \right] \right\}$$
$$\geqslant P_\theta^\infty \left\{ \bigcup_{r=1}^{\infty} A_r \right\} = 1.$$

This completes the proof of the theorem.

Remarks. Suppose (Θ, d) is a locally compact metric space and every sequence of AMLE is strongly consistent. Theorem 3.4.1 implies that relation (3.4.16) holds for every compact neighborhood M_θ of θ and hence for every neighborhood. Hence the posterior distribution is concentrated around θ a.s. under the conditions (A_1)–(A_4), which in turn implies that Bayes estimators are consistent. Results in this section are due to Strassen (1981a).

3.5 MAXIMUM LIKELIHOOD ESTIMATION

Let Θ be a measurable subset of R^k and P_θ be a probability measure on $(\Omega, \mathscr{B})$ for each $\theta \in \Theta$. Let $\rho(\theta, \phi)$ be the affinity

$$\int \left\{ \frac{dP_\theta}{d\mu} \frac{dP_\phi}{d\mu} \right\}^{1/2} d\mu,$$

where μ dominates P_θ and P_ϕ as defined in Section 1.3. It is easy to check that $\rho(\theta, \phi)$ is a covariance kernel on Θ. One can define a stochastic process $\{\xi(\theta), \theta \in \Theta\}$ such that $E\xi(\theta)\xi(\phi) = \rho(\theta, \phi)$. Then $E|\xi(\theta) - \xi(\phi)|^2 = H^2(\theta, \phi)$, where we write $H(\theta, \phi)$ for the Hellinger distance $H(P_\theta, P_\phi)$ as defined in Section 1.3. For instance, $\xi(\theta)$ could be chosen to be a Gaussian process with mean zero and covariance kernel $\rho(\theta, \phi)$. Let $\| \xi(\theta) - \xi(\phi) \|^2 = E|\xi(\theta) - \xi(\phi)|^2$. The following theorem gives sufficient conditions for the process $\xi(\theta)$, $\theta \in \Theta$, to be differentiable in quadratic mean.

Theorem 3.5.1. Let S be a measurable set of points $\theta \in \Theta$ such that

$$\overline{\lim_{|\tau| \to 0}} |\tau|^{-1} \| \xi(\theta + \tau) - \xi(\theta) \| < \infty. \tag{3.5.1}$$

Then, for a.e. every $\theta \in S$, $\xi(\theta)$ is differentiable in quadratic mean, that is,

$$|\tau|^{-1} |\xi(\theta + \tau) - \xi(\theta)| \xrightarrow{\text{q.m.}} \xi'(\theta) \quad \text{as } |\tau| \to 0 \tag{3.5.2}$$

for some process $\xi'(\theta)$ (here "a.e." is with respect to Lebesgue measure on R^k).

For a sketch of the proof, see Le Cam (1970).

Let us now suppose that $P_\theta \ll \mu$, μ σ-finite, such that

$$\mu(A) = 0 \Leftrightarrow \sup_\theta P_\theta(A) = 0 \tag{3.5.3}$$

Define

$$\xi(\theta) = \left(\frac{dP_\theta}{d\mu} \right)^{1/2}. \tag{3.5.4}$$

Then $E|\xi(\theta)|^2 = 1$ and $E|\xi(\theta) - \xi(\phi)|^2 = H^2(\theta, \phi)$. Here $E(\cdot)$ denotes expectation taken with respect to a measure μ. Define

$$X_\theta(\phi) = \left(\frac{dP_\phi}{dP_\theta} \right)^{1/2} - 1 \tag{3.5.5}$$

and

$$Y_\theta(\phi) = X_\theta(\phi) - E_\theta X_\theta(\phi), \tag{3.5.6}$$

where E_θ denotes the expectation under the measure P_θ. Let J_θ be the indicator function of the set $\{\omega:\xi(\theta,\omega)\neq 0\}$. Then

$$(3.5.7)\quad \rho(\theta_1,\theta_2)=\int_\Omega\left\{\frac{dP_{\theta_1}}{d\mu}\frac{dP_{\theta_2}}{d\mu}\right\}^{1/2}d\mu$$

$$=\int_\Omega J_\theta\left\{\frac{dP_{\theta_1}}{d\mu}\frac{dP_{\theta_2}}{d\mu}\right\}^{1/2}d\mu+\int_\Omega(1-J_\theta)\left\{\frac{dP_{\theta_1}}{d\mu}\frac{dP_{\theta_2}}{d\mu}\right\}^{1/2}d\mu$$

$$=E[1+X_\theta(\theta_1)][1+X_\theta(\theta_2)]+\beta_\theta(\theta_1,\theta_2),$$

where $\beta_\theta(\theta_1,\theta_2)$ denotes the second integral. Note that

$$(3.5.8)\qquad E_\theta X_\theta(\phi)=\rho(\phi,\theta)-1=-\tfrac{1}{2}H^2(\theta,\phi)$$

and

$$(3.5.9)\qquad \operatorname{cov}_\theta[X_\theta(\phi_1),X_\theta(\phi_2)]=E_\theta[Y_\theta(\phi_1)Y_\theta(\phi_2)]$$

$$=\rho(\phi_1,\phi_2)-\rho(\phi_1,\theta)\rho(\phi_2,\theta)-\beta_\theta(\phi_1,\phi_2).$$

Suppose the process $X_\theta(\phi)$ is differentiable in quadratic mean at $\phi=\theta$ with derivative $\nabla X_\theta(\theta)$. It is easy to check that

$$(3.5.10)\qquad E_\theta[\phi'\nabla X_\theta(\theta)]=-\tfrac{1}{2}\lim_{\varepsilon\to 0}\varepsilon^{-1}H^2(\theta,\theta+\varepsilon\phi)$$

and

$$(3.5.11)\qquad E_\theta[\phi_1'\nabla X_\theta(\theta)\nabla X_\theta(\theta)\phi_2]=\lim_{\varepsilon\to 0}\varepsilon^{-2}E_\theta[X_\theta(\theta+\varepsilon\phi_1)X_\theta(\theta+\varepsilon\phi_2)]$$

for all ϕ_1,ϕ_2 such that $\theta+\varepsilon\phi_i\in\theta$, $i=1,2$.

Proposition 3.5.2. Suppose the process $\xi(\phi)$, $\phi\in\Theta$, defined by (3.5.4) is differentiable in quadratic mean at $\phi=\theta$ with derivative $\nabla\xi(\theta)$ and $\Gamma(\theta)=E_\theta[\nabla\xi(\theta)\nabla\xi(\theta)']$. Then

(3.5.12) (i) $X_\theta(\phi)$ and $Y_\theta(\phi)$ are differentiable in quadratic mean at $\phi=\theta$,

(ii) $\nabla X_\theta(\theta)=\nabla Y_\theta(\theta)$, and

(iii) $E_\theta\nabla X_\theta(\theta)=0$, $E_\theta\nabla X_\theta(\theta)\nabla X_\theta(\theta)'=\Gamma(\theta)$.

Proof. Since $Y_\theta(\phi)=X_\theta(\phi)+\tfrac{1}{2}H^2(\theta,\phi)$, parts (i) and (ii) will be proved if we show that $X_\theta(\phi)$ is differentiable in quadratic mean at $\phi=\theta$. It is also clear from (3.5.10) that $E_\theta \nabla X_\theta(\theta) = \mathbf{0}$ when it exists. Define $J_\theta = \{\omega : \xi(\theta, \omega) \neq 0\}$ as before. Then

$$\xi(\mathbf{t}) - \xi(\theta) - (\mathbf{t} - \theta)' \nabla \xi(\theta) = J_\theta[\xi(\mathbf{t}) - \xi(\theta) - (\mathbf{t} - \theta)' \nabla \xi(\theta)] \\ + (1 - J_\theta)[\xi(\mathbf{t}) - (\mathbf{t} - \theta)' \nabla \xi(\theta)]$$

and hence, by hypothesis,

$$\| \mathbf{t} - \theta \|^{-1} \| J_\theta[\xi(\mathbf{t}) - \xi(\theta) - (\mathbf{t} - \theta)' \nabla \xi(\theta)] \| \to 0.$$

But

$$\| \mathbf{t} - \theta \|^{-2} \int_\Omega J_\theta \left| \frac{\xi(\mathbf{t})}{\xi(\theta)} - 1 - (\mathbf{t} - \theta)' \frac{\nabla \xi(\theta)}{\xi(\theta)} \right|^2 \xi^2(\theta) d\mu$$
$$= \| \mathbf{t} - \theta \|^{-2} E_\theta | X_\theta(\mathbf{t}) - (\mathbf{t} - \theta)' \nabla X_\theta(\theta) |^2$$

since $\nabla X_\theta(\theta) = \nabla \xi(\theta)/\xi(\theta)$ whenever $\xi(\theta) \neq 0$. Hence $X_\theta(\phi)$ is differentiable in quadratic mean at θ. In fact

$$\nabla \xi(\theta) = \xi(\theta) \nabla X_\theta(\theta) + (1 - J(\theta)) \xi(\theta)$$

and (iii) follows from the fact that $\{P_\theta, \theta \in \Theta\}$ is equivalent to μ.

Proposition 3.5.3. Suppose the process ξ defined by (3.5.4) is differentiable in quadratic mean at θ. Let $\{X_j, j \geqslant 1\}$ be i.i.d. as the process $X \equiv X_\theta$ induced by P_θ as given by (3.5.5). Let ∇X_j be the derivative of X_j at θ in quadratic mean and define

$$V_n = n^{-1/2} \sum_{j=1}^{n} \nabla X_j(\theta). \tag{3.5.13}$$

If $\| \tau_n \|$, $n \geqslant 1$, is bounded in R^k, then

$$\sum_{j=1}^{n} X_j(\theta + \tau_n n^{-1/2}) - \tau_n' V_n + \tfrac{1}{2} \tau_n' \Gamma(\theta) \tau_n \xrightarrow{\text{q.m.}} 0 \quad \text{as } n \to \infty \tag{3.5.14}$$

where $\Gamma(\theta) = E[\nabla \xi(\theta) \nabla \xi(\theta)']$.

Proof. Let $Y_j(\phi) = X_j(\phi) - E_\theta X_j(\phi)$. Proposition 3.5.2 implies that Y_j is differentiable in quadratic mean at $\phi = \theta$ and the derivative is $\nabla X_j(\theta)$. Note that $X_j(\theta) = 0$ by the definition of X_θ. Hence

$$Y_j(\theta + \mathbf{t}) = \mathbf{t}' \nabla X_j(\theta) + \| \mathbf{t} \| R_j(\mathbf{t}),$$

where $ER_j(\mathbf{t})=0$ and $ER_j^2(\mathbf{t})\to 0$ as $\|\mathbf{t}\|\to 0$. Therefore

$$\sum_{j=1}^{n} Y_j(\theta+\tau n^{-1/2})=\tau' V_n+\|\tau\| n^{-1/2}\sum_{j=1}^{n} R_j(\tau n^{-1/2}). \tag{3.5.15}$$

Note that $E_\theta[\sum_{j=1}^n R_j(\tau n^{-1/2})]=0$ and

$$\operatorname{Var}_\theta\left[\|\tau\| n^{-1/2}\sum_{j=1}^{n} R_j(\tau n^{-1/2})\right]=\|\tau\|^2 E_\theta[R_1^2(\tau n^{-1/2})]$$

and the last term tends to zero since τ is bounded. On the other hand,

$$\begin{aligned}\sum_{j=1}^{n} E_\theta X_j(\theta+\tau n^{-1/2}) &= -\frac{n}{2}H^2(\theta,\theta+\tau n^{-1/2}) \\ &= -\frac{n}{2}E|\xi(\theta+\tau n^{-1/2})-\xi(\theta)|^2,\end{aligned} \tag{3.5.16}$$

which tends to $-\frac{1}{2}E_\theta|\tau'\nabla\xi(\theta)|^2$ since ξ is differentiable in quadratic mean. This completes the proof of the lemma in view of (3.5.15).

As a consequence of Proposition 3.5.3, we have the following result.

Proposition 3.5.4. Suppose the conditions stated in Proposition 3.5.3 are satisfied. Let

$$Z_n(\tau)=\sum_{j=1}^{n} X_j(\theta+\tau n^{-1/2}). \tag{3.5.17}$$

Then the finite-dimensional distributions of the process Z_n converge weakly to the corresponding finite-dimensional distributions of a Gaussian process Z with mean $E_\theta[Z(\tau)]=-\frac{1}{2}\tau'\Gamma(\theta)\tau$ and covariance kernel $k(\mathbf{s},\mathbf{t})=\operatorname{cov}_\theta(Z(\mathbf{s}),Z(\mathbf{t}))=\mathbf{s}'\Gamma(\theta)\mathbf{t}$. Let

$$\Lambda_n(\theta,\mathbf{t})=\log\frac{dP_{\mathbf{t},n}}{dP_{\theta,n}}$$

be the logarithm of the likelihood ratio, where $P_{\mathbf{t},n}$ denotes the probability measure corresponding to n i.i.d. observations distributed as $P_{\mathbf{t}}$. Note that

$$\Lambda_n(\theta,\mathbf{t})=2\sum_{j=1}^{n}\log[1+X_j(\mathbf{t})], \tag{3.5.18}$$

where X_j is as defined in Proposition 3.5.3.

Proposition 3.5.5. Suppose ξ is differentiable in quadratic mean at $t=\theta$. Then, for every bounded sequence τ_n, $n \geqslant 1$, in R^k,

$$\tfrac{1}{2}\Lambda_n(\theta,\theta+\tau_n n^{-1/2}) - \sum_{j=1}^{n} X_j(\theta+\tau_n n^{-1/2}) + \tfrac{1}{2}\tau_n'\Gamma(\theta)\tau_n \to 0 \tag{3.5.19}$$

in $P_{\theta,n}$-probability as $n\to\infty$. In fact

$$\Lambda_n(\theta,\theta+\tau n^{-1/2}) \xrightarrow{\mathscr{L}} N(-2\tau'\Gamma(\theta)\tau,\, 4\tau'\Gamma(\theta)\tau). \tag{3.5.20}$$

Proof. Note that X is differentiable in quadratic mean since ξ is differentiable in quadratic mean by Proposition 3.5.2. In view of the representation (3.5.18) and applying Taylor expansion, we obtain (3.5.19) as in (3.5.14). Relation (3.5.20) is an immediate consequences of Proposition 3.5.4.

Let us again suppose that $\{P_\theta, \theta\in\Theta\}$ is a family of probability measures on a measurable space $(\Omega,\mathscr{B})$ and assume that $\{P_\theta,\theta\in\Theta\}$ is dominated by a σ-finite measure μ on $(\Omega,\mathscr{B})$. Let the density of P_θ with respect to μ be denoted by $f(x,\theta)$ and define $\Phi(x,\theta)=\log f(x,\theta)$. Suppose the following regularity conditions are satisfied.

(A) There is a vector-valued function $\boldsymbol{\phi}(x,\theta)$ and matrix-valued functions $\mathbf{B}(x,\theta)$ and $\mathbf{B}(x,\theta,\mathbf{t})$ such that

- **(i)** $\Phi(x,\theta+\mathbf{t})-\Phi(x,\theta)=\mathbf{t}'\boldsymbol{\phi}(x,\theta)-\tfrac{1}{2}\mathbf{t}'\mathbf{B}(x,\theta,\mathbf{t})\mathbf{t}$,
- **(ii)** $E_\theta\boldsymbol{\phi}(X,\theta)=0$, $\quad E_\theta\boldsymbol{\phi}(X,\theta)\boldsymbol{\phi}(X,\theta)'=\mathbf{M}(\theta)$,
- **(iii)** $\lim_{\varepsilon\to0}E_\theta\{\sup_{\|\mathbf{t}\|<\varepsilon}\|\mathbf{B}(x,\theta,\mathbf{t})-\mathbf{B}(x,\theta)\|\}=0$, and
- **(iv)** $\mathbf{C}(\theta)=E_\theta\mathbf{B}(X,\theta)$ exists.

Proposition 3.5.6. Let $X_i, i\geqslant 1$, be i.i.d. according to the probability measure P_θ. Define

$$\boldsymbol{\Delta}_n = n^{-1/2}\sum_{j=1}^{n}\boldsymbol{\phi}(X_j,\theta), \tag{3.5.21}$$

$$\Lambda_n(\theta,t)=\sum_{j=1}^{n}[\Phi(X_j,\mathbf{t})-\Phi(X_j,\theta)], \tag{3.5.22}$$

and

$$S_n(\mathbf{t})=\Lambda_n(\theta,\theta+\mathbf{t})-n^{1/2}\mathbf{t}'\boldsymbol{\Delta}_n+\tfrac{1}{2}n\mathbf{t}'\mathbf{C}(\theta)\mathbf{t}. \tag{3.5.23}$$

Suppose the process $S_n(\mathbf{t})$ is separable. Then there is a function $\eta(\|\mathbf{t}\|)\to 0$ as $\|\mathbf{t}\|\to 0$ such that, for every $\varepsilon>0$,

$$P\{|S_n(\mathbf{t})|\leqslant n[\varepsilon+\eta(\|\mathbf{t}\|)]\,\|\mathbf{t}\|^2 \quad \text{for all } \mathbf{t} \text{ with } \theta+\mathbf{t}\in\Theta\} \tag{3.5.24}$$

tends to 1 as $n\to\infty$.

Remarks. It can be shown that the condition (A) implies that the corresponding process ξ is differentiable in quadratic mean at $t=\theta$ if $\mathbf{C}(\theta)=\mathbf{M}(\theta)$. In fact, a weaker version of (A) will imply the same [see Le Cam (1970, p. 818)]. In particular relations (3.5.19) and (3.5.20) hold in view of Proposition 3.5.5.

Proof. Note that

$$\begin{aligned} n^{-1}S_n(\mathbf{t}) = n^{-1}\sum_{j=1}^{n} \mathbf{t}'[\mathbf{B}(X_j,\theta)-\mathbf{C}(\theta)]\mathbf{t} \\ -n^{-1}\sum_{j=1}^{n} \mathbf{t}'[\mathbf{B}(X_j,\theta,\mathbf{t})-\mathbf{B}(X_j,\theta)]\mathbf{t}. \end{aligned} \tag{3.5.25}$$

Let

$$D(x,\|\mathbf{t}\|)=\sup_{s}\{\|\mathbf{B}(x,\theta,\mathbf{s})-\mathbf{B}(x,\theta)\|:\|\mathbf{s}\|\leqslant\|\mathbf{t}\|\} \tag{3.5.26}$$

and

$$\eta_1(\|\mathbf{t}\|)=E_\theta D(X,\|\mathbf{t}\|). \tag{3.5.27}$$

Assumption (A) (iii) implies that $\eta_1(\|\mathbf{t}\|)\to 0$ as $\|\mathbf{t}\|\to 0$. Furthermore,

$$\begin{aligned} n^{-1}|S_n(\mathbf{t})| \leqslant \left|n^{-1}\sum_{j=1}^{n} \mathbf{t}'[\mathbf{B}(X_j,\theta)-\mathbf{C}(\theta)]\mathbf{t}\right| \\ +\|\mathbf{t}\|^2 n^{-1}\sum_{j=1}^{n} D(X_j,\|\mathbf{t}\|). \end{aligned} \tag{3.5.28}$$

By the SLLN for i.i.d. random variables, the first term on the right-hand side of the inequality (3.5.28) is bounded by $\frac{1}{2}\varepsilon\|\mathbf{t}\|^2$ and the second term tends a.s. $\eta_1(\|\mathbf{t}\|)$. Define

$$\eta(\|\mathbf{t}\|)=\lim_{\tau\to 0}\eta_1(\|\mathbf{t}\|+|\tau|)$$

whenever $\eta(\|\mathbf{t}\|)$ is finite and $\eta(\|\mathbf{t}\|)=\infty$ otherwise. Observe that $\eta(\|\mathbf{t}\|)$ is monotone. The function $\eta(\|\mathbf{t}\|)$ has only a finite number of jumps larger than $\varepsilon/4$ in the set where $\eta(\|\mathbf{t}\|)<\infty$. Therefore one can find values $0=\tau_0\leqslant\tau_1\leqslant\cdots\leqslant\tau_m$ that divide the range of values of $\|\mathbf{t}\|$ where $\eta(\|\mathbf{t}\|)<\infty$ into intervals such that the oscillation of $\eta(\|\mathbf{t}\|)$ in each interval $[\tau_{i-1},\tau_i]$ is utmost $\varepsilon/4$. The result now follows since

$$D(x,\|\mathbf{t}\|)\leqslant D(x,\tau_i)\quad\text{whenever } \tau_{i-1}\leqslant\|\mathbf{t}\|\leqslant\tau_i.$$

Theorem 3.5.7 Suppose condition (A) holds and the matrix $\mathbf{C}(\theta)$ given in (A)(iv) is nonsingular. Then there exists an $\alpha>0$ with the following property. Let $\hat{\theta}_n\in\theta$ such that

$$\sum_{i=1}^{n}\Phi(X_j,\hat{\theta}_n)\geqslant\sup_{\mathbf{t}\in\theta}\left\{\sum_{j=1}^{n}\Phi(X_j,\mathbf{t}):\|\mathbf{t}-\theta\|<\alpha\right\}-\frac{1}{n}. \tag{3.5.29}$$

Define T_n to be any element of θ such that

$$\begin{aligned}&n^{1/2}(T_n-\theta)'\boldsymbol{\Delta}_n-\tfrac{1}{2}n(T_n-\theta)'\mathbf{C}(\theta)(T_n-\theta)\\&\quad\geqslant\sup_{\theta+\mathbf{t}\in\theta}\{n^{1/2}\mathbf{t}'\boldsymbol{\Delta}_n-\tfrac{1}{2}n\mathbf{t}'\mathbf{C}(\theta)\mathbf{t}\}-\frac{1}{n}.\end{aligned} \tag{3.5.30}$$

Then

(3.5.31) (i) $n^{1/2}(\hat{\theta}_n-\theta)=O_p(1)$,

(ii) $n^{1/2}(T_n-\hat{\theta}_n)\to 0$ in P_θ-probability, and

(iii) $n^{1/2}(\hat{\theta}_n-\theta)\xrightarrow{\mathscr{P}}N(\mathbf{0},\mathbf{C}^{-1}\mathbf{M}\mathbf{C}^{-1})$ provided $\theta\in\Theta^0$. Here $\mathbf{C}=\mathbf{C}(\theta)$ and $\mathbf{M}=\mathbf{M}(\theta)$.

Proof. Choose α small so that

$$\eta(\|\mathbf{t}\|)\|\mathbf{t}\|^2\leqslant\tfrac{1}{8}\mathbf{t}'\mathbf{C}(\theta)\mathbf{t}\quad\text{for }\|\mathbf{t}\|\leqslant\alpha,$$

where η is as defined in Proposition 3.5.6. Then

$$\Lambda_n(\theta,\theta+\mathbf{t})\leqslant n^{1/2}\mathbf{t}'\boldsymbol{\Delta}_n-\tfrac{1}{4}n\mathbf{t}'\mathbf{C}(\theta)\mathbf{t}$$

for all $\|\mathbf{t}\|\leqslant\alpha$ by Proposition 3.5.6 with probability tending to 1. Hence

$$n^{1/2}(\hat{\theta}_n-\theta)'\boldsymbol{\Delta}_n-\tfrac{1}{4}[n(\hat{\theta}_n-\theta)'\mathbf{C}(\theta)(\hat{\theta}_n-\theta)]^{-1}\geqslant-1/n$$

by the definition of $\hat{\theta}_n$. In other words,

$$[n^{1/2}(\hat{\theta}_n-\theta)-2\mathbf{C}^{-1}\mathbf{\Delta}_n]'\mathbf{C}[n^{1/2}(\hat{\theta}_n-\theta)-2\mathbf{C}^{-1}\mathbf{\Delta}_n]\leqslant 4\mathbf{\Delta}_n'\mathbf{C}^{-1}\mathbf{\Delta}_n+\frac{1}{n},$$

where $\mathbf{C}\equiv\mathbf{C}(\theta)$. Since

$$\mathbf{\Delta}_n\xrightarrow{\mathscr{L}}N(\mathbf{0},\mathbf{M})$$

under P_θ in view of (A)(ii), it follows that $n^{1/2}(\hat{\theta}_n-\theta)=O_p(1)$. Similar argument shows that $n^{1/2}(T_n-\theta)=O_p(1)$. Choose b large so that

$$P_\theta[n^{1/2}\,\|\hat{\theta}_n-\theta\|\geqslant b]\quad\text{and}\quad P_\theta[n^{1/2}\,\|T_n-\theta\|\geqslant b]$$

are as small as desired. Define

$$\theta_n=(\mathbf{t}:\mathbf{t}\in\Theta, n^{1/2}\,\|\mathbf{t}-\theta\|<b\}.$$

Then

$$|\Lambda_n(\theta,\theta+\tau n^{-1/2})-\tau'\mathbf{\Delta}_n+\tfrac{1}{2}\tau'\mathbf{C}\tau|\leqslant\varepsilon\,\|\tau\|^2$$

for all τ such that $\|\tau\|\leqslant b$ and $\theta+\tau\in\Theta_n$ with probability approaching 1 by Proposition 3.5.6. Hence

$$\begin{aligned}&n^{1/2}(\hat{\theta}_n-\theta)'\Delta_n-\tfrac{1}{2}n(\hat{\theta}_n-\theta)'\mathbf{C}(\hat{\theta}_n-\theta)\\&\geqslant n^{1/2}(T_n-\theta)'\mathbf{\Delta}_n-\tfrac{1}{2}n(T_n-\theta)\mathbf{C}(T_n-\theta)-n^{-1}-\varepsilon\end{aligned}$$

with probability approaching 1. Therefore

$$n^{1/2}(T_n-\hat{\theta}_n)\to 0\quad\text{in}\quad P_\theta\text{-probability}.$$

But

$$n^{1/2}(T_n-\theta)\xrightarrow{\mathscr{L}}N(0,\mathbf{C}^{-1}\mathbf{M}\mathbf{C}^{-1})$$

since

$$\mathbf{\Delta}_n\xrightarrow{\mathscr{L}}N(\mathbf{0},\mathbf{M}).$$

Therefore

$$n^{1/2}(\hat{\theta}_n-\theta)\xrightarrow{\mathscr{L}}N(\mathbf{0},\mathbf{C}^{-1}\mathbf{M}\mathbf{C}^{-1}).$$

Throughout this section, our approach to the asymptotic theory of maximum likelihood estimation is based on the work of Le Cam (1970, 1973a). The standard set of assumptions in the literature [cf. Cramer (1946) and Rao (1974)] involve the existence of two or three derivatives of the function

dP_t/dP_θ and some additional conditions. The role of these conditions is of purely technical character. In general the derivatives of the function may fail to exist at a point, which may depend on θ or $\{P_t, t \in \Theta\}$ may not be mutually absolutely continuous. Le Cam (1970) proposed an alternative set of conditions that involve differentiability in quadratic mean of $(dP_t/dP_\theta)^{1/2}$. These conditions are weaker than Cramer's conditions whereas the Cramer's conditions imply Le Cam's conditions. Conditions (A) stated earlier do resemble Cramer's conditions. We now state an alternative weaker set of sufficient conditions that imply asymptotic normality of the maximum likelihood estimator in the case when $\theta \subset R$. The multiparameter case is not amenable to this approach. An alternative set of conditions guaranteeing the asymptotic normality of MLE is given in Section 3.6. Suppose the following conditions hold.

(B) (i) The set Θ is an interval in R. For each $\theta \in \Theta$, let P_θ be a probability measure. Let $H(t,\theta) \equiv H(P_t, P_\theta)$ be the Hellinger distance between P_t and P_θ as usual. Let

$$X(t) = \left(\frac{dP_t}{dP_\theta}\right)^{1/2} - 1,$$

and

$$Y(t) = X(t) - E_\theta X(t).$$

(We assume that suitable versions of dP_t/dP_θ are chosen so that the process $Y(t)$ is separable.)

(ii) The process X is continuous in quadratic mean.

(iii) Let $s(\tau)$ be the variation of the process Y on $[\theta, \tau]$. Assume that $s(\tau) < \infty$ for some $\tau > \theta$ and there exists $\alpha > 0$ such that

$$\alpha^2 = \varliminf_{t \to \theta} \frac{H^2(\theta, t)}{s^2(t)}.$$

Let $X_j, j \geqslant 1$, be i.i.d. as X. Define

$$\Lambda_n(\theta, t) = \log \frac{dP_{t,n}}{dP_{\theta,n}} = 2 \sum_{j=1}^{n} \log[1 + X_j(t)] \tag{3.5.32}$$

as stated in (3.5.18).

Lemma 3.5.8. Suppose conditions (B)(i) and (B)(ii) hold. Let

$$Z_n(t) = n^{-1/2} \sum_{j=1}^{n} Y_j(t) \tag{3.5.33}$$

and $J \subset \Theta$ be an interval such that $\theta \in J$ and the variation of Y on J is equal to a finite number L. Then the relations

(3.5.34) (i) $\sup_{t\in J} |Z_n(t)| \leqslant B,$

(ii) $\Lambda_n(\theta, t) \leqslant -nH^2(\theta, t) + 2Bn^{1/2}$ for $t \in J$,

(iii) $J \cap \{t : \Lambda_n(\theta, t) \geqslant 0\} \subset \{t : n^{1/2}H^2(\theta, t) \leqslant 2B\}$

hold with probability greater than $1 - 24(L^2B^{-2})$.

Proof. Observe that conditions (B)(ii) and the separability of the process Y_j imply that the process Y_j over the interval J has continuous sample paths by Proposition 1.10.21 and hence Z_n has continuous sample paths on J and hence separable. Proposition 1.10.20 implies (i). Note that

$$\begin{aligned}\Lambda_n(\theta, t) &= 2\sum_{j=1}^{n} \log[1 + X_j(t)]\\ &\leqslant 2\sum_{j\leqslant n} X_j(t)\\ &= -nH^2(\theta, t) + 2n^{1/2}Z_n(t)\\ &\leqslant -nH^2(\theta, t) + 2Bn^{1/2}\end{aligned}$$

for $t \in J$ and (iii) is a consequence of (ii).

Lemma 3.5.9. Suppose that assumptions (B)(i)–(B)(iii) hold. Then there exists $\delta > 0$ such that for any $\varepsilon > 0$, if $a_n > \theta$ and $s(a_n) = 16n^{-1/2}(\alpha^2\varepsilon)^{-1}$, then

(3.5.35) $$\Lambda_n(\theta, t) \leqslant -\tfrac{1}{2}n\alpha^2 s^2(t) + 2n^{1/2}Z_n(a_n), \qquad t \in [a_n, \theta + \delta),$$

with probability greater than $1 - (216)\varepsilon^2$.

Proof. Let us consider the process $\{Y(t)\beta(t),\ \theta < t < \theta + \delta\}$, where $\beta(t)$ is a continuous positive decreasing function. The variation of the process $Y\beta$ [see (1.10.20) for the definition] over an interval $[a, b) \subset (\theta, \theta + \delta)$ is smaller than

$$\begin{aligned}M &= \int_a^b \beta(t)\,ds(t) - \int_a^b \|Y(t) - Y(a)\|\,d\beta(t)\\ &\leqslant \int_a^b \beta(t)\,ds(t) - \int_a^b s(t)\,d\beta(t)\end{aligned}$$

and hence, if we choose $\beta(t) = [s(t)]^{-2}$, then

$$M \leqslant 3 \int_a^b [s(t)]^{-2}\, ds(t) \leqslant \frac{3}{s(a)}.$$

Therefore, it follows from Lemma 3.5.8 that

$$P[|Z_n(t) - Z_n(a)| \leqslant Bs^2(t) \quad \text{for all } t \in [a, b)] \geqslant 1 - (216)(s^2(a)B^2)^{-1}.$$

In particular, it follows that

$$\Lambda_n(\theta, t) \leqslant -nH^2(\theta, t) + 2n^{1/2}Z_n(a) + 2Bn^{1/2}s^2(t)$$

for all $t \in [a, b)$ with probability greater than

$$1 - (216)(s^2(a)B^2)^{-1}.$$

Assumption (B)(iii) implies that there exists $\delta > 0$ such that

$$H^2(\theta, t) \geqslant \tfrac{3}{4}\alpha^2 s^2(t), \qquad t \in [\theta, \theta + \delta).$$

Let $B = \frac{1}{8}\alpha^2 n^{1/2}$. Then

$$s^2(a)B^2 = \tfrac{1}{64}\alpha^4 n s^2(a)$$

and

$$\Lambda_n(\theta, t) \leqslant -\tfrac{1}{2}\alpha^2 n s^2(t) + 2n^{1/2}Z_n(a), \qquad t \in [\theta, \theta + \delta],$$

with probability greater than

$$1 - (216)(\tfrac{1}{64}\alpha^4 n s^2(a))^{-1}.$$

Given $\varepsilon > 0$, choose a_n such that $\frac{1}{64}\alpha^4 n s^2(a_n) = \varepsilon^{-2}$ or equivalently

$$s(a_n) = 16n^{-1/2}(\varepsilon\alpha^2)^{-1}.$$

This proves the lemma.

Lemma 3.5.10. Suppose that assumptions (B)(i)–(B)(iii) hold and suppose that $s(\tau) > 0$ for $\tau \in (\theta, \theta + \delta)$. Let $\varepsilon > 0$, and define $a_n > \theta$ such that

$$s(a_n) = 16n^{-1/2}(\alpha^2\varepsilon)^{-1}.$$

as in Lemma 3.5.9. Then

(3.5.36) $$P[\Lambda_n(\theta, t) < 0 \quad \text{for all} \quad t \in [a_n, \theta + \delta)] \geqslant 1 - (232)\varepsilon^2.$$

Proof. In view of Lemma 3.5.9, this lemma holds if $2n^{1/2}Z_n(a_n) \leqslant \frac{1}{2}n\alpha^2 s^2(a_n)$. Note that

(3.5.37) $$P\{|Z_n(a_n)| \geqslant (4\varepsilon)^{-1}s(a_n)\} \leqslant 16\varepsilon^2$$

from Chebyshev's inequality and the choice of a_n. Hence, with probability greater than $1 - (232)\varepsilon^2$. we have

$$\begin{aligned}\Lambda_n(\theta, t) &< \frac{2n^{1/2}}{4\varepsilon} s(a_n) - \tfrac{1}{2}n\alpha^2 s^2(t) \\ &< \frac{1}{2}\left\{\frac{n^{1/2}}{\varepsilon} s(a_n) - n\alpha^2 s^2(a_n)\right\} \\ &= \frac{1}{2}\left\{\frac{16}{\alpha^2\varepsilon^2} - \alpha^2\left(\frac{16}{\alpha^2\varepsilon}\right)^2\right\} \\ &< 0.\end{aligned}$$

Theorem 3.5.11. In addition to assumptions (B)(i)–(B)(iii), suppose that the following identifiability condition holds:

(B) (iv) $t \neq \theta \Leftrightarrow P_t \neq P_\theta$.

Let $J = [\theta, b)$ be an interval on which the process Y has bounded variation and suppose there exists $\varepsilon > 0$ such that the set

$$J \cap \{t : H^2(\theta, t) \leqslant \varepsilon\}$$

is compact. Then the probability that there exists $\hat{\theta}_n \in J$ such that

(3.5.38) $$\Lambda_n(\theta, \hat{\theta}_n) = \sup\{\Lambda_n(\theta, t) : t \in J\}$$

tends to 1 as $n \to \infty$ and if $\theta_n^* \in J$ is such that $\Lambda_n(\theta, \theta_n^*) \geqslant 0$, then $n^{1/2}s(\theta_n^*) = O_p(1)$. In particular, $n^{1/2}s(\hat{\theta}_n) = O_p(1)$.

Proof. Let $\theta_n^* \in J$ such that $\Lambda_n(\theta, \theta_n^*) \geqslant 0$. Lemma 3.5.8 implies that $n^{1/2}H^2(\theta, \theta_n^*) \leqslant 2B$ with probability greater than $1 - (24)L^2B^{-2}$, where L is the variation of Y on J. Hence $n^{1/2}H^2(\theta, \theta_n^*) = O_p(1)$. In particular, the probability that $\theta_n^* \in K = J \cap [t: H^2(\theta, t) \leqslant \varepsilon]$, which is compact, tends to 1 as $n \to \infty$. Since $\Lambda_n(\theta, t)$ is continuous in t and K is compact, there exists a measurable $\hat{\theta}_n$ such that $\hat{\theta}_n \in K$ and

$$\Lambda_n(\theta, \hat{\theta}_n) = \sup\{\Lambda_n(\theta, t), t \in J\}.$$

The identifiability condition (B)(iv) implies that

$$\inf\{H^2(\theta, t): t \in K \cap [a, b]\} > 0$$

whenever $\theta < a$. Therefore for any $\delta > 0$,

$$P[\theta_n^* \in [\theta, \theta + \delta)] \to 1 \quad \text{as } n \to \infty \tag{3.5.39}$$

and it follows that $n^{1/2} s(\theta_n^*) = O_p(1)$ from (B)(iii) in view of Lemma 3.5.9 and the inequality (3.5.37). This completes the proof of Theorem 3.5.11.

We now state and prove the main result regarding the existence and asymptotic normality of MLE.

Theorem 3.5.12. Let $\theta \in \Theta^0$, where Θ is an interval in R. Suppose the process $\xi = (dP_\theta / d\mu)^{1/2}$ has bounded variation on Θ and $K = \{t: H^2(\theta, t) \leqslant \varepsilon\}$ is compact for some $\varepsilon > 0$. Furthermore, assume that conditions (B)(i), (B)(ii), and (B)(iv) are satisfied and that ξ is differentiable in quadratic mean at θ. Then, with probability tending to 1 there exists a measurable maximum likelihood estimator $\hat{\theta}_n$ and

$$n^{1/2}(\hat{\theta}_n - \theta) \xrightarrow{\mathscr{L}} N\left(0, \frac{1}{4\sigma^2(\theta)}\right) \tag{3.5.40}$$

provided

$$0 < \sigma(\theta) = \overline{\lim_{\tau \to 0}} |\tau|^{-1} H[\theta, \theta + \tau] < \infty \tag{3.5.41}$$

and θ is a Lebesgue point of σ in the sense that

$$\lim_{\tau \to 0} \frac{1}{\tau} \int_{\theta - \tau}^{\theta + \tau} |\sigma(t) - \sigma(\theta)| \, dt = 0. \tag{3.5.42}$$

Proof. Note that the variation of the process ξ in $[\theta, t)$ is

$$v(t) = \int_\theta^t \sigma(\tau) \, d\tau.$$

Furthermore, the variation of the process X or Y is smaller than that of ξ, and hence

$$s(t) \leqslant \int_{\theta}^{t} \sigma(\tau)\,d\tau \leqslant (t-\theta)\sigma(\theta) + \int_{\theta}^{t} |\sigma(\tau)-\sigma(\theta)|\,d\tau.$$

Therefore

$$\varliminf_{t\to 0} \frac{H(\theta,\theta+t)}{s(\theta+t)} \geqslant \frac{1}{\sigma(\theta)} \varliminf_{t\to 0} \frac{H(\theta,\theta+t)}{|t|}$$

and condition (B)(iii) holds.

For any $k \geqslant 1$, cover the compact set K by intervals $[a_j, a_{j+1}]$ of length 2^{-k}. Since K is compact, a finite number of such intervals of length 2^{-k} are needed to cover K. Consider the first index j such that $\Lambda_n(\theta, t)$ reaches its maximum in $[a_j, a_{j+1}]$. Let $T_k = a_j$. Since the Λ_n are continuous, as the partitions of K become finer and finer, the function T_k will converge to a choice of $\hat{\theta}_n$ a.s. as $k \to \infty$. Hence this choice of $\hat{\theta}_n$ gives a *measurable* maximum likelihood estimator.

Since $n^{1/2}s(\hat{\theta}_n) = O_p(1)$ and the variation $s(t)$ is bounded by $|t-\theta|$ by a factor in a neighborhood of θ, one can restrict attention to interval of the type $(\theta - an^{-1/2}, \theta + an^{-1/2})$ for maximization of the likelihood. Let

$$W_n(\tau) = \tfrac{1}{2}\Lambda_n(\theta, \theta + \tau n^{-1/2}) \tag{3.5.43}$$

and

$$V_n(\tau) = \sum_{j=1}^{n} X_j(\theta + \tau n^{-1/2}) \tag{3.5.44}$$

for $\tau \in [-a, a]$.

Since ξ is differentiable in quadratic mean, it follows that $V_n(\tau)$ is asymptotically normal by Proposition 3.5.4. In fact the measures generated by the processes $(W_n, n \geqslant 1\}$ on $C[-a, a]$ form a tight family (see Proposition 1.10.21). Hence the processes $\{W_n, n \geqslant 1\}$ converge weakly to a Gaussian process on $[-a, a]$. In fact

$$\sup_{|\tau|\leqslant a} |W_n(\tau) - V_n(\tau) + \tfrac{1}{2}\operatorname{var}_\theta[V_n(\tau)]| \xrightarrow{p} 0 \tag{3.5.45}$$

(see Proposition 3.5.5). Let

$$\Delta_n = n^{-1/2} \sum_{j=1}^{n} X_j^{(1)}(\theta),$$

where $X_j^{(1)}(\theta)$ is the derivative in q.m. of $X_j(t)$ at $t = \theta$ and observe that

$$\sup_{|\tau| \leqslant a} |V_n(\tau) + \tfrac{1}{2} n H^2(\theta, \theta + \tau n^{-1/2}) - \tau \Delta_n| \xrightarrow{p} 0 \tag{3.5.46}$$

and

$$\sup_{|\tau| \leqslant a} |\text{var}_\theta[V_n(\tau)] - \sigma^2(\theta)\tau^2| \to 0$$

since $E_\theta[X_1^{(1)}(\theta)]^2 = \sigma^2(\theta)$. Hence

$$\sup_{|\tau| \leqslant a} |W_n(t) - \tau \Delta_n + \tfrac{1}{2}[\sigma^2(\theta) + \sigma^2(\theta)]\tau^2| \xrightarrow{p} 0,$$

which proves that

$$n^{1/2}(\hat{\theta}_n - \theta) - [\sigma^2(\theta) + \sigma^2(\theta)]^{-1} \Delta_n \xrightarrow{p} 0. \tag{3.5.47}$$

This in turn proves the result since $\Delta_n \xrightarrow{\mathscr{L}} N(0, \sigma^2(\theta))$. This completes the proof of the main result of this section.

Remarks. Le Cam (1973a) discussed sufficient conditions on the sample functions for the differentiability in quadratic mean of the process $\xi(\theta) = (dP_\theta/d\mu)^{1/2}$, $\theta \in \Theta \subset R$, where Θ is an interval. We state these results now.

A real-valued function f defined on an interval I is said to satisfy *Lusin's condition* if, for every set $S \subset I$ that has Lebesgue measure zero, thc range $f(S)$ has also Lebesgue measure zero.

Suppose ξ is a process defined on an interval I such that

$$\sigma^2(t) = \overline{\lim_{\tau \to 0}} \frac{1}{|\tau|^2} E|\xi(t + \tau) - \xi(t)|^2 < \infty. \tag{3.5.48}$$

Further suppose that

$$\int_I \sigma(t)\, dt < \infty. \tag{3.5.49}$$

Then the process ξ is differentiable with derivative $\xi^{(1)}$ in quadratic mean at almost all points of I and, for any $\theta \in I$,

$$\xi(t) = \xi(\theta) + \int_\theta^t \xi^{(1)}(\tau)\, d\tau, \qquad t \in I.$$

Consider a version $\xi(\omega, t)$ of ξ with continuous sample paths and let $\xi^{(1)}(\omega, t)$ be the derivative evaluated at ω. Then, except for a fixed null set,

$$\xi(\omega, t) = \xi(\omega, \theta) + \int_\theta^t \xi^{(1)}(\omega, \tau)\, d\tau, \tag{3.5.50}$$

and hence the sample paths of ξ are almost surely absolutely continuous, which implies that they satisfy Lusin's condition.

Conversely, let ξ be a process defined on an interval I. Define, for any fixed t,

$$\dot{\xi}(\omega, t) = \lim_{|\tau|\to 0} \frac{1}{\tau}[\xi(\omega, t+\tau) - \xi(\omega, \mathrm{t})] \tag{3.5.51}$$

if the limit exists on the right-hand side and $\dot{\xi}(\omega, t) = 0$ otherwise. Let

$$s^2(t) = E|\dot{\xi}(\omega, t)|^2. \tag{3.5.52}$$

Proposition 3.5.13. Suppose the sample paths of the process ξ are continuous and satisfy Lusin's condition. Furthermore, assume that $s(\cdot)$ is locally integrable. Then the process ξ is differentiable in quadratic mean with derivative $\xi^{(1)}$ in the class of $\dot{\xi}$ and $\sigma(t) = s(t)$ for almost all t, where $\sigma(t)$ is as defined by (3.5.48). Furthermore, these properties hold at least at all $t \in I$, where ξ is differentiable in probability and t is a Lebesgue point of S.

For proof of this result and related discussion, see Le Cam (1973a). The following example due to Le Cam (1973a) indicates the necessity of the assumption of continuity of sample paths. Let $\eta(\omega, \theta)$ be the function defined by

$$\eta(\omega, \theta) = \begin{cases} 1 & \text{if } |\omega - \theta| \leqslant \frac{1}{2} \\ 0 & \text{otherwise.} \end{cases}$$

Then $\eta^{(1)}(\omega, \theta) = 0$ for all (ω, θ) except at the points $(\omega - \frac{1}{2}, \omega)$ and $(\omega + \frac{1}{2}, \omega)$. The sample paths of η satisfy Lusin's condition since the range of η consists of only two points. However, η is not differentiable in quadratic mean.

Remarks. Maximum likelihood estimator of a parameter $\theta \in \Theta \subset R^k$ may be consistent over all of Θ but may fail to be consistent over $\Theta_0 \subset \Theta$ if there is additional information that indicates that $\theta \in \Theta_0$. Le Cam (1973a) indicates other situations when MLE is not necessarily a good estimator and suggests some competitors that still have all the properties of MLE like asymptotic sufficiency and asymptotic normality. We briefly indicate the construction of such estimators. For details, see Le Cam (1974).

Let $\tilde{\theta}_n$ be an estimator of θ such that $\delta_n^{-1}(\tilde{\theta}_n - \theta) = O_p(1)$. Discretize $\tilde{\theta}_n$ as follows. For every $n \geqslant 1$, pave the space R^k with cubes of diameter δ_n. Choose the center of thc cube closest to $\tilde{\theta}_n$ as θ_n^*. Let $\{\boldsymbol{\varepsilon}_j, 1 \leqslant j \leqslant k\}$ be a basis for R^k and $\Lambda_n(\mathbf{t}, \mathbf{s})$ be the log-likelihood ratio of $P_{\mathbf{t},n}$ with respect to $P_{\mathbf{s},n}$ as before. We assume that, for all bounded sequences $\{(\mathbf{s}_n, \mathbf{t}_n)\}$ with $\mathbf{s}_n, \mathbf{t}_n$ in R^k,

$$\Lambda_n(\theta + (\mathbf{t}_n + \mathbf{s}_n)\delta_n, \theta + \mathbf{s}_n\delta_n) - \Lambda_n(\theta + \mathbf{t}_n\delta_n, \theta) + \mathbf{t}_n'\boldsymbol{\Gamma}\mathbf{s}_n \xrightarrow{p} 0 \tag{3.5.53}$$

for some covariance matrix $\boldsymbol{\Gamma}$. Compute the differences

$$\Lambda_n(\theta_n^* + \delta_n(\boldsymbol{\varepsilon}_j + \boldsymbol{\varepsilon}_i), \theta_n^* + \delta_n\boldsymbol{\varepsilon}_i) - \Lambda_n(\theta_n^* + \delta_n\boldsymbol{\varepsilon}_i, \theta^*) \tag{3.5.54}$$

as if θ_n^* were the true value of θ. Note that these differences can be used to estimate $\boldsymbol{\varepsilon}_j'\boldsymbol{\Gamma}\boldsymbol{\varepsilon}_i$ for $1 \leqslant i \leqslant k$ in view of (3.5.53). Denote this estimator by $\boldsymbol{\varepsilon}_j'\hat{\boldsymbol{\Gamma}}_i\boldsymbol{\varepsilon}_i$. Define the vector W_n by the relations

$$\boldsymbol{\varepsilon}_i'\mathbf{W}_n = \Lambda_n(\theta_n^* + \delta_n\boldsymbol{\varepsilon}_i, \theta^*) + \tfrac{1}{2}\boldsymbol{\varepsilon}_i'\hat{\boldsymbol{\Gamma}}_n\boldsymbol{\varepsilon}_i, \qquad 1 \leqslant i \leqslant k.$$

Note that

$$\mathbf{t}'\mathbf{W}_n - \tfrac{1}{2}\mathbf{t}'\hat{\boldsymbol{\Gamma}}_n\mathbf{t} \tag{3.5.55}$$

is a quadratic function. Let $\hat{\mathbf{t}}_n$ be the vector in R^k at which (3.5.55) is maximized. Define

$$T_n = \theta_n^* + \delta_n\hat{\mathbf{t}}_n. \tag{3.5.56}$$

Le Cam (1974, Section 12) proved that $\{T_n\}$ is asymptotically sufficient and asymptotically normal. We omit the details.

3.6 MAXIMUM LIKELIHOOD ESTIMATION OF A VECTOR PARAMETER

In Section 3.5, we discussed asymptotic properties of the MLE under conditions weaker than the standard Cramer conditions. However, the conditions proposed therein go through mainly in the case of a one-dimensional parameter. This is because the sufficient conditions usually proposed to study fluctuations of random fields are *too strong* for use in statistical investigations. For instance, condition (B_4) in Section 3.3 is of a similar nature. An alternative set of conditions for the study of weak convergence of log-likelihood ratio random field are discussed by Inagaki and Ogata (1975). We now discuss their results. The regularity conditions assumed here are similar to those in Huber (1967) in the i.i.d. case and Prakasa Rao (1972) for discrete-time stationary Markov processes. We confine our attention to the i.i.d. case.

Let $\mathbf{X}_i$, $1 \leqslant i \leqslant n$, be i.i.d. p-dimensional random vectors with probability measure P_θ, $\theta \in \Theta \subset R^k$. Suppose $P_\theta \ll \mu$, μ σ-finite. Let

$$f(\mathbf{x}, \theta) = \frac{dP_\theta}{d\mu}(\mathbf{x})$$

and

$$Z_n(h) = \prod_{i=1}^{n} \{f(\mathbf{X}_i, \theta_0 + \mathbf{h}n^{-1/2})/f(\mathbf{X}_i, \theta_0)\} \tag{3.6.1}$$

for θ_0 and $\theta_0 + \mathbf{h}n^{-1/2} \in \Theta$ where θ_0 is arbitrary but fixed element in Θ. Suppose the following regularity conditions hold.

(A)

(A_1) $\Theta \subset R^k$ and the true parameter $\theta_0 \in \Theta^0$.

(A_2) $\theta_1 \neq \theta_2 \Leftrightarrow P_{\theta_1} \neq P_{\theta_2}$.

(A_3) $f(\mathbf{x}, \theta)$ is continuous in θ a.e. $[\mu]$ $\mathbf{x}$.

(B) Given $\theta_0 \in \Theta^0$, there exists a neighborhood $U = U_0 = U_{d_0}(\theta_0) = \{\theta : \|\theta - \theta_0\| \leqslant d_0\}$ satisfying the following conditions. (Here $\|\theta\| = \max\{|\theta^{(i)}|, 1 \leqslant i \leqslant k\}$, where $\theta = (\theta^{(1)}, \ldots, \theta^{(k)})'$).

(B_1) For any $\theta \in U_0$, $f(\mathbf{x}, \theta)$ has a common support and for a.e. $[\mu]\mathbf{x}$ $\log f(\mathbf{x}, \theta)$ is continuously differentiable at $\theta \in U_0$ with

$$\boldsymbol{\eta}(\mathbf{x}, \theta) = \frac{\partial}{\partial \theta} \log f(\mathbf{x}, \theta) \equiv \nabla \log f(\mathbf{x}, \theta) = \left(\frac{\partial}{\partial \theta^{(1)}}, \ldots, \frac{\partial}{\partial \theta^{(k)}}\right)' \log f(\mathbf{x}, \theta).$$

(B_2) For each $\theta \in U_0$, $\boldsymbol{\eta}(\mathbf{x}, \theta)$ is a measurable function of $\mathbf{x}$. Let

$$\boldsymbol{\lambda}(\theta) = E_{\theta_0} \boldsymbol{\eta}(\mathbf{x}, \theta)$$

and

$$u(\mathbf{x}, \theta, d) = \sup_{\|\tau - \theta\| \leqslant d} \|\boldsymbol{\eta}(\mathbf{x}, \tau) - \boldsymbol{\eta}(\mathbf{x}, \theta)\|.$$

(B_3) For all $\theta \in U_0, \boldsymbol{\lambda}(\theta) < \infty$, and $\boldsymbol{\Gamma}(\theta) = E_{\theta_0}\{\boldsymbol{\eta}(\mathbf{x}, \theta)\boldsymbol{\eta}(\mathbf{x}, \theta)'\}$ exists and is continuous at θ_0. Furthermore $\boldsymbol{\Gamma}(\theta_0)$ is positive definite.

(B_4) Suppose $\boldsymbol{\lambda}(\theta)$ is differentiable at $\theta = \theta_0$ with derivative

$$\boldsymbol{\Lambda}(\theta_0) = \left.\frac{\partial \boldsymbol{\lambda}(\theta)}{\partial \theta}\right|_{\theta=\theta_0} = \left(\left(\frac{\partial \lambda^{(i)}(\theta)}{\partial \theta^{(j)}}\right)\right)_{k \times k}$$

and

$$-\boldsymbol{\Lambda}(\theta_0) = \boldsymbol{\Gamma}(\theta_0).$$

(B_5) There exist positive constants b_1 and b_2 such that

$$E_{\theta_0} u(\mathbf{X}, \theta, d) \leqslant b_1 d \quad \text{for} \quad \|\theta - \theta_0\| + d \leqslant d_0, d > 0,$$

and

$$E_{\theta_0} u^2(\mathbf{X}, \theta, d) \leqslant b_2 d \quad \text{for} \quad \|\theta - \theta_0\| + d \leqslant d_0, d > 0.$$

(C) Let

$$\delta(\theta_1, \theta_2) = \frac{\|\theta_1 - \theta_2\|}{1 + \|\theta_1 - \theta_2\|}$$

and suppose $(\bar{\Theta}, \delta)$ is a metric space satisfying the following conditions:

(C_1) $(\bar{\Theta}, \delta)$ is Bahadur compactification of Θ in the sense that

(i) $\bar{\Theta}$ is compact,
(ii) Θ is everywhere dense in $\bar{\Theta}$,
(iii) If

$$g(\mathbf{x}, \bar{\theta}, d) = \sup\{f(\mathbf{x}, \theta) : \theta \in \Theta,\ \delta(\theta, \bar{\theta}) < d\} \text{ for } \bar{\theta} \in \bar{\Theta} \text{ and } \delta(\theta_0, \bar{\theta}) < 1,$$

and

$$g(\mathbf{x}, \theta_\infty, d) = \sup\{f(\mathbf{x}, \theta) : \theta \in \Theta,\ \delta(\theta_0, \theta) > 1 - d\}$$
$$\text{for } \theta_\infty \in \bar{\Theta} \text{ with } \delta(\theta_0, \theta_\infty) = 1,$$

then, for each $\bar{\theta} \in \bar{\Theta}$, there exists $d_1 = d_1(\bar{\theta}) > 0$ such that $g(\mathbf{x}, \bar{\theta}, d)$ is measurable for each $0 < d \leqslant d_1$, and for each $\bar{\theta} \in \bar{\Theta}$,

$$\int_{R^p} g(\mathbf{x}, \bar{\theta}, 0)\, \mu(d\mathbf{x}) \leqslant 1$$

where

$$g(\mathbf{x}, \bar{\theta}, 0) = \lim_{d \to 0} g(\mathbf{x}, \bar{\theta}, d).$$

(C_2) $\int_{R^p} |g(\mathbf{x}, \bar{\theta}, 0) - f(\mathbf{x}, \theta_0)| \mu(d\mathbf{x}) > 0$ if $\bar{\theta} \neq \theta_0$.

(C_3) For every $\bar{\theta} \in \bar{\Theta}$, there exists $d = d(\bar{\theta})$, $0 < d \leqslant d_1$ such that

$$E_{\theta_0}\left[\log^+ \frac{g(\mathbf{X}, \bar{\theta}, d)}{f(\mathbf{X}, \theta_0)}\right] < \infty.$$

Here

$$\log^+ x = \begin{cases} \log x & \text{if } \log x > 0 \\ 0 & \text{if } \log x \leqslant 0. \end{cases}$$

(C_4) Given $t > 0$ and $\bar{\theta} \in \bar{\Theta}$, there exists $d = d(t, \bar{\theta})$, $0 < d \leqslant d_1$, such that

$$E_{\theta_0}\left[\left\{\frac{g(\mathbf{X}, \bar{\theta}, d)}{f(\mathbf{X}, \theta_0)}\right\}^t\right] < \infty.$$

(C_5) Given $t > 0$ and $\theta_\infty \in \bar{\Theta}$ with $\delta(\theta_\infty, \theta_0) = 1$, there exists $\alpha = \alpha(t) > 0$ such that

$$\overline{\lim_{d \to 0}} \frac{1}{d^\alpha} E_{\theta_0}\left[\frac{g(\mathbf{X}, \theta_\infty, d)}{f(\mathbf{X}, \theta_0)}\right]^t < \infty.$$

Assumptions A are of the standard type, assumptions B refer to the local behavior of the likelihood at θ_0, and assumptions C deal with global behavior of likelihood over Θ.

Proposition 3.6.1. Under assumptions (A) and (B),

$$\lim_{\varepsilon \to 0} \frac{1}{\varepsilon^2} \int_{R^p} [f^{1/2}(\mathbf{x}, \theta_0 + \varepsilon \mathbf{h}) - f^{1/2}(\mathbf{x}, \theta_0)]^2 \mu(d\mathbf{x}) = \tfrac{1}{4} \mathbf{h}' \mathbf{\Gamma}(\theta_0) \mathbf{h} \tag{3.6.2}$$

for every $\mathbf{h} \in R^k$ such that $\theta_0 + \varepsilon \mathbf{h} \in U_0$.

Proof. In view of assumption (B_1),

$$\lim_{\varepsilon \to 0} \frac{1}{\varepsilon} [f^{1/2}(\mathbf{x}, \theta_0 + \varepsilon \mathbf{h}) - f^{1/2}(\mathbf{x}, \theta_0)] = h' \nabla f(\mathbf{x}, \theta_0) [2 f(\mathbf{x}, \theta_0)]^{-1/2}$$

$$= \tfrac{1}{2} \mathbf{h}' \boldsymbol{\eta}(\mathbf{x}, \theta_0) f^{1/2}(\mathbf{x}, \theta_0).$$

Hence, by assumption (B_3) and Fatou's lemma, it follows that

$$\begin{aligned}
&\lim_{\varepsilon\to 0}\frac{1}{\varepsilon^2}\int_{R^p}[f^{1/2}(\mathbf{x},\theta_0+\varepsilon\mathbf{h})-f^{1/2}(\mathbf{x},\theta_0)]^2\mu(d\mathbf{x})\\
&\geqslant \int_{R^p}\lim_{\varepsilon\to 0}\frac{1}{\varepsilon^2}[f^{1/2}(\mathbf{x},\theta_0+\varepsilon\mathbf{h})-f^{1/2}(\mathbf{x},\theta_0)]^2\,\mu(d\mathbf{x})\\
&=\frac{1}{4}\int_{R^p}[\mathbf{h}'\boldsymbol{\eta}(\mathbf{x},\theta_0)\boldsymbol{\eta}(\mathbf{x},\theta_0)'\mathbf{h}]f(\mathbf{x},\theta_0)\,\mu(d\mathbf{x})\\
&=\tfrac{1}{4}\mathbf{h}'\boldsymbol{\Gamma}(\theta_0)\mathbf{h}.
\end{aligned}$$

On the other hand,

$$f^{1/2}(\mathbf{x},\theta_0+\varepsilon\mathbf{h})-f^{1/2}(\mathbf{x},\theta_0)=\frac{1}{2}\int_0^{\varepsilon}[\mathbf{h}'\boldsymbol{\eta}(\mathbf{x},\theta_0+t\mathbf{h})]f^{1/2}(\mathbf{x},\theta_0+t\mathbf{h})\,dt,$$

and hence, by Fubini's theorem, we have

$$\begin{aligned}
&\int_{R_p}[f^{1/2}(\mathbf{x},\theta_0+\varepsilon\mathbf{h})-f^{1/2}(\mathbf{x},\theta_0)]^2\,\mu(d\mathbf{x})\\
&=\int_{R^p}\mu(d\mathbf{x})\left\{\frac{1}{2}\int_0^{\varepsilon}\mathbf{h}'\boldsymbol{\eta}(\mathbf{x},\theta_0+t\mathbf{h})f^{1/2}(\mathbf{x},\theta_0+t\mathbf{h})\right\}\\
&\leqslant\int_{R^p}\mu(d\mathbf{x})\left\{\frac{\varepsilon}{4}\int_0^{\varepsilon}\mathbf{h}'\boldsymbol{\eta}(\mathbf{x},\theta_0+t\mathbf{h})\boldsymbol{\eta}'(\mathbf{x},\theta_0+t\mathbf{h})\mathbf{h}f(\mathbf{x},\theta_0+t\mathbf{h})\,dt\right\}\\
&\qquad\text{(by Cauchy–Schwarz inequality)}\\
&=\frac{\varepsilon}{4}\int_0^{\varepsilon}\mathbf{h}'\boldsymbol{\Gamma}(\theta_0+t\mathbf{h})\mathbf{h}\,dt.
\end{aligned}$$

Hence, by assumption ($\mathbf{B}_3$), it follows that

$$\begin{aligned}
&\overline{\lim_{\varepsilon\to 0}}\frac{1}{\varepsilon^2}\int[f^{1/2}(\mathbf{x},\theta_0+\varepsilon\mathbf{h})-f^{1/2}(\mathbf{x},\theta_0)]^2\,\mu(d\mathbf{x})\\
&\leqslant\overline{\lim_{\varepsilon\to 0}}\frac{1}{4\varepsilon}\int_0^{\varepsilon}\mathbf{h}'\boldsymbol{\Gamma}(\theta_0+t\mathbf{h})\mathbf{h}\,dt=\tfrac{1}{4}\mathbf{h}'\boldsymbol{\Gamma}(\theta_0)\mathbf{h}.
\end{aligned}$$

This proves the proposition.

In particular

$$\rho(P_{\theta_0},P_{\theta_0+\varepsilon\mathbf{h}})=1-\tfrac{1}{8}\mathbf{h}'\boldsymbol{\Gamma}(\theta_0)\mathbf{h}\varepsilon^2(1+o(1)) \tag{3.6.3}$$

as $\varepsilon \to 0$, where $\rho(P, Q)$ is the affinity between the probability measures P and Q. Furthermore, this proposition implies that $f^{1/2}(\mathbf{x}, \theta)$ is differentiable in quadratic mean at θ_0. Therefore, by Proposition 3.5.4, we have the following result.

Theorem 3.6.2. Under assumptions (A) and (B), the finite-dimensional distributions of the random field $Z_n(\mathbf{h})$ defined by (3.6.1) converge to the corresponding finite-dimensional distributions of the random field

$$Z(\mathbf{h}) = \exp\{\mathbf{h}'\Gamma(\theta_0)^{1/2}\xi - \tfrac{1}{2}\mathbf{h}'\Gamma(\theta_0)\mathbf{h}\}, \tag{3.6.4}$$

where ξ is $N_k(\mathbf{0}, \mathbf{I})$.

Inagaki and Ogata (1975) proved the following lemmas, leading to the tightness of the family of measures generated by processes $\{Z_n, n \geq 1\}$. We omit the proofs.

Lemma 3.6.3. Under assumptions (A) and (B), for any $M > 0$ and $\varepsilon > 0$,

$$\lim_{n\to\infty} P_{\theta_0}\left\{\sup_{\|\mathbf{h}\| \leq M}\left|L_n(\mathbf{h}) - \mathbf{h}'\frac{1}{\sqrt{n}}\sum_{i=1}^{n}\eta(\mathbf{X}_i, \theta_0) + \tfrac{1}{2}\mathbf{h}'\Gamma(\theta_0)\mathbf{h}\right| > \varepsilon\right\} = 0, \tag{3.6.5}$$

where $L_n(\mathbf{h}) = \log Z_n(\mathbf{h})$.

Lemma 3.6.4. Under assumptions (A) and (B), there exist positive numbers $d, 0 < d \leq d_0$, and $c_1 > 0$ such that forall $\mathbf{h}$ such that

$$\|\mathbf{h}n^{-1/2}\| < d,$$

$$P_{\theta_0}\{Z_n(\mathbf{h}) > e^{-c_1\|\mathbf{h}\|^2}\} \leq e^{-c_1\|\mathbf{h}\|^2}. \tag{3.6.6}$$

Lemma 3.6.5. Suppose assumptions (A) and (B) hold. Choose $d > 0$ and $c_1 > 0$ as in Lemma 3.6.3. Then there exists $c_2 > 0$ such that

$$P_{\theta_0}\left\{\sup_{l \leq \|\mathbf{h}\| \leq l+1} Z_n(\mathbf{h}) > e^{-c_1 l^2}\right\} < c_2 l^{-2} \tag{3.6.7}$$

whenever l is any positive integer and $l + 1 < \sqrt{n}\,d$.

Lemmas 3.6.4 and 3.6.5 describe the behavior of the log-likelihood function locally. The global behavior is given by the following results.

Lemma 3.6.6. Suppose assumptions (A) and (C) hold. Then, for any $d>0$ and $M>0$, there exist positive numbers c_3 and an integer $n_0>0$ such that, for all $n \geqslant n_0$, $d \leqslant \|\mathbf{h}n^{-1/2}\| \leqslant M$,

$$P_{\theta_0}\left\{\sup_{\|\mathbf{h}\|\geqslant l} Z_n(\mathbf{h}) > e^{-c_3 l^2}\right\} \leqslant e^{-c_3 l^2}, \qquad l \geqslant 1. \tag{3.6.8}$$

Lemma 3.6.7. Suppose assumptions (A) and (C) hold. Then, for any $N>0$, there exist $M>0$ and an integer $n_0>0$ such that, for all $n \geqslant n_0$ and $l \geqslant M\sqrt{n}$,

$$P_{\theta_0}\left\{\sup_{\|\mathbf{h}\|\geqslant l} Z_n(\mathbf{h}) > \frac{1}{l^N}\right\} \leqslant \frac{1}{l^N}, \qquad l \geqslant 1. \tag{3.6.9}$$

Combining the results in Lemmas 3.6.4–3.6.7, one can obtain the following theorem.

Theorem 3.6.8. Suppose assumptions (A), (B), and (C) hold. Then, for any $N>0$, there exist $c_0>0$ and an integer $n_0>0$ (depending on N possibly) such that, for all $n \geqslant n_0$,

$$P_{\theta_0}\left\{\sup_{l\leqslant\|\mathbf{h}\|\leqslant l+1} Z_n(\mathbf{h}) > \frac{1}{l^N}\right\} \leqslant \frac{c_0}{l^2} \tag{3.6.10}$$

and

$$P_{\theta_0}\left\{\sup_{\|\mathbf{h}\|\geqslant M} Z_n(\mathbf{h}) > \frac{1}{M^N}\right\} \leqslant \frac{c_0}{M^N} \tag{3.6.11}$$

for $l \geqslant 1$ and $M \geqslant 1$.

Define

(3.6.12)

$$\bar{Z}_n(\mathbf{h}) = \begin{cases} Z_n(\mathbf{h}) & \text{if } \theta_0 + \mathbf{h}n^{-1/2} \in \Theta \\ \prod_{i=1}^{n} \{g(\mathbf{X}_i, \theta_0 + \mathbf{h}n^{-1/2}, 0)/f(\mathbf{X}_i, \theta_0)\} & \text{if } \theta_0 + \mathbf{h}n^{-1/2} \in \bar{\Theta} \\ 0 & \text{if } \theta_0 + \mathbf{h}n^{-1/2} \in \bar{\Theta}_n = \{\theta : \delta(\theta, \bar{\Theta}) = n^{-1/2}\} \\ \text{continuous} & \text{otherwise.} \end{cases}$$

In view of Theorem 3.6.8, it follows that the sample paths of the process $\bar{Z}_n(\mathbf{h})$ belong to $C_0(R^k)$ with probability 1, where $C_0(R^k)$ is the space of continuous functions f on R^k such that $\lim_{\|\mathbf{h}\|\to\infty} f(\mathbf{h})=0$. Furthermore, the following theorem implies the tightness of the family of measures generated by the random fields $\bar{Z}_n$ over $C_0(R^k)$.

Theorem 3.6.9. Suppose assumptions (A), (B), and (C) hold. Then, for any $\varepsilon>0$,

(3.6.13) $$\lim_{d\to 0}\overline{\lim_{n\to\infty}}\, P_{\theta_0}\left\{\sup_{\|\mathbf{h}_1-\mathbf{h}_2\|<d}|Z_n(\mathbf{h}_1)-Z_n(\mathbf{h}_2)|>\varepsilon\right\}=0$$

This theorem follows from Theorem 3.6.8 and Lemma 3.6.3 in view of the inequality

$$\sup_{\substack{\|\mathbf{h}_1-\mathbf{h}_1\|<d\\ \|\mathbf{h}\|\leqslant M}}|Z_n(\mathbf{h}_1)-Z_n(\mathbf{h}_2)|\leqslant \sup_{\|\mathbf{h}\|\leqslant M} Z_n(\mathbf{h}) \sup_{\|\mathbf{h}_1-\mathbf{h}_2\|<d}|L_n(\mathbf{h}_1)-L_n(\mathbf{h}_2)|.$$

As a consequence of Theorems 3.6.2 and 3.6.9, we obtain the following result. Observe that the sample paths of the process $\bar{Z}_n$ are bounded and continuous on the compactification $(\bar{R}^k,\delta)$ with probability 1 and $\bar{Z}_n(0)=1$. The following theorem is now a consequence of Straf (1970, p. 207) [cf. Prakasa Rao (1975a)] (see Section 1.10).

Theorem 3.6.10. Suppose assumptions (A), (B), and (C) hold. Then the sequence of random fields $\{\bar{Z}_n(\mathbf{h})\}$ converge in distribution to the random field $Z(\mathbf{h})$ given by

(3.6.14) $$Z(\mathbf{h})=\exp\{\mathbf{h}'\boldsymbol{\Gamma}(\theta_0)^{1/2}\xi-\tfrac{1}{2}\mathbf{h}'\boldsymbol{\Gamma}(\theta_0)\mathbf{h}\},$$

where ξ is $N_k(\mathbf{0},\mathbf{I})$. In particular, if ψ_n is a measurable functional on $C_0(R^k)$ continuously converging to ψ, that is,

$$f_n\to f\in C_0(R^k)\Rightarrow\psi_n(f_n)\to\psi(f),$$

then

(3.6.15) $$\lim_{n\to 0} P_{\theta_0}\{\psi_n(\bar{Z}_n)\leqslant x\}=P_{\theta_0}\{\psi(Z)\leqslant x\},\qquad x\in R.$$

Example 3.6.1. As an application of Theorem 3.6.10, we now obtain the asymptotic distribution of the maximum likelihood estimator $\hat{\theta}_n$ defined by

(3.6.16) $$g_n(\mathbf{X},\hat{\theta}_n)=\sup\{g_n(\mathbf{X},\theta):\theta\in\bar{\Theta}\},$$

where

$$(3.6.17) \qquad g_n(\mathbf{X}, \theta) = \prod_{i=1}^{n} g(\mathbf{X}_i, \theta, 0) = \begin{cases} \prod_{i=1}^{n} f(\mathbf{X}_i, \theta) & \text{for } \theta \in \Theta \\ \prod_{i=1}^{n} g(\mathbf{X}_i, \theta, 0) & \text{for } \theta \in \bar{\Theta} - \Theta. \end{cases}$$

Let

$$\Delta_{\mathbf{y}} = \prod_{r=1}^{k} [-\infty, y^{(r)}] \quad \text{where } \mathbf{y} = (y^{(1)}, \ldots, y^{(k)})' \in R^k$$

and define

$$\psi_{\mathbf{y}}(z) = \sup\{|z(\mathbf{h})| : \mathbf{h} \in \Delta_{\mathbf{y}}\}$$

and

$$\Psi_{\mathbf{y}}(\mathbf{z}) = \sup\{|z(\mathbf{h})| : \mathbf{h} \notin \Delta_{\mathbf{y}}\}$$

for $z \in C_0(R^k)$. Note that

$$\psi_{\mathbf{y}}(\bar{Z}_n) \geqslant \Psi_{\mathbf{y}}(\bar{Z}_n) \quad \text{iff} \quad \sqrt{n}(\hat{\theta}_n - \theta_0) \in \Delta_{\mathbf{y}}$$

and

$$\psi_{\mathbf{y}}(Z) \geqslant \Psi_{\mathbf{y}}(Z) \quad \text{iff} \quad \mathbf{\Gamma}(\theta_0)^{-1/2} \xi \in \Delta_{\mathbf{y}}.$$

Furthermore $\psi_{\mathbf{y}} - \Psi_{\mathbf{y}}$ is a continuous functional on $C_0(R^k)$. Hence, by Theorem 3.6.10, it follows that

$$\begin{aligned} P_{\theta_0}\{\sqrt{n}(\hat{\theta}_n - \theta_0) \in \Delta_{\mathbf{y}}\} &= P_{\theta_0}\{\psi_{\mathbf{y}}(\bar{Z}_n) - \Psi_{\mathbf{y}}(\bar{Z}_n) \geqslant 0\} \to P\{\psi_{\mathbf{y}}(Z) - \Psi_{\mathbf{y}}(Z) \geqslant 0\} \\ &= P\{\mathbf{\Gamma}(\theta_0)^{-1/2} \xi \in \Delta_{\mathbf{y}}\} = N_k(\mathbf{0}, \mathbf{\Gamma}^{-1}(\theta_0))(\mathbf{y}) \end{aligned}$$

as $n \to \infty$, proving that

$$(3.6.18) \qquad \sqrt{n}(\hat{\theta}_n - \theta_0) \xrightarrow{\mathscr{L}} N_k(\mathbf{0}, \mathbf{\Gamma}^{-1}(\theta_0)).$$

Remarks. For other applications of Theorem 3.6.10, see Inagaki and Ogata (1975). The idea of using the likelihood ratio process for the study of asymptotic distribution of MLE is due to Chernoff and Rubin (1956). Prakasa Rao (1966, 1968) applied the theory of weak convergence of the likelihood ratio process for deriving the asymptotic properties of MLE in a nonregular case. Ibragimov and Hasminskii (1972) developed the method independently in their study of asymptotic theory of MLE. Ibragimov and Hasminskii (1981) gives a comprehensive survey of their work. See also Basawa and Prakasa Rao (1980b) and Kutoyants (1984). In the next section, we discuss briefly the work of Prakasa Rao (1968).

3.7 MAXIMUM LIKELIHOOD ESTIMATION IN A NONREGULAR CASE

In this section, we consider the problem of the estimation of the location of the cusp of a continuous density. As we shall see below, this problem is not amenable to the standard methods, because the likelihood function is not differentiable. However, one can use the theory of weak convergence for the study of the log-likelihood ratio process. Results in this section are due to Prakasa Rao (1966, 1968). For a general discussion of problem of this type, see Ibragimov and Hasminskii (1981) and Polfeldt (1970).

Consider the family of densities $\{f(x,\theta),\ \theta\in\Theta\}$, $\Theta=(a,b)$, where $-A<a<b<A$, $A>0$, and

$$\text{(3.7.0)}\quad \text{(i)}\quad \log f(x,\theta)=\begin{cases}\varepsilon(x,\theta)|x-\theta|^{\lambda}+g(x,\theta) & \text{for } |x|\leqslant A\\ g(x,\theta) & \text{for } |x|>A,\end{cases}$$

$$\text{(ii)}\quad \varepsilon(x,\theta)=\begin{cases}\beta(\theta) & \text{if } x<0\\ \gamma(\theta) & \text{if } x>0,\end{cases}$$

$$\text{(iii)}\quad 0<\lambda<\tfrac{1}{2}.$$

Suppose the family $\{f(x,\theta),\theta\in\Theta\}$ satisfies the following regularity conditions.

(R_1) For each $\theta\neq\theta_0\in[a,b]$, there exists $\delta(\theta,\theta_0)>0$ such that

$$E_{\theta_0}[\sup\{\log f(X,\phi)-\log f(X,\theta_0):|\phi-\theta|\leqslant\delta(\theta,\theta_0)\}]<0.$$

(R_2) For all θ,θ_0 in $[a,b]$, $\partial g(x,\theta)/\partial\theta$ and $\partial^2 g(x,\theta)/\partial\theta^2$ exist,

$$E_{\theta_0}\left|\left.\frac{\partial g(X,\theta)}{\partial\theta}\right|_{\theta=\theta_0}\right|<\infty,$$

and

$$\sup_{\theta\in\Theta} E_{\theta_0}\left|\frac{\partial^2 g(X,\theta)}{\partial\theta^2}\right|\leqslant K_1(\theta_0)<\infty.$$

(R_3) For every $\theta_0 \in [a,b]$,

$$E_{\theta_0}\left[\frac{\partial \log f(X,\theta)}{\partial \theta}\bigg|_{\theta=\theta_0}\right]=0.$$

(R_4) For every $\theta_0 \in [a,b]$, there exists $0 < K_2(\theta_0) < \infty$ such that

$$|f(x,\theta_0)-f(\theta_0,\theta_0)| \leqslant K_2(\theta_0)|x-\theta_0|^{\lambda}, \qquad -A \leqslant x \leqslant A.$$

(R_5) $\beta(\theta)$ and $\gamma(\theta)$ are twice differentiable with bounded second derivatives.

It can be shown that a MLE $\hat{\theta}_n$ based on i.i.d. observations X_i distributed with density $f(x,\theta)$ is strongly consistent under condition (R_1). The method is essentially the same as that of Wald (1949).

Let us now consider the log-likelihood ratio process

$$L_n(\theta)-L_n(\theta_0)=\sum_{i=1}^{n}{}^{*}[\varepsilon(X_i,\theta)|X_i-\theta|^{\lambda}-\varepsilon(X_i,\theta_0)|X_i-\theta_0|^{\lambda}] + \sum_{i=1}^{n}[g(X_i,\theta)-g(X_i,\theta_0)] \tag{3.7.1}$$

where $\sum_{i=1}^{n}*$ denotes that the sum is extended over those X_i for which $|X_i-\theta_0| \leqslant A$. Since our interests center on obtaining the limiting distribution of the MLE $\hat{\theta}_n$, if any, without loss of generality, we can and will assume that $\theta_0=0$. The following estimates can be obtained on the expectation and variance of $L_n(\theta)-L_n(0)$. We omit the proofs. For details, see Prakasa Rao (1966, 1968).

Lemma 3.7.1. Let $\beta=\beta(0)$, $\gamma=\gamma(0)$, and $f=f(0,0)$. Under conditions (R_1)–(R_5), the following relations hold:

$$E_0[L_n(\theta)-L_n(0)] = -nCf|\theta|^{2\lambda+1}[1+o(1)], \tag{3.7.2}$$

$$\text{Var}_0[L_n(\theta)-L_n(0)] = 2nCf|\theta|^{2\lambda+1}[1+o(1)], \tag{3.7.3}$$

and in general, for any θ and ϕ in $[a,b]$,

$$E_0[L_n(\theta)-L_n(0)] \leqslant -nH|\theta|^{2\lambda+1} \tag{3.7.4}$$

and

(3.7.5) $$\mathrm{Var}_0[L_n(\theta)-L_n(\phi)]\leqslant nQ|\theta-\phi|^{2\lambda+1},$$

where

(3.7.6) $$C=\Gamma(\lambda+1)\Gamma(\tfrac{1}{2}-\lambda)[2^{2\lambda+1}\pi^{1/2}(2\lambda+1)]^{-1}[\beta^2+\gamma^2-2\beta\gamma\cos\pi\lambda]$$

and H, Q are nonnegative constants independent of θ, ϕ, and n.

As a consequence of Lemma 3.7.1, we obtain the following proposition, which enables us to conclude that the probability that the log-likelihood ratio $L_n(\theta)-L_n(0)$ attains its maximum outside the interval $[-\tau n^{-\rho}, \tau n^{-\rho}]$, where $\rho=(2\lambda+1)^{-1}$, approaches zero as $n\to\infty$ and $\tau\to\infty$. More precisely, we have the following proposition.

Proposition 3.7.2. There exists $\eta>0$ such that

(3.7.7) $$\lim_{\tau\to\infty}\overline{\lim_{n\to\infty}}\, P_0\left[\sup_{|\theta|>\tau n^{-\rho}}\left\{\frac{L_n(\theta)-L_n(0)}{n|\theta|^{2\lambda+1}}\right\}\geqslant-\eta\right]=0.$$

Proof. Let $M_n(\theta)=L_n(\theta)-L_n(0)$. Since $M_n(\theta)$ is continuous in θ, it is sufficient to prove that there exists an $\eta>0$ such that

(3.7.8) $$\lim_{\tau\to\infty}\overline{\lim_{n}}\, P_0\left[\sup_{|\theta_{ijk}|>\tau n^{-\rho}}\left\{\frac{M_n(\theta)}{n|\theta_{ijk}|^{2\lambda+1}}\right\}\geqslant-\eta\right]=0,$$

where $\{\theta_{ijk}\}$ is dense in the set $\{\theta:|\theta|>\tau n^{-\rho}\}$. Let

$$\theta_{ijk}=\tau n^{-\rho}2^{i+k2^{-j}},\quad i\geqslant 0,\quad j\geqslant 0,\quad 0\leqslant k<2^j.$$

Clearly the set $\{\theta_{ijk}\}$ is dense in $\{\theta:\theta>\tau n^{-\rho}\}$. Let $\zeta=2\lambda+1$, and define

(3.7.9) $$T_n(\theta)=M_n(\theta)-E_0[M_n(\theta)]-nH\theta^{\zeta},$$

where H is as given in Lemma 3.7.1. In view of Lemma 3.7.1, it is easy to check that

(3.7.10) (i) $E_0[T_n(\theta_{i00})]=-H\tau^{\zeta}2^{i\zeta}$,

(ii) $E_0[T_n(\theta_{i,j,2k+1})-T_n(\theta_{i,j-1,k})]=0, \qquad 0\leqslant k<2^{j-1}-1$,

(iii) $\mathrm{Var}_0[T_n(\theta_{i00})]\leqslant Q\tau^{\zeta}2^{i\zeta}$,

(iv) $\mathrm{Var}_0[T_n(\theta_{i,j,2k+1})-T_n(\theta_{i,j-1,k})]\leqslant Q\tau^{\zeta}(2\log 2)^{\zeta}2^{(i-j)\zeta}$.

For any $0<\eta<H$, it is clear that

$$P_0\left[\sup_{\theta_{ijk}>\tau n^{-\rho}}\frac{M_n(\theta)}{n\theta_{ijk}^{\zeta}}\geqslant-\eta\right]\leqslant P_0\left[\sup_{\theta_{ijk}>\tau n^{-\rho}}\frac{T_n(\theta)}{n\theta_{i00}^{\zeta}}\geqslant-\eta\right].$$

Let $0<\xi<H, p_j=\delta 2^{-\lambda j/2}$, where $0<\delta<\xi(2^{\lambda/2}-1)$ and $\eta=\xi-\delta(2^{\lambda/2}-1)^{-1}$. Then

$$\begin{aligned}P_0\left[\sup_{\theta_{ijk}>\tau n^{-\rho}}\frac{T_n(\theta)}{n\theta_{i00}^{\zeta}}\geqslant-\eta\right]&\leqslant\sum_{i=0}^{\infty}P_0[T_n(\theta_{i00})\geqslant-n\xi\theta_{i00}^{\zeta}]\\&\quad+\sum_{i=0}^{\infty}\sum_{j=1}^{\infty}P_0[T_n(\theta_{i,j,2k+1})-T_n(\theta_{i,j-1,k})\geqslant np_j\theta_{i00}^{\zeta}]\\&\leqslant Q\tau^{-\zeta}(1-2^{-\zeta})^{-1}[(H-\xi)^{-2}+\delta^{-2}\\&\quad\cdot(2\log 2)^{\zeta}(2^{\lambda+1}-2)^{-1}]\end{aligned}$$

by (3.7.10). Hence

$$\lim_{\tau\to\infty}\overline{\lim_{n}}\,P_0\left[\sup_{\theta_{ijk}>\tau n^{-\rho}}\frac{M_n(\theta_{ijk})}{n\theta_{ijk}^{\zeta}}\geqslant-\eta\right]=0. \tag{3.7.11}$$

Similar analysis proves that

$$\lim_{\tau\to\infty}\overline{\lim_{n}}\,P_0\left[\sup_{\theta_{ijk}<\ \tau n^{-\rho}}\frac{M_n(\theta_{ijk})}{n|\theta_{ijk}|^{\zeta}}\geqslant-\eta\right]=0. \tag{3.7.12}$$

Combining these two relations, the proposition is proved.

In view of the Proposition 3.7.2, the log-likelihood ratio $M_n(0)$ has the global maximum in the interval $[-\tau n^{-\rho},\tau n^{-\rho}]$ with probability approaching 1 as $n\to\infty$ and $\tau\to\infty$. For any such $\tau>0$, let

$$X_n(t)=M_n(tn^{-\rho})\quad\text{for}\quad t\in[-\tau,\tau] \tag{3.7.13}$$

and

$$A_n(t)=X_n(t)-E_0[X_n(t)]. \tag{3.7.14}$$

Let X be a Gaussian process on $[-\tau,\tau]$ with

$$\begin{aligned}&\text{(i)}\quad E[X(t)]=-Cf|t|^{\zeta},\\&\text{(ii)}\quad \operatorname{cov}[X(t_1),X(t_2)]=Cf[|t_1|^{\zeta}+|t_2|^{\zeta}-|t_1-t_2|^{\zeta}],\end{aligned} \tag{3.7.15}$$

where C is as given in (3.7.6), $f=f(0,0)$ and $\zeta=2\lambda+1$. We can assume that the sample paths of the process X are continuous with probability 1 since

$$E|X(t_1) - X(t_2)|^2 \leqslant Cf|t_1 - t_2|^{\zeta}, \qquad t_1, t_2 \in [-\tau, \tau] \tag{3.7.16}$$

and $\zeta = 2\lambda + 1 > 1$ (cf. Proposition 1.10.8). Let

$$A(t) = X(t) - E[X(t)]. \tag{3.7.17}$$

It can be checked that the finite-dimensional distributions of the process $\{A_n(t), t \in [-\tau, \tau]\}$ converge weakly to the corresponding finite-dimensional distributions of the Gaussian process $\{A(t), t \in [-\tau, \tau]\}$. Furthermore,

$$E_0[A_n(t_1) - A_n(t_2)]^2 < Q|t_1 - t_2|^{\zeta} \tag{3.7.18}$$

for all $t_1, t_2 \in [-\tau, \tau]$ and for all $n \geqslant 1$ by Lemma 3.7.1 where $\zeta = 2\lambda + 1$. Note that the processes A_n and A have continuous sample paths with probability 1. Hence, by Proposition 1.10.12, it follows that the sequence of measures μ_n induced by A_n on $C[-\tau, \tau]$ converges weakly to the measure μ induced by A. In other words, the processes A_n converge in distribution to the process A on $C[-\tau, \tau]$. It is easy to see that $E_0[X_n(t)]$ converges uniformly to $E[X(t)]$ for $t \in [-\tau, \tau]$. Hence, by Slutsky theorem for random elements on complete separable metric spaces, it follows that the sequence of processes X_n converge in distribution to the process X.

Theorem 3.7.3. The sequence of processes X_n converges in distribution to the process X.

For any $x \in C[-\tau, \tau)$, let $g(x)$ denote the point at which x attains its maximum over $[-\tau, \tau]$. If $x_n \to x$, x_n, $x \in C[-\tau, \tau]$, and x has a unique maximum, then it is easy to see that $g(x_n) \to g(x)$. Since the process X has continuous sample paths and the finite-dimensional distributions of $X(t)$, $t \in [-\tau, \tau]$, are absolutely continuous, it follows that the set of the discontinuities of g has zero measure with respect to the measure induced by X on $C[-\tau, \tau]$. Hence we have the following theorem by Proposition 1.10.12.

Theorem 3.7.4. The distribution of the location of the maximum of the log-likelihood ratio $M_n(0)$ over $[-\tau n^{-\rho}, \tau n^{-\rho}]$ converges weakly to the distribution of the location of the maximum of the process X over $[-\tau, \tau]$.

It can be shown that the process X has its maximum in a finite interval $[-\tau, \tau]$ with probability approaching 1 as $|\tau| \to \infty$. In fact

$$P\left[\overline{\lim_{n \to \infty}} \frac{X(\tau)}{Cf|\tau|^{\zeta}} \leqslant -1\right] = 1$$

[cf. Prakasa Rao (1966, 1968)]. Combining Proposition 3.7.1 and Theorem 3.7.4 with the remark made above, we have the following result.

Theorem 3.7.5. Under conditions (R_1)–(R_5), the maximum likelihood estimator $\hat{\theta}_n$ is strongly consistent and

$$n^{\rho}(\hat{\theta}_n - \theta_0) \xrightarrow{\mathcal{L}} Z,$$

where Z has the distribution of the location of the maximum of Gaussian process X with mean and covariance function defined by (3.7.15) where $\rho = (2\lambda + 1)^{-1}$.

Remarks. Observe that $\rho > \frac{1}{2}$ when $0 < \lambda < \frac{1}{2}$. Hence the asymptotic variance of $\hat{\theta}_n$ is $O(n^{-2\rho})$. Such estimators are said to be *hyperefficient.* In the classical or standard setup, as in Section 3.5, the asymptotic variance is $O(n^{-1})$, which is of higher order.

The limiting process X discussed above occurs in the nonregular context in view of non-LAN of the likelihood ratio. Properties of this process are discussed in Pflug (1982). Pflug (1983) also studies limiting log-likelihood ratio process for density families with jumps. He shows that the process is a compound Poisson process generalizing the work of Ibragimov and Hasminskii (1981).

The approach to maximum likelihood theory via weak convergence has been used in the nonparametric context in Prakasa Rao (1969, 1970). For some details, see Prakasa Rao (1983b).

3.8 BERNSTEIN–VON MISES THEOREM

One of the fundamental results in the asymptotic theory of inference is concerned with the approach of the posterior density to the normal density. This result was proved first in Le Cam (1956) and later with a different proof in Le Cam (1958). Special cases of results of this type were first proved by Bernstein and von Mises. Our approach is from Borwanker et al. (1971). We consider the usual case of i.i.d. observations. The discussion here is akin to the work of Bickel and Yahav (1969). Prakasa Rao (1974) generalized these results to arbitrary discrete-time stochastic processes [cf. Basawa and Prakasa Rao (1980a, b)]. The Bernstein–von Mises theorem for a class of diffusion processes and diffusion fields was proved in Prakasa Rao (1981, 1984a).

Let X_i, $1 \leqslant i \leqslant n$, be i.i.d. random variables defined on a measurable space

$(\Omega, \mathscr{B})$ with probability measure P_θ, $\theta \in \Theta \subset R$. Suppose the following regularity conditions hold.

(R_1) The parameter space Θ is an open interval in R. Λ is a prior probability measure on $(\Theta, \mathscr{F})$ when $\mathscr{F}$ is the σ-algebra of Borel subsets of Θ and Λ is absolutely continuous with respect to the Lebesgue measure on R.

(R_2) Suppose $P_\theta \ll \mu$, μ σ-finite on $(\Omega, \mathscr{B})$. Let

$$f(x, \theta) = \frac{dP_\theta}{d\mu}(x).$$

Let $h(x, \theta) = \log f(x, \theta)$. Suppose $\partial h/\partial\theta$ and $\partial^2 h/\partial\theta^2$ exist and are continuous in θ for x a.e. $[\mu]$.

(R_3) For every $\theta \in \Theta$, there exists $\eta(\theta) > 0$ such that

$$E\left[\sup\left\{\left|\frac{\partial^2 h(x,\theta)}{\partial\theta^2}\right| : |\theta - \theta'| < \eta(\theta), \theta' \in \Theta\right\}\right] < \infty.$$

(R_4) For every $\theta \in \Theta$ and any $\varepsilon > 0$,

$$-\infty < E_\theta[\sup\{h(X, \theta') - h(X, \theta) : |\theta' - \theta| \geqslant \varepsilon, \theta' \in \Theta\}] < 0.$$

(R_5) Let

$$i(\theta) \equiv -E_\theta\left[\frac{\partial^2 h(X, \theta)}{\partial\theta^2}\right], \qquad \theta \in \Theta.$$

It is clear that $i(\theta) < \infty$ for all $\theta \in \Theta$. Suppose that $i(\theta) > 0$ and $i(\theta)$ is continuous in θ.

(R_6) Let θ_0 denote the true parameter and $P_0 = P_{\theta_0}$. Let $K(\cdot)$ be non-negative function satisfying the following conditions: there exists $0 < \varepsilon < i_0 \equiv i(\theta_0)$ such that

$$M(\theta_0) = \left(\frac{i_0}{2\pi}\right)^{1/2} \int_{-\infty}^{\infty} K(t) \exp[-(i_0 - \varepsilon)t^2/2]\, dt < \infty.$$

Consider a compact neighborhood U_{θ_0} of θ_0 and an estimator $\hat{\theta}_n = \hat{\theta}_n(X_1, \ldots, X_n)$ such that

$$L_n(\hat{\theta}_n) = \sup_{\theta \in U_{\theta_0}} L_n(\theta),$$

where $L_n(\theta) = \sum_{i=1}^{n} \log f(X_i, \theta)$ is the log-likelihood function. It is clear that one can choose a measurable estimator $\hat{\theta}_n$ since $f(x, \theta)$ is continuous in θ and measurable in x and U_{θ_0} is compact.

(R_7) For every $h > 0$ and every $\delta > 0$,

$$e^{-n\delta} \int_{|t|>h} K(n^{1/2}t)\lambda(\hat{\theta}_n + t)\, dt \to 0 \quad \text{a.e. } [P_{\theta_0}].$$

(R_8) The prior density λ is continuous and positive in an open neighborhood of the true parameter θ_0.

Theorem 3.8.1. Under assumptions (R_1)–(R_5), there exists a compact neighborhood U_{θ_0} of θ_0 such that

(i) $\hat{\theta}_n \to \theta_0$ a.s.,

(ii) $\left.\dfrac{\partial \log L_n(\theta)}{\partial \theta}\right|_{\theta=\hat{\theta}_n} = 0$ for $n \geq N$, N depending on $\omega \in \Omega$, and

(iii) $n^{1/2}(\hat{\theta}_n - \theta_0) \xrightarrow{\mathscr{L}} N(0, i_0^{-1})$.

The proof of this theorem in the more general case of Markov processes is given in Borwanker et al. (1971). We omit it.

Let $f_n(\theta \mid x_1, \ldots, x_n)$ denote the posterior density of θ given the observations $x_1, \ldots, x_n$ with prior probability density λ. Let

$$f_n^*(t \mid x_1, \ldots, x_n) = n^{-1/2} f_n(\theta \mid x_1, \ldots, x_n).$$

Note that $f_n^*(t \mid x_1, \ldots, x_n)$ is the posterior density of $n^{1/2}(\theta - \hat{\theta}_n)$, where $\hat{\theta}_n$ is an estimator satisfying (R_7) and the conditions in Theorem 3.8.1.

Theorem 3.8.2. Under assumptions (R_1)–(R_8),

$$\lim_n \int_{-\infty}^{\infty} K(t) \left| f_n^*(t \mid X_1, \ldots, X_n) - \left(\frac{i_0}{2\pi}\right)^{1/2} \exp(-\tfrac{1}{2}t^2) \right| dt = 0 \quad \text{a.s. } [P_0]. \tag{3.8.1}$$

We prove this result after proving a series of lemmas.

Define

$$v_n(t) = \exp\left[\sum_{i=1}^{n} \{ h(X_i, \hat{\theta}_n + tn^{-1/2}) - h(X_i, \hat{\theta}_n) \} \right] \tag{3.8.2}$$

and

$$(3.8.3)\qquad C_n = \int_{-\infty}^{\infty} v_n(t)\lambda(\hat{\theta}_n + tn^{-1/2})\,dt.$$

It is clear that

$$(3.8.4)\qquad f_n^*(t|X_1,\ldots,X_n) = C_n^{-1} v_n(t)\lambda(\hat{\theta}_n + tn^{-1/2}).$$

Lemma 3.8.3. Suppose conditions (R_1) to (R_5) hold. Then

(i) For every $0 < \varepsilon < i_0$, there exists $\delta_0 > 0$ and integer $N \geq 1$ such that

$$(3.8.5)\qquad v_n(t) \leq \exp[-\tfrac{1}{2}t^2(i_0 - \varepsilon)] \quad \text{for } |t| \leq \delta_0 n^{1/2} \text{ and } n \geq N;$$

(ii) for every $\delta > 0$, there exists $\varepsilon > 0$ and integer $N \geq 1$ such that

$$(3.8.6)\qquad \sup_{|t| > \delta n^{1/2}} v_n(t) \leq \exp[-\tfrac{1}{4}n\varepsilon] \quad \text{for } n \geq N;$$

(iii) for every fixed t,

$$(3.8.7)\qquad \lim_{n\to\infty} v_n(t) = \exp[-\tfrac{1}{2}i_0 t^2] \quad \text{a.s. } [P_0].$$

Proof. Note that

$$(3.8.8)\qquad \log v_n(t) = \sum_{i=1}^{n} [h(X_i, \hat{\theta}_n + tn^{-1/2}) - h(X_i, \hat{\theta}_n)]$$
$$= n^{-1/2} t \sum_{i=1}^{n} \left.\frac{\partial h(X_i,\theta)}{\partial\theta}\right|_{\theta=\hat{\theta}_n} + \frac{t^2}{2n}\sum_{i=1}^{n} \left.\frac{\partial^2 h(X_i,\theta)}{\partial\theta^2}\right|_{\theta=\theta_n'},$$

where $|\theta_n' - \hat{\theta}_n| \leq |t|n^{-1/2}$. The first term on the right-hand side is zero a.s. for $n \geq N_1$ (say) by Theorem 3.8.1. The second term can be written in the form

$$(3.8.9)\qquad \frac{t^2}{2n}\sum_{i=1}^{n} \left.\frac{\partial^2 h(X_i,\theta)}{\partial\theta^2}\right|_{\theta=\theta_0} + \frac{t^2}{2n}\left[\sum_{i=1}^{n}\left\{\left.\frac{\partial^2 h(X_i,\theta)}{\partial\theta^2}\right|_{\theta=\theta_n'} - \left.\frac{\partial^2 h(X_i,\theta)}{\partial\theta^2}\right|_{\theta=\theta_0}\right\}\right].$$

The first term in expression (3.8.9) converges a.s. $[P_{\theta_0}]$ to $-\frac{1}{2}i_0 t^2$ by the SLLN. Hence, for $0 < \varepsilon < i_0$,

$$(3.8.10)\qquad \frac{t^2}{2n}\sum_{i=1}^{n} \left.\frac{\partial^2 h(X_i,\theta)}{\partial\theta^2}\right|_{\theta=\theta_0} < \frac{t^2}{2}\left(-i_0 + \frac{\varepsilon}{2}\right)$$

for $n \geqslant N_2$, where we may take $N_2 \geqslant N_1$. Let $0 < \delta < \frac{1}{2}\eta(\theta_0)$, where $\eta(\theta_0)$ is as given by (R_3). In view of Theorem 3.8.1, $|\hat{\theta}_n - \theta_0| < \delta$ and $|\theta_n' - \hat{\theta}_n| \leqslant |t|n^{-1/2} \leqslant \delta$ for $n \geqslant N_3$ (say). We can assume again that $N_3 \geqslant N_2$. Hence, for $n \geqslant N_3$, $|\theta_n' - \theta_0| \leqslant 2\delta$ and

$$(3.8.11)\qquad n^{-1}\sum_{i=1}^{n}\left\{\frac{\partial^2 h(X_i,\theta)}{\partial\theta^2}\bigg|_{\theta=\theta_0'} - \frac{\partial^2 h(X_i,\theta)}{\partial\theta^2}\bigg|_{\theta=\theta_0}\right\}$$
$$\leqslant n^{-1}\sum_{i=1}^{n}\sup_{|\theta-\theta_0|\leqslant 2\delta}\left\{\frac{\partial^2 h(X_i,\theta)}{\partial\theta^2} - \frac{\partial^2 h(X_i,\theta)}{\partial\theta^2}\bigg|_{\theta=\theta_0}\right\}$$

which in turn tends to

$$(3.8.12)\qquad E\left\{\sup_{|\theta-\theta_0|\leqslant 2\delta}\left[\frac{\partial^2 h(X,\theta)}{\partial\theta^2} - \frac{\partial^2 h(X,\theta)}{\partial\theta^2}\bigg|_{\theta=\theta_0}\right]\right\}$$

by (R_3). Condition (R_2) implies that the expression inside the expectation given in (3.8.12) tends to zero a.s. $[P_0]$, and, furthermore, it is bounded by an integrable function

$$(3.8.13)\qquad 2\sup_{|\theta-\theta_0|\leqslant\eta(\theta_0)}\left|\frac{\partial^2 h(x,\theta)}{\partial\theta^2}\right|$$

by (R_3). Hence, by the dominated convergence theorem, there exists $0 < \delta_0 < \frac{1}{2}\eta(\theta_0)$ such that

$$(3.8.14)\qquad E\left\{\sup_{|\theta-\theta_0|\leqslant 2\delta_0}\left[\frac{\partial^2 h(X,\theta)}{\partial\theta^2} - \frac{\partial^2 h(X,\theta)}{\partial\theta^2}\bigg|_{\theta=\theta_0}\right]\right\} < \frac{\varepsilon}{4}.$$

Combining the relations (3.8.8)–(3.8.14), we have

$$(3.8.15)\qquad \log v_n(t) < -\tfrac{1}{2}t^2(i_0 - \varepsilon) \quad \text{for } |t| \leqslant n^{1/2}\delta_0, \qquad n > N,$$

proving (3.8.5). We now prove (3.8.6). Note that

$$(3.8.16)\qquad n^{-1}\log v_n(t) = n^{-1}\sum_{i=1}^{n}[h(X_i,\hat{\theta}_n + tn^{-1/2}) - h(X_i,\theta_0)]$$
$$+ n^{-1}\sum_{i=1}^{n}[h(X_i,\theta_0) - h(X_i,\hat{\theta}_n)].$$

Suppose that $|\hat{\theta}_n - \theta_0| < \delta/2$ for $n \geqslant N_4$. Then $|tn^{-1/2}| > \delta$ implies that

$|\hat{\theta}_n + tn^{-1/2} - \theta_0| > \delta/2$. Hence, for $n \geqslant N_4$,

$$(3.8.17) \qquad n^{-1} \sum_{i=1}^{n} [h(X_i, \hat{\theta}_n + tn^{-1/2}) - h(X_i, \theta_0)] \leqslant n^{-1} \sum_{i=1}^{n} \sup_{|\theta - \theta_0| > \delta/2} \{h(X_i, \theta) - h(X_i, \theta_0)\}$$

and the last term tends a.s. $[P_0]$ to

$$(3.8.18) \qquad E\left[\sup_{|\theta - \theta_0| > \delta/2} \{h(X, \theta) - h(X, \theta_0)\}\right] < 0$$

by (R_4). Furthermore,

$$(3.8.19) \quad n^{-1} \sum_{i=1}^{n} \{h(X_i, \theta_0) - h(X_i, \hat{\theta}_n)\} = \frac{1}{n}(\theta_0 - \hat{\theta}_n) \sum_{i=1}^{n} \left.\frac{\partial h(X_i, \theta)}{\partial \theta}\right|_{\theta = \hat{\theta}_n} + \frac{1}{2n}(\theta_0 - \hat{\theta}_n)^2 \sum_{i=1}^{n} \left.\frac{\partial^2 h(X_i, \theta)}{\partial \theta^2}\right|_{\theta = \theta_n'},$$

which, by arguments given earlier and the fact that $\hat{\theta}_n \to \theta_0$ a.s., converges to zero a.s. P_0. Now let ε be such that

$$(3.8.20) \qquad 0 < \varepsilon < -E\left[\sup_{|\theta - \theta_0| > \delta/2} \{h(X, \theta) - h(X, \theta_0)\}\right].$$

Combining (3.8.16)–(3.8.19), it follows that, for $|t| > n^{1/2}\delta$ and $n \geqslant N = \max(N_4, N_3)$,

$$(3.8.21) \qquad \log v_n(t) \leqslant -\tfrac{1}{4} n\varepsilon,$$

which proves (3.8.6). For any fixed t and $\varepsilon > 0$, choose $\varepsilon_1 > 0$ such that

$$(3.8.22) \qquad 0 < (t^2/2)\varepsilon_1 < \varepsilon.$$

It is easy to check, from (3.8.8)–(3.8.11), that

$$(3.8.23) \qquad |\log v_n(t) + \tfrac{1}{2} t^2 i_0| < \tfrac{1}{2} t^2 \varepsilon_1 < \varepsilon$$

for $n \geqslant \max(N, t^2/\delta^2)$. This proves (3.8.7).

This completes the proof of Lemma 3.8.3.

Lemma 3.8.4. Under assumptions (R_1)–(R_8), there exists $\delta_0 > 0$ such that

$$\lim_{n\to\infty}\int_{|t|\leqslant\delta_0 n^{1/2}} K(t)|v_n(t)\lambda(\hat\theta_n + tn^{-1/2}) - \lambda(\theta_0)\exp(-\tfrac{1}{2}i_0t^2)|\,dt = 0 \quad \text{a.s. } [P_0]. \tag{3.8.24}$$

Proof. Observe that

$$\begin{aligned}\int_{|t|\leqslant\delta_0 n^{1/2}} & K(t)|v_n(t)\lambda(\hat\theta_n + tn^{-1/2}) - \lambda(\theta_0)\exp(-\tfrac{1}{2}i_0t^2)|\,dt \\ &\leqslant \int_{|t|\leqslant\delta_0 n^{1/2}} K(t)\lambda(\theta_0)|v_n(t) - \exp(-\tfrac{1}{2}i_0t^2)|\,dt \\ &\quad + \int_{|t|\leqslant\delta_0 n^{1/2}} K(t)v_n(t)|\lambda(\theta_0) - \lambda(\hat\theta_n + tn^{-1/2})|\,dt.\end{aligned} \tag{3.8.25}$$

Let $\varepsilon > 0$ such that

$$\int_{-\infty}^{\infty} K(t)\exp\{-\tfrac{1}{2}(i_0 - \varepsilon)t^2\}\,dt < \infty. \tag{3.8.26}$$

This can be done by (R_6). In view of Lemma 3.8.3, there exists $\delta_1 > 0$, and an integer $N \geqslant 1$ such that

$$v_n(t) \leqslant \exp[-(i_0 - \varepsilon)\tfrac{1}{2}t^2], \qquad |t| \leqslant \delta_1 n^{1/2}, \qquad n \geqslant N. \tag{3.8.27}$$

In view of (3.8.7), applying the dominated convergence theorem, it follows that

$$\int_{|t|\leqslant\delta_1 n^{1/2}} K(t)\lambda(\theta_0)|v_n(t) - \exp(-\tfrac{1}{2}i_0t^2)|\,dt \to 0 \qquad \text{a.s. } [P_0]. \tag{3.8.28}$$

On the other hand,

$$\begin{aligned}\int_{|t|\leqslant\delta_1 n^{1/2}} & K(t)v_n(t)|\lambda(\theta_0) - \lambda(\hat\theta_n + tn^{-1/2})|\,dt \\ &\leqslant \sup_{|\theta-\theta_0|\leqslant\delta_1}|\lambda(\theta) - \lambda(\theta_0)|\int_{|t|\leqslant\delta_1 n^{1/2}} K(t)\exp[-(i_0 - \varepsilon)\tfrac{1}{2}t^2]\,dt.\end{aligned} \tag{3.8.29}$$

For a given η, we can choose $0<\delta_0\leqslant\delta_1$ such that

$$(3.8.30)\qquad \sup_{|\theta-\theta_0|\leqslant\delta}|\lambda(\theta)-\lambda(\theta_0)|\int_{|t|\leqslant\delta_0 n^{1/2}}K(t)\exp[-(i_0-\varepsilon)\tfrac{1}{2}t^2]\,dt<\eta.$$

Relations (3.8.23), (3.8.28), and (3.8.30) prove the lemma.

Lemma 3.8.5. Under assumptions (R_1)–(R_8), for every $\delta>0$,

$$(3.8.31)\qquad \lim_{n\to\infty}\int_{|t|>\delta n^{1/2}}K(t)|v_n(t)\lambda(\hat{\theta}_n+tn^{-1/2})-\lambda(\theta_0)\exp(-\tfrac{1}{2}i_0t^2)|\,dt=0 \quad \text{a.s. } [P_0].$$

Proof. It is clear that

$$\begin{aligned}\int_{|t|>\delta n^{1/2}}&K(t)|v_n(t)\lambda(\hat{\theta}_n+tn^{-1/2})-\lambda(\theta_0)\exp(-\tfrac{1}{2}i_0t^2)|\,dt\\ &\leqslant\int_{|t|>\delta n^{1/2}}K(t)v_n(t)\lambda(\hat{\theta}_n+tn^{-1/2})\,dt+\int_{|t|>\delta n^{1/2}}K(t)\lambda(\theta_0)\exp(-\tfrac{1}{2}i_0t^2)\,dt\\ &\leqslant e^{-n\varepsilon/4}\int_{|t|>\delta n^{1/2}}\lambda(\hat{\theta}_n+tn^{-1/2})\,dt+\lambda(\theta_0)\int_{|t|>\delta n^{1/2}}K(t)\exp(-\tfrac{1}{2}i_0t^2)\,dt\end{aligned}$$

by (3.8.6) and the terms on the right hand side of the inequality tend to zero a.s. $[P_0]$ by (R_7) and (R_6).

We now prove the main theorem.

Proof of Theorem 3.8.2. In view of Lemmas 3.8.4 and 3.8.5, we obtain that

$$(3.8.32)\qquad \lim_{n\to\infty}\int K(t)|v_n(t)\lambda(\hat{\theta}_n+tn^{-1/2})-\lambda(\theta_0)\exp(-\tfrac{1}{2}i_0t^2)|\,dt=0 \text{ a.s. } [P_0].$$

Choosing $K(t)\equiv 1$, which satisfies conditions (R_6) and (R_7) trivially, we get

$$\begin{aligned}(3.8.33)\qquad C_n&=\int_{-\infty}^{\infty}v_n(t)\lambda(\hat{\theta}_n+tn^{-1/2})\,dt\to\lambda(\theta_0)\int_{-\infty}^{\infty}\exp(-\tfrac{1}{2}i_0t^2)\,dt\\ &=\lambda(\theta_0)\left(\frac{2\pi}{i_0}\right)^{1/2}.\end{aligned}$$

Hence

$$(3.8.34)\quad \int K(t)\left| f_n^*(t|X_1,\ldots,X_n) - \left(\frac{i_0}{2\pi}\right)^{1/2} \exp(-\tfrac{1}{2}i_0 t^2)\right| dt$$

$$\leqslant \int K(t)|C_n^{-1}\lambda(\hat{\theta}_n + tn^{-1/2}) - C_n^{-1}\lambda(\theta_0)\exp(-\tfrac{1}{2}i_0 t^2)|\,dt$$

$$+ \int K(t)\left| C_n^{-1}\lambda(\theta_0) - \left(\frac{i_0}{2\pi}\right)^{1/2}\right| \exp(-\tfrac{1}{2}i_0 t^2)\,dt$$

and the last two terms tend to zero a.s. $[P_0]$ by (3.8.22) and (3.8.23) completing the proof of Theorem 3.8.2.

As a corollary of Theorem 3.8.2, we obtain the following result, which generalizes the theorem of Bernstein and von Mises.

Theorem 3.8.6. Suppose conditions (R_1)–(R_5) hold. Let the prior density $\lambda(\cdot)$ satisfy (R_8) and

$$(3.8.35)\qquad \int_{-\infty}^{\infty} |\theta|^m \lambda(\theta)\,d\theta < \infty$$

for some integer $m \geqslant 0$. Then

$$(3.8.36)\quad \lim_{n\to\infty} \int_{-\infty}^{\infty} |t|^m \left| f_n^*(t|X_1,\ldots,X_n) - \left(\frac{i_0}{2\pi}\right)^{1/2} \exp(-\tfrac{1}{2}i_0 t^2)\right| dt = 0$$

a.s. $[P]$.

The result follows by taking $K(t) = |t|^m$, $m \geqslant 0$, and verifying (R_6) and (R_7). We omit the details. For $m = 0$, we obtain the classic form of the Bernstein–von Mises theorem, which asserts that the normalized posterior density converges to normal density in L_1-mean.

We have discussed the scalar parameter case. The multidimensional parameter case can be carried out without any difficulty.

As an application of Theorem 3.8.6, we now derive the asymptotic properties of regular Bayes estimators, that is, estimators $T_n = T_n(X_1,\ldots,X_n)$ that minimize the posterior risk $R_n(\psi) = \int L(\theta,\psi) f_n(\theta|X_1,\ldots,X_n)\,d\theta$, where $L(\theta,\psi)$ is a loss function. We assume that there exists a measurable regular Bayes estimator.

Theorem 3.8.7. Let conditions (R_1)–(R_5) and (R_8) hold and T_n be a regular

Bayes estimator with respect to a nonnegative loss function $L(\theta, \psi)$ of the type $L(\theta, \psi) = l(\theta - \psi)$, where $l(t_1) \geqslant l(t_2)$ if $t_1 \geqslant t_2 \geqslant 0$ or $t_1 \leqslant t_2 \leqslant 0$. Furthermore, suppose that there exist a nonnegative sequence $\{a_n\}$ and functions $K(\cdot)$, $G(\cdot)$ such that

(3.8.37) (i) $G(\cdot)$ satisfies (R_6) and (R_7),

(ii) $a_n l(tn^{-1/2}) \leqslant G(t)$ for all t and n,

(iii) $a_n l(tn^{-1/2}) \to K(t)$ uniformly on compact sets, and

(iv) $\int K(t+m)\exp(-\frac{1}{2}i_0 t^2)dt$ has a strict minimum at $m = 0$.

Then

(3.8.38) (i) $T_n \to \theta_0$ a.s. $[P_{\theta_0}]$,

(ii) $n^{1/2}(\theta_0 - T_n) \xrightarrow{\mathscr{L}} N(0, i_0^{-1})$,

(iii) $a_n B_n(T_n) \to \left(\frac{i_0}{2\pi}\right)^{1/2} \int_{-\infty}^{\infty} K(t)\exp(-\frac{1}{2}i_0 t^2)dt$ a.s. $[P_{\theta_0}]$.

Proof. It is sufficient to prove that $n^{1/2}(\hat{\theta}_n - T_n) \to 0$ a.s. $[P_{\theta_0}]$ and that

$$a_n B_n(\hat{\theta}_n) \to \left(\frac{i_0}{2\pi}\right)^{1/2} \int_{-\infty}^{\infty} K(t)\exp(-\tfrac{1}{2}i_0 t^2)dt$$

from which (3.8.38) follows easily in view of Theorem 3.8.1. Note that

$$\begin{aligned}
\text{(3.8.39)}\quad \overline{\lim_n}\, a_n B_n(T_n) &\leqslant \overline{\lim_n}\, a_n B_n(\hat{\theta}_n) \quad \text{(since } T_n \text{ is Bayes)} \\
&\leqslant \overline{\lim_n} \int_{-\infty}^{\infty} a_n l(\theta - \hat{\theta}_n) f_n(\theta \mid X_1, \ldots, X_n)d\theta \\
&= \overline{\lim_n} \int_{-\infty}^{\infty} a_n l(tn^{-1/2}) f_n^*(t \mid X_1, \ldots, X_n)dt \\
&\leqslant \lim_n \int_{-\infty}^{\infty} |a_n l(tn^{-1/2}) - K(t)| \\
&\quad \cdot \left| f_n^*(t \mid X_1, \ldots, X_n) - \left(\frac{i_0}{2\pi}\right)^{1/2} \exp(-\tfrac{1}{2}i_0 t^2)\right| dt
\end{aligned}$$

$$+\overline{\lim_n}\left(\frac{i_0}{2\pi}\right)^{1/2}\int_{-\infty}^{\infty}|a_n l(tn^{-1/2})-K(t)|\exp(-\tfrac{1}{2}i_0t^2)dt$$

$$+\overline{\lim_n}\int_{-\infty}^{\infty}K(t)f_n^*(t|X_1,\ldots,X_n)dt.$$

The first term on the right-hand side of (3.8.39) is smaller than

$$\overline{\lim_n}\int_{-\infty}^{\infty}2G(t)\left|f_n^*(t|X_1,\ldots,X_n)-\left(\frac{i_0}{2\pi}\right)^{1/2}\exp(-\tfrac{1}{2}i_0t^2)\right|dt=0 \quad \text{a.s. } [P_{\theta_0}] \tag{3.8.40}$$

by Theorem 3.8.2. The second term converges to zero by the dominated convergence theorem. The last term converges to

$$\left(\frac{i_0}{2\pi}\right)^{1/2}\int_{-\infty}^{\infty}K(t)\exp(-\tfrac{1}{2}i_0t^2)dt. \tag{3.8.41}$$

Hence

$$\overline{\lim_n}\, a_nB_n(T_n)\leqslant\left(\frac{i_0}{2\pi}\right)^{1/2}\int K(t)\exp(-\tfrac{1}{2}i_0t^2)dt. \tag{3.8.42}$$

We now observe that $\overline{\lim}_n|U_n|<\infty$ a.s. where $n^{1/2}(\hat\theta_n-T_n)=U_n$. For, if not, then, for every $M>0$, there exists a measurable set A_M with $P_{\theta_0}(A_M)>0$ such that

$$|U_n|>M \quad \text{infinitely often for } \omega\in A_M.$$

Without loss of generality, assume that $U_n>M$ infinitely often. Let $\{U_{n_j}\}$ be a subsequence such that $U_{n_j}>M$ for all $j\geqslant 1$. Then

$$\begin{aligned}
a_{n_j}B_{n_j}(T_{n_j}) &= \int a_{n_j}l((t+U_{n_j})n_j^{-1/2})f_{n_j}^*(t|X_1,\ldots,X_{n_j})dt\\
&\geqslant \int_{|t|\leqslant M} a_{n_j}l((t+U_{n_j})n_j^{-1/2})f_{n_j}^*(t|X_1,\ldots,X_{n_j})dt\\
&\geqslant \int_{|t|\leqslant M} a_{n_j}l((t+M)n_j^{-1/2})f_{n_j}^*(t|X_1,\ldots,X_n)dt\\
&\to \int_{|t|\leqslant M} K(t+M)\left(\frac{i_0}{2\pi}\right)^{1/2}\exp(-\tfrac{1}{2}i_0t^2)dt
\end{aligned}$$

by Theorem 3.8.2. Since $K(t+M)I_{[|t|\leqslant M]}$ is nondecreasing in M for each fixed t,

$$\lim_{M\to\infty}\int_{|t|\leqslant M} K(t+M)\left(\frac{i_0}{2\pi}\right)^{1/2}\exp(-\tfrac{1}{2}i_0t^2)dt$$
$$= K(\infty) > \int_{-\infty}^{\infty}\left(\frac{i_0}{2\pi}\right)^{1/2} K(t)\exp(-\tfrac{1}{2}i_0t^2)dt$$

by (3.8.37)(iv). Hence, for large M, on a set of positive probability

$$\underline{\lim}_n\, a_nB_n(T_n) > \left(\frac{i_0}{2\pi}\right)^{1/2}\int_{-\infty}^{\infty} K(t)\exp(-\tfrac{1}{2}i_0t^2)dt$$
$$\geqslant \overline{\lim}_n\, a_nB_n(\hat{\theta}_n),$$

contradicting the fact that T_n is Bayes estimator. Hence

(3.8.43) $$\overline{\lim}_n |U_n| < \infty \quad \text{a.s.}$$

Let $\varepsilon > 0$ and B_M be the event that $P(B_M) > 1-\varepsilon$ and $|U_n| \leqslant M$ for $\omega\in B_M$. For any $\omega\in B_M$, $\{U_n\}$ is a bounded sequence. Let m be a limit point of U_n. If possible, suppose $m \neq 0$. Let $\{U_{n_j}\}$ be a subsequence of U_n converging to m. Then

$$\underline{\lim}_j\, a_{n_j}B_{n_j}(T_{n_j}) \geqslant \underline{\lim}_j \int_{-T_0}^{T_0} a_{n_j} l\left(\frac{t+U_{n_j}}{n_j^{1/2}}\right) f^*_{n_j}(t|x_1,\ldots,x_{n_j})dt$$
$$\geqslant \int_{-T_0}^{T_0} \underline{\lim}_j\, a_{n_j} l\left(\frac{t+U_{n_j}}{n_j^{1/2}}\right) f^*_{n_j}(t|x_1,\ldots,x_{n_j})dt$$
$$= \left(\frac{i_0}{2\pi}\right)^{1/2}\int_{-T_0}^{T_0} K(t+M)\exp(-\tfrac{1}{2}i_0t^2)dt$$

by (3.8.38)(iii). Hence, for T_0 large,

(3.8.44) $$\underline{\lim}_j\, a_{n_j}B_{n_j}(T_{n_j}) > \left(\frac{i_0}{2\pi}\right)^{1/2}\int_{-\infty}^{\infty} K(t)\exp(-\tfrac{1}{2}t^2i_0)dt - \eta$$

by (3.8.38)(iv). Since (3.8.44) holds for every $\eta > 0$, we have

$$\underline{\lim}_j\, a_{n_j}B_{n_j}(T_{n_j}) > \left(\frac{i_0}{2\pi}\right)^{1/2}\int_{-\infty}^{\infty} K(t)\exp(-\tfrac{1}{2}t^2i_0)dt,$$

which is impossible by (3.8.42). Hence $m=0$ and $n^{1/2}(T_n-\hat{\theta}_n)\to 0$ a.s. $[P_{\theta_0}]$. Furthermore,

$$\underline{\lim}_{j}\, a_n B_n(T_n)=\underline{\lim}_{j}\, a_n B_n(\hat{\theta}_n)=\left(\frac{i_0}{2\pi}\right)^{1/2}\int K(t)\exp(-\tfrac{1}{2}i_0 t^2)dt.$$

Remarks. Hipp and Michel (1976) obtained the rate of convergence of posterior distribution to normal distribution improving over an earlier result of Strasser (1975a, b). Strasser (1977) obtained improved bounds for the difference between Bayes and maximum likelihood estimators. He proved that $n|T_n-\hat{\theta}_n|=O_p(1)$ under some conditions. These results were discussed in Prakasa Rao (1978a, b; 1979) for discrete-time stationary Markov processes. We shall not discuss these results here.

3.9 GLOBAL ASYMPTOTIC LOWER BOUNDS FOR RISK OF ESTIMATORS

In this section, we investigate the local and global asymptotic lower bounds for risk functions of estimators. Our approach is that of Strasser (1978, 1982).

Let $(\Omega_n, \mathscr{A}_n)$, $n\geqslant 1$, be a sequence of measurable spaces and $\Theta\subset R^k$ be open. Let $P_{\theta,n}$ be a probability measure on $(\Omega_n, \mathscr{A}_n)$ for every $\theta\in\Theta$ and suppose that for any fixed $A\in A_n$ and $n\geqslant 1$, $P_{\theta,n}(A)$ is a Borel-measurable function of θ.

For any two probability measures P and Q defined on a measurable space $(\Omega, \mathscr{A})$, let dP/dQ denote the Radon–Nikodým derivative of the absolutely continuous component of P with respect to Q.

Definition 3.9.1. Let τ be a probability measure on Θ. The sequence $\{P_{\theta,n}: \theta\in\Theta\}$ is said to satisfy *local asymptotic normality* (*LAN*) *in τ-measure* if there exists

- **(i)** $\delta_n\downarrow 0$,
- **(ii)** $\mathbf{\Gamma}(\theta)$, a symmetric $k\times k$ matrix positive definite a.e. τ,
- **(iii)** a measurable function $\mathbf{\Delta}_n:\Theta\times\Omega_n\to R^k$ such that

$$\int f(\cdot,\theta)d\mathscr{L}(\mathbf{\Delta}_n(\theta)|P_{\theta,n})\to\int f(\cdot,\theta)dN(\mathbf{0},\mathbf{\Gamma}(\theta)) \tag{3.9.1}$$

in τ-measure for all measurable functions $f:R^k\times\Theta\to R$ such that for any given θ, $f(\cdot,\theta)\in C_{00}(R^k)$ [here $C_{00}(R^k)$ denotes the space of continuous functions with compact support, $\mathscr{L}(\Delta_n(\theta)|P_{\theta,n})$ denotes the distribution of $\mathbf{\Delta}_n(\theta)$ under $P_{\theta,n}$, and $N(\mathbf{0},\mathbf{\Gamma}(\theta))$ denotes the k-variate normal distribution with mean vector zero and covariance matrix $\mathbf{\Gamma}(\theta)$], and

(iv) there exists a measurable function $z_n:\Theta \times R^k \times \Omega_n \to R$ such that

(3.9.2) $$\lim_{n\to\infty} z_n(\cdot,\mathbf{t})=0 \quad \text{in} \int_{\Theta} P_{\theta,n}\tau(d\theta)\text{-measure for } \mathbf{t}\in R^k$$

and

(3.9.3) $$\frac{dP_\theta+\delta_n\mathbf{t},n}{dP_{\theta,n}}=\exp\{\mathbf{t}'\mathbf{\Delta}_n(\theta)-\tfrac{1}{2}\mathbf{t}'\mathbf{\Gamma}(\theta)\mathbf{t}+Z_n(\theta,\mathbf{t})\}$$

for $\theta\in\Theta$, $\mathbf{t}\in R^k$, $n\geqslant 1$.

Definition 3.9.2. Suppose the sequence $\{P_{\theta,n},\theta\in\Theta\}$ is LAN in τ-measure. A sequence $\{\mathbf{T}_n\}$ is said to be *similar to logarithm of likelihood ratio (SLLR) in τ-measure* if

(3.9.4) $$\lim_{n\to\infty}\int_{R^k} P_{\theta,n}\{\|\delta_n^{-1}(T_n-\theta)-\mathbf{\Gamma}^{-1}(\theta)\mathbf{\Delta}_n(\theta)\|>\varepsilon\}\tau(d\theta)=0.$$

Let $W:R^k\to[0,1]$ be a loss function such that $1-W\in C_0(R^k)$, the uniform hull of $C_{00}(R^k)$. Suppose further that for every $N(\mathbf{0},\mathbf{A})$ and every probability measure P

(3.9.5) $$\int W dN(\mathbf{0},\mathbf{A})\leqslant\int W d(N(\mathbf{0},\mathbf{A})*P).$$

It can be checked that the loss function $W(\mathbf{t})=L(\|\mathbf{t}\|)$, where L is non-decreasing, satisfies (3.8.5). Here $N(\mathbf{0},\mathbf{A})$ denotes normal probability measure with mean vector $\mathbf{0}$ and covariance matrix $\mathbf{A}$ as before.

The following theorem gives a lower bound for the risk of any sequence of estimators $\{\Psi_n\}$. Let λ denote the Lebesgue measure on R^k.

Theorem 3.9.1. Suppose $\tau\ll\lambda$ and $\{P_{\theta,n},\theta\in\Theta\}$ is LAN in τ-measure. Then, for every sequence of estimators $\{\Psi_n\}$, the following inequality holds for any $g\in L_1^+(\tau)$:

(3.9.6) $$\varliminf_{n\to\infty}\iint W(\delta_n^{-1}(\theta-\Psi_n))g(\theta)dP_{\theta,n}\tau(d\theta)\geqslant\int W(\cdot)g(\theta)d\mathbf{N}(\theta)\tau(d\theta)$$

where $\mathbf{N}(\theta)$ denotes $N(0,\mathbf{\Gamma}^{-1}(\theta))$.

Before we prove this theorem, we prove a convolution theorem. An earlier result of this kind is in Hajek (1970). The convolution theorem gives a characterization of all the possible limiting distributions of the sequence $\{\Psi_n\}$ under some conditions.

Suppose τ is a probability measure on Θ and the sequence $\{P_{\theta,n}:\theta\in\Theta\}$ is LAN in τ-measure. Let $\{\Psi_n\}$ be any sequence of estimators.

Define

(3.9.7) $$R_n(\theta, \mathbf{t}) = \mathscr{L}(\mathbf{t} - \delta_n^{-1}(\Psi_n - \theta) | P_{\theta+\delta_n\mathbf{t},n})$$

and

(3.9.8) $$S_n(\theta, \mathbf{t}) = \mathscr{L}(\mathbf{\Gamma}^{-1}(\theta)\mathbf{\Delta}_n(\theta) - \delta_n^{-1}(\Psi_n - \theta) | P_{\theta+\delta_n\mathbf{t},n})$$

whenever $\theta + \delta_n\mathbf{t}\in\theta$. Let $D_\alpha = \{\mathbf{t}\in R^k: \|\mathbf{t}\| \leqslant \alpha\}$, $\alpha \geqslant 0$. Let

(3.9.9) $$R_{n,\alpha}(\theta) = \frac{1}{\lambda(D_\alpha)}\int_{D_\alpha} R_n(\theta, \mathbf{t})\,d\mathbf{t}$$

and

(3.9.10) $$S_{n,\alpha}(\theta) = \frac{1}{\lambda(D_\alpha)}\int_{D_\alpha} S_n(\theta, \mathbf{t})\,d\mathbf{t}.$$

Theorem 3.9.2. Under the assumptions stated above, if $\{\Psi_n\}$ is any sequence of estimators, then

(3.9.11) $$\lim_{\alpha\to\infty} \overline{\lim_{n}} \int \| R_{n,\alpha}(\theta) - N(\theta) * S_{n,\alpha}(\theta) \| \tau(d\theta) = 0,$$

where $\|P - Q\|$ denotes the variation norm between the two probability measures P and Q and $*$ denotes the convolution operation.

Since the proof of this theorem is long, we state two lemmas before giving a proof of it. For proofs of these lemmas, see Strasser (1978).

Lemma 3.9.3. Suppose τ is a probability measure on Θ and $\{P_{\theta,n}:\theta\in\Theta\}$ is LAN in τ-measure. Then there exist

(i) $K_n \uparrow \infty$,
(ii) measurable functions $C_n(\cdot, \mathbf{t}):\Theta \to R$ such that

(3.9.12) $$\lim_{n\to\infty} \int \sup_{\|\mathbf{t}\| \leqslant a} |C_n(\theta, \mathbf{t}) - 1| \tau(d\theta) = 0 \text{ for every } a > 0,$$

(iii) probability measures $Q_n(\theta, t)$ on $(\Omega_n, \mathscr{A}_n)$ such that

(a) $Q_n(\theta, \mathbf{t}) \ll P_{\theta,n}$ for all $\theta \in \Theta$ and

$$\text{(b)}\quad \frac{dQ_n(\theta,\mathbf{t})}{dP_{\theta,n}} = C_n(\theta,\mathbf{t})\exp[\mathbf{t}'\mathbf{\Delta}_n^*(\theta) - \tfrac{1}{2}\mathbf{t}'\mathbf{\Gamma}(\theta)\mathbf{t}], \tag{3.9.13}$$

where

$$\begin{aligned} \mathbf{\Delta}_n^* &= \mathbf{\Delta}_n \quad \text{if } \|\mathbf{\Delta}_n\| \leqslant K_n \\ &= 0 \quad \text{if } \|\mathbf{\Delta}_n\| > K_n \end{aligned} \tag{3.9.14}$$

and

(c) $Q_n(\theta, \mathbf{t})$ approximates $P_{\theta+\delta_n\mathbf{t},n}$ in the sense that

$$\lim_{n\to\infty} \int_\Theta \|P_{\theta+\delta_n\mathbf{t},n} - Q_n(\theta,\mathbf{t})\| \,\tau(d\theta) = 0, \qquad \mathbf{t}\in R^k. \tag{3.9.15}$$

Lemma 3.9.6. Suppose $\tau(\cdot)$ is a probability measure on Θ and the sequence $\{P_{\theta,n}, \theta\in\Theta\}$, $n \geqslant 1$, is LAN in τ-measure. Then

$$\lim_{n\to\infty} \int_\Theta |\mathscr{L}(\mathbf{\Delta}_n(\theta) | P_{\theta+\delta_n\mathbf{t},n}) - \mathbf{N}(\mathbf{\Gamma}(\theta)\mathbf{t}, \mathbf{\Gamma}(\theta))| \,\tau(d\theta) = 0 \tag{3.9.16}$$

weakly for every $\mathbf{t}\in R^k$.

The next two lemmas are the main tools for proving Theorem 3.8.2.

Lemma 3.9.5. Let P be a probability measure on R^k and $\delta_\mathbf{t}$ denote the probability measure degenerate at $\mathbf{t}\in R^k$. Let $D_\alpha = \{\mathbf{s}\in R^k : \|\mathbf{s}\| \leqslant \alpha\}$ and λ be the Lebesgue measure on R^k. Then

$$\lim_{\alpha\to\infty} \frac{1}{\lambda(D_\alpha)} \int_{D_\alpha} (P * \delta_\mathbf{t})(D_\alpha^c)\, d\mathbf{t} = 0, \tag{3.9.17}$$

where $*$ denotes convolution.

Proof. Let $E_\alpha = \{\mathbf{s}\in R^k : \|\mathbf{s}\| \leqslant \alpha - \sqrt{\alpha}\}$. Then

$$D_{\sqrt{\alpha}} + E_\alpha \subset D_\alpha$$

and hence

$$D_{\sqrt{\alpha}} \subset D_\alpha - \mathbf{t} \quad \text{for every } \mathbf{t}\in E_\alpha.$$

Therefore

$$\frac{1}{\lambda(D_\alpha)}\int_{D_\alpha} P(D_\alpha-\mathbf{t})\,d\mathbf{t} \geqslant \frac{1}{\lambda(D_\alpha)}\int_{E_\alpha} P(D_\alpha-\mathbf{t})\,d\mathbf{t}$$

$$\geqslant \frac{\lambda(E_\alpha)}{\lambda(D_\alpha)} P(D_{\sqrt{\alpha}})$$

$$\geqslant \frac{\lambda(E_\alpha)}{\lambda(D_\alpha)} - P(D^{c}_{\sqrt{\alpha}})$$

and hence

$$\lim_{\alpha\to\infty}\frac{1}{\lambda(D_\alpha)}\int_{D_\alpha}(P*\delta_{\mathbf{t}})(D_\alpha^{c})\,d\mathbf{t} \leqslant \lim_{\alpha\to\infty} P(D^{c}_{\sqrt{\alpha}}) + \lim_{\alpha\to\infty}\left(1-\frac{\lambda(E_\alpha)}{\lambda(D_\alpha)}\right),$$

which implies the result.

Lemma 3.9.6. Let $(\Omega, \mathscr{A}, \mu)$ be a σ-finite measure space and P be a probability measure absolutely continuous with respect to the Lebesgue measure λ. Let $dP/d\lambda = h$ and $\mathbf{X}, \mathbf{Y}$ be k-dimensional random vectors defined on $(\Omega, \mathscr{A})$. Define

$$Q_{\mathbf{t}}(A) = C(\mathbf{t})\int_A h(\mathbf{t}-\mathbf{X}(\omega))\,\mu(d\omega)$$

for any $A\in\mathscr{A}$ where $C(\mathbf{t})$ is a known function. Let

$$R_{\mathbf{t}} = \mathscr{L}(\mathbf{t}-\mathbf{Y}\,|\,Q_{\mathbf{t}}), \qquad S_{\mathbf{t}} = \mathscr{L}(\mathbf{X}-\mathbf{Y}\,|\,Q_{\mathbf{t}})$$

for $\mathbf{t}\in R^k$. Furthermore, define

$$\bar{R}_\alpha = \frac{1}{\lambda(D_\alpha)}\int_{D_\alpha} R_{\mathbf{t}}\,d\mathbf{t}, \qquad \bar{S}_\alpha = \frac{1}{\lambda(D_\alpha)}\int_{D_\alpha} S_{\mathbf{t}}\,d\mathbf{t},$$

where $D_\alpha = \{\mathbf{s}\in R^k : \|\mathbf{s}\| \leqslant \alpha\}$. Then

$$\|\bar{R}_\alpha - N*\bar{S}_\alpha\| \leqslant \min\{1, 2\sup_{\mathbf{t}\in D_\alpha}|C(\mathbf{t})-1|\} + \frac{1}{\lambda(D_\alpha)}\int_{D_\alpha}(P*\mathscr{L}(X\,|\,Q_{\mathbf{t}}))(D_\alpha^{c})d\mathbf{t}. \tag{3.9.18}$$

Proof. Define

$$\bar{Q}_\alpha(A \times B) = \frac{1}{\lambda(D_\alpha)} \int_{D_\alpha \cap B} \bar{Q}_t(A)dt, \qquad A \in \mathscr{A}, \quad B \in \mathscr{B}^k, \tag{3.9.19}$$

where $\mathscr{B}^k$ is the σ-algebra of Borel subsets of R^k. Note that

$$\begin{aligned}\bar{R}_\alpha(B) = \frac{1}{\lambda(D_\alpha)} \int_{D_\alpha} R_t(B)\,dt &= \bar{Q}_\alpha\{\omega, \mathbf{t}) : \mathbf{t} - \mathbf{Y}(\omega) \in B\} \\ &= \int_\Omega \bar{Q}_\alpha(B + Y(\omega) | \mathscr{A})(\omega) \bar{Q}_\alpha(d\omega)\end{aligned}$$

and hence

$$\begin{aligned}&|\bar{R}_\alpha(B) - \int_\Omega (P * \delta_{X(\omega)})(B + Y(\omega)) \bar{Q}_\alpha(d\omega)| \\ &\leqslant \int_\Omega \| \bar{Q}_\alpha(\cdot | \mathscr{A})(\omega) - (P * \delta_{X(\omega)}) \| \bar{Q}_\alpha(d\omega).\end{aligned}$$

Observe that

$$\begin{aligned}\int_\Omega (P * \delta_{\mathbf{X}(\omega)})(P + \mathbf{Y}(\omega)) \bar{Q}_\alpha(d\omega) &= \int_\Omega [P(B + \mathbf{Y}(\omega)) - \mathbf{X}(\omega)]\, \bar{Q}_\alpha(d\omega) \\ &= (P * \bar{S}_\alpha)(B).\end{aligned}$$

Therefore

$$\| \bar{R}_\alpha - P * \bar{S}_\alpha \| \leqslant \int_\Omega \| \bar{Q}_\alpha(\cdot | \mathscr{A})(\omega) - \mathbf{P} * \delta_{\mathbf{X}(\omega)} \| \bar{Q}_\alpha(d\omega). \tag{3.9.20}$$

Note that, for any two probability measures μ and ν on A with $\mu \ll \nu$ and for every $A \in \mathscr{A}$ with $\mu(A) > 0$,

$$\| \mu(\cdot | A) - \nu \| \leqslant 2 \min \left\{ 1, \sup_{\omega \in A} \left| \frac{d\mu}{d\nu} - 1 \right| \right\} + 2\nu(A^c). \tag{3.9.21}$$

Clearly, from the definition of Q_t and Q_α, it follows that

$$\bar{Q}_\alpha(B | \mathscr{A})(\omega) = \frac{\int_{B \cap D_\alpha} C(\mathbf{t})(P * \delta_{\mathbf{X}(\omega)})\,(d\mathbf{t})}{\int_{D_\alpha} C(\mathbf{t})(P * \delta_{\mathbf{X}(\omega)})\,(d\mathbf{t})}. \tag{3.9.22}$$

Applying inequality (3.9.21) with $\mu = \bar{Q}_\alpha$ and $\nu = P * \delta_{\mathbf{X}(\omega)}$, it follows that

(3.9.23)
$$\| \bar{Q}_\alpha(\cdot \mid \mathscr{A})(\omega) - P * \delta_{\mathbf{X}(\omega)} \| \leqslant 2 \min \left\{ 1, \sup_{\mathbf{t} \in D_\alpha} | C(\mathbf{t}) - 1 | \right\} + 2(P * \delta_{\mathbf{X}(\omega)})(D_\alpha^c).$$

The result stated in the lemma now follows from inequalities (3.9.20) and (3.9.23) and the observation

(3.9.24)
$$\int_\Omega (P * \delta_{\mathbf{X}(\omega)})(D_\alpha^c) Q_{\mathbf{t}}(d\omega) = \int_\Omega P(D_\alpha^c - \mathbf{X}(\omega))\, Q_{\mathbf{t}}(d\omega) = (P * \mathscr{L}(\mathbf{X} \mid Q_{\mathbf{t}}))(D_\alpha^c).$$

This completes the proof of Lemma 3.9.6.

We now prove the convolution theorem (Theorem 3.9.2).

Proof of Theorem 3.9.2. Let

(3.9.25) $$U_n(\theta, \mathbf{t}) = \mathscr{L}(\mathbf{t} - \delta_n^{-1}(\Psi_n - \theta) \mid Q_n(\theta, \mathbf{t})),$$

(3.9.26) $$V_n(\theta, \mathbf{t}) = \mathscr{L}(\mathbf{\Gamma}^{-1}(\theta)\mathbf{\Delta}_n(\theta) - \delta_n^{-1}(\Psi_n - \theta) \mid Q_n(\theta, \mathbf{t})),$$

(3.9.27) $$U_{n,\alpha}(\theta) = \frac{1}{\lambda(D_\alpha)} \int_{D_\alpha} U_n(\theta, \mathbf{t})\, d\mathbf{t},$$

(3.9.28) $$V_{n,\alpha}(\theta) = \frac{1}{\lambda(D_\alpha)} \int_{D_\alpha} V_n(\theta, \mathbf{t})\, d\mathbf{t}.$$

Observe that $U_n(\theta, \mathbf{t})$ and $R_n(\theta, \mathbf{t})$ [as given by (3.9.7)] are the distributions of the same random variable under $P_{\theta + \delta_n \mathbf{t}, n}$ and $Q_n(\theta, \mathbf{t})$, respectively. Similar statement holds for V_n and S_n given by (3.9.8).

Lemma 3.9.3 implies that

(3.9.29) (i) $$\lim_{n \to \infty} \int_\Theta \| R_{n,\alpha}(\theta) - U_{n,\alpha}(\theta) \| \tau(d\theta) = 0$$

and

(3.9.30) (ii) $$\lim_{n \to \infty} \int_\Theta \| S_{n,\alpha}(\theta) - V_{n,\alpha}(\theta) \| \tau(d\theta) = 0.$$

Hence the theorem holds provided

$$\lim_{\alpha\to\infty}\overline{\lim_{n}}\int_{\Theta}\|U_{n,\alpha}(\theta)-N(\theta)*V_{n,\alpha}(\theta)\|\,\tau(d\theta)=0. \tag{3.9.31}$$

Let $\theta\in\Theta$ be fixed and $n\geqslant 1$. Applying Lemma 3.9.6 with

$$Q_{\mathbf{t}}=Q_n(\theta,\mathbf{t}),\qquad \mathbf{X}=\mathbf{\Gamma}^{-1}(\theta)\mathbf{\Delta}_n(0),\qquad \mathbf{Y}=\delta_n^{-1}(\Psi_n-\theta),$$
$$P=\mathbf{N}(\theta)=\mathbf{N}(\mathbf{0},\mathbf{\Gamma}^{-1}(\theta)),$$

we have

$$\|U_{n,\alpha}(\theta)-\mathbf{N}(\theta)*V_{n,\alpha}(\theta)\|\leqslant\min\left\{1,2\sup_{\mathbf{t}\in D_\alpha}|C_n(\theta,\mathbf{t})-1|\right\}$$
$$+\frac{1}{\lambda(D_\alpha)}\int_{D_\alpha}(\mathbf{N}(\theta)*\mathscr{L}(\mathbf{\Gamma}^{-1}(\theta)\mathbf{\Delta}_n(\theta)|Q_n(\theta,\mathbf{t}))(D_\alpha^{\mathrm{c}})\,d\mathbf{t}.$$

Applying Lemmas 3.9.3(ii) and 3.9.4, we obtain

$$\overline{\lim_{n}}\int\|U_{n,\alpha}(\theta)-\mathbf{N}(\theta)*V_{n,\alpha}(\theta)\|\,\tau(d\theta) \tag{3.9.32}$$
$$\leqslant\frac{1}{\lambda(D_\alpha)}\int\int_{D_\alpha}(\mathbf{N}(\theta)*\mathbf{N}(\theta)*\delta_t)(D_\alpha^{\mathrm{c}})\,d\mathbf{t}\,\tau(d\theta),$$

and the last term tends to zero as $\alpha\to\infty$ by Lemma 3.9.5. This completes the proof of the Theorem 3.9.2.

Let (T,τ) be a locally compact space with countable base with the associated Borel σ-algebra $\mathscr{B}$. Let $(\Omega,\mathscr{A},\mu)$ be a σ-finite measure space and $M_1(\mathscr{B})$ be the family of substochastic measures on $\mathscr{B}$. Let $P:\Omega\to M_1(\mathscr{B})$ and C_{00} be the space of continuous function on T vanishing outside compact sets. Suppose that the functional

$$\iint_{T\times\Omega} f(t)P(\omega)(dt)\,g(\omega)\mu(d\omega)$$

for $f\in C_{00}$ and $g\in L_1(\mu)$ is continuous on $C_{00}\otimes L_1(\mu)$. It can be shown that the space $C_{00}\otimes L_1(\mu)$ endowed with this requirement is metrizable and compact.

The following theorem generalizes a result of Le Cam (1973c, 1974). It is known as *Le Cam's invariance theorem.* We omit the proof.

Theorem 3.9.7. Let τ be a probability measure on $\Theta \subset R^k$. Suppose $\tau \ll \lambda$ where λ is the Lebesgue measure on R^k. Furthermore, suppose that $\{F_n, n \geqslant 1\}$ is a sequence of kernels from $\Theta \times R^k$ into $M_1(\mathscr{B})$ such that $F_n \to F$ in the space $C_{00}(R^k) \otimes L_1(\tau \otimes \lambda)$. Suppose there is a sequence $\delta_n \downarrow 0$ such that $F_n(\theta + \delta_n \mathbf{r}, \mathbf{t}) = F_n(\theta, \mathbf{r} + \mathbf{t})$ for all $\theta \in \Theta, \mathbf{r}, \mathbf{t} \in R^k$, $n \geqslant 1$. Then, for τ-almost all $\theta \in \Theta$,

$$\int F(\theta, \mathbf{t}) h(\mathbf{t})\, d\mathbf{t} = \int F(\theta, \mathbf{t} + \mathbf{r}) h(\mathbf{t})\, d\mathbf{t} \tag{3.9.33}$$

for all $\mathbf{r} \in R^k$ and $h \in L_1(\lambda)$.

As a corollary of Theorems 3.9.2 and 3.9.7, we have the following result.

Corollary 3.9.8. Suppose the assumptions in Theorem 3.9.2 hold and $\tau \ll \lambda$. Let $\{\Psi_n\}$ be a sequence of estimators. Then, for every subsequence $\{n_j\}$ such that $\{R_{n_j}\}$ and $\{S_{n_j}\}$ are $C_{00}(R^k) \otimes L_1(\tau \otimes \lambda)$-convergent [where R_n and S_n are as defined by (3.9.7) and (3.9.8)],

$$\lim_{\alpha \to \infty} \lim_{n} (R_n(\cdot, 0) - N * S_{n,\alpha}) = 0 \tag{3.9.34}$$

on $C_{00}(R^k) \otimes L_1(\tau)$.

Proof. In view of Theorem 3.9.2, it is sufficient to prove that

$$\lim_{n \to \infty} (R_n(\cdot, 0) - R_{n,\alpha}) = 0 \tag{3.9.35}$$

for every $\alpha > 0$ on $C_{00}(R^k) \otimes L_1(\tau)$.

Let R be the limit of $\{R_{n_j}\}$. It can be shown that for every $\mathbf{t} \in R^k$ and $\varepsilon > 0$, there exists $\eta(\varepsilon, \mathbf{t}) > 0$ such that

$$\lim_{n \to \infty} \int_{R^k} \| P_{\theta + \delta_n \mathbf{s}} - P_{\theta + \delta_n \mathbf{t}} \| \, \tau(d\theta) \leqslant \varepsilon \quad \text{provided } \| \mathbf{s} - \mathbf{t} \| \leqslant \eta(\varepsilon, \mathbf{t})$$

by using Lemmas 3.9.3 and 3.9.4 [see Strasser (1978, Lemma 1)]. In view of Theorem 3.9.7, it follows that

$$\int R(\theta, 0) g(\theta)\, \tau(d\theta) = \int R(\theta, \mathbf{t}) g(\theta)\, \tau(d\theta)$$

for all $g \in L_1(\tau)$ and $\mathbf{t} \in R^k$. This proves the result.

We now obtain the main result concerning lower bounds for risks of estimators.

Proof of Theorem 3.9.1. Without loss of generality we assume that $g(\theta) \equiv 1$. We have to show that

$$\lim_{n_j} \int\int W \, dR_{n_j}(\theta, 0) \, \tau(d\theta) \geqslant \int\int W \, dN(\theta) \, \tau(d\theta)$$

for every subsequence $\{n_j\}$ such that $\{R_{n_j}\}$ and $\{S_{n_j}\}$ are $C_{00} \otimes L_1(\tau \otimes \lambda^k)$-convergent. For such a sequence, Corollary 3.9.8 and the property of loss function W show that

$$\int\int W \, dN(\theta) \, \tau(d\theta) \leqslant \lim_{\alpha \to \infty} \lim_{n_j} \int\int W \, d(N * S_{n_j,\alpha})(\theta) \, \tau(d\theta)$$

$$= \lim_{n_j} \int\int W \, dR_n(\theta, 0) \, \tau(d\theta).$$

This completes the proof of Theorem 3.9.1.

Remark 1. Suppose the loss function $W(\cdot)$ has the additional property that

$$\int W \, dW(0, A) < \int W \, d(N(0, A) * P) \tag{3.9.35}$$

whenever $P \neq \delta_0$. It can be shown [cf. Strasser (1978, p. 38)] that equality occurs in Theorem 3.9.1 if and only if $\{\Psi_n\}$ is SLLR in τ-measure (cf. Definition 3.9.2). We omit the details. In particular, it follows that if τ is the prior, then the corresponding Bayes estimators are SLLR in τ-measure provided the loss function W satisfies (3.9.35).

Remark 2. For an alternative proof of Theorem 3.9.2, see Jeganathan (1981).

We have discussed global asymptotic properties of risk function of estimators till now. We shall briefly discuss local asymptotic minimax properties of estimators in the next section.

3.10 LOCAL ASYMPTOTIC LOWER BOUNDS FOR RISKS OF ESTIMATORS

Let T be an arbitrary set. An *experiment* is a triple $(\Omega, \mathscr{A}, \mathscr{P})$ where $(\Omega, \mathscr{A})$ is a measurable space and $\mathscr{P} = \{P_t, t \in \mathbf{T}\}$ is a family of probability measures over $(\Omega, \mathscr{A})$. $\mathbf{T}$ is the parameter space. The collection of all experiments is denoted by $\mathscr{E}(T)$.

Two experiments $E_i = (\Omega_i, \mathscr{A}_i, \{P_{i,\mathbf{t}}:\mathbf{t}\in\mathbf{T}\})$, $i = 1, 2$, are said to be *equivalent* if, for every $\mathbf{s}\in\mathbf{T}$ and every finite subset $\mathbf{T}_0 \subset \mathbf{T}$,

$$\mathscr{L}\left(\left\{\frac{dP_{1,\mathbf{t}}}{P_{1,\mathbf{s}}}\right\}_{\mathbf{t}\in\mathbf{T}_0}\middle| P_{1,\mathbf{s}}\right) = \mathscr{L}\left(\left\{\frac{dP_{2,\mathbf{t}}}{dP_{2,\mathbf{s}}}\right\}_{\mathbf{t}\in\mathbf{T}_0}\middle| P_{2,\mathbf{s}}\right).$$

(Here dP/dQ denotes the Radon–Nikodým derivative of the absolutely continuous component of P with respect to Q and $\mathscr{L}(X|P)$ denote the distribution of X under P). We denote by $E_1 \sim E_2$ if E_1 and E_2 are equivalent.

Suppose $\mathbf{T} \subset R^k$. Let $\mathscr{B}_{\mathbf{T}}$ be the Borel σ-algebra of $\mathbf{T}$. An *estimator* for the experiment E is a stochastic kernel $\rho:\Omega \times \mathscr{B}_{\mathbf{T}} \to [0, 1]$. Let $\mathscr{R}(E, \mathbf{T})$ be the set of all such estimators and $V:R^k \to [0, \infty)$ be a bounded continuous *loss function*. The *risk* ρ for given V and $\mathbf{t}\in\mathbf{T}$ is

$$V_\rho P_{\mathbf{t}} = \int\int V(\mathbf{x} - \mathbf{t})\, \rho(\omega, d\mathbf{x})\, P_{\mathbf{t}}(d\omega). \tag{3.10.0}$$

Let $\mathbf{K}$ be a compact subset of $\mathbf{T}$. The *minimax* risk corresponding to $(V, \mathbf{K})$ is

$$\inf_{\rho\in\mathscr{R}(E,\mathbf{T})} \sup_{\mathbf{t}\in\mathbf{K}} V_\rho P_t. \tag{3.10.1}$$

Let $\mathbf{v}$ be a set of bounded continuous functions V and $\mathscr{K}$ be the family of all compact subsets of T. The minimax risk over $(\mathbf{v}, \mathscr{K})$ for ρ is

$$A \equiv \sup_{V\in\mathbf{v}} \sup_{\mathbf{K}\in\mathscr{K}} \inf_{\rho\in\mathscr{R}(E,\mathbf{T})} \sup_{\mathbf{t}\in\mathbf{K}} V_\rho P_{\mathbf{t}}, \tag{3.10.2}$$

which can be written in the form

$$\inf_{\rho\in\mathscr{R}(E,\mathbf{T})} \sup_{\substack{V\in\mathbf{v}, K\in\mathscr{K} \\ t\in\mathbf{K}}} V_\rho P_{\mathbf{t}} = \inf_{\rho\in\mathscr{R}(E,\mathbf{T})} \sup_{\mathbf{t}\in\mathbf{T}} W_\rho P_{\mathbf{t}} \tag{3.10.3}$$

by the minimax theorem, where W is the upper envelope of the family $\mathbf{v}$. W need not be bounded and continuous. We call an estimator $\rho\in\mathscr{R}(E, \mathbf{T})$ *minimax* for $\mathbf{v}$ if

$$\sup_{\mathbf{t}\in\mathbf{T}} W_\rho P_{\mathbf{t}} = \inf_{\sigma\in\mathscr{R}(E,\mathbf{T})} \sup_{\mathbf{t}\in\mathbf{T}} W_\sigma P_{\mathbf{t}}. \tag{3.10.4}$$

An *experiment* $E = (\Omega, \mathscr{A}, \{P_{\mathbf{t}}:\mathbf{t}\in\mathbf{T}\})$ is said to be *translation invariant* if

$E \simeq (\Omega, \mathscr{A}, \{P_{\mathbf{t}+\mathbf{s}}:\mathbf{t}\in\mathbf{T}\})$ for every $\mathbf{s}\in\mathbf{T}$. An *estimator* $\rho\in\mathscr{R}(E,\mathbf{T})$ is said to be *translation invariant* if

(3.10.5)
$$\mathscr{L}(\rho(\cdot, B)|P_0) = \mathscr{L}(\rho\cdot, B+\mathbf{s})|P_{\mathbf{s}}).$$

for every $B\in\mathscr{B}_{\mathbf{T}}$ and $\mathbf{s}\in\mathbf{T}$.

It is clear that every location parameter experiment is translation invariant. The converse is not true. However, as can be seen from Theorem 3.9.7 (Le Cam, 1973c), in asymptotic theory, the limit experiment is translation invariant is general.

We assume the set $\mathscr{P} = \{P_{\mathbf{t}}:\mathbf{t}\in\mathbf{T}\}$ is separable so that product-measurable densities exist (Pfanzagl, 1969b). Let $\mathbf{T} = R^k$.

An estimator $\rho\in\mathscr{R}(E,\mathbf{T})$ is a *Pitman estimator with respect to a given measurable loss function* $W:T\to[0,\infty)$ if, for every $\mathbf{s}\in\mathbf{T}$,

(3.10.6)
$$\int\int W(\mathbf{x}-\mathbf{t})\rho(\cdot, d\mathbf{x})\frac{dP_{\mathbf{t}}}{dP_{\mathbf{s}}}\lambda(d\mathbf{t}) = \inf_{\mathbf{y}\in\mathbf{T}}\int W(\mathbf{y}-\mathbf{t})\frac{dP_{\mathbf{t}}}{dP_{\mathbf{s}}}\lambda(d\mathbf{t}), \quad P_{\mathbf{s}}-\text{a.e.}$$

where $\lambda(\cdot)$ is the Lebesgue measure on R^k. It can be checked that this definition agrees with the earlier definition given in Section 3.2 under some conditions on the loss function W. We assume that the following conditions are satisfied.

(A) $W:R^k\to[0,\infty)$ is a measurable loss function such that $W(0)=0$. Furthermore, the following conditions hold.

- **(i)** W is *separating* that is, there exists $a<\sup W$ such that the set $[\mathbf{x}:W(\mathbf{x})\leqslant a]$ is a neighborhood of zero.
- **(ii)** W is *level-compact*, that is, $[\mathbf{x}:W(\mathbf{x})\leqslant a]$ is compact for every $a<\sup W$.
- **(iii)** W is of *order* $p\geqslant 0$, that is, $W(\mathbf{t})\leqslant C_1\|\mathbf{t}\|^p + C_2$, $\mathbf{t}\in R^k$, for some constants C_1 and C_2.

(B) Let $E = (\Omega, \mathscr{A}, \{P_{\mathbf{t}}:\mathbf{t}\in\mathbf{T}\})$ be the experiment. Then

- **(i)** E is *continuous* that is, $P_{\mathbf{t}}(\cdot)$ is continuous in $\mathbf{t}$ for the variational distance $\|P-Q\| = \sup_{A\in\mathscr{A}}|P(A)-Q(A)|$.
- **(ii)** E is *p-times $\lambda(\cdot)$-integrable*, that is for every $\mathbf{s}\in\mathbf{T}$ and $\varepsilon>0$ there exists a compact set $\mathbf{K}\subset\mathbf{T}$ and a test function $\phi:\Omega\to[0,1]$ such that

(3.10.7)
$$E_{\mathbf{s}}\phi > 1-\varepsilon \quad \text{and} \quad \int_{K^c} E_{\mathbf{t}}\phi\,\|\mathbf{t}\|^p\,\lambda(d\mathbf{t}) < \varepsilon.$$

(C) The experiment $E=(\Omega, \mathscr{A}, \{P_{\mathbf{t}}:\mathbf{t}\in\mathbf{T}\})$ and the loss function $W:R^k\to[0,\infty)$ satisfy the *uniqueness condition*, that is the function

$$\int W(\mathbf{x}-\mathbf{t})\frac{dP_{\mathbf{t}}}{dP_{\mathbf{s}}}(\omega)\,\lambda(d\mathbf{t})$$

attains its infimum at a single point for $P_{\mathbf{s}}$—almost all $\omega\in\Omega$ and every $\mathbf{s}\in\mathbf{T}$.

The following theorem is due to Strasser (1982). We omit the proof [cf. Strasser (1982, Proposition 1.6)].

Theorem 3.10.1. Let $\mathbf{T}=R^k$. Suppose that an experiment E is continuous, translation invariant, and p-times λ-integrable. Furthermore, suppose that the loss function W is separating, level-compact, and of order p. Then

(i) there exist Pitman estimators,

(ii) every translation invariant estimator is a minimax estimator, and, furthermore

(iii) if the uniqueness condition is satisfied, then every Pitman estimator is translation invariant.

Let $\{\mathbf{T}_n\}$ be a sequence of open subsets of R^k-increasing to $\mathbf{T}=R^k$ as $n\to\infty$. Let $E_n=(\Omega, \mathscr{A}_n, \{P_{n,\mathbf{t}}:\mathbf{t}\in\mathbf{T}_n\})$, $n\geqslant 1$, be a sequence of experiments. E_n is said to *converge weakly* in f.d.d. (finite dimensional distributions) to E if for every $\mathbf{s}\in\mathbf{T}$ and every *finite* subset $\mathbf{T}_0\subset\mathbf{T}$,

$$\mathscr{L}\left(\left\{\frac{dP_{n,\mathbf{t}}}{dP_{n,\mathbf{s}}}\right\}_{\mathbf{t}\in\mathbf{T}_0}\middle|P_{n,\mathbf{s}}\right)\to\mathscr{L}\left(\left\{\frac{dP_{n,\mathbf{t}}}{dP_{n,\mathbf{s}}}\right\}_{\mathbf{t}\in\mathbf{T}_0}\middle|P_{\mathbf{s}}\right) \tag{3.10.8}$$

weakly as $n\to\infty$, Define $\mathbf{v}$ and $\mathscr{K}$ as before, and let A be the minimax risk of E for $\mathbf{v}$ as defined by (3.10.2). A sequence of estimators $\rho_n\in\mathscr{R}(E_n,\mathbf{T})$, $n\geqslant 1$, is *asymptotically minimax* for $\mathbf{v}$ if

$$\overline{\lim_{n\to\infty}}\sup_{\mathbf{t}\in\mathbf{K}} V_{\rho_n}P_{n,\mathbf{t}}\leqslant A$$

for $V\in\mathbf{v}, \mathbf{K}\in\mathscr{K}$.

The problem is to obtain estimators that are asymptotically minimax. Hajek (1972) studied the problem in case the family $\{P_{n,\mathbf{t}}\}$ is LAN. We first extend the notion of Pitman estimator.

Suppose $\{P_{n,t}, \mathbf{t}\in\mathbf{T}_n\}$, $n\geq 1$, is a separable family. Let $W: R^k \to [0, \infty)$ be a measurable loss function. A sequence of estimators $\rho_n \in \mathscr{R}(E_n, \mathbf{T})$ is a sequence of *asymptotic Pitman estimators* if for every $\varepsilon > 0$ and every $\mathbf{s}\in\mathbf{T}$,

$$\lim_{n\to\infty} P_{n,s}\left\{\int\int W(\mathbf{x}-\mathbf{t})\rho_n(\cdot, d\mathbf{x})\frac{dP_{n,t}}{dP_{n,s}}\lambda(d\mathbf{t}) > \inf_{\mathbf{y}\in\mathbf{T}_n}\int W(\mathbf{y}-\mathbf{t})\frac{dP_{n,t}}{dP_{n,s}}\lambda(d\mathbf{t})+\varepsilon\right\}=0. \tag{3.10.9}$$

Let $E_n = (\Omega_n, \mathscr{A}_n, \{P_{n,t}, \mathbf{t}\in\mathbf{T}_n\})$, $n\geq 1$, be a sequence of experiments. $\{E_n\}$ is said to be *equicontinuous* if the functions $\mathbf{t}\to P_{n,t}$ $\mathbf{t}\in\mathbf{T}_n, n\geq 1$, are equicontinuous in the variational distance. The sequence is said to be *uniformly p-times λ-integrable* if for every $\mathbf{s}\in\mathbf{T}$ and every $\varepsilon>0$ there exists a compact set $\mathbf{K}\subset T$ and a sequence $\{\phi_n\}$ of test functions $\phi_n:\Omega_n\to[0, 1]$ such that

$$E_{n,s}\phi_n > 1-\varepsilon \quad\text{and}\quad \int_{\mathbf{T}_n\setminus\mathbf{K}} E_{n,t}\phi_n\|\mathbf{t}\|^p\lambda(d\mathbf{t}) < \varepsilon \tag{3.10.10}$$

as soon as $\mathbf{K}\subset\mathbf{T}_n$.

Theorem 3.10.2. Suppose $\{E_n\}$ is equicontinuous and uniformly p-times λ-integrable. Furthermore, suppose that W is separating, level-compact, and of order p. Let $\{\rho_n, n\geq 1\}$ be a sequence of asymptotic Pitman estimators for $\{E_n\}$ and W and ρ be a Pitman estimate for E and W. If $E_n\to E$ weakly in f.d.d. and if E and W satisfy the uniqueness condition, then

$$\mathscr{L}(\rho_n|P_{n,s})\to\mathscr{L}(\rho|P_s) \quad \text{weakly,}\quad \mathbf{s}\in\mathbf{T}. \tag{3.10.11}$$

It can be checked that if $\{E_n\}$ is equicontinuous and uniformly p-times λ-integrable and if $E_n\to E$ weakly, then E is continuous and p-times λ-integrable. Note that the theorem implies that the distribution of ρ_n under $P_{n,s}$ converges weakly to the distribution of ρ under P_s. We sketch the proof [cf. Strasser (1982, Proposition 1.10)] later in this section.

As a consequence of Theorems 3.10.1 and 3.10.2, we have the following main theorem showing that Pitman estimators are asymptotically minimax.

Theorem 3.10.3. Let $\mathbf{v}$ be a set of bounded and continuous loss functions with upper envelope W. Suppose the condition stated in Theorems 3.10.1 and 3.10.2 hold. Then $\{\rho_n\}$ is asymptotically minimax for $\mathbf{v}$.

Proof. Since $\{E_n\}$ is equicontinuous, Theorem 3.10.2 shows that

$$\lim_{n\to\infty} \sup_{\mathbf{t}\in\mathbf{K}} |V_{\rho_n} P_{n,\mathbf{t}} - V_\rho P_{\mathbf{t}}| = 0$$

for every $V \in \mathbf{v}$ and $\mathbf{K} \in \mathscr{K}$, where $V_\rho P_{\mathbf{t}}$ is given by (3.9.0). Hence

$$\lim_{n\to\infty} \sup_{\mathbf{t}\in\mathbf{K}} V_{\rho_n} P_{n,\mathbf{t}} = \sup_{\mathbf{t}\in\mathbf{K}} V_\rho P_{\mathbf{t}} \leqslant \sup_{\mathbf{t}\in\mathbf{T}} W_\rho P_{\mathbf{t}} = A,$$

where W is the upper envelope of the family $\mathbf{v}$.

We now discuss implications of these results in the classical setup.

Let the parameter space $\Theta \subset R^k$ be open. Consider a sequence of experiments $F_n = (\Omega_n, \mathscr{A}_n, \{P_{n\theta}; \theta\in\Theta\})$ and $\delta_n \downarrow 0$. Let

$$E_{n,\theta} = (\Omega_n, \mathscr{A}_n, \{P_{n,\theta+\delta_n\mathbf{t}} : \mathbf{t}\in\mathbf{T}_n(\Theta)\}), \tag{3.10.12}$$

where $\mathbf{T}_n(\Theta) = \{\mathbf{t}\in R^k : \theta + \delta_n \mathbf{t} \in \Theta\}$, $n \geqslant 1$. $\{E_{n,\theta}, n \geqslant 1\}$ is a sequence of localized experiments around θ. In order to study the asymptotic properties of estimators, we fix $\theta\in\Theta$ and $\delta_n\downarrow 0$ and study the limiting behavior of localized experiments $\{E_{n,\theta}\}$. $\{\delta_n\}$ is chosen so that

$$0 < \underline{\lim}_n \| P_{n,\theta} - P_{n,\theta+\delta_n\mathbf{t}_n} \| \leqslant \overline{\lim}_n \| P_{n,\theta} - P_{n,\theta+\delta_n\mathbf{t}_n} \| < 1 \tag{3.10.13}$$

whenever $\{\mathbf{t}_n\}$ is bounded and bounded away from zero.

Example 3.10.1. Suppose X_i, $i \geqslant 1$, are i.i.d. random variables with probability measure P_θ, $\theta\in\Theta$, defined on $(\Omega, \mathscr{A})$. Let $F_n = (\Omega^n, \mathscr{A}^n, \{P_\theta^n : \theta\in\Theta\})$. Suppose the Hellinger distance $H(P_\sigma, P_\tau)$ satisfies the following condition: for every $\theta\in\Theta$, there is a neighborhood U_θ of θ such that

$$c_1 |\sigma - \tau|^a \leqslant H(P_\sigma, P_\tau) \leqslant c_2 |\sigma - \tau|^a, \tag{3.10.14}$$

where c_1 and c_2 are constants depending on U_θ and θ. In this case $\delta_n = n^{-(1/2a)}$ (see Section 3.7) is the right norming sequence.

Example 3.10.2. Let $F = \{R, \mathscr{B}, \{P_\theta, \theta\in R\}\}$, where

$$\frac{dP_\theta}{d\lambda}(x) = C(\alpha)\exp(|-x-\theta)|^\alpha), \qquad \alpha > 0.$$

A sample choice of $\delta_n \downarrow 0$ in this case is

$$\delta_n = \begin{cases} n^{-1/2} & \text{if} \quad \alpha > \frac{1}{2} \\ (n \log n)^{-1/2} & \text{if} \quad \alpha = \frac{1}{2} \\ n^{-1/(2\alpha+1)} & \text{if } 0 < \alpha < \frac{1}{2}. \end{cases}$$

If $\alpha \geqslant \frac{1}{2}$, then the sequence $\{E_{n,\theta}\}$ converges weakly in f.d.d. to a Gaussian shift on $(R, \mathscr{B})$. If $0 < \alpha < \frac{1}{2}$, then the sequence $\{E_{n,\theta}\}$ converges weakly in f.d.d. to an experiment E characterized by the property that the log-likelihood ratio process $\log(dP_t/dP_0)$, $t \in R$, is a Gaussian process with mean $-(c/2)|t|^{2\alpha+1}$ and covariance $(c/2)(|s|^{2\alpha+1} + |t|^{2\alpha+1} - |s-t|^{2\alpha+1})$, where c is a constant depending on α [cf. Prakasa Rao (1968), Section 3.7].

Example 3.10.3. Let $\mathbf{\Gamma}$ be a positive definite $k \times k$ matrix. Then $G = (R^k, \mathscr{B}^k, \{N(\mathbf{\Gamma t}, \mathbf{\Gamma}) : \mathbf{t} \in R^k\})$ is a Gaussian shift with covariance $\mathbf{\Gamma}$. The sequence $\{F_n, n \geqslant 1\} \subset \mathscr{E}(\Theta)$ is locally asymptotically normal (LAN) at $\theta \in \Theta$ if $E_{n,\theta} \to G$ weakly in f.d.d.

We now sketch a proof of Theorem 3.10.2. We first state a couple of lemmas. For proofs, see Strasser (1982).

Lemma 3.10.3. Let $\{E_n\} = (\Omega_n, \mathscr{A}_n, \{P_{n,t} : \mathbf{t} \in \mathbf{T}_n\}), \mathbf{T}_n$ open, and $\mathbf{T}_n \uparrow \mathbf{T}$ be equicontinuous and uniformly p-times λ-integrable. Suppose $W(\cdot)$ is separating, level-compact, and of order $p \geqslant 0$. Define $W_a(\cdot) = \min(a, W(\cdot))$ for any $a \geqslant 0$. Then, for every $\mathbf{s} \in \mathbf{T}$ and sufficiently small $\varepsilon > 0$, there exist $b_\varepsilon > 0$, $A_{n,\varepsilon} \in \mathscr{A}_n$, $n \geqslant 1$, and a compact set $\mathbf{K}_\varepsilon \subset \mathbf{T}$ such that

(3.10.15) (i) $P_{n,s}(A_{n,\varepsilon}) > 1 - \varepsilon, \qquad n \geqslant 1,$

(ii) $$\inf_{\mathbf{x} \in \mathbf{T} \setminus K_\varepsilon} \int W_a(\mathbf{x} - \mathbf{t}) F_n(d\mathbf{t}) > \inf_{\mathbf{y} \in \mathbf{T}_n} \int W_a(\mathbf{y} - \mathbf{t}) F_n(d\mathbf{t}) + \varepsilon \quad \text{on } A_{n\varepsilon}$$

provided $a \in [b_\varepsilon, \infty]$ and $\mathbf{K}_\varepsilon \subset \mathbf{T}_n$, where $F_n(\cdot)$ denotes the posterior distributon of E_n, namely.

$$F_n(\cdot, B) = \int_B \frac{dP_{n,t}}{dP_{n,s}} \lambda(d\mathbf{t}) \bigg/ \int_{\mathbf{T}_n} \frac{dP_{n,t}}{dP_{n,s}} \lambda(d\mathbf{t}), \tag{3.10.16}$$

for every Borel set B in R^k and $\mathbf{s} \in \mathbf{T}$.

Remarks. Under the conditions of Lemma 3.10.3, it can be checked that a sequence of estimators $\rho_n \in \mathscr{R}(E_n, \mathbf{T})$, $n \geqslant 1$, is a sequence of asymptotic Pitman estimators iff for every $\varepsilon > 0$ and every $\mathbf{s} \in \mathbf{T}$

$$\lim_{n\to\infty} \left\{ P_{n,\mathbf{s}} \int\int W(\mathbf{x}-\mathbf{t})\rho_n(d\mathbf{x},\cdot)F_n(d\mathbf{t}) > \inf_{\mathbf{y}\in\mathbf{T}_n} \int W(\mathbf{y}-\mathbf{t})F_n(d\mathbf{t}) + \varepsilon \right\} = 0. \tag{3.10.17}$$

Furthermore, for every $\varepsilon > 0$ and $\mathbf{s} \in \mathbf{T}$,

$$\lim_{n\to\infty} \int_{\Omega_n} \rho_n(\cdot, \mathbf{T}\backslash \mathbf{B}_n^\varepsilon) dP_{n,\mathbf{s}} = 0, \tag{3.10.18}$$

where

$$B_n^\varepsilon(\omega_n) = \left\{ \mathbf{x} \in \mathbf{T} : \int W(\mathbf{x}-\mathbf{t})F_n(d\mathbf{t}) \leqslant \inf_{\mathbf{y}\in\mathbf{T}_n} \int W(\mathbf{y}-\mathbf{t})F_n(d\mathbf{t}) + \varepsilon \right\} \tag{3.10.19}$$

for $\omega_n \in \Omega_n$, $n \geqslant 1$.

Lemma 3.10.4. Let $\{E_n\}$ be equicontinuous and uniformly p-times λ-integrable. Suppose $W(\cdot)$ is separating, level-compact, and of order $p \geqslant 0$. Define

$$X_n(\mathbf{r}) = \int W(\mathbf{r}-\mathbf{t})F_n(d\mathbf{t}), \qquad \mathbf{r} \in \mathbf{T}_n, \quad n \geqslant 1. \tag{3.10.20}$$

Assume that $E_n \to E$ weakly in f.d.d. Then, for every $\mathbf{s} \in \mathbf{T}$ and every finite set $\boldsymbol{\alpha} \subset \mathbf{T}$,

$$\mathscr{L}((X_n(\mathbf{r}))_{\mathbf{r}\in\boldsymbol{\alpha}} | P_{n,\mathbf{s}}) \xrightarrow{\mathrm{w}} \mathscr{L}((X_n(\mathbf{r}))_{\mathbf{r}\in\boldsymbol{\alpha}} | P_{\mathbf{s}}). \tag{3.10.21}$$

We now prove Theorem 3.10.2.

Proof. Let F be the posterior distribution of E. Define $\mathbf{B}_n^\varepsilon$ as in (3.10.19) and $\mathbf{B}^\varepsilon$ with $\mathbf{T}_n$ replaced by $\mathbf{T}$ and F_n replaced by F in (3.10.19).

Let $\mathbf{s} \in \mathbf{T}$ and ρ_n, $n \geqslant 1$, be a sequence of asymptotic Pitman estimators. Let $\mathbf{K} \subset \mathbf{T}$ be compact. If $\mathbf{s} \in \mathbf{T}_n$, $\mathbf{K} \subset \mathbf{T}_n$, then for any $\varepsilon > 0$,

$$\int \rho_n(\cdot, \mathbf{K}) dP_{n,\mathbf{s}} \leqslant \int \rho_n(\cdot, \mathbf{K} \cap \mathbf{B}_n^\varepsilon) dP_{n,\mathbf{s}} + \int \rho_n(\cdot, \mathbf{T}_n \backslash \mathbf{B}_n^\varepsilon) dP_{n,\mathbf{s}} \tag{3.10.22}$$

and hence, by (3.10.18),

$$\overline{\lim_{n\to\infty}} \int \rho_n(\cdot, \mathbf{K})dP_{n,\mathbf{s}} \leqslant \overline{\lim_{n\to\infty}} P_{n,\mathbf{s}}\{P_n(\cdot, \mathbf{K}\cap\mathbf{B}_n^\varepsilon) > 0\} \tag{3.10.23}$$

$$\leqslant \overline{\lim_{n\to\infty}} P_{n,\mathbf{s}}(\mathbf{K}\cap\mathbf{B}_n^\varepsilon \neq \phi).$$

Suppose we prove that

$$\overline{\lim_{n\to\infty}} P_{n,\mathbf{s}}(\mathbf{K}\cap\mathbf{B}_n^\varepsilon \neq \phi) \leqslant \lim_{\varepsilon\downarrow 0} \overline{\lim_{n\to\infty}} P_{n,\mathbf{s}}(\mathbf{K}\cap\mathbf{B}_n^\varepsilon \neq \phi) \tag{3.10.24}$$

$$\leqslant P_{\mathbf{s}}(\mathbf{B}^0\cap\mathbf{K} \neq \phi).$$

Then it follows that

$$\overline{\lim_{n\to\infty}} \int \rho_n(\cdot, \mathbf{K})\,dP_{n,\mathbf{s}} \leqslant P_{\mathbf{s}}(\mathbf{B}^0\cap\mathbf{K} \neq \phi). \tag{3.10.25}$$

Let $\rho\in\mathscr{R}(E,\mathbf{T})$ be a Pitman estimator. Since card $(B^0(\omega)) = 1$ $P_{\mathbf{s}}$-a.e. by assumption, for every $\mathbf{s}\in\mathbf{T}$, it follows that

$$P_{\mathbf{s}}(\mathbf{B}^0\cap\mathbf{K}) \neq \phi) = P_{\mathbf{s}}(\rho\in\mathbf{K}), \qquad \mathbf{s}\in\mathbf{T}. \tag{3.10.26}$$

Relations (3.10.25) and (3.10.26) prove the theorem. We now prove (3.10.24). Let $\mathbf{s}\in\mathbf{T}$ and $\mathbf{K}\subset\mathbf{T}$ be compact as before. Compactness of $\mathbf{K}$ implies that

$$\bigcap_{\varepsilon>0} (\mathbf{B}^\varepsilon\cap\mathbf{K} \neq \phi) = (\mathbf{B}^0\cap\mathbf{K} \neq \phi)$$

and hence

$$\lim_{\varepsilon\downarrow 0} P_{\mathbf{s}}(\mathbf{B}^\varepsilon\cap\mathbf{K} \neq \phi) = P_{\mathbf{s}}(\mathbf{B}^0\cap\mathbf{K} \neq \phi). \tag{3.10.27}$$

Therefore the inequalities in (3.10.24) will hold clearly provided there exists $\varepsilon_{\mathbf{K}} > 0$ such that

$$\overline{\lim_{n\to\infty}} P_{n.\mathbf{s}}(\mathbf{B}_n^\varepsilon\cap\mathbf{K} \neq \phi) \leqslant P_{\mathbf{s}}(\mathbf{B}^\varepsilon\cap\mathbf{K} \neq \phi) + 2\varepsilon, \qquad 0 \leqslant \varepsilon \leqslant \varepsilon_{\mathbf{K}}. \tag{3.10.28}$$

Define $X_n(\mathbf{r})$ as in (3.10.20) and

$$X(\mathbf{r}) = \int W(\mathbf{r}-\mathbf{t})F(d\mathbf{t}), \qquad \mathbf{r}\in\mathbf{T}. \tag{3.10.29}$$

In view of Lemma 3.10.3, for sufficiently small $\varepsilon > 0$, there exists a compact set $\mathbf{K}_\varepsilon \subset \mathbf{T}$ such that

$$P_{n,\mathbf{s}}\left\{\inf_{\mathbf{r}'\in\mathbf{K}_\varepsilon} X_n(\mathbf{r}') > \inf_{\mathbf{r}''\in\mathbf{T}_n} X_n(\mathbf{r}'') + \varepsilon\right\} > 1 - \varepsilon \quad \text{if } \mathbf{K}_\varepsilon \subset \mathbf{T}_n. \tag{3.10.30}$$

Since $E_n \to E$ weakly in f.d.d., it follows that E is continuous. Furthermore, since $\{E_n\}$ are equicontinuous and uniformly λ-integrable, it can be shown that E is λ-integrable. Hence, by Lemma 3.9.3, it follows that there exists a compact set $\mathbf{K} \subset \mathbf{T}$ such that

$$P_{\mathbf{s}}\left\{\inf_{\mathbf{r}'\in\mathbf{K}} X(\mathbf{r}') > \inf_{\mathbf{r}''\in\mathbf{T}} X(\mathbf{r}'') + \varepsilon\right\} > 1 - \varepsilon. \tag{3.10.31}$$

Assume without loss of generality that $\mathbf{K}_\varepsilon$ is decreasing for $\varepsilon > 0$ and $\varepsilon_{\mathbf{K}}$ be such that $\mathbf{K} \subset \mathbf{K}_\varepsilon$ if $0 < \varepsilon < \varepsilon_{\mathbf{K}}$. Therefore

$$P_{n,\mathbf{s}}(\mathbf{B}_n^\varepsilon \subset \mathbf{K}_\varepsilon) > 1 - \varepsilon \quad \text{if } \mathbf{K}_\varepsilon \subset \mathbf{T}_n \tag{3.10.32}$$

and

$$P_{\mathbf{s}}(\mathbf{B}^\varepsilon \subset \mathbf{K}_\varepsilon) > 1 - \varepsilon \tag{3.10.33}$$

for all $0 < \varepsilon < \varepsilon_{\mathbf{K}}$. Note that

$$[\mathbf{B}_n^\varepsilon \cap K \neq \phi] \Leftrightarrow \inf_{\mathbf{K}} \mathbf{X}_n \leqslant \inf_{\mathbf{K}_\varepsilon} X_n + \varepsilon \quad \text{if } \mathbf{B}_n^\varepsilon \subset \mathbf{K}_\varepsilon \subset \mathbf{T}_n \tag{3.10.34}$$

and

$$[\mathbf{B}^\varepsilon \cap K \neq \phi] \Leftrightarrow \inf_{\mathbf{K}} X \leqslant \inf_{\mathbf{K}_\varepsilon} X + \varepsilon \quad \text{if } \mathbf{B}^\varepsilon \subset \mathbf{K}_\varepsilon. \tag{3.10.35}$$

Therefore, by Lemma 3.10.4, it follows that

$$\begin{aligned}
\overline{\lim_n}\, P_{n,\mathbf{s}}(\mathbf{B}_n^\varepsilon \cap \mathbf{K} \neq \phi) &\leqslant \varepsilon + \overline{\lim_n}\, P_{n,\mathbf{s}}\left\{\inf_{\mathbf{K}} X_n \leqslant \inf_{\mathbf{K}_\varepsilon} X_n + \varepsilon\right\} \\
&\leqslant \varepsilon + P_{\mathbf{s}}\left\{\inf_{\mathbf{K}} X \leqslant \inf_{\mathbf{K}_\varepsilon} X_n + \varepsilon\right\} \\
&\leqslant 2\varepsilon + P_{\mathbf{s}}(\mathbf{B}^\varepsilon \cap \mathbf{K} \neq \phi),
\end{aligned} \tag{3.10.36}$$

which proves (3.10.28) as required. This completes the proof of Theorem 3.10.2.

3.11 MAXIMUM LIKELIHOOD ESTIMATION FOR STOCHASTIC PROCESSES

We now digress from the classical model of i.i.d. observations to stochastic processes as investigations into large sample inference for stochastic processes bring out some special problems and results which are not present in the classical case. We refer the reader to Billingsley (1961), Prakasa Rao (1974), Basawa and Prakasa Rao (1980a, b), Grenander (1981), Basawa and Scott (1983), Kutoyants (1984), and other references cited at the end for statistical inference for stochastic processes. Our approach here is that of Sweeting (1980, 1983). The major difference from the classic case is that the averaged conditional information does not in general behave asymptotically like a constant, for instance, in branching processes. Such processes are termed nonergodic [cf. Basawa and Scott (1983)].

Let $\{(\Omega_t, \mathscr{A}_t), t \in T\}$ be a family of measurable spaces where $t \in T$ is a discrete or continuous index. Let P_θ^t be a probability measure defined on $(\Omega_t \mathscr{A}_t)$, $\theta \in \Theta \subset R^k$, $k \geqslant 1$. Suppose that the following conditions hold.

(A_1) Θ is open and $P_\theta^t \ll \lambda_t$, where λ_t is a σ-finite measure on $(\Omega_t, \mathscr{A}_t)$. Let

$$p_t(\theta) = \frac{dP_\theta^t}{d\lambda_t} \quad \text{and} \quad l_t(\theta) = \log p_t(\theta). \tag{3.11.0}$$

Clearly $l_t(\theta) < \infty$ a.e. $[\lambda_t]$. Assume that

(A_2) (*Differentiability*) Second-order partial derivatives of $p_t(\theta)$ exist and are continuous a.e. for every $\theta \in \Theta$. Let

$$\mathbf{U}_t(\theta) = \nabla l_t(\theta) = \left(\frac{\partial l_t(\theta)}{\partial \theta^{(1)}}, \ldots, \frac{\partial l_t(\theta)}{\partial \theta^{(k)}}\right), \tag{3.11.1}$$

where $\theta = (\theta^{(1)}, \ldots, \theta^{(k)})'$, and

$$-\mathscr{I}_t(\theta) = \nabla^2 l_t(\theta) = \left(\left(\frac{\partial^2 l_t(\theta)}{\partial \theta^{(i)} \partial \theta^{(j)}}\right)\right) \tag{3.11.2}$$

be the matrix of second-order partial derivatives. $\mathscr{I}_t(\theta)$ is the (*random*) *information matrix*.

Denote by the symbol $\to_{\mathrm{u}}$ the uniform convergence on compact subsets of Θ, $\xrightarrow{\mathscr{L}}_{\mathrm{u}}$ will be the uniform weak convergence (see Section 1.10) and $\xrightarrow{p}_{\mathrm{u}}$ will denote the uniform convergence in probability. For any matrix $\mathbf{A}$, let

$$|\mathbf{A}| = (\operatorname{tr} \mathbf{A}'\mathbf{A})^{1/2}$$

and $\mathbf{A}_n \to \mathbf{A}$ if $|\mathbf{A}_n - \mathbf{A}| \to 0$. $\mathbf{A} > 0$ means $\mathbf{A}$ is positive definite and $\mathbf{A}^{1/2}$ denotes the symmetric positive definite square root of $\mathbf{A}$ when $\mathbf{A} > 0$. Let $\mathbf{I}_k$ be the identity matrix of order $k \times k$.

Suppose the information matrix $\mathscr{I}_t(\theta)$ satisfies the following conditions.

($\mathbf{A}_3$) (*Growth and convergence*) There exist nonrandom square matrices $A_t(\theta)$ of order $k \times k$ continuous in θ such that

(3.11.3) (i) $\mathbf{A}_t(\theta)^{-1} \to_{\mathrm{u}} 0$,

(ii) $\mathbf{W}_t(\theta) \equiv \{\mathbf{A}_t(\theta)^{-1}\} \mathscr{I}_t(\theta) \{\mathbf{A}_t(\theta)^{-1}\}' \xrightarrow{\mathscr{L}}_{\mathrm{u}} \mathbf{W}(\theta)$,

(iii) $P[\mathbf{W}(\theta) > 0] = 1$.

(A_4) (*Continuity*) For all $c > 0$

(i) $\sup\{|\mathbf{A}_t(\theta)^{-1}\mathbf{A}_t(\phi) - \mathbf{I}_k| : |\mathbf{A}_t(\theta)'(\phi - \theta) \leqslant c)\} \xrightarrow{p}_{\mathrm{u}} 0$,

(ii) $\sup|\{\mathbf{A}_t(\theta)^{-1}\}[\mathscr{I}_t(\Gamma) - \mathscr{I}_t(\theta)]\{\mathbf{A}_t(\theta)^{-1}\}'| \xrightarrow{p}_{\mathrm{u}} 0$,

where the supremum is taken over the set

$$|\mathbf{A}_t(\theta)'(\theta_i - \theta)| \leqslant c, \qquad 1 \leqslant i \leqslant k,$$

for fixed $\theta_i \in \Theta, 1 \leqslant i \leqslant k$, and $\mathscr{I}_t(\Gamma)$ is the matrix $\mathscr{I}_t(\theta)$, where row i is evaluated at θ_i for $1 \leqslant i \leqslant k$.

We shall comment on these conditions later in this section. We first prove a lemma.

Lemma 3.11.1. Suppose condition (A_3) holds. Let G_θ be the distribution function of $\mathbf{W}(\theta)$. Then $G_\theta(\cdot)$ is continuous in $\theta \in \Theta$.

Proof. Let $G_{t,\theta}$ be the distribution function of $\mathbf{W}_t(\theta)$ defined by (3.11.3). Since $\mathbf{W}_t(\theta) \xrightarrow{\mathscr{L}}_{\mathrm{u}} \mathbf{W}(\theta)$, it is sufficient to show that $G_{t,\theta}$ is continuous in θ for each t. Let $g \in C_{\mathrm{u}}(\mathscr{X})$ the space of real bounded uniformly continuous functions on $\mathscr{X} = R^{k^2}$ and $\theta_n \to \theta$ with $\theta_n \in \Theta$ and $\theta \in \Theta$. Then

$$\begin{aligned}\left|\int g\,dG_{t,\theta_n} - \int g\,dG_{t,\theta}\right| &= \left|\int g[\mathbf{W}_t(\theta_n)]\,dP_{t,\theta_n} - \int g[\mathbf{W}_t(\theta)]dP_{t,\theta}\right| \\ &\leqslant \|g\| \int |p_t(\theta_n) - p_t(\theta)|\,d\lambda(t) \\ &\quad + \int |g(\mathbf{W}_t(\theta_n)) - g(\mathbf{W}_t(\theta))|\,dP_{t,\theta}.\end{aligned}$$

The first term on the right hand side of the inequality tends to zero as $n \to \infty$ since $p_t(\theta)$ is continuous a.e. by Scheffe's theorem and the second term tends to zero by the bounded convergence theorem since $\mathbf{W}_t(\theta)$ is continuous a.e. (λ_t) and $g \in C_u(\mathscr{X})$. Let

$$\text{(3.11.4)} \qquad \mathbf{X}_t(\theta) = \{\mathbf{A}_t(\theta)\}^{-1}\mathbf{U}_t(\theta), \qquad Y_t(\theta) = \mathbf{A}_t(\theta)'(\hat{\theta}_t - \theta),$$

$$\text{(3.11.5)} \qquad \mathbf{Z}_t(\theta) = \{\mathbf{W}_t(\theta)\}^{1/2}\, Y_t(\theta),$$

where $\hat{\theta}_t$ is any local maximum of $l_t(\theta)$. If none exists, set $\hat{\theta}_t = +\infty$.

Theorem 3.11.2. Under conditions (A_1)–(A_4).

$$\text{(3.11.6)} \qquad (\mathbf{X}_t(\theta), \mathbf{W}_t(\theta)) \xrightarrow{\mathscr{L}}_{u} (\mathbf{W}(\theta)^{1/2}\mathbf{Z}, \mathbf{W}(\theta))$$

where $\mathbf{Z}$ is $\mathbf{N}_k(\mathbf{0}, \mathbf{I}_k)$ independent of $\mathbf{W}(\theta)$, $\theta \in \Theta$.

Proof. Let $\theta_t, \theta \in \Theta$, such that $\theta_t \to \theta$ as $t \to \infty$. For simplicity, write $\mathbf{W}_t = \mathbf{W}_t(\theta_t)$, $\mathbf{A}_t = \mathbf{A}_t(\theta_t)$, $\mathbf{X}_t = \mathbf{X}_t(\theta_t)$. Observe that $\mathbf{A}_t^{-1} \to 0$ from (A_3). We will omit θ from the argument when it is clear from the context. Let $\theta \in R^k$ and

$$\psi_t = \theta_t + (\mathbf{A}_t^{-1})'\theta.$$

Note that $\psi_t \to \theta$ since $\mathbf{A}_t^{-1} \to 0$ and $\theta_t \to \theta$. Furthermore, $\psi_t \in \Theta$ for t large (say) $t > t_0$. Hereafter, assume that $t > t_0$. Note that

$$\text{(3.11.7)} \quad l_t(\psi_t) = l_t(\theta_t) + (\psi_t - \theta_t)'\nabla l_t(\theta_t) - \tfrac{1}{2}(\psi_t - \theta_t)'\mathscr{I}_t(\phi_t)(\psi_t - \theta_t),$$

where $\phi_t = \alpha_t\theta_t + (1 - \alpha_t)\psi_t$, $0 < \alpha_t < 1$, and α_t is random. Let

$$\mathbf{V}_t = \{\mathbf{A}_t^{-1}\}\mathscr{I}(\phi_t)\{\mathbf{A}_t^{-1}\}'.$$

By exponentiating on both sides of (3.11.7), we have

$$\text{(3.11.8)} \qquad e^{\theta'\mathbf{X}_t}p_t(\theta_t) = e^{(1/2)\theta'\mathbf{V}_t\theta}p_t(\psi_t).$$

As a consequence of (A_3) and (A_4), we have

$$\text{(3.11.9)} \qquad \{\mathbf{A}_t(\theta')\}^{-1}\mathscr{I}_t(\Gamma)\{\mathbf{A}_t(\theta')^{-1}\}' \xrightarrow{p}_{u} \mathbf{W}(\theta)$$

when $|\{\mathbf{A}_t(\theta)\}'(\theta' - \theta)| \leqslant c$ and θ_i, $1 \leqslant i \leqslant k$, are random vectors such that

$|\{A_t(\theta)\}'(\theta_i - \theta)| \leqslant c$ with $P(W(\theta) > 0) = 1$. Hence

$$\mathbf{V}_t \xrightarrow{\mathscr{L}} \mathbf{W} \tag{3.11.10}$$

both under $\{\theta_t\}$ and under $\{\psi_t\}$. Let $0 < \varepsilon < 1$ and choose K such that

$$\mathrm{P}(|\mathbf{W}| \geqslant K) \leqslant \varepsilon \quad \text{and} \quad P(|\mathbf{W}| = K) = 0$$

Since $\mathbf{V}_t \xrightarrow{\mathscr{L}} \mathbf{W}$ and the set $[|\mathbf{W}| < K]$ is a G_θ-continuity set, it follows that

$$P^t_{\theta_t}(|\mathbf{V}_t| < K) \to P(|\mathbf{W}| < K). \tag{3.11.11}$$

Let Q^t be the probability measure $P^t_{\theta_t}$ conditional on $[|\mathbf{V}_t| < K]$. Q^t has the density

$$q_t = \begin{cases} p_t(\theta_t)/P^t_{\theta_t}(|\mathbf{V}_t| < K) & \text{if } |\mathbf{V}_t| < K \\ 0 & \text{otherwise.} \end{cases} \tag{3.11.12}$$

Let E^*_t denote the expectation under Q^t. Let $h(\cdot)$ be a bounded function on M_k, the space of all $k \times k$ matrices, continuous for $|A| < K$ with $h(A) = 0$ for $|A| \geqslant K$. Then relation (3.11.9) implies that

$$E^*_t\{h(\mathbf{V}_t)e^{\theta'\mathbf{X}_t}\} = E\{h(\mathbf{V}_t)e^{(1/2)\theta'\mathbf{V}_{t}s}\}/P^t_{\theta_t}(|\mathbf{V}_t| < K). \tag{3.11.13}$$

The term on the right hand side of (3.11.13) tends to

$$E\{h(\mathbf{W})e^{(1/2)\theta'\mathbf{W}\theta}\}/P(|\mathbf{W}| < K) \tag{3.11.14}$$

from (3.11.11) and the fact that $h(\mathbf{w})e^{(1/2)\theta'\mathbf{w}\theta}$ is a G_θ-continuous function [cf. Billingsley (1968, Theorem 5.2)]. The term in (3.11.14) is equal to

$$E^*\{h(\mathbf{W})e^{(1/2)\theta'\mathbf{W}\theta}\}, \tag{3.11.15}$$

where E^* denotes the expectation given that $|\mathbf{W}| < K$. But

$$E^*\{h(\mathbf{W})e^{(1/2)\theta'\mathbf{W}\theta}\} = E^*\{h(\mathbf{W})e^{\theta'\mathbf{W}^{1/2}\mathbf{Z}}\}, \tag{3.11.16}$$

where Z is $\mathbf{N}_k(0, \mathbf{I}_k)$ independent of W. By the uniqueness of the bilateral Laplace transform and the weak compactness theorem, it follows that

$$(\mathbf{X}_t, \mathbf{V}_t) \xrightarrow{\mathscr{L}} (\mathbf{W}^{1/2}\mathbf{Z}, \mathbf{W}) \tag{3.11.17}$$

conditional on $|\mathbf{W}| < K$ with respect to the family (Q^t) of distributions. Since ε is arbitrary, we have

$$(3.11.18) \qquad (\mathbf{X}_t, \mathbf{V}_t) \xrightarrow{\mathscr{L}} (\mathbf{W}^{1/2}\mathbf{Z}, \mathbf{W})$$

and hence

$$(3.11.19) \qquad (\mathbf{X}_t, \mathbf{W}_t) \xrightarrow{\mathscr{L}} (\mathbf{W}^{1/2}\mathbf{Z}, \mathbf{W})$$

since

$$\mathbf{V}_t - \mathbf{W}_t \xrightarrow{p} 0$$

with respect to $\{P^t_{\theta_t}\}$. Since (3.11.19) holds for arbitrary $\theta_t \to \theta$ and since $\{\mathbf{W}(\theta)^{1/2}\mathbf{Z}, \mathbf{W}(\theta)\}$ has a distribution continuous in θ by Lemma 3.11.1, it follows that

$$(3.11.20) \qquad (\mathbf{X}_t(\theta), \mathbf{W}_t(\theta)) \xrightarrow{\mathscr{L}}_{\mathrm{u}} (\{\mathbf{W}(\theta)\}^{1/2}\mathbf{Z}, \mathbf{W}(\theta)).$$

This completes the proof of Theorem 3.11.2.

Proposition 3.11.3. Suppose conditions (A_1)–(A_4) hold. Then there exists a local maximum $\hat{\theta}_t$ of $l_t(\theta)$ such that

$$\mathbf{Y}_t(\theta) = \{\mathbf{A}_t(\theta)\}'(\hat{\theta}_t - \theta)$$

is uniformly stochastically bounded, that is, for every $\varepsilon > 0$ and compact set $K \subset \theta$ there exist $c > 0$ and t_0 such that

$$(3.11.21) \qquad P_\theta(|\mathbf{Y}_t(\theta)| > c) < \varepsilon$$

for all $t > t_0$ and $\theta \in K$.

Proof. Let $\theta_t, \theta \in \Theta$, such that $\theta_t \to \theta$ as $t \to \infty$. Define

$$(3.11.22) \qquad S_t = \{\phi \in R^k : |A_t'(\phi - \theta_t)| = c\}, \qquad c > 0.$$

It is easy to check that if

$$(3.11.23) \qquad (\phi - \theta_t)'\nabla l_t(\phi) < 0$$

for all $\phi \in S_t$, then there exists a local maximum $\hat{\theta}_t$ of $l_t(\theta)$ and $|\mathbf{Y}_t(\theta_t)| \leqslant c$ [cf. Aitchinson and Silvey (1958)]. Let

$$\pi_t = P^t_{\theta_t}\left(\sup_{\phi \in S_t} (\phi - \theta_t)' \nabla l_t(\phi) \geqslant 0 \right). \tag{3.11.24}$$

We shall prove that

$$\lim_{c \to \infty} \overline{\lim_{t \to \infty}}\, \pi_t = 0, \tag{3.11.25}$$

which implies the result. Note that

$$(\phi - \theta_t)' \nabla l_t(\phi) = (\phi - \theta_t)' \nabla l_t(\theta_t) - (\phi - \theta_t)' \mathscr{I}_t(\Gamma)(\phi - \theta_t), \tag{3.11.26}$$

where $\Gamma = (\theta_1, \ldots, \theta_k)$ and $|A_t'(\theta_i - \theta_t)| \leqslant c, 1 \leqslant i \leqslant k$. Hence

$$\pi_t \leqslant P^t_{\theta_t}\left(\sup_{|\mathbf{x}|=1} \mathbf{x}'\mathbf{X}_t \geqslant c \inf_{|\mathbf{x}|=1} \mathbf{x}'\mathbf{U}_t\mathbf{x} \right), \tag{3.11.27}$$

where $\mathbf{U}_t = \mathbf{A}_t^{-1} \mathscr{I}_t(\Gamma) \mathbf{A}_t^{-1}$. But $\inf_{|\mathbf{x}|=1} \mathbf{x}' U_t \mathbf{x} = \mu_t$, the least eigenvalue of U_t. Relation (3.11.9) and the uniformity in convergence imply that

$$\mu_t \xrightarrow{\mathscr{L}} \mu(\theta), \tag{3.11.28}$$

where $\mu(\theta)$ is the least eigenvalue of $\mathbf{W}(\theta)$. But, for every $\varepsilon > 0$,

$$\pi_t \leqslant P^t_{\theta_t}(|\mathbf{X}_t| \geqslant c\mu_t) \leqslant P^t_{\theta_t}(|\mathbf{X}_t| \geqslant c\varepsilon) + P^t_{\theta_t}(\mu_t \leqslant \varepsilon). \tag{3.11.29}$$

Since $X_t \xrightarrow{\mathscr{L}} \{\mathbf{W}(\theta)\}^{1/2}\mathbf{Z}$ by Theorem 3.11.2 and $\mu_t \xrightarrow{\mathscr{L}} \mu(\theta)$, it follows that

$$\begin{aligned} \overline{\lim_{t \to \infty}}\, \pi_t &\leqslant P\{|\{\mathbf{W}(\theta)\}^{1/2}\mathbf{Z}| \geqslant c\varepsilon\} + P\{\mu(\theta) \leqslant \varepsilon\} \\ &\to P(\mu(\theta) \leqslant \varepsilon) \end{aligned} \tag{3.11.30}$$

as $c \to \infty$ by Theorem 2.1 in Billingsley (1968). Let $\varepsilon \to 0$. It follows that

$$\lim_{c \to \infty} \overline{\lim_{t \to \infty}}\, \pi_t = 0. \tag{3.11.31}$$

This completes the proof.

Theorem 3.11.4. Suppose conditions (A_1)–(A_4) hold. Then there exists a local maximum $\hat{\theta}_t$ of $l_t(\theta)$ with probability tending to 1 such that

$$\mathbf{X}_t(\theta) - \mathbf{W}_t(\theta)\mathbf{Y}_t(\theta) \xrightarrow{p}_u 0. \tag{3.11.32}$$

Proof. Let $F_t = [\hat{\theta}_t < \infty]$. Observe that

$$\mathbf{U}_t(\theta) = \mathscr{I}_t(\Gamma)(\hat{\theta}_t - \theta),$$

where $\Gamma = (\theta_1, \ldots, \theta_k)$. Here θ_i, $1 \leqslant i \leqslant k$, are as given in (A_4). Hence

$$\mathbf{X}_t(\theta) = \mathbf{W}_t(\theta, \hat{\theta}_t)\mathbf{Y}_t(\theta),$$

where

$$\mathbf{W}_t(\theta, \hat{\theta}_t) = \{\mathbf{A}_t(\theta)\}^{-1}\mathscr{I}_t(\Gamma)\{\mathbf{A}_t(\theta)\}^{-1'}.$$

Since $\mathbf{A}_t(\theta)'(\theta_i - \theta)$, $1 \leqslant i \leqslant k$, are uniformly stochastically bounded, it follows from (A_4) that

$$\mathbf{W}_t(\theta, \hat{\theta}_t) - \mathbf{W}_t(\theta) \xrightarrow{p}_u 0.$$

Since $\mathbf{Y}_t(\theta)$ is uniformly stochastically bounded, the theorem holds conditionally on the set F_t. Hence the result is true since $P_\theta^t(F_t) \to_u 1$.

As a consequence of Theorems 3.11.2 and 3.11.4 and the continuous mapping theorem [see Proposition 1.10.11 or Billingsley (1968)] it follows that

$$(\{\mathbf{W}_t(\theta)\}^{1/2}\mathbf{Y}_t(\theta), \mathbf{W}_t(\theta)) \xrightarrow{\mathscr{L}}_u (\mathbf{Z}, \mathbf{W}(\theta)), \tag{3.11.33}$$

where $\mathbf{Z}$ is $\mathbf{N}_k(\mathbf{0}, \mathbf{I}_k)$. In particular, we have

$$\mathbf{Z}_t = (\mathbf{W}_t(\theta))^{1/2}\mathbf{Y}_t(\theta) \xrightarrow{\mathscr{L}}_u \mathbf{Z}, \tag{3.11.34}$$

or in other words,

$$\mathbf{W}_t(\theta)^{1/2}\mathbf{A}_t(\theta)'(\hat{\theta}_t - \theta) \xrightarrow{\mathscr{L}}_u \mathbf{Z}. \tag{3.11.35}$$

In view of (A_3), it follows that

$$\mathbf{Y}_t(\theta) = \mathbf{A}_t(\theta)'(\hat{\theta}_t - \theta) \xrightarrow{\mathscr{L}} \{\mathbf{W}(\theta)\}^{-1/2}\mathbf{Z}. \tag{3.11.36}$$

Remarks. In applications, $\mathbf{A}_t(\theta)\mathbf{A}_t(\theta)' = E_\theta \mathscr{I}_t(\theta)$ and one can choose $\mathbf{A}_t(\theta) = \{E_\theta \mathscr{I}_t(\theta)\}^{1/2}$. The conditions assumed above do not involve the existence of $E_\theta \mathscr{I}_t(\theta)$. Note that (A_4) and Proposition 3.11.3 show that

$$\{\mathbf{A}_t(\theta)\}^{-1}\mathbf{A}_t(\hat{\theta}_t) \xrightarrow{\mathrm{p}}_{\mathrm{u}} \mathbf{I}_k$$

and hence $\mathbf{A}_t(\theta)$ may be replaced by $\mathbf{A}_t(\hat{\theta}_t)$ in the asymptotic analysis discussed earlier.

In the case of ergodic models, that is, when $\mathbf{W}(\theta)$ is nonrandom, condition (A_3) holds if $E\{\mathscr{I}_t(\theta)\} \to_{\mathrm{u}} \mathbf{W}(\theta)$ and $\operatorname{Var}\{\mathscr{I}_t(\theta)\} \to_{\mathrm{u}} 0$. For the discrete ergodic case, see Prakasa Rao (1974) and Weiss (1973). As a simple example of a nonergodic model, we now discuss the problem of estimation of the mean in a supercritical Galton–Watson branching process. It can be checked that

$$\mathscr{I}_n(\theta) = \phi(\theta) \sum_{i=0}^{n-1} X_i - \frac{\partial \phi}{\partial \theta} \sum_{i=1}^{n} (X_i - \theta X_{i-1}),$$

where $\phi(\theta) = \sigma^2(\theta)$ is the offspring distribution variance and X_i is the size of the ith generation. The results stated in this section hold with $A_n(\theta)^2 = \theta^n$. For more details in inference for branching processes, see Basawa and Prakasa Rao (1980b). Another example of a nonergodic model is that of pure-birth process. The problem is the estimation of the birth parameter λ [cf. Keiding (1974)]. Here $\mathscr{I}_t(\lambda) = \lambda^{-2}(X_t - X_0)$, where X_t is the population size at time t and the results given above hold with $A_t(\lambda)^2 = e^{\lambda t}$.

We shall now discuss the optimality property of the estimator $\hat{\theta}_t$ obtained above, following Sweeting (1983). We assume that conditions (A_1)–(A_4) hold as before. Hence

$$\mathbf{Y}_t(\theta) = \mathbf{A}_t(\theta)'(\hat{\theta}_t - \theta) \xrightarrow{\mathscr{L}}_{\mathrm{u}} \{\mathbf{W}(\theta)\}^{-1/2}\mathbf{Z}. \tag{3.11.37}$$

Let $F_\theta(\cdot)$ be the distribution function of $\{\mathbf{W}(\theta)\}^{-1/2}\mathbf{Z}$. Let $\mathscr{R}$ be a class of sets J in R^k that are convex and symmetric about the origin. In view of (3.11.37), it follows that

$$P_\theta^t\{\mathbf{Y}_t(\theta) \in J\} \to_{\mathrm{u}} F_\theta\{J\} \tag{3.11.38}$$

by Lemma 3.10.1 and Proposition 1.10.2. Let $\mathscr{C}$ be the class of estimators T_t such that

$$\mathbf{A}_t(\theta)'(T_t - \theta) \xrightarrow{\mathscr{L}}_{\mathrm{u}} T, \tag{3.11.39}$$

where T has some distributions, say L_θ. The form of L_θ is not specified. We now show that $\hat{\theta}_t$ is optimal in the class $\mathscr{C}$ in the following sense.

Theorem 3.11.5. Suppose conditions (A_1)–(A_4) hold. Then, for any $(T_t)\in\mathscr{C}$,

$$L_\theta\{J\} \leqslant F_\theta\{J\} \tag{3.11.40}$$

for all $J\in\mathscr{R}$ and $\theta\in\Theta$.

Remarks. In other words, the maximum likelihood estimator is efficient in the sense of having asymptotically maximum probability of concentration about the true parameter among the class of all estimators that have limiting distribution in a uniform sense as specified by (3.11.39).

Proof. Let $\theta_0\in\Theta$ and denote $\mathbf{A}_t=\mathbf{A}_t(\theta_0)$, $\mathscr{I}_t=\mathscr{I}_t(\theta_0)$, and $\mathbf{W}_t=\mathbf{W}_t(\theta_0)$ as before. Let $h>0$ and $H_t=\{\theta:|\mathbf{A}_t'(\theta-\theta_0)|<h\}$. Since

$$\mathbf{A}_t'(T_t-\theta)\xrightarrow{\mathscr{L}}{}_{\mathrm{u}}T, \tag{3.11.41}$$

where T has the distribution L_θ, it follows that

$$P_\theta^t\{\mathbf{A}_t'(T_t-\theta)\in J\}-P_{\theta_0}^t\{\mathbf{A}_t'(T_t-\theta_0)\in J\}\to 0 \tag{3.11.42}$$

uniformly for $\theta\in H_t$ and $J\in\mathscr{R}$ [note that the distribution of $\mathbf{A}_t'(\theta)(T_t-\theta)$ is continuous in θ for fixed t since $A_t(\theta)$ is continuous and hence the distribution of T, namely, L_θ is continuous in θ by uniform convergence]. Furthermore, conditions (A_3)–(A_4) imply the following for any given $\varepsilon>0$. Condition (A_3) implies that there exists t_0 such that for all $t>t_0$ and $\theta\in H_t$,

$$P_\theta^t\left\{\sup_{\theta'\in H_t}|\mathbf{A}_t^{-1}[\mathscr{I}_t(\theta')-\mathscr{I}_t]\mathbf{A}_t^{-1'}|>\frac{\varepsilon}{h^2}\right\}<\varepsilon. \tag{3.11.43}$$

Conditions (A_3) and (A_4) and the inequality (3.11.43) prove that

$$P_\theta^t\{|\mathbf{W}_t|>0)>1-\varepsilon. \tag{3.11.44}$$

Since $Y_t(\theta)$ is uniformly stochastically bounded by Proposition 3.10.3, it follows that there exists $M>0$ and $b_t\to\infty$ such that for all θ satisfying $|\mathbf{A}_t'(\theta-\theta_0)|\leqslant b_t$ and $t>t_1$,

$$P_\theta^t(|\mathbf{A}_t'(\hat{\theta}_t-\theta)|>M)<\varepsilon. \tag{3.11.45}$$

In view of (3.11.42) to (3.11.45), we obtain the result as follows.

Suppose $J \in \mathscr{R}$ and J is bounded. Let λ_t be the Lebesgue measure of $H_t \subset R^k$. In view of (3.10.42), it follows that

$$\text{(3.11.46)} \quad \lim_t \lambda_t^{-1} \int_{H_t} P_\theta^t(\mathbf{A}_t'(T_t - \theta) \in J)\, d\theta = \lim_t P_{\theta_0}^t(A_t'(T_t - \theta_0) \in J).$$

In particular (3.11.46) holds for $\hat{\theta}_t$ too. In order to prove the theorem, it is sufficient to show that

$$\text{(3.11.47)} \quad \lambda_t^{-1} \int_{H_t} P_\theta^t(\mathbf{A}_t'(T_t - \theta) \in J)\, d\theta \leqslant e^{4\varepsilon} \lambda_t^{-1} \int_{H_t} P_\theta^t(\mathbf{A}_t'(\hat{\theta}_t - \theta) \in J)\, d\theta + 4\varepsilon$$

in view of (3.11.46). Let $B(0, z) \supset J$, where $B(0, z)$ denotes the sphere in R^k with center at 0 and radius z. Let

$$\text{(3.11.48)} \quad D_t = \{\omega_t \colon \sup_{\theta' \in H_t} |\mathbf{A}_t^{-1}[\mathscr{I}_t(\theta') - \mathscr{I}_t]\mathbf{A}_t^{-1}| \leqslant \varepsilon h^{-2}; |\mathbf{W}_t(\theta_0)| > 0;$$
$$\cdot |\mathbf{A}_t'(\hat{\theta} - \theta_0)| \leqslant h - z\}$$

and

$$\text{(3.11.49)} \qquad B_t(d) = H_t \cap \{\mathbf{A}_t'(d - \theta) \in J\}.$$

Note that

$$\text{(3.11.50)} \quad \lambda_t^{-1} \int_{H_t} P_\theta^t(\mathbf{A}_t'(T_t - \theta) \in J)\, d\theta \leqslant \lambda_t^{-1} \int_{D_t} \int_{B_t(T_t)} p_t(\theta)\, d\theta\, d\lambda_t$$
$$+ \lambda_t^{-1} \int_{H_t} P_\theta^t(D_t^c)\, d\theta$$
$$= K_1 + K_2 \quad \text{(say)}.$$

For $\theta \in H_t$ and $x \in D_t$,

$$\text{(3.11.51)} \qquad p_t(\theta) = p_t(\hat{\theta}_t) \exp\{-\tfrac{1}{2}(\theta - \hat{\theta}_t)' \mathscr{I}_t(\theta - \hat{\theta}_t) + \delta_t\},$$

where $|\delta_t| \leqslant 2\varepsilon$. By Proposition 1.14.17 (Anderson, 1955), it can be seen that the integral of

$$\exp\{-\tfrac{1}{2}(\theta - \hat{\theta}_t)' \mathscr{I}_t(\theta - \hat{\theta}_t)\}$$

over the set $\mathbf{A}_t'(\mathbf{d} - \theta) \in J$ is maximized when $\mathbf{d} = \hat{\theta}_t$ and hence the integral is also maximum when $\mathbf{d}=\hat{\theta}_t$ for $\theta \in H_t$, since $|\mathbf{A}_t'(\hat{\theta}_t-\theta_0)| \leqslant h-z$. Therefore

(3.11.52) $$K_1 \leqslant e^{4\varepsilon}\lambda_t^{-1}\int_{H_t} P_\theta^t(\mathbf{A}_t'(\hat{\theta}_t-\theta)\in J)\,d\theta.$$

Let $J_t=\{\theta:|\mathbf{A}_t'(\theta-\theta_0)| \leqslant (1-\varepsilon)^{1/k}h\}$. It can be checked from relations (3.11.43)–(3.11.45) that

(3.11.53) $$P_\theta^t(D_t^c) < 3\varepsilon$$

for $\theta \in J_t$. On the other hand,

(3.11.54) $$\lambda_t^{-1}\int_{H_t \setminus J_t} d\theta < \varepsilon.$$

Relations (3.11.53) and (3.11.54) show that

(3.11.55) $$K_2 < 4\varepsilon.$$

Combining (3.11.50) with (3.11.52) and (3.11.55), we have the required result for bounded intervals J.

Suppose J is unbounded. Let $K>0$ such that

(3.11.56) $$P_\theta^t(|\mathbf{A}_t(\theta)'(T_t-\theta)|>K)<\varepsilon$$

for all $t \in T$. Let $\tilde{J}=J\cap\{|\mathbf{x}| \leqslant K\}$. We have seen above that the result holds for $\tilde{J}$ and it can be checked that

$$L_\theta\{J\} \leqslant F_\theta\{J\}+\varepsilon.$$

Since ε is arbitrary, we have the main result. This completes the proof of Theorem 3.11.5.

As a consequence of the above theorem, the following corollary can be obtained.

Corollary 3.11.6. Suppose conditions (A_1)–(A_4) hold. Furthermore, suppose that $E(\{\mathbf{W}(\theta)\}^{-1})<\infty$. Let $\mathbf{M}(L_\theta)$ denote the asymptotic mean-square error matrix of L_θ whenever it exists, that is,

$$\mathbf{M}(L_\theta)=\int \mathbf{x}\mathbf{x}'\,dL_\theta(\mathbf{x}).$$

Then $\mathbf{M}(L_\theta) - \mathbf{M}(F_\theta)$ is nonnegative definite where $F_\theta(\cdot)$ is the distribution function of $\mathbf{W}(\theta)^{-1/2}\mathbf{Z}$ and $\mathbf{Z}$ is $\mathbf{N}_k(0, \mathbf{I}_k)$.

Proof. Note that F_θ has mean $\mathbf{0}$ and covariance matrix $E\{\mathbf{W}(\theta)\}^{-1}$. The corollary follows from the following relations.

Let $\mathbf{X} \sim L_\theta$ and $\mathbf{Y} \sim F_\theta$ and $\mathbf{x} \in R^k$. Let $L_\theta^{\mathbf{x}}$ and $F_\theta^{\mathbf{x}}$ be the distribution functions of $|\mathbf{x}'\mathbf{X}|^2$ and $|\mathbf{x}'\mathbf{Y}|^2$, respectively. Then

$$\begin{aligned} E|\mathbf{x}'\mathbf{X}|^2 &= \int_0^\infty u\,dL_\theta^{\mathbf{x}}(u) = \int_0^\infty (1 - L_\theta^{\mathbf{x}}(u))\,du \\ &\geqslant \int_0^\infty (1 - F_\theta^{\mathbf{x}}(u))\,du = E|\mathbf{x}'\mathbf{Y}|^2. \end{aligned}$$

3.12 CHANGE-POINT PROBLEM

Suppose we want to test the hypothesis

$$H_0: \quad X_1, \ldots, X_n \text{ i.i.d } N(\theta, 1)$$

against the alternative

$$H_1: \quad \text{there exists } 1 \leqslant k \leqslant n-1$$

such that $X_1, \ldots, X_t$ i.i.d. $N(\theta_1, 1)$ and $X_{k+1}, \ldots, X_n$ i.i.d. $N(\theta_2, 1)$ with $\theta_1 \neq \theta_2$. Examples of such problems arise in many practical situations.

The likelihood ratio test for testing H_0 versus H_1 is

$$L_n = \frac{\sup_k \sup_{\theta_1, \theta_2} g(X_1, \ldots, X_k; \theta_1) g(X_{k+1}, \ldots, X_n; \theta_2)}{\sup_\theta g(X_1, \ldots, X_n; \theta)}, \tag{3.12.0}$$

where $g(X, \ldots, X_n; \theta)$ is the joint density of $(X_1, \ldots, X_n)$ under $N(\theta, 1)$.

It can be shown that $L_n \to \infty$ a.s. under H_0 unlike the classic i.i.d. case when $-2\log L_n$ is asymptotically χ^2 under H_0 under some regularity conditions. In order to study the existence and identification of a change point and its asymptotic behaviour, we now extend the weak convergence properties of the likelihood ratio processes by adding a time parameter to the classical one discussed earlier, for instance, in Section 3.6. This type of result is necessary when the order in which observations are obtained is important. We shall indicate only the main results with some applications. For details, see Deshayes and Picard (1983, 1984a, b).

Let $\phi(\cdot)$ be a piecewise continuous function on $[0,1]$ such that $\phi(0)=0$, $\phi(1)=1$, and $\psi(t)=\phi(t)/\sqrt{t}$ is nondecreasing on $[0,1]$. Let C_ϕ denote the set of functions $z(\theta,t)$ continuous on $R^p\times[0,1]$ with $z(\theta,0)\equiv 1$ and

$$\lim_{N\to\infty}\sup_{D_M} z(\theta,t)=0, \tag{3.12.1}$$

where

$$D_M=\{(\theta,t):\phi(t)\|\theta\|>M\}. \tag{3.12.2}$$

Let z_n and z belong to C_ϕ. z_n is said to converge to z in C_ϕ if for every $\varepsilon>0$ and every compact $K\subset R^p$, there exists n_0 such that for $n\geqslant n_0$,

$$\sup_{K\times[0,1]}|z_n-z|\leqslant\varepsilon$$

and for every $\varepsilon>0$, there exists M_0, beyond which, for every n

$$\sup_{D_M}|z_n|\leqslant\varepsilon.$$

The space C_ϕ is metrizable under the metric

$$d(z_1,z_2)=\sum_{j=0}^{\infty}2^{-j}\sup_{\substack{j\leqslant|\theta|<j+1\\0\leqslant t\leqslant 1}}|z_1-z_2|+\lim_{M\to\infty}\sup_{D_M}|z_1-z_2|. \tag{3.12.3}$$

Furthermore, by Ascoli-Arzela's theorem, a set $J\subset C_\phi$ is relatively compact if and only if

(i) for every $\varepsilon>0$ and every compact set K, there exists $\delta>0$ such that

$$\sup_{z\in J}\sup_{\substack{(\theta_i,t_i)\in K\times[0,1]\\|(\theta_1,t_1)-(\theta_2,t_2)|\leqslant\delta}}|z(\theta_1,t_1)-z(\theta_2,t_2)|\leqslant\varepsilon \tag{3.12.4}$$

and

(ii) for every $\varepsilon>0$, there exists an M such that

$$\sup_{z\in J}\sup_{D_M}|z'|\leqslant\varepsilon. \tag{3.12.5}$$

Let $X_1,X_2,\ldots,X_n$ be i.i.d. random variables taking values in $(\mathscr{X},\mathscr{B})$ with probability measure $P_\theta, \theta \in \Theta$ open in R^p. Suppose $P_\theta \ll \mu$, μ σ-finite. Let

$$\frac{dP_\theta}{d\mu}(x) = g(x, \theta).$$

For every $\theta \in R^p$, let $Z_n(\theta, \cdot)$ be the polygonal line joining the points $(0, 1)$ and

$$\left(\frac{k}{n}, \left[\prod_{i=1}^{k} \frac{g(X_i, \theta_0 + \theta n^{-1/2})}{g(X_i, \theta_0)}\right]\right), \qquad 1 \leqslant k \leqslant n,$$

where $\theta_0 \in \Theta$ is fixed. Suppose the following regularity conditions hold.

(A) (i) The function $g(x, \theta)$ is absolutely continuous in θ on R^p for x a.e. $[\mu]$.

(3.12.6) (ii) $$\mathbf{I}(\theta) = \int_{\mathscr{X}} \nabla g^{1/2}(x, \theta) \nabla g^{1/2}(x, \theta)' \, d\mu(x)$$

is continuous on R^p and positive definite at θ_0. (Here ∇h is the gradient of h with respect to θ.)

(iii) For all $\delta > 0$ and $\mathbf{w} \in R^p$ such that $\|\mathbf{w}\| = 1$,

(3.12.7) $$\lim_{\varepsilon \to 0} \varepsilon^{-p} \int_{\|\mathbf{v} - \theta_0\| \leqslant \varepsilon} d\lambda(\mathbf{v}) \int \nabla g^{1/2}(\mathbf{x}, \mathbf{v})' \nabla g^{1/2}(x, \mathbf{v}) \, d\mu(x) = 0$$
$$\cdot \left\{ x : \left| \log \frac{g(x, \theta_0 + \varepsilon \mathbf{w})}{g(x, \theta_0)} \right| > \delta \right\}$$

where λ is the Lebesgue measure on R^p.

For every nonempty subset u of $\{1, 2, \ldots, p\}$, denote by ∂_u the derivative operator with respect to the coordinates of indices belonging to u. If $u = \{j_1, \ldots, j_{l(u)}\}$, then

$$\partial_u = \frac{\partial^{l(u)}}{\partial \theta^{(j_1)} \ldots \partial \theta^{(j_{l(u)})}}.$$

(B) There exist $a > 0$, an integer $m \geqslant 1$, and an even integer q such that $mq > p$ and

(3.12.8) $$\sup_{\theta \in R^p} (1 + \|\theta\|^a)^{-1} \int_{\mathscr{X}} |\partial_u g^{1/q}(x, \theta_0 + \theta)|^q \, d\mu(x) < \infty$$

for every subset u of $\{1, 2, \ldots, p\}$ such that its cardinality $l(u) \leqslant m$.

(C) There exists $\eta > 0$ such that

$$\overline{\lim_{|\theta| \to \infty}} \|\theta\|^{\eta} \rho(\theta_0, \theta_0 + \theta) < \infty, \tag{3.12.9}$$

where ρ denotes the affinity (see Section 1.3)

$$\rho(\theta_0, \theta_0 + \theta) = \int [g(x, \theta_0) g(x, \theta_0 + \theta)]^{1/2} \, d\mu(x). \tag{3.12.10}$$

(D) For every $\varepsilon > 0$,

$$\lim_{M \to \infty} P_{\theta_0} \left\{ \sup_{\|\theta\| > M} \frac{g(X, \theta_0 + \theta)}{g(X, \theta_0)} > \varepsilon \right\} = 0. \tag{3.12.11}$$

(E) The Hessian matrix $\ddot{L}(X, \theta)$ of $L(X, \theta) = \log g(X, \theta_0 + \theta)$ in θ is continuous at $\theta = 0$ almost surely and there exists $\delta > 0$ such that

$$E_{\theta_0} \left\{ \sup_{|\theta| \leqslant \delta} |\ddot{L}(X, \theta)| \right\} < \infty. \tag{3.12.12}$$

If $\Theta \neq R^p$, then the process Z_n is defined a priori on the set $\mathscr{D}_n = \{\theta : \theta_0 + \theta n^{-1/2} \in \Theta\}$ only, but we set

$$Z_n(\theta, 0) = 1 \quad \text{for } \theta \in R^p, \tag{3.12.13}$$

$$Z_n\left(\theta, \frac{k}{n}\right) = 0$$

for θ belonging to the complement of a neighborhood of $\mathscr{D}_n$ for $k = 1, 2, \ldots, n$ and it is extended to $R^p \times [0, 1]$ by continuity. Hereafter we suppose that $\Theta = R^p$.

The following theorem is due to Deshayes and Picard (1984). We omit the proof.

Theorem 3.12.1. Suppose the function ϕ satisfies

$$\int_0^1 \left[\frac{\phi(t)}{t}\right]^2 dt < \infty$$

and conditions (A)–(E) hold. Then the sequence of processes Z_n converge weakly under P_{θ_0} on the space C_ϕ to the process Z given by

$$Z(\theta, t) = \exp\left\{\theta'\mathbf{I}(\theta_0)^{1/2}\mathbf{W}_p(t) - \frac{t}{2}\theta'\mathbf{I}(\theta_0)\theta\right\}, \tag{3.12.14}$$

where $\mathbf{W}_p(\cdot)$ is the Brownian motion of p dimensions on $[0,1]$ with independent components.

Let us now suppose that there is a change ("rupture") in the model at an instant k so that $X_1, \ldots, X_k$ are i.i.d. P_{θ_1}, and $X_{k+1}, \ldots, X_n$ are i.i.d. P_{θ_2} with $\theta_1 \neq \theta_2$. Suppose that k, θ_1, and θ_2 are unknown and the problem is the estimation of the three parameters k, θ_1, and θ_2. In order to study the asymptotic behavior, we assume that the (instant of rupture) change point varies with n so that

(3.12.15) (i) $k = k_n \to \infty$ and $n - k_n \to \infty$.

Furthermore, assume that θ_1 and θ_2 depend on n so that

(ii) $\lim_{n\to\infty} \theta_1^{(n)} = \lim_{n\to\infty} \theta_2^{(n)} = \theta_0$ and

(iii) there exists $0 < \alpha < 1$ such that, for $\mathbf{d}^{(n)} = \theta_2^{(n)} - \theta_1^{(n)}$,

$$\left\{\frac{k_n(n-k_n)}{n}\right\}^\alpha \mathbf{d}^{(n)\prime}\mathbf{I}(\theta_0)\mathbf{d}^{(n)} \to \infty \text{ as } n \to \infty.$$

Theorem 3.12.2. Suppose $X_1, \ldots, X_{k_n}$ are i.i.d. with probability measure $P_{\theta_1^{(n)}}$ and $X_{k_n+1}, \ldots, X_n$ are i.i.d. with probability measure $P_{\theta_2^{(n)}}$. Let $\hat{\theta}_{1n}, \hat{\theta}_{2n}$, and $\hat{k}_n$ denote the maximum likelihood estimators of θ_1, θ_2, and k, respectively. Suppose conditions (A)–(E) and (3.12.1) hold. Then $\hat{\theta}_{1n}, \hat{\theta}_{2n}$, and $\hat{k}_n$ are asymptotically independent and

(3.12.16) (i) $k_n^{1/2}(\hat{\theta}_{1n} - \theta_1^{(n)}) \xrightarrow{\mathscr{L}} \mathbf{N}(\mathbf{0}, \mathbf{I}(\theta_0)^{-1})$,

(ii) $(n-k_n)^{1/2}(\hat{\theta}_{2n} - \theta_2^{(n)}) \xrightarrow{\mathscr{L}} \mathbf{N}(\mathbf{0}, \mathbf{I}(\theta_0)^{-1})$,

(iii) $\psi(n)(\hat{k}_n - k_n) \xrightarrow{\mathscr{L}} \xi_1$,

where

$$\psi(n) = \frac{k_n}{n}\left(1 - \frac{k_n}{n}\right)\mathbf{d}^{(n)\prime}\mathbf{I}(\theta_0)\mathbf{d}^{(n)} \tag{3.12.17}$$

and ξ_1 is a random variable with density

$$\tfrac{3}{2}e^{|s|}\int_{(3/\sqrt{2})|s|} e^{-u^2/2}\,du - \frac{1}{2}\int_{(1/2)\sqrt{|s|}} (2\pi)^{-1/2}e^{-u^2/2}\,du. \tag{3.12.18}$$

We shall now show that Theorem 3.12.2 is a consequence of Theorem 3.12.1 under the model

$$H_1(\theta_1^{(n)}, k_n, \theta_2^{(n)}):\quad X_1,\ldots,X_{k_n}$$

are i.i.d. $P_{\theta_1^{(n)}}$ and $X_{k_n+1},\ldots,X_n$ are i.i.d. $P_{\theta_2^{(n)}}$ with an appropriate scale change in time. Change

$$\theta_1 \rightsquigarrow \theta_1^{(n)} + \frac{\theta_1^*}{\sqrt{k_n}},\qquad \theta_2 \rightsquigarrow \theta_2^{(n)} + \frac{\theta_2^*}{\sqrt{k_n}},\qquad k \rightsquigarrow k_n + \frac{\tau}{\psi_n} = k(\tau) \tag{3.12.19}$$

in Theorem 3.12.1. The likelihood ratio process $Y_n(\theta_1^*, \tau, \theta_2^*)$ to be studied is then defined for θ_1^* and θ_2^* and determined by linearization between the points τ corresponding to $k(\tau) = 1, 2, \ldots, n$,

$$Y_n(\theta_1^*, \tau, \theta_2^*) = \frac{dP_{H_1(\theta_1^{(n)}+\theta_1^*/\sqrt{k_n},\,k(\tau),\,\theta_2^{(n)}+\theta_2^*/\sqrt{k_n})}}{dP_{H_1(\theta_1^{(n)},k_n,\theta_2^{(n)})}}. \tag{3.12.20}$$

Let $C_0(R^{2k+1})$ be the space of real-valued continuous functions on R^{2k+1} that tend to zero at infinity. Let C_0 be equipped with the supremum norm. We first prove the following theorem concerning the weak convergence of the process Y_n.

Theorem 3.12.3. The sequence of processes Y_n converge weakly in $C_0(R^{2k+1})$ under $H_1(\theta_1^{(n)}, k_n, \theta_2^{(n)})$ to the process Y defined by

$$\log Y(\theta_1^*, \tau, \theta_2^*) = \theta_1^{*\prime}\zeta_1 + W(\tau) + \theta_2^{*\prime}\zeta_2 - \tfrac{1}{2}\theta_1^{*\prime}\mathbf{I}(\theta_0)\theta_1^* - \frac{|\tau|}{2} - \tfrac{1}{2}\theta_2^{*\prime}\mathbf{I}(\theta_0)\theta_2^*, \tag{3.12.21}$$

where ζ_1 and ζ_2 are i.i.d. $\mathbf{N}_p(0, \mathbf{I}(\theta_0))$ and independent of $W(\cdot)$, which is Brownian motion on R tied down at 0.

Remark. Theorem 3.12.2 is a consequence of Theorem 3.12.3 from the standard theory of weak convergence (Billingsley, 1968) and from a result in

Shepp (1979) on the joint density of the maximum and its location for a Wiener process with drift.

Proof of Theorem 3.12.3. From the symmetry in the problem, we can suppose that $\tau \leqslant 0$ [and hence $k(\tau) \leqslant k_n$]. For $\tau \leqslant 0$, the term $Y_n(\theta_1^*, \tau, \theta_2^*)$ can be written as the product of the two terms

$$Z_{k_n}\left(\theta_1^*, \frac{k(\tau)}{k_n}\right)\tilde{Z}_{n-k_n}(\theta_2^*, 1) \tag{3.12.22}$$

and

$$\bar{Y}_{k_n}(\theta_2^*, \tau) \equiv \frac{Z_{k_n}(\theta_2^*\sqrt{k_n/(n-k_n)} + \sqrt{k_n}\mathbf{d}^{(n)}, 1)}{Z_{k_n}(\theta_2^*\sqrt{k_n/(n-k_n)} + \sqrt{k_n}\mathbf{d}^{(n)}, k(\tau)/k_n)}, \tag{3.12.23}$$

where the sequence Z_k is constructed from an independent sample of size k_n under the law $P_{\theta_1^{(n)}}$ and $\tilde{Z}_{n-k_n}$ is constructed from an independent sample of size $n-k_n$ under the law $P_{\theta_2^{(n)}}$. The main effect of this partition is to separate the parameters (θ_1^*, θ_2^*) to one part and τ to the other part. The following lemma describes the asymptotic behavior of $\bar{Y}_n$.

Lemma 3.12.4. (i) For every compact $K \subset R^k$ and for every $\tau_1 < 0$ and $\varepsilon > 0$, there exists $n_0 \geqslant 1$ such that, for every $n \geqslant n_0$,

$$P_{\theta_1^{(n)}}\left\{\sup_{\theta \in K}\sup_{\tau < \tau_1}|\log \bar{Y}_n(\theta, \tau) - \log \bar{Y}_n(0, \tau)| \geqslant \varepsilon\right\} \leqslant \varepsilon; \tag{3.12.24}$$

(ii) for every $\tau_1 < 0$ and $\varepsilon > 0$, there exist $A > 1$ and integer $n_0 \geqslant 1$ such that, for every $n \geqslant n_0$,

$$P_{\theta_1^{(n)}}\left\{\sup_{\tau < \tau_1}\sup_{\|\theta\| \leqslant 2\sqrt{n-k_n}\|\mathbf{d}^{(n)}\|}\bar{Y}_n(\theta, \tau) > A\right\} \leqslant \varepsilon; \tag{3.12.25}$$

(iii) the sequence $\bar{Y}_n(0, \tau)$ converges weakly, under $\mathscr{P}_{\theta_1}(n)$ on $C(R^-)$ (equipped with the uniform convergence on compact sets), to the process

$$W(\tau) - \frac{|\tau|}{2}; \tag{3.12.26}$$

(iv) for every $\varepsilon > 0$ and $N > 0$, there exists $\tau_1 < 0$ such that

$$P_{\theta_1^{(n)}}\left\{\sup_{\tau \leqslant \tau_1}\sup_{\|\theta\| \leqslant 2\sqrt{n-k_n}\|\mathbf{d}^{(n)}\|}\bar{Y}_n(\theta, \tau) > \varepsilon\right\} \leqslant \varepsilon, \tag{3.12.27}$$

(3.12.28) $$P_{\theta_1^{(n)}}\left\{\sup_{k(\tau)\leqslant(k_n/2)}\ \sup_{\|\theta\|\leqslant 2\sqrt{n-k_n}\,\|\mathbf{d}^{(n)}\|} \bar{Y}_n(\theta,\tau) > k_n^{-N}\right\} \leqslant \varepsilon,$$

and

(v) for every sequence a_n satisfying $\lim_{n\to\infty}(\log k_n)a_n^{-1} < \infty$,

(3.12.29) $$P_{\theta_1^{(n)}}\left\{\sup_{\tau\leqslant 0}\ \sup_{\theta\in R^k} \log \bar{Y}_n(\theta,\tau) > a_n\right\} \leqslant \varepsilon.$$

For proof of this lemma, see Deshayes and Picard (1984).

Remarks. In order to obtain weak convergence under the topology of uniform convergence on compact sets, it is sufficient to combine (i) and (iii) of Lemma 3.12.4. The following lemma allows us to replace $Z_{k_n}(\theta_1^*, k(\tau)/n)$ by $Z_{k_n}(\theta_1^*, 1)$ in studying the asymptotic behavior of the log-likelihood ratio process, and the lemma is a consequence of Theorem 3.12.1.

Lemma 3.12.5. For every compact set $K \subset R^k$ and for every $\tau_1 > 0$ and $\varepsilon > 0$, there exists n_0 such that for $n > n_0$,

(3.12.30) $$P\left\{\sup_{k(\tau_1)\leqslant j\leqslant k_n}\ \sup_{\theta\in K}\left| Z_{k_n}\left(\theta, \frac{j}{k_n}\right) - Z_{k_n}(\theta, 1)\right| > \varepsilon\right\} \leqslant \varepsilon.$$

Lemma 3.12.6 (Deshayes and Picard, 1984). For every $A > 2$ and $N > 0$, there exist $B > 0$, $\gamma > 0$, and an integer $n(N)$ beyond which

$$P\left\{\sup_{\|\theta\|>A} Z_n(\theta, 1) \geqslant \varepsilon\right\} \leqslant B\varepsilon^{-\gamma}A^{-N}.$$

In order to obtain the weak convergence of Y_n in $C_0(R^{2k+1})$, we separate the parameters (θ_1^*, θ_2^*) and τ and prove the following lemma.

Lemma 3.12.7. (i) For every $\varepsilon > 0$, there exists $\tau_1 < 0$ such that

(3.12.31) $$P\left\{\sup_{\tau\leqslant\tau_1}\ \sup_{(\theta_1^*,\theta_2^*)\in R^{2k}} Y_n(\theta_1^*, \tau, \theta_2^*) > \varepsilon\right\} \leqslant \varepsilon,$$

and

(ii) for every $\varepsilon > 0$, there exists $M > 0$ such that

(3.12.32) $$P\left\{\sup_{\tau_1\leqslant\tau\leqslant 0}\ \sup_{|(\theta_1^*,\theta_2^*)|\geqslant M} Y_n(\theta_1^*, \tau, \theta_2^*) > \varepsilon\right\} \leqslant \varepsilon.$$

Proof. In order to prove inequality (3.12.31), we divide the domain $[\tau \leqslant \tau_1] \times [(\theta_1^*, \theta_2^*) \in R^{2k}]$ into three sets:

(3.12.33)

$$\text{(i)} \quad D_1 = \left\{ (\theta_1^*, \theta_2^*, \tau) : \| \theta_2^* \| < 2\sqrt{n - k_n} \, \| \mathbf{d}^{(n)} \|, \quad \frac{k_n}{2} \leqslant k(\tau) \leqslant k(\tau_1) \right\}$$

$$\text{(ii)} \quad D_2 = \left\{ (\theta_1^*, \theta_2^*, \tau) : \| \theta_2^* \| < 2\sqrt{n - k_n} \, \| \mathbf{d}^{(n)} \|, \quad 1 \leqslant k(\tau) \leqslant \frac{k_n}{2} \right\}$$

$$\text{(iii)} \quad D_3 = \left\{ (\theta_1^*, \theta_2^*, \tau) : \| \theta_2^* \| > 2\sqrt{n - k_n} \, \| \mathbf{d}^{(n)} \|, \quad 1 \leqslant k(\tau) \leqslant k(\tau_1) \right\}.$$

Note that on set D_1, the processes $Z_{k_n}(\theta_1^*, k(\tau)/k_n)$ and $\tilde{Z}_{n-k_n}(\theta_2^*, 1)$ are bounded in probability by Theorem 3.12.1. The asymptotic behavior of $\bar{Y}_n(\theta_2^*, \tau)$ is governed by the inequality (3.12.24). On D_2, the process $\tilde{Z}_{n-k_n}(\theta_2^*, 1)$ is bounded in probability. Theorem 3.12.1 implies that, if $\{a_n\}$ is a sequence such that $\lim(\log k_n) a_n^{-1} < \infty$, then the process

$$\log Z_{k_n}\left(\theta_1^*, \frac{k(\tau)}{k_n} \right)$$

is smaller than a_n on the set D_2 with probability larger than $1 - \varepsilon$, the remaining then reduces to $\bar{Y}_n(\theta_2^*, \tau)$ and one can use inequality (3.12.25). On the set D_3, Theorem 3.12.1 and inequality (3.12.27) give the bound in probability for the product

$$Z_{k_n}\left(\theta_1^*, \frac{k(\tau)}{k_n} \right) \bar{Y}_n(\theta_2^*, \tau). \tag{3.12.34}$$

The remaining part reduces to $\tilde{Z}_{n-k_n}(\theta_2^*, 1)$, and one can use Lemma 3.12.6. This prove (3.12.31).

In order to show inequality (3.12.32), divide the domain into two sets:

(3.12.35)

$$B_1 = \left\{ (\theta_1^*, \theta_2^*, \tau) : \tau_1 \leqslant \tau \leqslant 0, |(\theta_1^*, \theta_2^*)| > M, \left\| \left(\frac{k_n}{n - k_n} \right)^{1/2} \theta_2^* + k_n^{1/2} \mathbf{d}^{(n)} \right\| \right.$$
$$\left. \leqslant 2 k_n^{1/2} \| \mathbf{d}^{(n)} \| \right\},$$

and

$$B_2 = \left\{ (\theta_1^*, \theta_2^*, \tau) \colon \tau_1 \leqslant \tau \leqslant 0, \left\| \left(\frac{k_n}{n - k_n} \right)^{1/2} \theta_2^* + k_n^{1/2} \mathbf{d}^{(n)} \right\| > 2k_n^{1/2} \| \mathbf{d}^{(n)} \| \right\}. \tag{3.12.36}$$

On set B_1, the process $\bar{Y}_n(\theta_2^*, \tau)$ is bounded in probability by inequality (3.12.25) and the remaining term reduces to

$$Z_{k(n)}\left(\theta_1^*, \frac{k(\tau)}{k_n} \right) \quad \text{or} \quad \tilde{Z}_{n-k_n}(\theta_2^*, 1) \tag{3.12.37}$$

according as $\| \theta_1^* \|$ or $\| \theta_2^* \|$ is larger than M, and one can use Theorem 3.12.1. On B_2, the process $\bar{Y}_n(\theta_2^*, \tau)$ is controlled by inequality (3.12.27), the process $Z_{k_n}(\theta_1, k(\tau)/k_n)$ is bounded in probability (by Theorem 3.12.1), and the balance reduces to $\tilde{Z}_{n-k_n}(\theta_2^*, 1)$ and we can use Lemma 3.12.6. This completes the proof of Lemma 3.12.7 and hence of Theorem 3.12.3 since convergence of finite-dimensional distribution is easy to check.

Remark. Theorem 3.12.3 can also be used for obtaining the asymptotic properties of Bayes estimators of parameters θ_1, θ_2, and k as in the classical problem [cf. Basawa and Prakasa Rao (1980b) and Kutoyants (1984)].

Suppose the prior information on the parameters θ_1, k, θ_2 is given by a prior density $q(\theta_1, k/n, \theta_2)$ on $R^p \times [0, 1] \times R^p$, which is continuous and strictly positive with limit point $(\theta_0, t_0, \theta_0)$ for the sequence $(\theta_1^{(n)}, k_n/n, \theta_2^{(n)})$. The Bayes estimators for each of the parameters under $H_1(\theta_1^{(n)}, k_n, \theta_2^{(n)})$ can be obtained with the help of the likelihood process Y_n. Suppose the loss function is quadratic. Then the Bayes estimators are given by

$$\text{(i)} \quad \tilde{\theta}_1 = A^{-1} \int \theta_1 Q(\theta_1, t, \theta_2; \mathbf{X})\, d\theta_1\, dt\, d\theta_2, \tag{3.12.38}$$

$$\text{(ii)} \quad \tilde{k} = nA^{-1} \int tQ(\theta_1, t, \theta_2; \mathbf{X})\, d\theta_1\, dt\, d\theta_2,$$

$$\text{(iii)} \quad \tilde{\theta}_2 = A^{-1} \int \theta_2 Q(\theta_1, t, \theta_2; \mathbf{X})\, d\theta_1\, dt\, d\theta_2,$$

where

$$A = \int Q(\theta_1, t, \theta_2; \mathbf{X})\, d\theta_1\, dt\, d\theta_2 \tag{3.12.39}$$

and $Q(\theta_1, t, \theta_2; \mathbf{X})$ denotes the joint density

$$q(\theta_1, t, \theta_2) \prod_{i=1}^{[nt]} g(X_i, \theta_1) \prod_{i=[nt]+1}^{n} g(X_i, \theta_2). \tag{3.12.40}$$

The parametrization (3.12.19) allows us to write the Bayes estimators as

$$\theta_1^{(n)} + \frac{\tilde{\theta}_1^*}{k_n^{1/2}}, \qquad k_n + \frac{\tilde{\tau}^*}{\psi_n^{1/2}}, \qquad \theta_2^{(n)} + \frac{\tilde{\theta}_2^*}{(n-k_n)^{1/2}}, \tag{3.12.41}$$

where

(3.12.42)

$$\tilde{\theta}_1^* = \frac{\int \theta_1^* q\left(\theta_1^{(n)} + \frac{\theta_1^*}{k_n^{1/2}}, \frac{k_n}{n} + \frac{\tau}{n\psi_n}, \theta_2^{(n)} + \frac{\theta_2^*}{(n-k_n)^{1/2}}\right) Y_n(\theta_1^*, \tau, \theta_2^*)\, d\theta_1^*\, d\tau\, d\theta_2^*}{\int q\left(\theta_1^{(n)} + \frac{\theta_1^*}{k_n^{1/2}}, \frac{k_n}{n} + \frac{\tau}{n\psi_n}, \theta_2^{(n)} + \frac{\theta_2^*}{(n-k_n)^{1/2}}\right) Y_n(\theta_1^*, \tau, \theta_2^*)\, d\theta_1^*\, d\tau\, d\theta_2^*}$$

and $\tilde{\theta}_2^*$ and $\tilde{\tau}^*$ are defined similarly.

It is clear now that Theorem 3.12.1 can be used to derive the limit laws for the Bayes estimator.

Theorem 3.12.8. Suppose the conditions of Theorem 3.11.2 hold. Under the sequence of hypothesis $H_1(\theta_1^{(n)}, k_n, \theta_2^{(n)})$, the Bayesian estimators are asymptotically independent and

(3.12.43) (i) $k_n^{1/2}(\tilde{\theta}_1 - \theta_1^{(n)}) \xrightarrow{\mathscr{L}} \mathbf{N}_p(\mathbf{0}, \mathbf{I}(\theta_0)^{-1})$,

(ii) $(n-k_n)^{1/2}(\tilde{\theta}_2 - \theta_2^{(n)}) \xrightarrow{\mathscr{L}} \mathbf{N}_p(\mathbf{0}, \mathbf{I}(\theta_0)^{-1})$

and

(3.12.44) (iii) $\psi(n)(\tilde{k} - k_n) \xrightarrow{\mathscr{L}} \xi_2$,

where ξ_2 is the random variable

$$\frac{\int \tau e^{W(\tau) - |\tau|/2}\, d\tau}{\int e^{W(\tau) - |\tau|'2}\, d\tau}. \tag{3.12.45}$$

It is known (Ibragimov and Hasminskii, 1981) that

$$E(\xi_2^2) = 19.5 \quad \text{and} \quad E\xi_1^2 = 26, \tag{3.12.46}$$

where ξ_1 is defined in Theorem 3.12.2. Hence the Bayes estimator for the change point has strictly smaller asymptotic variance than the maximum likelihood estimator.

Remarks. Deshayes and Picard (1984) studied the likelihood ratio tests for the existence of a change point using Theorem 3.12.1. We do not discuss the details here.

REFERENCES

Aitchinson, J. and Silvey, S. D. (1958). Maximum likelihood estimation of parameters subject to restraints. *Ann. Math. Statist.* **29**, 813–828.

Anderson, T. W. (1955). The integral of a symmetric unimodal function. *Proc. Amer. Math. Soc.*, **6**, 170–176.

Basawa, I. V. and Prakasa Rao, B. L. S. (1980a). Asymptotic inference for stochastic processes. *Stoch. Proc. and Their. Appl.* **10**, 221–254.

Basawa, I. V. and Prakasa Rao, B. L. S. (1980b). *Statistical Inference for Stochastic Processes*, Academic Press, London.

Basawa, I. V, and Scott, D. J. (1983). *Asymptotic Optimal Inference for Non-ergodic Models.* Lecture Notes in Statistics, No. 17, Springer-Verlag, New York.

Bickel, P. and Yahav, J. (1969). Some contributions to the asymptotic theory of Bayes solutions. *Z. Wahr. verw. Geb.* **11**, 257–276.

Billingsley, P. (1961). *Statistical Inference for Markov Processes*, University of Chicago Press, Chicago.

Billingsley, P. (1968). *Convergence of Probability Measures*, Wiley, New York.

Borwanker, J. D., Kallianpur, G., Prakasa Rao, B. L. S. (1971). The Bernstein–von Mises theorem for Markov processes. *Ann. Math. Statist.* **43**, 1241–1253.

Chernoff, H. and Rubin, H. (1956). The estimation of the location of a discontinuity of a density. *Proc. Third. Berkeley Symp. Math. Statist.* **1**, 19–37.

Cramer, H. (1946). *Mathematical Methods of Statistics*, Princeton University Press, Princeton, New Jersey.

Deshayes, J. and Picard, D. (1983). Ruptures de modeles enstatistiques, theses d'Etat. Université Paris-Sud.

Deshayes, J. and Picard, D. (1984a). Principle d'invariance sur le processes de vraisemblance. *Ann. Inst. Henri Poincaré*, **20**, 1–20.

Deshayes, J. and Picard, D. (1984b). Lois asymptotiques des tests et estimateurs de rupture dans un modele statistique classique. *Ann. Inst. Henri Poincaré*, **20**, 309–327.

Dharmadhikari, S. W. and Jogdev, K. (1969). Bounds on moments of certain random variables. *Ann. Math. Statist.* **40**, 1506–1509.

Grenander, U. (1981). *Abstract Inference*, Wiley, New York.

Hajek, J. (1970). A characterization of limiting distributions of regular estimates. *Z. Wahr. verw. Geb.* **14**, 323–330.

Hajek, J. (1972). Local asymptotic minimax and admissibility in estimation. *Proc. 6th Berkeley Symp. Math. Statist. Prob.* **1**, 175–194.

Hall, P. and Heyde, C. (1980). *Martingale Limit Theory and Its Applications*, Academic Press, New York.

Hipp, C. and Michel, R. (1976). On the Bernstein–von Mises approximation of posterior distributions. *Ann. Statistics* **4**, 972–980.

Huber, P. J. (1967). The behavior of maximum likelihood estimators under non-standard conditions. *Proc. 5th Berkeley Symp. Math. Statist. Prob.* **1**, 221–233.

Ibragimov, I. A. and Hasminskii, R. Z. (1972). Asymptotic behaviour of statistical estimators in the smooth case I. Study of the likelihood ratio. *Theor. Prob. and Appl.* **17**, 445–462.

Ibragimov, I. A. and Hasminskii, R. Z. (1981) *Statistical Estimation-Asymptotic Theory*, Springer-Verlag, New York.

Inagaki, N. and Ogata, Y. (1975). The weak convergence of the likelihood ratio random fields and its applications. *Ann. Inst. Statist. Math.* **27**, 391–419.

Jeganathan, P. (1981). On a decomposition of the limit distribution of a sequence of estimators. *Sankhya Ser. A.* **43**, 26–36.

Keiding, N. (1974). Estimation in the birth process. *Biometrika* **61**, 71–80.

Kutoyants, Yu. A. (1984). *Parameter Estimation for Stochastic Processes* (Translated from Russian and Edited by B. L. S. Prakasa Rao), Heldermann-Verlag, Berlin.

Landers, D. (1968). Existing Und Konsistenz von Maximum Likelihood Schatzern. Thesis, University of Cologne.

Le Cam, L. (1956). On the asymptotic theory of estimation and testing of hypotheses. *Proc. 3rd Berkeley Symp. Math. Statist. and Prob.* **1**, 129–156.

Le Cam, L. (1958). Les proprietes asymptotiques des solutions de Bayes. *Publ. Inst. Statist. Univ. Paris* **7**, 18–35.

Le Cam, L. (1970). On the assumptions used to prove asymptotic normality of maximum likelihood estimators. *Ann. Math. Statist.* **41**, 802–828.

Le Cam, L. (1973a). Asymptotic normality of experiments defined by independent and identically distributed variables. Tech. Report CRM-313, Centre de Récherches Mathématiques. Université de Montréal.

Le Cam, L. (1973b). Convergence of estimates under dimensionality restrictions *Ann. Statistics* **1**, 38–53.

Le Cam. L. (1973c). Sur les constraints imposees par les passages a la limite usuels in statistique. Inter. Stat. Inst. Congress, Vienna.

Le Cam, L. (1974). *Notes on Asymptotic Methods in Statistical Decision Theory.* CRM-245, Centre de Récherches Mathématiques, Université de Montréal.

Loeve, M. (1963). *Probability Theory*, Von Nostrand, Princeton.

Michel, R. and Pfanzagl, J. (1971). The accuracy of the normal approximation for minimum contrast estimate. *Z. Wahr. verw. Geb.* **18**, 73–84.

Mishra, M. N. and Prakasa Rao, B. L. S. (1985a). Asymptotic study of the maximum likelihood estimation for nonhomogeneous diffusion processes. *Statistics & Decisions* **3**, 193–203.

Mishra M. N. and Prakasa Rao, B. L. S. (1985b). On the Berry–Esseen bound for maximum likelihood estimator for non-homogeneous diffusion processes. *Sankhya Ser. A.* **47**, 392–398.

Pfaff, T. (1982). Quick consistency of quasi maximum likelihood estimators. *Ann. Statistics* **10**, 990–1005.

Pfanzagl, J. (1969a). On the measurability and consistency of minimum contrast estimates. *Metrika* **14**, 249–272.

Pfanzagl, J. (1969b). On the existence of product measurable densities. *Sankhya* **31**, 13–18.

Pfanzagl, J. (1971). The Berry–Esseen bound for minimum contrast estimates. *Metrika* **17**, 82–91.

Pflug, Georg Ch. (1982). A statistically important Gaussian process. *Stoch. Proc. and Their Appl.* **13**, 45–57.

Pflug, Georg Ch. (1983). The limiting log-likelihood process for discontinuous density families. *Z. Wahr. verw. Geb.* **64**, 15–35.

Polfeldt, T. (1970). Asymptotic results in non-regular estimation. *Skand. Aktuar. Suppl.* **1–2**, 1–78.

Prakasa Rao, B. L. S. (1966). Asymptotic distributions in some nonregular statistical problems. Ph.D. Thesis, Michigan State University.

Prakasa Rao, B. L. S. (1968). Estimation of the location of the cusp of a continuous density. *Ann. Math. Statist.* **39**, 76–87.

Prakasa Rao, B. L. S. (1969). Estimation of a unimodal density. *Sankhya Ser. A.* **31**, 23–36.

Prakasa Rao, B. L. S. (1970). Estimation for distributions with monotone failure rate. *Ann. Math. Statist.* **41**, 507–519.

Prakasa Rao, B. L. S. (1972). Maximum likelihood estimation for Markov processes. *Ann. Inst. Statist. Math.* **24**, 333–345.

Prakasa Rao, B. L. S. (1973). On the rate of convergence of estimators for Markov processes. *Z. Wahr. verw. Geb.* **26**, 141–152.

Prakasa Rao, B. L. S. (1974). Statistical Inference for stochastic processes. Tech. Repo. CRM-465, Centre de Récherches Mathématiques, Université de Montréal.

Prakasa Rao, B. L. S. (1975a). Tightness of probability measures generated by stochastic processes on metric spaces. *Bull. Inst. Math. Acad. Sinica.* **3**, 353–367.

Prakasa Rao, B. L. S. (1975b). On the Berry–Esseen bound for minimum contrast estimators in the independent not identically distributed case. *Metrika* **22**, 225–239.

Prakasa Rao, B. L. S. (1978a). Rate of convergence of Bernstein–von Mises approximation for Markov processes. *Serdica* **4**, 36–42.

Prakasa Rao, B. L. S. (1978b). On some problems of inference for Markov processes. *Proc. Gen. Math. Seminar. Univ. of Patras* **4**, 35–49.

Prakasa Rao, B. L. S. (1979). The equivalence between (modified) Bayes estimator and maximum likelihood estimator for Markov processes. *Ann. Inst. Statist. Math.* **31**, 499–513.

Prakasa Rao, B. L. S. (1981). The Bernstein–von Mises theorem for a class of diffusion processes. *Teor. Sluch. Proc.* **9**, 95–104. (In Russian.)

Prakasa Rao, B. L. S. (1982a). Maximum probability estimation for diffusion processes. *Statistics and Probability: Essays in Honour of C. R. Rao.* (Ed. G. Kallianpur et al.), North-Holland, Amsterdam.

Prakasa Rao, B. L. S. (1982b). Remarks on linear parametric inference for stochastic processes. *Statistics and Decisions* **1**, 39–55.

Prakasa Rao, B. L. S. (1983a). Asymptotic theory for non-linear least squares estimators for diffusion processes. *Math. Operations–Forch Statist. Ser. Statistics* **14**, 195–209.

Prakasa Rao B. L. S. (1983b). *Nonparametric Functional Estimation*, Academic Press, New York.

Prakasa Rao, B. L. S. (1983d). Maximum likelihood estimator for Markov processes with nuisance parameters (preprint).

Prakasa Rao, B. L. S. (1984a). On Bayes estimation for diffusion fields. In *Statistics: Applications and New Directions* (Ed. J. K. Ghosh et al.), Statistical Publishing Society, Calcutta.

Prakasa Rao, B. L. S. (1984b). Law of iterated logarithm for fluctuation of posterior distributions for a class of diffusion processes and a sequential test of power one (preprint).

Prakasa Rao, B. L. S. (1985). Estimation of the drift for diffusion processes. *Statistics* **16**, 263–275.

Prakasa Rao, B. L. S. and Rubin, H. (1968). A property of log-likelihood ratio process for Gaussian processes. *Ann. Inst. Statist. Math.* 311–314.

Prakasa Rao, B. L. S. and Rubin, H. (1981). Asymptotic theory of estimator in non-linear stochastic differential equations. *Sankhya Ser. A.* **43**, 170–189.

Prasad, M. S. and Prakasa Rao B. L. S. (1976). Maximum likelihood estimation for dependent random variables. *J. Indian Stat. Assoc.* **14**, 75–97.

Rao, C. R. (1974). *Linear Statistical Inference and its Applications*, Wiley, New York.

Schwarz, L. (1965). On Bayes procedures. *Z. Wahr. verw. Geb.* **4**, 10–26.

Shepp, L. A. (1979). The joint density of the maximum and its location for a Wiener process with drift. *J. Appl. Probab.* **16**, 423–427.

Straf, M. L. (1970). Weak convergence of random processes with several parameters. *Proc. 6th Berkeley Symp. Math. Statist. Prob.* **2**, 187–221.

Strasser, H. (1973). On Bayes estimates. *J. Multivariate Anal.* **3**, 293–310.

Strasser, H. (1975a). Asymptotic properties of posterior distributions. *Z. Wahr. verw. Geb.* **35**, 269–282.

Strasser, H. (1975b). The asymptotic equivalence of Bayes and maximum likelihood estimation. *J. Multivariate Anal.* **5**, 206–226.

Strasser, H. (1977). Improved bounds for the equivalence of Bayes and maximum likelihood estimation. *Theory of Prob. and Its Appl.* **22**, 349–361.

Strasser, H. (1978). Global asymptotic properties of risk functions in estimation. *Z. Wahr. verw. Geb.* **45**, 35–48.

Strasser, H. (1981a). Consistency of maximum likelihood and Bayes estimates. *Ann. Statistics* **9**, 1107–1113.

Strasser, H. (1981b). Convergence of estimates. Part I. *J. Multivariate Anal.* **11**, 127–151.

Strasser, H. (1981c). Convergence of estimates. Part II. *J. Multivariate Anal.* **11**, 152–172.

Strasser, H. (1982). Local asymptotic minimax properties of Pitman estimates. *Z. Wahr. verw. Geb.* **60**, 223–247.

Sweeting, T. J. (1980). Uniform asymptotic normality of the maximum likelihood estimator. *Ann. Statistics* **8**, 1375–1381.

Sweeting, T. J. (1983). On estimator efficiency in stochastic processes. *Stoch. Proc. and Their Appl.* **15**, 93–98.

Wald, A. (1949). Note on the consistency of the maximum likelihood estimate. *Ann. Math. Statist.* **20**, 595–601.

Weiss, L. (1973). Asymptotic properties of maximum likelihood estimators in some nonstandard cases II. *J. Amer. Statist. Assoc.* **68**, 428–430.

CHAPTER 4

Linear Parametric Inference

4.1 LINEAR PARAMETRIC ESTIMATION

In Chapter 3, we studied asymptotic properties of Bayes and maximum likelihood estimators. It is well known that computation of these estimators is difficult in practice, and the alternative method used for estimation in most of the problems is the so-called method of moments. However, unlike the maximum likelihood theory, no optimality properties can be described for the estimators obtained by the method of moments. We now describe a more general method of estimation and study the asymptotic and optimality properties of estimators. The discussion here is based on the work of Klebanov and Melamed (1979) and Prakasa Rao (1982). Linear parametric inference for arbitrary stochastic processes is discussed in Prakasa Rao (1982).

Let $(\Omega, \mathscr{B})$ be a measurable space and $\{P_\theta, \theta \in \Theta\}$ be a family of probability measures defined on $(\Omega, \mathscr{B})$, where Θ is a closed interval in R. Let $X_i, i \geqslant 1$, be i.i.d. random variables distributed as P_θ.

(A_1) Let $\psi_i, 1 \leqslant i \leqslant k$, be real-valued functions defined on Ω such that the system $\{1, \psi_1, \ldots, \psi_k\}$ is linearly independent, and suppose there exists $0 < \delta < 1$ such that

(4.1.0) $$E_\theta |\psi_i(X)|^{2+\delta} < \infty, \qquad 1 \leqslant i \leqslant k, \qquad \theta \in \Theta.$$

Then

(4.1.1) $$\gamma_i(\theta) = E_\theta \psi_i(X), \qquad \gamma_{ij}(\theta) = E_\theta \psi_i(X)\psi_j(X), \quad 1 \leqslant i, j \leqslant k$$

are finite. Assume that

(A_2) $\gamma_i(\theta)$ is differentiable with respect to θ and

$$\inf_{\theta\in\Theta}|\gamma_i^{(1)}(\theta)|>0, \qquad 1\leqslant i\leqslant k, \tag{4.1.2}$$

where $\gamma_i^{(1)}(\theta)$ is the derivative of $\gamma_i(\theta)$.

Since $\gamma_i(\theta)$ is continuous in θ by (A_2) and Θ is an interval, it follows that $\gamma_i(\theta)$ is a connected subset of R. We assume hereafter that

(A_3) $\gamma_i(\Theta)$ is a neighborhood of $\gamma_i(\theta)$ for all $\theta\in\Theta^\circ$, the interior of Θ. Let $\sigma_{ij}(\theta)=\mathrm{cov}_\theta[\psi_i(X),\psi_j(X)]$ and $\sigma_i^2(\theta)\equiv\sigma_{ii}(\theta)$. Assume that

(A_4) (i) $\sigma_i^2(\theta)>0, 1\leqslant i\leqslant k$, and $\Delta_\theta=((\delta_{ij}(\theta)))$ is nonsingular, where

$$\delta_{ij}(\theta)=\frac{\sigma_{ij}(\theta)}{\gamma_i^{(1)}(\theta)\gamma_j^{(1)}(\theta)}, \tag{4.1.3}$$

(ii) $\delta_{ij}(\theta)$ are continuous in θ for $1\leqslant i,j\leqslant k$.

It is clear from the fact that X_i, $1\leqslant i\leqslant n$, are i.i.d., that

$$\left(\frac{1}{\sqrt{n}}\sum_{r=1}^{n}\psi_j(X_r)-E_\theta\psi_j(X_r), 1\leqslant j\leqslant k\right)\xrightarrow{\mathscr{L}}\mathbf{N}_k(\mathbf{0},\boldsymbol{\Sigma}_\theta), \tag{4.1.4}$$

where

$$\boldsymbol{\Sigma}_\theta=((\sigma_{ij}(\theta))). \tag{4.1.5}$$

Let

$$U_{jn}=\frac{1}{n}\sum_{r=1}^{n}\psi_j(X_r), \qquad 1\leqslant j\leqslant k, \tag{4.1.6}$$

and

$$\tilde{U}_{jn}=\begin{cases}U_{jn} & \text{if } U_{jn}\in\gamma_j(\Theta)\\ V_{jn} & \text{if } U_{jn}\notin\gamma_j(\Theta),\end{cases} \tag{4.1.7}$$

where $V_{jn}\in\gamma_j(\Theta)$ and

$$|V_{jn}-\gamma_j(\theta)|\leqslant|U_{jn}-\gamma_j(\theta)|, \qquad \theta\in\Theta, \quad 1\leqslant j\leqslant k. \tag{4.1.8}$$

Note that

$$0 \leqslant P_\theta(U_{jn} \neq \bar{U}_{jn}) \leqslant P_\theta(U_{jn} \notin \gamma_j(\Theta)). \tag{4.1.9}$$

But

$$U_{jn} \to \gamma_j(\theta) \quad \text{a.s. } [P_\theta] \text{ as } n \to \infty \tag{4.1.10}$$

by the SLLN. Hence

$$P_\theta(U_{jn} \neq \tilde{U}_{jn}) \to 0 \quad \text{as } n \to \infty \tag{4.1.11}$$

for $1 \leqslant j \leqslant k$ and $\theta \in \Theta^0$. As a consequence of (4.1.4) and (4.1.11), we obtain that

$$\tilde{U}_{jn} \to \gamma_j(\theta) \quad \text{a.s. } [P_\theta] \text{ as } n \to \infty \tag{4.1.12}$$

and

$$(\sqrt{n}(\tilde{U}_{jn} - \gamma_j(\theta)), 1 \leqslant j \leqslant k) \xrightarrow{\mathscr{L}} \mathbf{N}_k(\mathbf{0}, \boldsymbol{\Sigma}_\theta), \qquad \theta \in \Theta^0. \tag{4.1.13}$$

Furthermore,

$$\begin{aligned} E_\theta|\sqrt{n}(\tilde{U}_{jn} - \gamma_j(\theta))|^{2+\delta} &\leqslant E_\theta|\sqrt{n}(U_{jn} - \gamma_j(\theta))|^{2+\delta} \\ &\leqslant C_\theta \operatorname{Var}_\theta[\sqrt{n}(U_{jn} - \gamma_j(\theta))]^{1+\delta/2} \end{aligned}$$

for some constant $C_\theta > 0$ [cf. Ibragimov and Linnik (1971, p. 340)]. But $\operatorname{Var}_\theta[U_{jn}] = (1/n)\sigma_j^2(\theta)(1 + o(1))$. Hence

$$E_\theta|\sqrt{n}(\tilde{U}_{jn} - \gamma_j(\theta))|^{2+\delta} = O(1). \tag{4.1.14}$$

Relations (4.1.13) and (4.1.14) prove that

$$\lim_{n\to\infty} E_\theta[\sqrt{n}(\tilde{U}_{in} - \gamma_i(\theta))\sqrt{n}(\tilde{U}_{jn} - \gamma_j(\theta))] = \sigma_{ij}(\theta), \qquad \theta \in \Theta^0, \tag{4.1.15}$$

by Proposition 1.2.11 under (A_1).

In view of condition (A_2), $\gamma_j(\cdot)$ possesses a well-defined inverse for $1 \leqslant j \leqslant k$. Let

$$\phi_{jn} = \gamma_j^{-1}(\tilde{U}_{jn}). \tag{4.1.16}$$

Note that

$$(4.1.17) \qquad |\phi_{jn}-\theta| = \left|\int_{\gamma_j(\theta)}^{\tilde{U}_{jn}} [\gamma_j^{-1}(u)]^{(1)}\,du\right|$$
$$\geqslant \sup_u |[\gamma_j^{-1}(u)]^{(1)}|\,|\tilde{U}_{jn}-\gamma_j(\theta)|$$
$$\leqslant C_\theta|\tilde{U}_{jn}-\gamma_j(\theta)|$$

for some $C_\theta > 0$ by (A_2). Therefore

$$(4.1.18) \qquad \phi_{jn}\to\theta \quad \text{a.s. } [P_\theta] \text{ as } n\to\infty$$

for $1\leqslant j\leqslant k$ and $\theta\in\Theta^0$. In fact

$$(4.1.19) \qquad E_\theta[\sqrt{n}|\phi_{jn}-\theta|^{2+\delta}] = O(1),$$

by relations (4.1.17) and (4.1.14). It can be seen now, from (4.1.13) (Rao, 1974, p. 387), that

$$(4.1.20) \qquad \{\sqrt{n}(\phi_{jn}-\theta), 1\leqslant j\leqslant k\} \xrightarrow{\mathscr{L}} \mathbf{N}_k(\mathbf{0}, \boldsymbol{\Delta}_\theta),$$

where $\boldsymbol{\Delta}_\theta$ is defined in (A_4). In fact,

$$(4.1.21) \qquad \lim_{n\to\infty} E_\theta[\sqrt{n}(\phi_{in}-\theta)\sqrt{n}(\phi_{jn}-\theta)] = \delta_{ij}(\theta).$$

Lemma 4.1.1. For any nonsingular matrix $\mathbf{B}$ of order $k\times k$,

$$(4.1.22) \qquad \inf_{\substack{\boldsymbol{\alpha}\in R^k \\ \boldsymbol{\alpha}'\mathbf{1}=1}} \boldsymbol{\alpha}'\mathbf{B}\boldsymbol{\alpha} = (\mathbf{1}'\mathbf{B}^{-1}\mathbf{1})^{-1},$$

and the minimum is attained at $\hat{\boldsymbol{\alpha}} = (\mathbf{1}'\mathbf{B}^{-1}\mathbf{1})^{-1}\mathbf{B}^{-1}\mathbf{1}$.

Remark. Since U_{jn} has expectation $\gamma_j(\theta)$, one can choose $\phi_{jn} = \gamma_j^{-1}(\tilde{U}_{jn})$ as an estimator of θ. The problem is to choose a "best" linear combination of ϕ_{jn}, $1\leqslant j\leqslant k$, as an estimator of θ. Heuristically speaking,

$$E_\theta\left[\sum_{j=1}^k \alpha_j\phi_{jn}\right] \simeq \left(\sum_{j=1}^k \alpha_j\right)\theta$$

and

$$\operatorname{Var}_\theta\left[\sum_{j=1}^k \alpha_j\phi_{jn}\right] \simeq \boldsymbol{\alpha}'\boldsymbol{\Delta}_\theta\boldsymbol{\alpha}.$$

Therefore the problem is to choose $\boldsymbol{\alpha} \in R^k$ so that $\boldsymbol{\alpha}'\Delta_\theta\boldsymbol{\alpha}$ is minimum subject to $\boldsymbol{\alpha}'\mathbf{1} = 1$.

The matrix Δ_θ is nonsingular by assumption (A_4). Define

$$C(\theta) = (C_1(\theta), \ldots, C_k(\theta)), \tag{4.1.23}$$
$$= (\mathbf{1}'\boldsymbol{\Delta}_\theta^{-1}\mathbf{1})^{-1}\boldsymbol{\Delta}_\theta^{-1}\mathbf{1}$$

and $\tilde{C}_j = C_j(\phi_{jn})$, $1 \leqslant j \leqslant k$, where ϕ_{jn} is as defined by (4.1.16). Let

$$\hat{\theta}_n = \sum_{j=1}^{n} \tilde{C}_j \phi_{jn}. \tag{4.1.24}$$

Note that $\hat{\theta}_n$ is a linear function of $\tilde{C}_j\phi_{jn}$, $1 \leqslant j \leqslant k$, but not of ϕ_{jn}, $1 \leqslant j \leqslant k$. Furthermore, it minimizes the asymptotic variance among all linear functions of ϕ_{jn} that are consistent for θ.

We now study the asymptotic and optimality properties of the estimator $\hat{\theta}_n$ motivated by the considerations discussed above. We first state a lemma.

Lemma 4.1.2. Under conditions (A_1)–(A_4), for any $\theta \in \Theta^0$,

$$\{\sqrt{n}(\tilde{C}_i\phi_{in} - C_i(\theta)\theta), 1 \leqslant i \leqslant k\} \to \mathbf{N}_k(\mathbf{0}, \mathbf{V}_\theta) \quad \text{as } n \to \infty, \tag{4.1.25}$$

where

(i) $\mathbf{V}_\theta = \mathbf{G}_\theta\boldsymbol{\Delta}_\theta\mathbf{G}_\theta'$,

(ii) $\mathbf{G}_\theta = \left(\left(\dfrac{\partial g_i}{\partial \phi_j}\bigg|_{\phi_j=\theta}\right)\right)$,

(iii) $g_i(\boldsymbol{\phi}) = C_i(\phi_i)\phi_i, \qquad 1 \leqslant i \leqslant k$,

(iv) $\sum_{i=1}^{k} C_i(\theta) = 1$.

This lemma is again a consequence of a general result on limit distributions for transformation of random variables [cf. Rao (1974, p. 387)] in view of (4.1.20). See Proposition 2.1.4.

The following theorem describes the asymptotic properties of the estimator $\hat{\theta}_n$.

Theorem 4.1.3. Suppose conditions (A_1)–(A_4) hold. Then

(4.1.26) (i) $\hat{\theta}_n \to \theta$ a.s. $[P_\theta]$ as $n \to \infty$ for all $\theta \in \Theta^0$,

(ii) $\sqrt{n}(\hat{\theta}_n - \theta) \xrightarrow{\mathscr{L}} \mathbf{N}(\mathbf{0}, \mathbf{1}' \mathbf{V}_\theta \mathbf{1})$, $\theta \in \Theta^0$,

and furthermore, if

(A_5) $C_i(\theta)$, $1 \leqslant i \leqslant k$, are differentiable in θ with bounded second derivatives where $\mathbf{C}(\theta) = (C_1(\theta), \ldots, C_k(\theta))$ is defined by (4.1.23), then

(4.1.27) (iii) $\lim_{n \to \infty} E_\theta[\sqrt{n}(\hat{\theta}_n - \theta)]^2 = \mathbf{1}'\mathbf{V}_\theta\mathbf{1}.$

Proof. Note that $\mathbf{C}(\theta)$ is continuous in Θ by (A_4). The result follows from the representation (4.1.24) of $\hat{\theta}_n$ and the fact that

$$\phi_{jn} \to \theta \quad \text{a.s. } [P_\theta] \quad \text{as } n \to \infty$$

as given by relation (4.1.18). This proves part (i). Observe that

$$\sqrt{n}(\hat{\theta}_n - \theta) = \sum_{j=1}^{k} \sqrt{n}(\bar{C}_j \phi_{jn} - C_j(\theta)\theta) \tag{4.1.28}$$

and hence

$$\sqrt{n}(\hat{\theta}_n - \theta) \xrightarrow{\mathscr{L}} N(0, \mathbf{1}'\mathbf{G}_\theta \mathbf{\Delta}_\theta \mathbf{G}'_\theta \mathbf{1}) \tag{4.1.29}$$

by Lemma 4.1.2. This proves part (ii). Note that

$$\begin{aligned} E_\theta|\sqrt{n}(\hat{\theta}_n - \theta)|^{2+\delta} &\leqslant K_\theta \sum_{i=1}^{k} E_\theta|\sqrt{n}(C_i(\phi_{in})\phi_{in} - C_i(\theta)\theta)|^{2+\delta} \\ &\leqslant K_\theta^* \sum_{i=1}^{k} E_\theta|\sqrt{n}(\phi_{in} - \theta)|^{2+\delta} \\ &= O(1). \end{aligned} \tag{4.1.30}$$

The first inequality is a consequence of the c_r inequality (Proposition 1.14.12). The second inequality follows from assumption (A_5) of the hypothesis by using the mean value theorem. The last equality follows from (4.1.19). Hence

$$\lim_{n \to \infty} E_\theta[\sqrt{n}(\hat{\theta}_n - \theta)]^2 = \mathbf{1}'\mathbf{V}_\theta\mathbf{1}. \tag{4.1.31}$$

Remark. Lemma 4.1.1 and the choice of $\hat{\theta}_n$ imply that

$$(\mathbf{1}'\boldsymbol{\Delta}_\theta\mathbf{1})^{-1} = \mathbf{1}'\mathbf{V}_\theta\mathbf{1}. \tag{4.1.32}$$

4.2 ASYMPTOTIC EFFICIENCY

We now study the optimality of the estimator $\hat{\theta}_n$ defined in (4.1.28) in a sense to be described.

Suppose the following additional regularity conditions hold:

(A_7) There exists a σ-finite measure μ dominating the family $\{P_\theta, \theta\in\Theta\}$ and $(dP_\theta/d\mu)(\omega) = f(\omega,\theta)$ is absolutely continuous in θ for ω a.e. $[\mu]$. Furthermore,

$$J = J(x,\theta) = \frac{1}{f(\omega,\theta)}\frac{df(\omega,\theta)}{d\theta} \in L^2(P_\theta), \qquad \theta\in\Theta.$$

(A_8) $$\frac{d}{d\theta}\int_\Omega f(\omega,\theta)\,d\mu(\omega) = \int_\Omega \frac{\partial f(\omega,\theta)}{\partial\theta}\,d\mu(\omega).$$

(A_9) $$\frac{d}{d\theta}\int_\Omega \psi_j(\omega) f(\omega,\theta)\,d\mu(\omega) = \int_\Omega \psi_j(\omega)\frac{\partial f(\omega,\theta)}{\partial\theta}\,d\mu(\omega).$$

Let $J_n(\mathbf{X},\theta) = \sum_{i=1}^n J(X_i,\theta)$ and χ be the linear space spanned by the set $\{1, C_1(\psi_{1n})\phi_{1n},\ldots, C_k(\phi_{kn})\phi_{kn}\}$. Define

$$J_n^*(\theta) = \hat{E}_\theta(J_n(\mathbf{X},\theta)\,|\,\chi), \tag{4.2.0}$$

where $\hat{E}_\theta(\cdot|\cdot)$ denotes the conditional expectation in the wide sense [cf. Doob (1953)]. $E_\theta(J_n^*(\theta)^2)$ is the information contained in the linear space χ.

Theorem 4.2.1 (Asymptotic Efficiency). Suppose conditions (A_1)–(A_9) hold. Then the estimator $\hat{\theta}_n$ is asymptotically efficient in the sense that its asymptotic variance is the inverse of the average limiting information contained in the linear space χ.

Proof. It can be checked that [cf. Prakasa Rao (1982)]

$$(n^{-1/2}J_n(\mathbf{X},\theta)), \sqrt{n}(\tilde{U}_{1n}-\gamma_1(\theta)),\ldots,\sqrt{n}(\tilde{U}_{kn}-\gamma_k(\theta)) \xrightarrow{\mathscr{L}} \mathbf{N}_{k+1}(\mathbf{0},\boldsymbol{\Sigma}_\theta^*), \tag{4.2.1}$$

where

$$\Sigma_\theta^* = \begin{bmatrix} I(\theta) & \gamma_1^{(1)}(\theta) & \cdots & \gamma_k^{(1)}(\theta) \\ \gamma_1^{(1)}(\theta) & & & \\ \vdots & & \Sigma_\theta & \\ \gamma_k^{(1)}(\theta) & & & \end{bmatrix}. \tag{4.2.2}$$

Here $I(\theta) = E_\theta[\partial \log f(X,\theta)/\partial\theta]^2$ and Σ_θ is as given by (4.1.5). In particular, it follows that

$$(n^{-1/2}J_n(\mathbf{X},\theta)), \sqrt{n}(C_1(\phi_{1n})\phi_{1n} - C_1(\theta)\theta), \ldots, \tag{4.2.3}$$
$$\sqrt{n}(C_k(\phi_{kn})\phi_{kn} - C_k(\theta)\theta) \xrightarrow{\mathscr{L}} \mathbf{N}_{k+1}(\boldsymbol{\theta}, \mathbf{V}_\theta^*),$$

where

$$\mathbf{V}_\theta^* = \begin{bmatrix} I(\theta) & C_1(\theta) + \theta C_1^{(1)}(\theta) & \cdots & C_k(\theta) + \theta C_k^{(1)}(\theta) \\ C_1(\theta) + \theta C^{(1)}(\theta) & & & \\ \vdots & & \mathbf{V}_\theta & \\ C_k(\theta) + \theta C_k^{(1)}(\theta) & & & \end{bmatrix}.$$

It follows again by assumptions (A_1)–(A_9) that

$$\lim_{n\to\infty} E_\theta[J_n(\tilde{C}_j\phi_{jn} - C_j(\theta)\theta)] = \theta C_j^{(1)}(\theta) + C_j(\theta), \qquad 1 \leqslant j \leqslant k, \tag{4.2.4}$$

for $\theta \in \Theta^0$. In particular, it follows that

$$\begin{aligned} \lim_{n\to\infty} \frac{1}{n} E_\theta(J_n - J_n^*)^2 &= \frac{\det \mathbf{V}_\theta^*}{\det \mathbf{V}_\theta} \\ &= I(\theta) - \mathbf{1}'\mathbf{G}_\theta \mathbf{V}_\theta^{-1}\mathbf{G}_\theta\mathbf{1} \\ &= I(\theta) - \mathbf{1}'\boldsymbol{\Delta}_\theta^{-1}\mathbf{1} \\ &= I(\theta) - (\mathbf{1}'\mathbf{V}_\theta\mathbf{1})^{-1}. \end{aligned} \tag{4.2.5}$$

The last equality follows from (4.1.32). Relation (4.2.5) follows from (4.2.3), assumption (A_5), and the formula for residual variance [cf. Cramer (1946, p. 305)] since $J_n^* = \hat{E}_\theta(J_n|\chi)$, where χ is spanned by $\{1, C_1(\phi_{1n})\phi_{1n}, \ldots, C_k(\phi_{kn})\phi_{kn}\}$.

Note that $J_n - J_n^*$ and J_n^* are uncorrelated. Hence

$$E_\theta(J_n^2) = E_\theta(J_n - J_n^*)^2 + E_\theta(J_n^*)^2,$$

which in turn proves that

$$\lim_{n\to\infty} \frac{1}{n} E_\theta(I_n^*)^2 = (\mathbf{1}'\mathbf{V}_\theta\mathbf{1})^{-1}$$

since $(1/n)E_\theta(J_n^2) = I(\theta)$ for all n. Relation (4.1.27) of Theorem 4.1.3(iii) proves that the asymptotic variance of $\hat{\theta}_n$ is the inverse of the average limiting information contained in the linear space χ.

Remarks. If $\psi_i(x) = x^i$, $1 \leqslant i \leqslant k$, then the method discussed here reduced to the method of moments. The result can be extended to multiparameter case [cf. Klebanov and Melamed (1979)].

REFERENCES

Cramer, H. (1946). *Mathematical Methods of Statistics*, Princeton University Press, Princeton, New Jersey.

Doob, J. L. (1953). *Stochastic Processes*, Wiley, New York.

Ibragimov, I. A. and Linnik, Yu. V. (1971). *Independent and Stationary Sequences of Random Variables*, Wolters Nordhoff, Groningen.

Klebanov, L. B. and Melamed, I. A. (1979). One linear method of estimation of parameters. *Asymptotic Statistics* (Ed. P. Mandl and M. Huskova), North-Holland, Amsterdam.

Prakasa Rao, B. L. S. (1982). Remarks on linear parametric inference for stochastic processes. *Statistics and Decisions* **1**, 39–55.

Rao, C. R. (1974). *Linear Statistical Inference and Its Applications*, 2nd ed., Wiley, New York.

CHAPTER 5

Martingale Approach to Inference

5.1 TESTS FOR GOODNESS OF FIT

In this section and in the following sections of this chapter, we discuss several nontraditional applications of the theory of martingales and semimartingales to the classical problems in mathematical statistics. Apart from the fact that they are discussed from a modern point of view, the approach adapted here brings in the interplay between the modern probabilistic concepts and classical statistical problems. Our approach in this chapter is based on the works of Khmaladze (1981), Aalen (1978, 1982) [cf. Prakasa Rao (1984)], Gill (1980), and others.

Let X_i, $1 \leqslant i \leqslant n$, be i.i.d. random variables with distribution function F. Let

(5.1.0) $$F_n(x) = \frac{1}{n}\sum_{i=1}^{n} I_{[X_i \leqslant x]}$$

be the empirical distribution function based on the sample X_i, $1 \leqslant i \leqslant n$, where I_A denotes the indicator function of the set A. Let

(5.1.1) $$V_n^F(x) = \sqrt{n}[F_n(x) - F(x)].$$

The weak convergence of the empirical processes $V_n^F(x), -\infty < x < \infty$, has been studied [cf. Billingsley (1968) and Parthasarathy (1967)] and the asymptotic distribution of the Kolmogorov–Smirnov statistic for testing goodness of fit has been obtained via the limiting process in the references cited above. We discuss an alternative approach to the problem due to Khmaladze (1981).

Let F be an absolutely continuous distribution function and $X_{(1)}, \ldots, X_{(n)}$ be the order statistics corresponding to the sample $X_1, \ldots, X_n$. Observe that $X_{(1)} < \cdots < X_{(n)}$ with probability 1. Note that

(5.1.2) $$F_n(x) = \frac{1}{n} \sum_{i=1}^{n} I_{[X_{(i)} \leqslant x]}.$$

In the representation (5.1.0), $F_n(x)$ is the sum of i.i.d. random variables $(1/n)I_{[X_i \leqslant x]}$, $1 \leqslant i \leqslant n$, whereas in (5.1.2), the random variables $I_{[X_{(i)} \leqslant x]}$, $1 \leqslant i \leqslant n$, are dependent. However, in the representation (5.1.2), $F_n(x)$ can be thought of as a point process with jumps at the points $X_{(i)}$, $1 \leqslant i \leqslant n$, and $F_n(x)$, $n \geqslant 1$, is a sub martingale for any fixed n since it is a nondecreasing process in x. We use this point of view in the rest of this chapter.

The problem now is to test the hypothesis that the true distribution function F belongs to a given family $\mathscr{F} = \{F(\cdot, \theta), \theta \in \Theta\}$, $\Theta \subset R^k$. Let

(5.1.3) $$V_n^{\theta}(x) = \sqrt{n}[F_n(x) - F(x, \theta)]$$

and

(5.1.4) $$V_n^{\hat{\theta}}(x) = \sqrt{n}[F_n(x) - F(x, \hat{\theta}_n)],$$

where $\hat{\theta}_n = \hat{\theta}_n(X_1, \ldots, X_n)$ is a suitable estimator.

Suppose θ is known. By making the transformation $t = F(x, \theta)$, the process (5.1.3) is transformed into

(5.1.5) $$V_n(t) = \sqrt{n}[F_n(t) - t], \qquad 0 \leqslant t \leqslant 1,$$

where $F_n(t)$, $0 \leqslant t \leqslant 1$, denotes the empirical distribution function of the observations $U_i = F(X_i, \theta)$, $1 \leqslant i \leqslant n$. $\{V_n(t), 0 \leqslant t \leqslant 1\}$ is called the *uniform empirical process*. It is well known that the limiting distribution of "smooth" functionals $\phi[V_n^{\theta}, F(\cdot, \theta)]$ of the empirical process V_n^{θ} (equivalently, those of functionals $\psi[V_n]$ of the uniform empirical process) are distribution free, that is, the limit distribution does not depend on $F(\cdot, \theta)$, and hence can be used for obtaining tests for goodness of fit, for instance, χ^2-test, Cramer–von Mises test, and Kolomogorov–Smirnov test.

The distribution-free property of the nonparametric tests is important in practice for constructing tests for goodness of fit.

Suppose θ is *unknown*. Let $\hat{\theta}_n = \hat{\theta}_n(X_1, \ldots, X_n)$ be an estimator of θ, and define the process $V_n^{\hat{\theta}}$ by (5.1.4). The problem now is whether one can construct functionals of the parametric empirical process $V_n^{\hat{\theta}}$ that depend on the family $\mathscr{F} = \{F(x,\theta):\theta\in\Theta\}$ in such a way that the asymptotic distribution does *not* depend on $\mathscr{F}$ under the hypothesis.

Let us suppose that the hypothesis $F\in\mathscr{F}$ holds and there exists $\theta\in\Theta$ such that $F(x)=F(x,\theta)$. Let $t=F(x,\theta)$, and transform the process (5.1.4) into

$$U_n(t)=\sqrt{n}[F_n(t)-G(t,\hat{\theta}_n)], \tag{5.1.6}$$

where $F_n(t)$ is the empirical distribution function of the random variables $F(X_i,\theta)$, $1\leqslant i\leqslant n$, and $G(t,\hat{\theta}_n)=F(x,\hat{\theta}_n)$, so that $G(t,\theta)=t$. Let us call $U_n(t)$ the parametric empirical process in this context. Suppose $\hat{\theta}_n$ admits the representation

$$\sqrt{n}(\hat{\theta}_n-\theta)=\int_0^1 \mathbf{e}(s,\theta)\,dV_n(s)+o_p(1), \tag{5.1.7}$$

where $V_n(t)=\sqrt{n}[F_n(t)-t]$. Then

$$U_n(t)=V_n(t)-\mathbf{g}(t,\theta)'\int_0^1 \mathbf{e}(s,\theta)\,dV_n(s)+r_n(t).$$

This follows by expanding $G(t,\hat{\theta}_n)$ in $\hat{\theta}_n$ in a neighborhood of the parameter θ up to a linear term. Here $\mathbf{g}(t,\theta)=\nabla G(t,\theta)$. It can be shown that the process $\{U_n(t),\ 0\leqslant t\leqslant 1\}$ converges weakly to the process $\{U(t),\ 0\leqslant t\leqslant 1\}$ under some regularity conditions (see Theorem 5.1.1) (Khmaladze, 1979) given by

$$U(t)=V(t)-\mathbf{g}(t,\theta)'\int_0^1 \mathbf{e}(s,\theta)\,dV(s),$$

where $V(\cdot)$ is a Brownian bridge. The distribution of U clearly depends on the specific value of the parameter θ, and hence it depends on the family $\mathscr{F}$ under consideration. The distribution of V does not depend on θ or on $\mathscr{F}$.

In (1.9.11) and (1.9.16) of Section 1.9, we have derived an alternative representation for $U(\cdot)$ that is a one-to-one transformation of $U(\cdot)$ into a Wiener process $W(\cdot)$. If this transformation is applied to the parametric empirical process $U_n(\cdot)$, then, under the hypothesis, the transformed process $W_n(\cdot)$ converges weekly in $L_2[0,1]$ to the Wiener process $W(\cdot)$. Hence the distribution of any functional of $U_n(\cdot)$ (not depending on $\mathscr{F}$) that can be represented as a fixed continuous functional on $L_2[0,1]$ of $W_n(\cdot)$ converges weakly to the corresponding functional of Wiener process (*not* depending on $\mathscr{F}$). Hence the limit distribution does not depend on $\mathscr{F}$ and is distribution free.

Suppose the following regularity conditions hold:

(A_1) For every $\theta\in\Theta$, $F(\cdot,\theta)$ is absolutely continuous and $F(\cdot,\theta_1)$ and

$F(\cdot, \theta_2)$ are mutually absolutely continuous for θ_1, θ_2 in Θ.

(A_2) For every $\theta \in \Theta$,

$$E_\theta[\mathbf{h}'(X, \theta, \varepsilon)\mathbf{h}(X, \theta, \varepsilon)] \to 0 \quad \text{as } \varepsilon \to 0, \tag{5.1.8}$$

where $\mathbf{h}' = (h_1, \ldots, h_k)$ and

$$h_i(x, \theta_0, \varepsilon) = \sup_{\|\theta - \theta_0\| < \varepsilon} \left| \frac{\partial \log f(x, \theta)}{\partial \theta_i} - \left. \frac{\partial \log f(x, \theta)}{\partial \theta_i} \right|_{\theta = \theta_0} \right|. \tag{5.1.9}$$

(A_3) For every $\theta \in \Theta$,

$$\mathscr{I}(\theta) = E_\theta[\nabla \log f(X, \theta)' \nabla \log f(X, \theta)] \tag{5.1.10}$$

is a finite and nonsingular matrix where $\nabla \log f = (\partial \log f/\partial \theta_1, \ldots, \partial \log f/\partial \theta_k)'$.

(A_4) There exists a k-dimensional vector $\mathbf{l}(\cdot, \theta)$ such that

$$\text{(i)} \quad E_\theta[\mathbf{l}'(X, \theta)\mathbf{l}(X, \theta)] < \infty \tag{5.1.11}$$

and

$$\text{(ii)} \quad E_\theta[\mathbf{l}(X, \theta) \nabla \log f(X, \theta)'] = \mathbf{I}_k, \tag{5.1.12}$$

where $\mathbf{I}_k$ is the identity matrix of order $k \times k$ and

(A_5) $\hat{\theta}_n$ is an estimator of θ such that

$$\text{(iii)} \quad \sqrt{n}(\hat{\theta}_n - \theta) = \int \mathbf{l}(x, \theta) V_n^\theta(dx) + o_p(1). \tag{5.1.13}$$

Let $\{V(t), 0 \leqslant t \leqslant 1\}$ denote the standard Brownian bridge on $[0, 1]$. Note that $V(t)$ can be represented as $W(t) - t\,W(1)$, where W is the standard Wiener process. Let

$$\mathbf{g}(t, \theta) = \int_0^t \mathbf{g}^{(1)}(s, \theta)\, ds, \qquad \mathbf{e}(t, \theta) = \mathbf{l}(x, \theta), \tag{5.1.14}$$

where $g^{(1)\prime} = (g_1^{(1)}, \ldots, g_k^{(1)})$ and

$$g_j^{(1)}(t, \theta) \equiv \frac{\partial \log f(x, \theta)}{\partial \theta_j} \quad \text{with } t = F(x, \theta). \tag{5.1.15}$$

The following theorem is due to Khmaladze (1979).

Theorem 5.1.1. Suppose conditions (A_1)–(A_5) hold. Then the process $\{U_n(t), 0 \leqslant t \leqslant 1\}$ defined by (5.1.7) converges weakly in $L_2[0,1]$ to the process $\{U(t), 0 \leqslant t \leqslant 1\}$, where

$$U(t) = V(t) - \mathbf{g}(t,\theta)' \int_0^t \mathbf{e}(s,\theta)\, V(ds). \tag{5.1.16}$$

Similar results for convergence in $C[0,1]$ (and $D[0,1]$) are obtained in Durbin (1973). Suppose ϕ and ψ are functionals satisfying the relation

$$\phi[V_n^\theta, F(\cdot,\theta)] = \psi(V_n), \tag{5.1.17}$$

where ψ is continuous on $L_2[0,1]$. Then it follows from Theorem 5.1.1 that

$$\phi[V_n^\theta, F(\cdot,\hat{\theta})] \xrightarrow{\mathscr{L}} \psi(u) \quad \text{as } n \to \infty. \tag{5.1.18}$$

However, note that the distribution of $\psi(u)$ *depends* on θ and hence the test is not distribution free. We shall indicate later some further conditions under which the limiting distribution does not depend on the parameter θ.

In order to indicate the new approach in a simple manner, we revert to the case when θ is known and suppress θ in the notation given below.

Given a process ξ on $[0,1]$, let $\mathscr{F}_t^\xi$ denote the σ-algebra generated by $\xi(s)$, $0 \leqslant s \leqslant t$. $\mathscr{F}^\xi = \{\mathscr{F}_t^\xi, 0 \leqslant t \leqslant 1\}$ is a flow of σ-algebras corresponding to the process ξ. The order statistics $X_{(1)}, \ldots, X_{(n)}$ are the stopping times with respect to the flow $\mathscr{F}^{F_n}$ since $[X_{(i)} \leqslant t] = [F_n(t) \geqslant i/n] \in \mathscr{F}_t^{F_n}$.

Note that the process $\{F_n(t), 0 \leqslant t \leqslant 1\}$ is a submartingale and a Markov process. Furthermore,

$$\begin{aligned} E[\Delta F_n(t) \mid \mathscr{F}_t^{F_n}] &= E[\Delta F_n(t) \mid F_n(t)] \\ &= \frac{1 - F_n(t)}{1-t} \Delta t, \qquad 0 \leqslant t < 1, \end{aligned} \tag{5.1.19}$$

where $\Delta F_n(t) = [F_n(t + \Delta t) - F_n(t)]$, $\Delta t \geqslant 0$. This follows from the fact that, given $F_n(t)$, $n\Delta F_n(t)$ is Binomial with $n[1 - F_n(t)]$ as the number of trials and $\Delta t/(1-t)$ as the probability of success. This suggests the Doob–Meyer decomposition for the submartingale $F_n(t)$, $0 \leqslant t \leqslant 1$, in the following way. Let

$$F_n(t) = \int_0^t \frac{1 - F_n(s)}{1-s}\, ds + M_n(t), \qquad 0 \leqslant t \leqslant 1. \tag{5.1.20}$$

It is easy to see that

$$E[M_n(t)|\mathscr{F}_s^{F_n}] = M_n(s), \qquad s \leqslant t, \tag{5.1.21}$$

and

$$\int_0^t \frac{1-F_n(s)}{1-s}\,ds \leqslant -\log(1-X_{(n)}). \tag{5.1.22}$$

Therefore $m_n(t)$ has a finite expectation for all $0 \leqslant t \leqslant 1$. Relation (5.1.21) implies that $\{M_n(t), \mathscr{F}_t^{F_n}, 0 \leqslant t \leqslant 1\}$ is a martingale. In other words, the identity (5.1.20) gives the Doob–Meyer decomposition for the submartingale $F_n(t)$. Let

$$V_n(t) = \sqrt{n}[F_n(t) - t] \tag{5.1.23}$$

as in (5.1.5). We have

$$V_n(t) = -\int_0^t \frac{V_n(s)}{1-s}\,ds + W_n(t), \tag{5.1.24}$$

where $\{W_n(t), \mathscr{F}_t^{V_n}, 0 \leqslant t \leqslant 1\}$ is a martingale. In fact, $\mathscr{F}_t^{F_n} = \mathscr{F}_t^{V_n}$, $0 \leqslant t \leqslant 1$,

$$\text{(i)} \quad W_n(t) = \sqrt{n}M_n(t), \tag{5.1.25}$$

$$\text{(ii)} \quad \sqrt{n}\int_0^t \frac{1-F_n(s)}{1-s}\,ds - \sqrt{n}t = -\int_0^t \frac{V_n(s)}{1-s}\,ds$$

from the representations (5.1.20) and (5.1.24). We have noted earlier that a Brownian bridge $\{V(t), 0 \leqslant t \leqslant 1\}$ can be represented in the form

$$V(t) = W(t) - tW(1), \tag{5.1.26}$$

where $W(\cdot)$ is the standard Wiener process. However, $V(t)$ can also be considered as a diffusion processes satisfying the stochastic differential equation

$$dV(t) = -\frac{V(t)}{1-t}\,dt + dW(t). \tag{5.1.27}$$

This representation is similar to (5.1.24). We shall now discuss the weak convergence of the sequence of processes $\{W_n(t), 0 \leqslant t \leqslant 1\}$ defined by (5.1.24) in $L_2[0,1]$. Note that

$$W_n(t) = V_n(t) + \int_0^t \frac{V_n(s)}{1-s}\,ds, \qquad 0 \leqslant t \leqslant 1. \tag{5.1.28}$$

Lemma 5.1.2. The sequence of processes $\{V_n(t), 0 \leqslant t \leqslant 1\}$ is weakly compact in $L_2[0,1]$ and, in fact, $V_n \xrightarrow{\mathscr{L}} V$ where V is given by (5.1.26).

Proof. See Khmaladze (1979).

Lemma 5.1.3. Let

$$Z_n(t) = \int_0^t \frac{V_n(s)}{1-s}\,ds, \qquad 0 \leqslant t \leqslant 1. \tag{5.1.29}$$

Then the sequence of process $\{Z_n, n \geqslant 1\}$ is weakly compact in $L_2[0,1]$.

Proof. Define the operator K on $L_2[0,1]$ by

$$K\phi(t) = \int_0^t \frac{\phi(s)}{1-s}\,ds. \tag{5.1.30}$$

It is sufficient to prove that the operator K is continuous in view of Lemma 5.1.2. Note that, for $\phi \in L_2[0,1]$,

$$\begin{aligned}
\|K\phi\|^2 &= \int_0^1 \left(\int_0^t \frac{\phi(s)}{1-s}\,ds\right)^2 dt \\
&= \left[-(1-t)\left(\int_0^t \frac{\phi(s)}{1-s}\,ds\right)^2\right]_{t=0}^{t=1} + 2\int_0^1 \left(\phi(t)\int_0^t \frac{\phi(s)}{1-s}\,ds\right) dt \\
&= 2\int_0^1 \left(\phi(t)\int_0^t \frac{\phi(s)}{1-s}\,ds\right) dt
\end{aligned}$$

by integration by parts. Furthermore,

$$\int_0^1 \left(\phi(t)\int_0^t \frac{\phi(s)}{1-s}\,ds\right) dt \leqslant \|K\phi\|\,\|\phi\|,$$

and hence

$$\|K\phi\|^2 \leqslant 2\|K\phi\|\,\|\phi\|$$

or equivalently

$$\|K\phi\| \leqslant 2\|\phi\|. \tag{5.1.31}$$

Therefore the norm of the operator does not exceed 2, and hence the operator K is continuous.

Proposition 5.1.4. The sequence of processes W_n, $n \geqslant 1$, defined by

$$W_n(t) = V_n(t) + \int_0^t \frac{V_n(s)}{1-s} ds, \qquad 0 \leqslant t \leqslant 1, \tag{5.1.32}$$

is weakly compact in $L_2[0,1]$.

Proof. This result follows as a consequence of Lemmas 5.1.2 and 5.1.3 and relation (5.1.32).

Theorem 5.1.5. The sequence of processes W_n defined by (5.1.32) converges weakly a standard Wiener process W in $L_2[0,1]$.

Proof. In view of Lemma 5.1.2 and Proposition 5.1.4, it follows that

$$W_n(t) = (I + K)V_n(t) \xrightarrow{\mathscr{L}} (I + K)V \tag{5.1.33}$$

since $I + K$ is a continuous operator. It is sufficient to show that $(I + K)V$ is a standard Wiener process. Let

$$b = (I + K)V. \tag{5.1.34}$$

Note that, for any $\phi \in L_2[0,1]$,

$$\begin{aligned}\int_0^1 \phi(t)b(t)\,dt &= \int_0^1 \phi(t)V(t)\,dt + \int_0^1 \left(\phi(t)\int_0^t \frac{V(s)}{1-s} ds\right) dt \\ &= \int_0^1 \left[\Phi(t) - \int_0^t \frac{\Phi(s)}{1-s} ds\right] dV(t),\end{aligned}$$

where

$$\Phi(t) = \int_t^1 \phi(s)\,ds$$

Hence

$$E\left[\int_0^1 \phi(t)b(t)\,dt\right] = 0 \tag{5.1.35}$$

and

$$\begin{aligned}\operatorname{Var}\left[\int_0^1 \phi(t)b(t)\,dt\right] &= \int_0^1 \left[\Phi(t) - \int_0^t \frac{\Phi(s)}{1-s}\,ds\right]^2 dt - \left\{\int_0^1 \left[\Phi(t) - \int_0^t \frac{\Phi(s)}{1-s}\,ds\right] dt\right\}^2 \\ &= \int_0^1 \left[\Phi(t) - \int_0^t \frac{\Phi(s)}{1-s}\,ds\right]^2 dt \\ &= \int_0^1 \Phi^2(t)\,dt.\end{aligned} \tag{5.1.36}$$

Since (5.1.35) and (5.1.36) hold for every $\phi \in L_2[0,1]$ and the process b is Gaussian from the fact $b = (I+K)V$, it follows that $\{b(t), t \geqslant 0\}$ is a standard Wiener process.

In general, the following result holds.

Theorem 5.1.6. Let $X_{1n}, \ldots, X_{nn}$ be i.i.d. random variables on $[0,1]$ with distribution function $A_n(t)$ for every $n \geqslant 1$. Let $F_n(t)$ be the empirical distribution function based on $X_{1n}, \ldots, X_{nn}$. Define

$$V_n^A(t) = \sqrt{n}[F_n(t) - A_n(t)] \tag{5.1.37}$$

and

$$W_n^A(t) = V_n^A(t) + \int_0^t \frac{V_n^A(s)}{1-s}\,ds. \tag{5.1.38}$$

Suppose the functions $a_n(t) = \sqrt{n}\,[A_n(t) - t]$ converge in $L_2[0,1]$ to a function $a(t)$ in $L_2[0,1]$. Then the sequence of processes W_n^A converge weakly to $W + c$ in $L_2[0,1]$ where W is the standard Wiener process and

$$c(t) = a(t) + \int_0^t \frac{a(s)}{1-s}\,ds, \qquad 0 \leqslant t \leqslant 1. \tag{5.1.39}$$

We omit the proof. For details, see Khmaladze (1981).

Let C_1 be the set of all functions of the form $\Phi(t) = \int_t^1 \phi(s)\,ds$ where $\phi \in L_2[0, 1]$. It is well known that the measures generated by the process W and $W + c$ are equivalent if and only if $c \in C_1$ [cf. Chatterjee and Mandrekar (1978)]. In order to study the problem of testing of hypotheses, it is important to know whether it is possible to distinguish the sequence of alternatives A_n, $n \geqslant 1$, from the hypothesis with asymptotic power 1. In other words, one should determine the functions $a(\cdot) \in L_2[0, 1]$ such that the corresponding $c(\cdot) \in L_2[0, 1] \backslash C_1$. Furthermore, it is crucial to know whether there exists $a(\cdot) \in L_2[0, 1]$, $a \not\equiv 0$, such that $c(\cdot) = 0$. These can be answered from the fact that the operator $I + K$ is one-to-one on $L_2[0, 1]$ and $(I + K)\phi \in C_1$ if and only if $\phi \in C_1$.

Instead of centering the process $F_n(t)$ by t as in (5.1.23), suppose we center it at tY_n, where Y_n is some random variable. Define

$$\text{(5.1.40)} \quad \text{(i)} \quad V_n(t, Y_n) = \sqrt{n}[F_n(t) - tY_n], \qquad 0 \leqslant t \leqslant 1,$$

$$\text{(ii)} \quad W_n(t) = V_n(t, Y_n) - \int_0^t \frac{V_n(1, Y_n) - V_n(s, Y_n)}{1 - s}\,ds.$$

Note that $W_n(t)$ *does not depend on* Y_n. If $Y_n \equiv 1$, we have $V_n(t, Y_n) = V_n(t)$ as discussed in (5.1.23). If $Y_n \equiv 0$, then

$$\text{(5.1.41)} \qquad W_n(t) = \sqrt{n}\left[F_n(t) - \int_0^t \frac{1 - F_n(s)}{1 - s}\,ds\right]$$

$$= n^{1/2} F_n(t) + n^{-1/2} \sum_{i=1}^n \log(1 - t \wedge X_i),$$

which is connected with the expression in (5.1.20). In view of Theorem 5.1.5, it follows that

$$\text{(5.1.42)} \qquad W_n(t) = n^{1/2} F_n(t) - n^{-1/2} \sum_{i=1}^n \log(1 - t \wedge X_i) \xrightarrow{\mathscr{L}} W(t).$$

Remarks. These arguments show that centering, in the definition of $V_n(t)$, can be done by any random variable tY_n instead of t (equivalently, instead of the distribution function specified by the hypothesis) so that the limiting distribution exists.

θ Unknown

We now consider the case of composite hypothesis. The problem is to test the hypothesis that the true distribution function $F \in \mathscr{F} = \{F(x, \theta), \theta \in \Theta\}$ $\Theta \subset R^k$. We study the properties of the process

(5.1.43) $$U_n(t) = V_n^{\hat{\theta}}(x) = \sqrt{n}[F_n(x) - F(x, \hat{\theta})]$$

as defined by (5.1.6), where $\hat{\theta}$ is an estimator for θ. Let $t = F(x, \theta)$ and

$$g_j^{(1)}(t, \theta) = \frac{\partial \log f(x, \theta)}{\partial \theta_j}, \qquad 1 \leqslant j \leqslant k,$$

as before, where $f(x, \theta)$ is the density of the distribution function $F(x, \theta)$. Condition (A_2) can also be written in the form

(A_2') $$\int_0^1 \left(\sup_{\phi: \|\phi - \theta\| < \varepsilon} \| \mathbf{g}^{(1)}(t, \phi) - \mathbf{g}^{(1)}(t, \theta) \| \right) dt \to 0 \quad \text{as } \varepsilon \to 0$$

for every $\theta \in \Theta$.

For studying the weak convergence of the process $U_n(t)$, we do not need (A_5), but the estimator $\hat{\theta}$ should satisfy the following condition:

(A_5') $\sqrt{n}(\hat{\theta}_n - \theta) = O_p(1)$.

Let

$$V_n(t) = \sqrt{n}[F_n(t) - t]$$

as before. Then

(5.1.44) $$U_n(t) = V_n(t) - \mathbf{g}(t, \theta)' \sqrt{n}(\hat{\theta}_n - \theta) + r_n(t),$$

where

(5.1.45) $$\int_0^1 [r_n^{(1)}(t)]^2 \, dt = o_p(1) \quad \text{as } n \to \infty$$

in view of (A_2').

In addition to (A_1), (A_2'), (A_3'), and (A_5') suppose that

(A_6') (i) the functions $g_j^{(1)}(t, \theta), 1 \leqslant j \leqslant k$, belong to $L_2[0, 1]$;

(ii) the functions $1, g_1^{(1)}(t, \theta), \ldots, g_k^{(1)}(t, \theta)$ are linearly independent in the neighborhood of $t = 1$. Let

(5.1.46) $$\Gamma = \left[\begin{array}{c|c} 1 & 0 \\ \hline 0 & \mathscr{I}(\theta) \end{array} \right],$$

where $\mathscr{I}(\theta)$ is as defined by (A_3), and let

$$\mathbf{g}(t) = \Gamma^{-1/2}\left(\frac{t}{\mathbf{g}(t, \theta)}\right). \tag{5.1.47}$$

Define

$$W_n(t) = U_n(t) - \int_0^t \left(\nabla\mathbf{g}(s)'\mathbf{C}^{-1}(s) \int_s^1 \nabla\mathbf{g}(\tau)\, dU_n(\tau) \right) ds \tag{5.1.48}$$

using the transformation derived in (1.9.16), where

$$\mathbf{C}(t) = \mathbf{I}_{k+1} - \mathbf{B}(t) \equiv \int_t^1 \nabla\mathbf{g}(\tau)\nabla\mathbf{g}(\tau)'d\tau \tag{5.1.49}$$

so that $\mathbf{C}(0) = \mathbf{I}_{k+1}$ and $\mathbf{C}(1) = \mathbf{0}$. Since

$$\nabla\mathbf{g}(t)' = \nabla\mathbf{g}(t)'[\mathbf{I}_{k+1} - \mathbf{B}(t)]^{-1} \int_t^1 \nabla\mathbf{g}(\tau)\nabla\mathbf{g}(\tau)'\, d\tau, \tag{5.1.50}$$

relation (5.1.45) implies that

$$W_n(t) = V_n(t) - \left(\int_0^t \left\{ \nabla\mathbf{g}(s)'\mathbf{C}^{-1}(s) \int_s^1 \nabla\mathbf{g}(\tau)dV_n(\tau) \right\} \right) ds + \varepsilon_n(t), \tag{5.1.51}$$

where

$$\varepsilon_n(t) = \int_0^t \left(\nabla\mathbf{g}(s)'\mathbf{C}^{-1}(s) \int_0^1 \nabla\mathbf{g}(\tau) r_n^{(1)}(\tau)d\tau \right) ds. \tag{5.1.52}$$

Lemma 5.1.7. Suppose (A'_6) holds. Then the operator M, defined by

$$\begin{aligned} M\phi(t) &= \int_0^t \left(\nabla\mathbf{g}(s)'\mathbf{C}^{-1}(s) \int_s^1 \nabla\mathbf{g}(\tau)\phi(\tau)d\tau \right) ds \\ &= \int_0^1 \left(\int_0^{t\wedge\tau} \nabla\mathbf{g}(s)'\mathbf{C}^{-1}(s)ds \right) \nabla\mathbf{g}(\tau)\phi(\tau)d\tau \end{aligned} \tag{5.1.53}$$

for $\phi \in L_2[0, 1]$, is a Hilbert–Schmidt operator.

Proof. Let

$$M(t, \tau) = \left(\int_0^{t\wedge\tau} \nabla\mathbf{g}(s)'\mathbf{C}^{-1}(s)ds \right) \nabla\mathbf{g}(\tau). \tag{5.1.54}$$

Then

$$\int_0^1 M^2(t,\tau)d\tau = \phi_1(\tau) + \int_0^t \phi_2^2(\tau)d\tau,$$

where

$$\phi_1(t) = \left(\int_0^t \nabla\mathbf{g}(s)'\mathbf{C}^{-1}(s)ds\right)\mathbf{C}(t)\left(\int_0^t \mathbf{C}^{-1}(s)\nabla\mathbf{g}(s)ds\right)$$

and

$$\phi_2(t) = \nabla\mathbf{g}(t)\int_0^t \nabla\mathbf{g}(s)'\mathbf{C}^{-1}(s)ds.$$

Let

$$\xi(t)' = \left(\int_0^t \nabla\mathbf{g}(s)'\mathbf{C}^{-1}(s)ds\right)(\mathbf{C}^{-1}(t) - \mathbf{I})^{-1},$$

where $\mathbf{I} \equiv \mathbf{I}_{k+1}$. Consider the function that is identically 1 on $[0, t]$. Then $\xi(t)$ is the vector of Fourier coefficients with respect to the function $\mathbf{C}^{-1}(s)\nabla\mathbf{g}(s)$. Hence

$$\begin{aligned}(5.1.55)\qquad t &\geqslant \xi(t)'[\mathbf{C}^{-1}(t) - \mathbf{I}]\xi(t)\\ &= \int_0^t \nabla\mathbf{g}(s)'\mathbf{C}^{-1}(s)ds[\mathbf{C}^{-1}(t) - \mathbf{I}]^{-1}\int_0^t \mathbf{C}^{-1}(s)\nabla\mathbf{g}(s)ds\\ &\geqslant \int_0^t \nabla\mathbf{g}(s)'\mathbf{C}^{-1}(s)ds\mathbf{C}(t)\int_0^t \mathbf{C}^{-1}(s)\nabla\mathbf{g}(s)ds = \phi_1(t),\end{aligned}$$

since

$$[\mathbf{C}^{-1}(t) - \mathbf{I}]^{-1} - \mathbf{C}(t) = \mathbf{C}^2(t)[\mathbf{I} - \mathbf{C}(t)]^{-1} > 0.$$

On the other hand,

$$(5.1.56)\qquad \int_0^t \phi_2^2(\tau)d\tau = -\phi_1(t) + 2\int_0^t \phi_2(\tau)d\tau$$

by integrating by parts. Hence

$$(5.1.57)\qquad \int_0^1 M^2(t,\tau)d\tau = 2\int_0^1 \phi_2(\tau)d\tau.$$

Let $\int_0^1 \phi_2^2(\tau)d\tau = a^2$ and $\phi_1(1) = b$. Relation (5.1.56) implies that $a^2 \leqslant -b + 2a$ and hence $a \leqslant 1 + \sqrt{1-b}$. But $b \leqslant 1$ since $\phi_1(t) \leqslant t$ by (5.1.55).

Hence

$$\int_0^1 M^2(t,\tau)d\tau \leqslant b + a^2 \leqslant b - b + 2a = 2a \leqslant 4. \tag{5.1.58}$$

This proves that the operator M is Hilbert–Schmidt.

Remarks. In particular, it also follows that

$$\int_0^1 \int_0^1 M^2(t,\tau)dt\,d\tau < \infty. \tag{5.1.59}$$

Lemma 5.1.8. Suppose assumptions (A_6') and relation (5.1.45) hold. Then

$$\int_0^1 \varepsilon_n^2(t)dt = o_p(1). \tag{5.1.60}$$

Proof. This lemma follows from the inequality

$$\begin{aligned}\sup_{0\leqslant t\leqslant 1} |\varepsilon_n(t)| &\leqslant \left\{\int_0^1 M^2(t,\tau)d\tau\right\}^{1/2} \|r_n^{(1)}\|_2\\ &\leqslant 2\|r_n^{(1)}\|_2\end{aligned}$$

and relation (5.1.45).

In view of Lemma 5.1.8, the weak convergence of $W_n(t)$ holds if it does for

$$\tilde{W}_n(t) = V_n(t) - \left\{\int_0^t \nabla\mathbf{g}(s)\mathbf{C}^{-1}(s)\left(\int_s^1 \nabla\mathbf{g}(\tau)dV_n(\tau)\right)ds\right\}, \quad 0 \leqslant t \leqslant 1. \tag{5.1.61}$$

Let $\mathscr{C}$ be the class functionals S defined on $L_2[0,1]$ for which, for any given $\varepsilon > 0$, there exists a function S_ε that is continuous on $L_2[0,1]$ and an integer $n(\varepsilon) \geqslant 1$ such that

$$\begin{aligned}&\text{(i)} \quad P\{|S(V) - S_\varepsilon(V)| > \varepsilon\} < \varepsilon,\\ &\text{(ii)} \quad P\{|S(V_n) - S_\varepsilon(V_n)| > \varepsilon\} < \varepsilon, \qquad n \geqslant n(\varepsilon).\end{aligned} \tag{5.1.62}$$

It is clear that, for any $S \in \mathscr{C}$,

$$S(V_n) \xrightarrow{\mathscr{L}} S(V) \quad \text{as } n \to \infty. \tag{5.1.63}$$

Let $X_{1n},\ldots,X_{nn}$ be i.i.d. random variables on [0, 1] with distribution function A_n. Let $F_n(t)$ be the empirical distribution function based on these random variable. Suppose that the distribution functions $A_n(t)$ are absolutely continuous with densities $a_n(t)$ with the property

(5.1.64) (i) $a_n(\mathrm{t}) = 1 + n^{-1/2}a^{(1)}(t) + \rho_n(t),$

where

(ii) $\sum_{i=1}^{n} \rho_n(U_n) = o_p(1)$

and U_i, $1 \leqslant i \leqslant n$, are i.i.d. random variables uniform on [0, 1]. Let

(5.1.65) $$V_n^A(t) = \sqrt{n}[F_n(t) - A_n(t)],$$

(5.1.66) $$Z_n^A(t) = \int_0^1 M(t, \tau)dV_n^A(\tau),$$

and

(5.1.67) $$\tilde{W}_n^A(t) = V_n^A(t) - Z_n^A(t).$$

Remark. Note that $\tilde{W}_n^A(t) = \tilde{W}_n(t)$ given by (5.1.61) when $A_n(t) \equiv t$.

Theorem 5.1.9. Suppose conditions (5.1.64) holds. Then

(5.1.68) $$\tilde{W}_n^A(\cdot) \xrightarrow{\mathscr{L}} W(\cdot) + c(\cdot),$$

where

(5.1.69) (i) $c(t) = a(t) + \int_0^1 M(t, \tau)a^{(1)}(\tau)d\tau, \qquad 0 \leqslant t \leqslant 1,$

(ii) $a(\cdot) = \int_0^t a^{(1)}(s)ds.$

Proof. It can be shown that

(5.1.70) $$V_n^A \xrightarrow{\mathscr{L}} V + a,$$

where V is a Brownian bridge on [0, 1] by using Theorem 5.1 in Chibisov

(1969). In fact, for any $S \in \mathscr{C}$,

$$S(V_n^A) \xrightarrow{\mathscr{L}} S(V+a). \tag{5.1.71}$$

Let C be any functional continuous on $L_2[0, 1]$. Define S by the relations

$$S(V_n^A) = C(Z_n^A) \quad \text{and} \quad S(V) = C(Z), \tag{5.1.72}$$

where

$$Z(t) = \int_0^1 M(t, \tau) dV(\tau), \qquad 0 \leqslant t \leqslant 1,$$

and $M(t, \tau)$ as defined by (5.1.54).

Suppose we show that $S \in \mathscr{C}$. Then it follows that

$$Z_n^A \xrightarrow{\mathscr{L}} Z + Ma^{(1)}, \tag{5.1.73}$$

where M is the operator as given in (5.1.53). Relations (5.1.70) and (5.1.72) prove that

$$W_n^A \xrightarrow{\mathscr{L}} W + c. \tag{5.1.74}$$

In view of Lemma 5.1.7, given $\varepsilon > 0$, there exists a kernel $M_\varepsilon(t, \tau)$ such that

(5.1.75) (i) $$\int_0^1 \int_0^1 [M(t, \tau) - M_\varepsilon(t, \tau)]^2 dt\, d\tau < \varepsilon,$$

(ii) $$\int_0^1 \int_0^1 \left[\frac{\partial}{\partial \tau} M_\varepsilon(t, \tau)\right]^2 dt\, d\tau < \infty.$$

Relation (5.1.75)(i) shows that

$$E \int_0^1 \left| Z(t) - \int_0^1 M_\varepsilon(t, \tau)\, dV(\tau) \right|^2 dt < \varepsilon$$

and

$$E \int_0^1 \left| Z_n(t) - \int_0^1 M_\varepsilon(t, \tau)\, dV_n(\tau) \right|^2 dt < \varepsilon.$$

Relation (5.1.75)(ii) implies that

$$\int_0^1 M_\varepsilon(t, \tau) dV(\tau) = \int_0^1 \frac{\partial}{\partial \tau} M_\varepsilon(t, \tau) V(\tau) d\tau \tag{5.1.76}$$

is a continuous mapping of $V(t)$. In particular, is S is a functional satisfying (5.1.72), then define

$$S_\varepsilon(\phi) = C(\zeta),$$

where

$$\zeta(t) = \int_0^1 \frac{\partial}{\partial \tau}(M(t, \tau))\phi(\tau)d\tau$$

and S_ε satisfies (5.1.62). This completes the proof of this theorem.

In view of the identity

$$\nabla \mathbf{g}(t)' = \nabla \mathbf{g}(t)'[\mathbf{I}_{k+1} - \mathbf{B}(t)]^{-1} \int_0^1 \nabla \mathbf{g}(\tau)\nabla \mathbf{g}(\tau)' d\tau, \tag{5.1.77}$$

the process $\{W_n^A(t), 0 \leqslant t \leqslant 1\}$ can also be expressed as

$$\tilde{W}_n^A(t) = \sqrt{n}\left[F_n(t) - \int_0^1 M(t, \tau) dF_n(\tau)\right]. \tag{5.1.78}$$

The function $M(t, \tau) = M(t, \tau, \theta)$ depends on $\mathbf{g}(t)$, which in turn depends on θ. If $\hat{\theta}$ is a $\sqrt{n}$-consistent estimator of θ, that is, $\sqrt{n}(\hat{\theta} - \theta) = O_p(1)$, then it can be shown that

$$\hat{\tilde{W}}_n^A(t) = \sqrt{n}\left[F_n(t) - \int_0^1 M(t, \tau, \hat{\theta})\, dF_n(\tau)\right] \to W(t) + c(t)$$

provided $M(t, \tau, \theta)$ is continuous in θ.

Remarks. Now suppose that ϕ is any functional of the parametric empirical process $V_n^{\hat{\theta}}$ defined by (5.1.4) and of the family $\mathscr{F}$ under the hypothesis such that

$$\phi[V_n^{\hat{\theta}}; \mathscr{F}] = \psi[\hat{W}_n],$$

where ψ is continuous $P_{[0,1]}$ a.e. in $L_2[0, 1]$. Then the limiting distribution of such functonal ϕ under the hypothesis does *not* depend on $\mathscr{F}$, and hence is useful for obtaining the tests for goodness of fit.

Before we conclude this section, we should like to mention that "almost sure" and "in probability" representations of the empirical process of appropriate Gaussian processes are obtained when unknown parameters of the underlying distribution function are estimated by Burke et al. (1978). Csörgö (1981) discussed the empirical characteristic processes (see Section 8.3) when parameters are estimated.

5.2 TESTS OF HOMOGENEITY

Let $X_{1n}, \ldots, X_{nn}$ be independent random variables, and let F_{in} be the distribution function of X_{in}. We should like to test the hypothesis

$$\text{H:}\quad F_{in} = F \quad \text{and} \quad F \in \mathscr{F}$$

for all i and n against the alternative hypothesis

$$\text{K:}\quad \text{There are values of } i \text{ for which } F_{in} \notin \mathscr{F}$$

Let

$$\hat{F}_{sn}(x) = \frac{1}{[sn]} \sum_{i \leqslant [sn]} I_{[X_{in} \leqslant x]}, \qquad s \in [0, 1], \quad x \in R, \quad n \geqslant 1. \tag{5.2.1}$$

$\hat{F}_{sn}(x)$ is called the *empirical field*. There is a one-to-one correspondence between the observations $X_{1n}, \ldots, X_{nn}$ and the sample paths of the empirical field $\hat{F}_{sn}$. This function is the analog of the empirical distribution function in the i.i.d. case for the case of independent but not necessarily identically distributed random variables.

Suppose $\mathscr{F} = \{F(x, \theta) : \theta \in \Theta\}$, $\Theta \subset R^k$. If Θ is a singleton consisting of θ, then one can consider functionals of the empirical field

$$Y_n^{\theta}(s, x) = \frac{[sn]}{\sqrt{n}} [\hat{F}_{sn}(x) - F(x, \theta)] \tag{5.2.2}$$

for testing H versus K. Using the transformation $t = F(x, \theta)$, we have the uniform empirical field

$$Y_n(s, t) = \frac{[sn]}{\sqrt{n}} [\hat{F}_{sn}(t) - t] = \frac{1}{\sqrt{n}} \sum_{i \leqslant [sn]} [I_{\{U_{in} \leqslant t\}} - t],$$

where $U_{in} = F(X_{in}, \theta)$ and I_A denotes the indicator function of the set A. If θ is unspecified and θ belongs to Θ, an open subset of R^k, then we consider

$$Z_n(s, t) = Y_n^{\hat{\theta}}(s, t) \tag{5.2.3}$$

with $t = F(x, \hat{\theta}_n)$, where $\hat{\theta}_n$ is an estimator of θ and use functionals of Z_n for testing purposes.

Example 5.2.1. Suppose X_{in}, $1 \leqslant i \leqslant m$, are i.i.d. as F and X_{jn}, $m+1 \leqslant j \leqslant n$, are i.i.d. as G and we want to test the hypothesis H: $F = G$ against the alternative K: $F \neq G$. Construct the empirical process

$$\sqrt{\frac{n}{m}}\, Y_n^F\left(\frac{m}{n}, x\right) = \frac{1}{\sqrt{m}} \sum_{i \leqslant m} [I_{\{X_{in} \leqslant x\}} - F(x)]$$

from the first m observations and the process

$$\sqrt{\frac{n}{n-m}}\left[Y_n^G(1, \cdot) - Y_n^G\left(\frac{m}{n}, \cdot\right)\right]$$

from the remaining $n - m$ observations. One can use the statistic

$$\sup_x \left[Y_n^F(1, x) - \frac{n}{m} Y_n^F\left(\frac{m}{n}, x\right)\right]$$

for testing homogeneity.

Example 5.2.2 (Change-Point Problem). Suppose X_{in}, $1 \leqslant i \leqslant n$, are independent random variables with distribution function F_{in} and we want to test the hypothesis H: $F_{in} = F \in \mathscr{F}$ for $1 \leqslant i \leqslant n$ against the alternative K: there is a number m such that $1 \leqslant m \leqslant n$ and $F_{in} \notin \mathscr{F}$ for $i > m$. A statistic for testing H versus K is $\eta(T_n) = \max|T_n(s)|$ [cf. Hinkley (1970)], where

$$T_n(s) = n^{-1/2} \sum_{i \leqslant [sn]} \left[h\left(\frac{i}{n}, X_{in}\right) - \int_{-\infty}^{\infty} h\left(\frac{i}{n}, x\right) F(dx, \hat{\theta})\right],$$

where $h(\sigma, x)$ is a function defined on $[0, 1] \times R$. But

$$T_n(s) = \int_0^s \int_{-\infty}^{\infty} h(u, x)\, Y_n^{\hat{\theta}}(du, dx),$$

where Y_n^{θ} is as defined by (5.2.2) and hence $\eta(T_n)$ is a functional of $Y_n^{\hat{\theta}}$.

We now discuss conditions for weak convergence of processes $Y_n^{\hat{\theta}}$ in $L_2[0.1] \times L_2[0, 1]$.

Suppose the following conditions hold:

(B_1) There exists $\theta \in \Theta$ such that the distribution function

$$A_{in}(t) = F_{in}[F^{-1}(t, \theta)] \quad \text{with } F(t, \theta) = \sup\{x: F(x, \theta) \leqslant t\}$$

satisfies the condition

$$a_n^{(1)}(s, t) \equiv \frac{1}{\sqrt{n}} \sum_{i \leqslant sn} [A_{in}^{(1)}(t) - 1]$$

converges in $L_2[0, 1] \times L_2[0, 1]$ to $a^{(1)}(s, t)$. Here $A_{in}^{(1)}(t)$ denotes the derivative of $A_{in}(t)$.

(B_2) $\hat{\theta}_n$ is a sequence of estimators such that the empirical field $Y_n^{\hat{\theta}}$ satisfies one of the following conditions:

(i) there is a vector function $\mathbf{l}(x, \theta)$ such that

(a) $E_\theta[\mathbf{l}(X, \theta)' \mathbf{l}(X, \theta)] < \infty$,

(b) $E_\theta[\mathbf{l}(X, \theta) \nabla \log f(X, \theta)'] = \mathbf{I}_k$,

and

(c) $\sqrt{n}(\hat{\theta}_n - \theta) = \int \mathbf{l}(x, \theta)\, Y_n^\theta(1, dx) + o_P(1) \quad \text{as } n \to \infty$

(ii) for a given s, the estimator $\hat{\theta}_{ns}$ depends on the first $[sn]$ random variables and

$$n^{1/2}(\hat{\theta}_{ns} - \theta) = s^{-1/2} \int \mathbf{l}(x, \theta)\, Y_n^\theta(s, dx) + \varepsilon_n(s),$$

where

$$\int_0^1 \varepsilon_n^2(s)\, ds = o_p(1)$$

as $n \to \infty$.

Let $Y(s, t)$ denote the Gaussian random field on $[0, 1]^2$ with mean zero and covariance function $(s_1 \wedge s_2)(t_1 \wedge t_2 - t_1 t_2)$. The trajectories of this random field are continuous and in particular belong to $L_2[0, 1] \times L_2[0, 1]$ a.s. Define

$$\bar{Z}(s, t) = \bar{\pi} Y(s, t) = Y(s, t) - s\mathbf{g}(t, \theta)' \int_0^1 \mathbf{e}(\tau, \theta)\, Y(1, d\tau) \tag{5.2.4}$$

and

$$Z(s,t) = \pi Y(s,t) = Y(s,t) - \mathbf{g}(t,\theta)' \int_0^1 \mathbf{e}(\tau,\theta)\, Y(s, d\tau), \tag{5.2.5}$$

where

$$\mathbf{g}^{(1)}(t,\theta) = \nabla \log f(x,\theta), \mathbf{g}(t,\theta) = \int_0^t \mathbf{g}^{(1)}(s,\theta)\, ds$$

and

$$\mathbf{e}(t,\theta) = \mathbf{l}(x,\theta), \qquad t = F(x,\theta)$$

as in Section 5.1. Let

$$a(s,t) = \int_0^t a^{(1)}(s,\tau)\, d\tau \tag{5.2.6}$$

The following theorem can be proved by methods analogous to those given in Section 5.1 [cf. Khmaladze (1979)].

Theorem 5.2.1. If (B_1) and (B_2) hold, then

$$\bar{Z}_n \xrightarrow{\mathscr{L}} \bar{Z} + \bar{c} \quad \text{or} \quad Z_n \xrightarrow{\mathscr{L}} Z + c \quad \text{as } n \to \infty, \tag{5.2.7}$$

where

$$\bar{c} = \bar{\pi} a \quad \text{and} \quad c = \pi a. \tag{5.2.8}$$

Since the distribution of the random fields Z and $\bar{Z}$ depend on $\mathbf{g}(t,\theta)$ and hence on $\mathscr{F}$, one cannot use functionals of $\bar{Z}_n$ or Z_n for testing purposes. However, we can consider

$$\tilde{W}_n(s,t) = \frac{1}{\sqrt{n}} \sum_{i \leq [sn]} [I_{\{U_{in} \leq t\}} - M(t, U_{in})] \tag{5.2.9}$$

in analogy with $\tilde{W}_n$ defined in (5.1.78), where $M(t,\tau)$ is as defined in Lemma 5.1.7 [equation (5.1.54)]. This is a transformation of Z_n.

It turns out the random field $\tilde{W}_n(s,t)$ converges to a Wiener random field under the hypothesis (for the definition, see Section 2.2).

A *standard Wiener random field* $W(\cdot,\cdot)$ is a Gaussian random field with mean zero and covariance function $(s_1 \wedge s_2)(t_1 \wedge t_2)$. Let $\mathscr{F}_{st}^W$ be the σ-algebra generated by $W(s',t')$ with $s' \leqslant s$ and $t' \leqslant t$ and $\zeta_{st} = \mathscr{F}^W_{s1} \vee \mathscr{F}^W_{1t}$. Let

$$\Delta^2_{s't'} W(s,t) = W(s',t') - W(s',t) - W(s,t') + W(s,t).$$

Then

$$E[\Delta^2_{s't'} W(s,t) \mid \zeta_{st}] = 0,$$

that is, W is a *strong martingale*. For any $t \in [0,1]$, the process $\{W(s,t), \mathscr{F}^W_{s1}\}$ and for any $s \in [0,1]$, the process $\{W(s,t), \mathscr{F}^W_{1t}\}$ are martingales and W is a *bimartingale*.

Let us consider again the random field $Z(s,t)$ as defined by (5.2.5). The process $\{Z(s,t), \mathscr{F}^z_{s1}\}$ is a martingale for fixed $t \in [0,1]$, but the process $\{Z(s,t), \mathscr{F}^z_{1t}\}$ is *not* a martingale for fixed $s \in [0,1]$. Let

$$W(s,t) = Z(s,t) - \int_0^1 M(t,\tau)\, Z(s, d\tau), \tag{5.2.10}$$

where M is as defined in (5.1.54). Then W is a Wiener random field as described above and the correspondence between the trajectories of W and of Z is one to one.

Theorem 5.2.2. Under the sequence of alternative hypotheses specified by (B_1),

$$\tilde{W}_n \xrightarrow{\mathscr{L}} W + m \quad \text{as } n \to \infty, \tag{5.2.11}$$

where

$$m(s,t) = c(s,t) - \int_0^1 M(t,\tau) c^{(1)}(s,\tau)\, d\tau \tag{5.2.12}$$

and the relation between m and c is one to one. Under the hypothesis,

$$W_n \xrightarrow{\mathscr{L}} W \quad \text{as } n \to \infty. \tag{5.2.13}$$

We omit the proof of this theorem. It is similar to that of Theorem 5.1.9. The function m can also be written in the form

$$m(s,t) = a(s,t) - \int_0^1 M(t,\tau) a^{(1)}(s,\tau)\, d\tau,$$

where $a^{(1)}(\cdot,\cdot)$ is as defined by (B_1). Note that if $F_{in} = F$ for $1 \leqslant i \leqslant n$, that is, the hypothesis is simple, then

$$\tilde{W}_n(s,t) = Y_n(s,t) - \int_0^1 M(t,\tau)\, Y_n(s, d\tau)$$

$$= n^{-1/2} \sum_{i \leqslant [sn]} I_{\{U_{in} \leqslant t\}} + n^{-1/2} \sum_{i \leqslant [sn]} \log(1 - U_{in} \wedge t)$$

and it is a strong martingale for any fixed n.

Suppose ϕ is a functional of the parametric empirical field $Y_n^{\hat{\theta}}$ and the family $\mathscr{F}$ and it satisfies the relation

$$\phi(Y_n^{\hat{\theta}}, \mathscr{F}) = \psi(\tilde{W}_n],$$

where ψ is a functional continuous $P_{[0,1]^2}$-a.e. on $L_2[0,1] \times L_2[0,1]$, where $P_{[0,1]^2}$ is the Wiener measure induced by the Wiener field W on $L_2[0,1] \times L_2[0,1]$. Under the hypothesis, the limiting distribution of the random field $\phi[Y_n^{\hat{\theta}}, \mathscr{F})$ does not depend on $\mathscr{F}$, and it can be used as a test statistic for testing homogeneity.

5.3 DECOMPOSABLE STATISTICS

Let $(\nu_{n0}, \ldots, \nu_{nN})$ be a multivariate random vector with multinomial distribution with probability vector $(p_{n0}, \ldots, p_{nN})$ with n denoting the number of trials. Let $h(x,t)$ be a function defined on $\mathscr{N} \times [0,1]$, where $\mathscr{N} = \{0, 1, 2, \ldots\}$. A statistic of the form

$$\sum_{i=1}^{N} h\left(\nu_{ni}, \frac{i}{N}\right)$$

is called a *decomposable statistic*. Examples of such statistics are maximum likelihood statistic with

$$h\left(\nu_{ni}, \frac{i}{N}\right) = \nu_{ni} \log\left(\frac{\nu_{ni}}{n p_{ni}}\right)$$

and the χ^2 statistic with

$$h\left(\nu_{ni}, \frac{i}{N}\right) = \frac{(\nu_{ni} - n p_{ni})^2}{n p_{ni}}.$$

Let

$$X_{nN}(s) = N^{-1/2} \sum_{i \leqslant [sN]} \left\{ h\left(v_{ni}, \frac{i}{N} \right) - Eh\left(v_{ni}, \frac{i}{N} \right) \right\} \tag{5.3.1}$$

for $0 \leqslant s \leqslant 1$. This process is a semimartingale (cf. Section 1.8) with respect to the flow of σ-algebras $\{\mathscr{F}_i^n, 1 \leqslant i \leqslant N\}$ where $\mathscr{F}_i^n$ is the σ-algebra generated by $v_{n0}, \ldots, v_{n,(i-1)}$. Khmaladze (1983) studied limit theorems for the martingale component of $X_{nN}(s)$ using limit theorems for semimartingales [cf. Shiryayev (1981) and Liptser and Shiryayev (1980)]. We discuss a special case of his result.

Let $Y_1, \ldots, Y_n$ be i.i.d. random variables with an absolutely continuous distribution function F and density f. Suppose that $F(0) = 0$, $F(1) = 1$, and $F(t) < 1$ for $t < 1$. Let $F_n(\cdot)$ be the empirical distribution function based on $Y_1, \ldots, Y_n$. Define

$$v_{ni} = n \Delta F_n\left(\frac{i}{n} \right) = n \left[F_n\left(\frac{i+1}{n} \right) - F_n\left(\frac{i}{n} \right) \right]. \tag{5.3.2}$$

v_{ni} is the normalized increment over the interval $[i/n, (i+1)/n]$. Let $\mathscr{F}_i^n$ denote the σ-algebra generated by $Y_1, \ldots, Y_{i-1}$. We assume that

(5.3.3) (i) $|h(x,t)| < ce^{ax}$ for some $c > 0$ and $a > 0$,

(ii) $h(x,t)$ is left continuous in t for each x, that is, $h(x,t) = h(x, t-)$.

In order to discuss the asymptotic behavior of the process $X_{nN}(\cdot)$, we assume that $N + 1 = n$ for simplicity and write X_n for $X_{n(n+1)}$ in (5.3.1). The Doob decomposition for the semimartingale $\{X_n(i/n), \mathscr{F}_{i+1}^n\}$ gives

$$X_n = K_n + W_n, \tag{5.3.4}$$

where

$$W_n(s) = n^{-1/2} \sum_{i \leqslant [sn]} \left\{ h\left(v_{ni}, \frac{i}{n} \right) - E\left[h\left(v_{ni}, \frac{i}{n} \right) \middle| \mathscr{F}_i^n \right] \right\} \tag{5.3.5}$$

and

$$\begin{aligned} K_n(s) &= n^{-1/2} \sum_{i \leqslant [sn]} \left\{ E\left[h\left(v_{ni}, \frac{i}{n} \right) \middle| \mathscr{F}_i^n \right] - E\left[h\left(v_{ni}, \frac{i}{n} \right) \right] \right\} \\ &= n^{-1} \sum_{i \leqslant [sn]} n^{1/2} \left\{ E\left[h\left(v_{ni}, \frac{i}{n} \right) \middle| \mathscr{F}_i^n \right] - E\left[h\left(v_{ni}, \frac{i}{n} \right) \right] \right\}. \end{aligned} \tag{5.3.6}$$

Note that

$$P[\nu_{ni}=x|\mathscr{F}_i^n]=b\left(x,\frac{\Delta F(i/n)}{1-F(i/n)},\quad n\left(1-F_n\left(\frac{i}{n}\right)\right)\right), \tag{5.3.7}$$

where

$$b(x,p,m)=\binom{m}{x}p^x(1-p)^{m-x}. \tag{5.3.8}$$

Let

$$\tau(s)=\int_0^s \sigma^2(t)\,dt, \tag{5.3.9}$$

where $\sigma^2(t)=\mathrm{Var}[h(Z,t)]$ and Z is Poisson with mean $f(t)$. Let W^τ denote the Wiener process on $[0,1]$ with respect to time τ.

Theorem 5.3.1. If $\|f\|=\sup|f(x)|<\infty$, then

$$W_n\xrightarrow{\mathscr{L}}W^\tau \tag{5.3.10}$$

in $D[0,1]$ as $n\to\infty$.

We first state a few technical lemmas. For proofs, see Khmaladze (1982, 1983).

Lemma 5.3.2. Let

$$b(x,p,m)=\binom{m}{x}p^x(1-p)^{m-x}\quad\text{and}\quad \pi(x,\lambda)=\frac{\lambda^x}{x!}e^{-\lambda}. \tag{5.3.11}$$

If $\lambda=mp$ and $h(x,t)$ satisfies (5.3.3), then there exists $c>0$ such that

$$\sum_{x\in A}|h(x,p)|\,b(x,p,m)\leqslant ce^{\lambda e^a}\sum_{x\in A}\pi(x,\lambda e^a). \tag{5.3.12}$$

Lemma 5.3.3. Let $p_n(t)$ and $m_n(t)$ be nonnegative functions for $t\in[0,1]$, $n\geqslant 1$, when $m_n(\cdot)$ is integer-valued. Suppose $[T_{1n},T_{2n}]\subset[0,1]$ and the following relations hold:

$$\text{(i)}\quad \lim_{n\to\infty} m_n=\infty\quad\text{where } m_n=\inf_{t\in[T_{1n},T_{2n}]} m_n(t), \tag{5.3.13}$$

(ii) $\overline{\lim_{n\to\infty}} \sup_{t\in[T_{1n},T_{2n}]} m_n(t)p_n(t) < \lambda < \infty.$

Let $\lambda_n(t) = m_n(t)p_n(t)$, and suppose $n = O(m_n)$ as $n \to \infty$ and (5.3.3) holds. Then

$$\sqrt{n} \sup_{t\in[T_{1n},T_{2n}]} \left| \sum_{x\geq 0} h(x,t)[b(x,p_n(t),m_n(t)) - \pi(x,\lambda_n(t))] \right| \to 0. \tag{5.3.14}$$

Lemma 5.3.4. Suppose λ_n and μ_n are functions defined on $[0,1]$ such that

(5.3.15) (i) $0 \leq \lambda_n(t) < \lambda, \quad 0 \leq \mu_n(t) < \lambda, \qquad t \in [T_{1n}, T_{2n}],$

(ii) $\lambda_n(t) \to \lambda(t), \quad \mu_n(t) \to \mu(t) \quad \text{as} \quad n \to \infty, \quad t \in [T_{1n}, T_{2n}].$

Furthermore, suppose that relation (5.3.3) holds. If (i) holds, then

$$\sup_{t\in[T_{1n},T_{2n}]} \left| \frac{1}{\lambda_n(t) - \mu_n(t)} \sum_{x\geq 0} h(x,t)[\pi(x,\lambda_n(t)) - \pi(x,\mu_n(t))] \right| \leq c(e^a + 1)\exp(\lambda e^a + 1); \tag{5.3.16}$$

and if (i) and (ii) hold, then, as $n \to \infty$,

$$\frac{1}{\lambda_n(t) - \mu_n(t)} \sum_{x\geq 0} h(x,t)[\pi(x,\lambda_n(t)) - \pi(x,\mu_n(t))] \to \sum_{x\geq 0} \Delta h(x,t)\pi(x,\lambda(t)). \tag{5.3.17}$$

Proof of Theorem 5.3.1. Let

$$\xi_{ni} = h\left(v_{ni}, \frac{i}{n}\right) \quad \text{and} \quad \eta_i = E\left[h\left(v_{ni}, \frac{i}{n}\right) \middle| \mathscr{F}_i^n \right]. \tag{5.3.18}$$

In view of Corollary 1.8.3 [cf. Liptser and Shiryayev (1980, Corollary 6)], the theorem follows if we prove the following assertions:

(a) for any $\varepsilon > 0$,

$$n^{-1} \sum_{i=1}^{n} E[(\xi_i - \eta_i)^2 I\{(\xi_i - \eta_i)^2 > \varepsilon n\} | \mathscr{F}_i^n] \xrightarrow{p} 0 \quad \text{as } n \to \infty \tag{5.3.19}$$

and

$$\text{(5.3.20) (b)} \quad n^{-1} \sum_{i \leq [sn]} E[(\xi_i - \eta_i)^2 | \mathscr{F}_i^n] \xrightarrow{p} \tau(s) \quad \text{as } n \to \infty$$

Let

$$\lambda_n(t) = n \frac{1 - \hat{F}_n(t)}{1 - F(t)} \Delta F(t), \tag{5.3.21}$$

where

$$\Delta F(t) = F(t + 1/n) - F(t). \tag{5.3.22}$$

Note that

$$\begin{aligned} \sup_{0 \leq t \leq 1} \lambda_n(t) &\leq \sup_{0 \leq t \leq 1} \frac{1 - \hat{F}_n(t)}{1 - F(t)} \|f\| = \lambda_n \quad \text{(say)} \\ &= o_p(1), \end{aligned} \tag{5.3.23}$$

since

$$R_n = \sup_{0 \leq t \leq 1} \frac{1 - \hat{F}_n(t)}{1 - F(t)} = o_p(1). \tag{5.3.24}$$

In view of Lemma 5.3.2 and relation (5.3.7), it follows that

$$E\left[\xi_i^2 I\left\{\xi_i^2 > \frac{\varepsilon n}{2}\right\} \middle| \mathscr{F}_i^n\right] \leq c^2 \exp(\lambda_n e^{2a}) \sum_{x \geq c_n} \pi(x, \lambda_n e^a), \tag{5.3.25}$$

where

$$c_n = (2a)^{-1} \log(n\varepsilon/2c^2). \tag{5.3.26}$$

Similarly, we have

$$\eta_i^2 E\left[I\left\{\xi_i^2 > \frac{\varepsilon n}{2}\right\} \middle| \mathscr{F}_i^n\right] \leq c^2 \exp(2\lambda_n e^a) \sum_{x \geq c_n} \pi(x, \lambda_n e^a) \tag{5.3.27}$$

and

$$\eta_i^2 I\left\{\eta_i^2 > \frac{\varepsilon n}{2}\right\} \leq E\{\xi_i^2 | \mathscr{F}_i^n\} I\left\{\eta_i^2 > \frac{\varepsilon n}{2}\right\} \leq c^2 \exp(\lambda_n e^{2a}) I(\lambda_n > a e^a c_n). \tag{5.3.28}$$

The expressions on the right-hand sides of the inequalities (5.3.25), (5.3.27), and (5.3.28) do not depend on i, and hence the inequalities hold uniformly in i. Furthermore, the expressions cited tend to zero in probability as $n \to \infty$. In view of the inequality

$$I\{(\xi_i - \eta_i)^2 > \varepsilon n\} \leqslant I\left\{\xi_i^2 > \frac{\varepsilon n}{2}\right\} + I\left\{\eta_i^2 > \frac{\varepsilon n}{2}\right\} \tag{5.3.29}$$

and the Cauchy–Schwarz inequality, it follows that

$$\max E[(\xi_i - \eta_i)^2 I\{(\xi_i - \eta_i)^2 > \varepsilon n\} | \mathscr{F}_i^n] = o_p(1) \quad \text{as } n \to \infty \tag{5.3.30}$$

proving in particular (5.3.19). We now prove (5.3.20). Let

$$\phi_n(t) = E\{(\xi_{[tn]} - \eta_{[tn]})^2 | \mathscr{F}_{[tn]}^n\}. \tag{5.3.31}$$

Since

$$\lambda_n([tb]/n) \to f(t) \quad \text{a.s. as } n \to \infty \tag{5.3.32}$$

and

$$h(x, [tn]/n) \to h(x, t) \tag{5.3.33}$$

by hypothesis, it follows from Lemmas 5.3.3 and 5.3.4 that

$$\phi_n(t) \to \sigma^2(t) \quad \text{a.s.} \tag{5.3.34}$$

for almost all $t < 1$. In view of Lemma 5.3.2 and the inequality

$$\begin{aligned} \sup_{0 \leqslant t \leqslant \tau} \lambda_n(t) &\leqslant \sup_{0 \leqslant t \leqslant T} \frac{1 - \hat{F}_n(t)}{1 - F(t)} \left(\| f \| + \frac{\varepsilon}{2} \right) \\ &\leqslant \| f \| + \varepsilon, \end{aligned} \tag{5.3.35}$$

which holds for large n and all $T < 1$, it is clear that

$$0 \leqslant \phi_n(t) \leqslant E\{\xi_{[tn]}^2 | \mathscr{F}_{[tn]}^n\} \leqslant c^2 \exp\{\| f \| e^{2a} + \varepsilon\} \tag{5.3.36}$$

for $t \leqslant T$ and sufficiently large n. Hence, by the dominated convergence theorem, almost surely,

$$\int_0^{t\wedge T} |\phi_n(s) - \sigma^2(s)|\,ds \to 0 \quad \text{as } n\to\infty \tag{5.3.37}$$

for $T<1$. However, for every t,

$$\text{(5.3.38)}\quad \text{(i)}\quad \phi_n(t) \leqslant c^2 \exp[R_n(e^{2a}\|f\| + \varepsilon)],$$

$$\text{(ii)}\quad \sigma^2(t) \leqslant c^2 \exp[\|f\|(e^{2a}-1)],$$

and hence, as $n\to\infty$ and $\varepsilon > 1-T$,

$$\begin{aligned}\text{(5.3.39)}\quad &\left|\int_0^t [\phi_n(s)-\sigma^2(s)]ds - \int_0^{t\wedge T}[\phi_n(s)-\sigma^2(s)]ds\right|\\ &\leqslant c^2(\exp[R_n(e^{2a}\|f\|+\varepsilon)] + \exp[\|f\|(e^a-1)])(1-T)\\ &\leqslant \varepsilon O_p(1)\end{aligned}$$

and the proof of the theorem is complete.

Let us now discuss heuristically the asymptotic behavior of the compensator K_n. Let $E_\lambda h(Z,t)$ be the expectation of $h(Z,t)$ when Z is Poisson with mean λ. The expression K_n given by (5.3.6) can be approximated by

$$\tilde{K}_n = \sqrt{n}\left[E_{\lambda_n(\frac{i}{n})} h\left(Z,\frac{i}{n}\right) - E_{\mu_n(\frac{i}{n})} h\left(Z,\frac{i}{n}\right)\right], \tag{5.3.40}$$

where

$$\text{(5.3.41)}\quad \text{(i)}\quad \lambda_n(t) = n\frac{1-\hat{F}_n(t)}{1-F(t)}\Delta F(t), \qquad \Delta F(t) = F\left(t+\frac{1}{n}\right) - F(t),$$

$$\text{(ii)}\quad \mu_n(t) = n\,\Delta F(t).$$

Since $\lambda_n(t)$ and $\mu_n(t)$ converge to $f(t)$ as $n\to\infty$, $\tilde{K}_n$ can be approximated by

$$K_n^* = \frac{d}{d\lambda}E_\lambda h\left(Z,\frac{i}{n}\right)\Big|_{\lambda=f(t)} \sqrt{n}\left[\lambda_n\left(\frac{i}{n}\right) - \mu_n\left(\frac{i}{n}\right)\right]. \tag{5.3.42}$$

But

$$\sqrt{n}[\lambda_n(t) - \mu_n(t)] = \frac{-V_n^F(t)}{1-F(t)} n\Delta F(t), \tag{5.3.43}$$

where

$$V_n^F(t) = \sqrt{n}[\hat{F}_n(t) - F(t)]. \tag{5.3.44}$$

Hence K_n can be approximated by the process

$$\bar{K}_n = \int_0^t \frac{V_n^F(s)}{1 - F(s)} E_{f(s)} h(Z, s) f(s) ds, \tag{5.3.45}$$

and one can obtain asymptotic behavior of K_n from that of $\bar{K}_n$, which in turn can be studied using the behavior of empirical process. The following theorem is proved in Khmaladze (1983). We omit the proof. We introduce some notation.

Let $\mathscr{K}$ be the operator defined by

$$\mathscr{K} x(t) = -\int_0^t k(s) x(s)\, ds, \tag{5.3.46}$$

where

$$k(s) = f(s) E_{f(s)} h(Z, s)(Z - f(s)) \tag{5.3.47}$$

and define the process

$$K = \mathscr{K} \frac{V}{1 - F}, \tag{5.3.48}$$

where $V \circ F = V$ denotes the Brownian bridge with respect to time $F(t)$.

Theorem 5.3.5. Suppose that $\| f \| < \infty$ and for any $\varepsilon > 0$ and $T_n = 1 - \varepsilon n^{-1/2}$,

$$n^{1/2}(1 - F(T_n)) \to \delta(\varepsilon) > 0 \quad \text{as } n \to \infty. \tag{5.3.49}$$

Then

$$K_n \xrightarrow{\mathscr{L}} K \quad \text{as } n \to \infty. \tag{5.3.50}$$

In order to study the limiting behavior of X_n defined by (5.3.4), it is not sufficient to obtain the limit theorems separately for the martingale W_n and compensator K_n. However, Khmaladze (1983) has proved that X_n converges in distribution to an Ito process with W^τ defined by Theorem 5.3.1 as its martingale component. We now state this result. For proof, see Khmaladze (1983).

Let (V, W) be a two-dimensional Gaussian process adapted to the flow of σ-algebra $\{\mathscr{F}_t^V, 0 \leqslant t \leqslant 1\}$ generated by the process V, where the covariance function of V is $F(t \wedge s) - F(t)F(s)$, the covariance function of W is $(t \wedge s)$, and the mutual covariance function of V and W is

$$EV(t)W(s) = [1 - F(t)] \int_0^{t \wedge s} \frac{k(y)}{1 - F(y)} dy, \tag{5.3.51}$$

where $k(\cdot)$ is as defined in (5.3.47).

Theorem 5.3.6. Suppose the conditions stated in Theorem 5.3.5 hold. Then

$$X_n \xrightarrow{\mathscr{L}} X \quad \text{as } n \to \infty, \tag{5.3.52}$$

where

$$X(t) = -\int_0^t k(s) \frac{V(s)}{1 - F(s)} ds + W(t), \qquad 0 \leqslant t \leqslant 1. \tag{5.3.53}$$

Remarks. As in Section 5.1, one can construct one-to-one mappings of the processes X_n into processes converging in distribution to a Wiener process. The usefulness of these mappings is that, in view of the one-to-one nature, distinguishing hypotheses on the distribution of the processes X_n and distinguishing them on the distribution of the transformed processes are equivalent. Since the limit distribution of the transformed processes are convenient to compute, one can distinguish the hypothesis in the transformed case much more easily.

The weak convergence of the process X_n is discussed in Theorem 5.3.6 under the null hypothesis. Khmaladze (1983) also discussed corresponding weak convergence of the process X_n under an alternative hypothesis contiguous to the null hypothesis. The techniques are essentially the same.

Note that the process X has the covariance function

$$\tau(t \wedge s) - \int_0^t k(y)\,dy \int_0^s k(y)\,dy.$$

Let

$$\tilde{X}(t) = \int_0^t \frac{k(y)}{\sigma^2(y)} dX(y). \tag{5.3.54}$$

Then covariance function of the process $\tilde{X}$ is $\rho(t \wedge s) - \rho(t)\rho(s)$, where

$$\rho(t) = \int_0^t r(s)\,ds, \qquad r(t) = \frac{k^2(t)}{\sigma^2(t)}. \tag{5.3.55}$$

Here $\tilde{X}$ is a Brownian bridge with respect to time ρ. In view of results in Section 1.9, $\tilde{X}$ has a stochastic differential of the form

$$dX(t) = -\frac{\tilde{X}(t)}{1-\rho(t)}\,d\rho(t) + d(\bar{W}\circ\rho)(t), \tag{5.3.56}$$

where $\bar{W}$ is a Wiener process with respect to time τ. Relations (5.3.54) and (5.3.56) prove that

$$dX(t) = -\frac{k(t)}{1-\rho(t)}\int_0^t \frac{k(s)}{\sigma^2(s)}\,dX(s) + d\bar{W}(t), \tag{5.3.57}$$

which is the stochastic differential equation satisfied by the process X. It can be seen that there is a one-to-one correspondence between the trajectories of X and W.

Let us consider the process $\bar{W}^n$-defined by

$$\Delta\bar{W}^n\left(\frac{i}{n}\right) = \Delta X_n\left(\frac{i}{n}\right) + \frac{k(i/n)}{n - \sum_{j\leqslant i} r(j/n)} \sum_{j\leqslant i} \frac{k(j/n)}{\sigma^2(j/n)}\,\Delta X_n\left(\frac{j}{n}\right), \tag{5.3.58}$$

where $\Delta Y(i/n) = Y[(i+1)/n] - Y(i/n)$ for any process Y in view of relation (5.3.57). It can be shown that if

$$X_n \xrightarrow{\mathscr{L}} X + m,$$

then

$$\bar{W}^n \xrightarrow{\mathscr{L}} W + m_1,$$

where

$$m_1^{(1)}(t) = m^{(1)}(t) + \frac{k(t)}{1-\rho(t)}\int_0^t \frac{k(s)}{\sigma^2(s)}\,m^{(1)}(s)\,ds. \tag{5.3.59}$$

Relation (5.3.58) gives the required transformation of the process X_n.

Observe that $r(t) \leqslant f(t)$ and hence $\rho(1) \leqslant 1$. If $\rho(1) \leqslant 1$, then one can give another transformation of the process X_n that is also one to one. We now describe these results.

Lemma 5.3.7. If $\rho(1) < 1$, then the process

$$b(t) = X(t) - \beta \int_0^t k(y)\,dy \int_0^1 \frac{k(y)}{\sigma^2(y)}\,dX(y) \tag{5.3.60}$$

with

$$\beta = \frac{1+\sqrt{1-\rho(1)}}{\rho(1)\sqrt{1-\rho(1)}} \quad \text{or} \quad \beta = \frac{1-\sqrt{1-\rho(1)}}{\rho(1)\sqrt{1-\rho(1)}} \tag{5.3.61}$$

is a Wiener process with respect to time τ and the correspondence between $b(\cdot)$ and $X(\cdot)$ is one to one.

Proof. The process $\tilde{X}$ can be represented in the form

$$\tilde{X}(t) = (b \circ \rho)(t) - \alpha\rho(t)(b \circ \rho)(1) \tag{5.3.62}$$

since $\tilde{X}$ is a Brownian bridge with respect to time ρ, where α is a solution of the equation

$$\rho(1)\alpha^2 - 2\alpha + 1 = 0. \tag{5.3.63}$$

If $\rho(1) < 1$, then $\alpha \neq 1$, and hence the representation (5.3.62) is invertible and

$$(b \circ \rho)(t) = \tilde{X}(t) - \beta\rho(t)\tilde{X}(1), \tag{5.3.64}$$

where $\beta = \alpha[\alpha\rho(1) - 1]^{-1}$. In view of (5.3.54) and the relation

$$(b \circ \rho)(t) = \int_0^t \frac{k(y)}{\sigma(y)}\,db(y), \tag{5.3.65}$$

we obtain the lemma.

Let us now consider the transformation

$$b_n(t) = X_n(t) - \frac{1}{n}\beta \sum_{1 \leqslant nt} k\left(\frac{i}{n}\right) \sum_{i \leqslant n} \frac{k(i/n)}{\sigma^2(i/n)} \Delta X_n\left(\frac{i}{n}\right). \tag{5.3.66}$$

It can be shown that if $X_n \xrightarrow{\mathscr{L}} X + m$, then $b_n \xrightarrow{\mathscr{L}} b + m^*$, where

(5.3.67)
$$m^*(t) = m(t) - \beta \int_0^t k(y)\,dy \int_0^1 \frac{k(y)}{\sigma^2(y)} m^{(1)}(y)\,dy.$$

Thus, relation (5.3.66) yields yet another transformation for the process X_n, which preserves the one-to-one property in case $\rho(1) < 1$.

Similar results can be obtained for statistics involving spacings as was done in Mnatsakanov (1983). Let $X_1, \ldots, X_n$ be i.i.d. with absolutely continuous distribution function F and density f. Denote by $X_{(1)}, \ldots, X_{(n)}$ the corresponding order statistics. Statistics of the form

$$\sum_{i=1}^{n} h\left(\Delta X_{(i)}, \frac{i}{n}\right)$$

play an important role in nonparametric statistics where $\Delta X_{(i)} = X_{(i+1)} - X_{(i)}$. Let $\mathscr{F}_i^{(n)}$ be the σ-algebra generated by $X_{(1)}, \ldots, X_{(i)}$, and define

$$Y_n(s) = n^{-1/2} \sum_{i \leqslant [sn]} \left\{ h\left(\Delta X_{(i)}, \frac{i}{n}\right) - Eh\left(\Delta X_{(i)}, \frac{i}{n}\right) \right\}.$$

Observe that

$$P\{\Delta X_{(i)} > \lambda \mid \mathscr{F}_i^{(n)}\} = \left\{ \frac{1 - F(\lambda + X_{(i)})}{1 - F(X_{(i)})} \right\}^{n-i},$$

and one can obtain the limit theorems for martingale component of $\{Y_n(i/n), \mathscr{F}_{i+1}^n\}$ as before [cf. Mnatsakanov (1983)].

5.4 CHI-SQUARE AND OMEGA-SQUARE STATISTICS

In Section 5.3 we described asymptotic behavior of the decomposible statistics, which include the χ^2 statistic in particular. Standard statistics of the χ^2 type are defined in terms of cells that are fixed prior to taking observations. Probabilities that observation fall into these cells might depend on unknown parameters. The asymptotic distribution of the χ^2 statistic depends on the type of estimators used for estimating these parameters. If maximum likelihood estimators are used assuming that the full sample is available, the asymptotic distribution of the χ^2 statistic need not be chi-square distribution, as shown by Chernoff and Lehmann (1954). In fact, if M cells are used and m parameters are estimated, then the asymptotic distribution under the null hypothesis is that of

(5.4.0)
$$\sum_{i=1}^{M-m-1} Z_i^2 + \sum_{i=M-m}^{M-1} \lambda_i Z_i^2,$$

where $Z_1, \ldots, Z_{M-1}$ are independent standard normal random variables and the λ's, which may *depend* on the parameters, lie between 0 and 1. In other words, the asymptotic distribution is not distribution-free, as mentioned in Section 5.3. In practice, the cell-boundaries are likely to depend on the estimated parameter values. For univariate observations, Roy (1956) obtained the asymptotic distribution of the χ^2 statistic under the null hypothesis and he proved that it is the same as (5.4.0) under some regularity conditions. He proved that the asymptotic distribution (5.4.0) is distribution-free in case the family under consideration is either location or scale parameter and the cell boundaries are properly chosen. Our approach here is due to Moore (1971) in the multiparameter case.

Let $\mathbf{X}_i$, $1 \leqslant i \leqslant n$, be i.i.d. k-dimensional random vectors with distribution function $F(\mathbf{x}, \theta)$, $\theta \in \Theta \subset R^p$. We assume that the following regularity conditions hold:

(A_0) Θ is open set;

(A_1) $F(\mathbf{x}, \theta)$ has a density $f(\mathbf{x}, \theta)$ that is continuous in $(\mathbf{x}, \theta)$ and continuously differentiable in θ;

(A_2) $$\frac{\partial}{\partial \theta_i} \int f(\mathbf{x}, \theta) d\mathbf{x} = \int \frac{\partial}{\partial \theta_i} f(\mathbf{x}, \theta) d\mathbf{x}, \qquad 1 \leqslant i \leqslant p;$$

(A_3) $$J_{ij}(\theta) = E_\theta \left[\frac{\partial \log f}{\partial \theta_i} \frac{\partial \log f}{\partial \theta_j} \right] < \infty$$

for $1 \leqslant i, j \leqslant p$ and the matrix $\mathbf{J}(\theta) = ((J_{ij}(\theta))$ is positive definite for all $\theta \in \Theta$.

Partition the x_i axis by functions of θ

$$-\infty \equiv \xi_{i0}(\theta) < \xi_{i1}(\theta) < \cdots < \xi_{i, v_i - 1}(\theta) < \xi_{i, v_i}(\theta) \equiv +\infty$$

for $1 \leqslant i \leqslant k$. This subdivision gives rectangular cells in R^k. Assume that

(A_4) $\dfrac{\partial \xi_{ij}}{\partial \theta_s}$ exist and are continuous in θ for $1 \leqslant i \leqslant k$, $1 \leqslant j \leqslant v_i$, and $1 \leqslant s \leqslant p$.

The partition of R_k has $M = \prod_{i=1}^k v_i$ cells. We index the cells of the partition by σ running from 1 to M in arbitrary but fixed assignment. Let $p_\sigma(\theta)$ be the probability that an observative falls into cell indexed by σ. It is easy to see that

$$p_\sigma(\theta) = \Delta_\sigma^\theta F(\mathbf{x}, \theta), \tag{5.4.1}$$

where Δ_σ^θ is the difference operator in R^k. We assume that

$$p_\sigma(\theta) > 0 \quad \text{for all } \theta \in \Theta.$$

(A_5) Suppose $M > p$ and the matrix $\mathbf{W} = ((w_{i\sigma}))_{p \times M}$ with

$$w_{i\sigma} = \frac{\partial}{\partial\theta} \Delta_\sigma^\theta F(\mathbf{x}, \theta), \qquad 1 \leqslant i \leqslant p, \quad 1 \leqslant \sigma \leqslant m, \tag{5.4.2}$$

has rank p.

Let N_σ be the number of observations $\mathbf{X}_1, \ldots, \mathbf{X}_n$ in the σth cell and $\hat{\theta}_n$ be a sequence of estimators of θ. We *allow* the observation to choose the cells by replacing θ in $\xi_{ij}(\theta)$ by $\hat{\theta}_n$ and consider the χ^2 statistic so obtained, namely,

$$T_n = \sum_{\sigma=1}^{M} \frac{[N_\sigma - np_\sigma(\hat{\theta}_n)]^2}{np_\sigma(\hat{\theta}_n)}. \tag{5.4.3}$$

Furthermore, we assume that

(A_6) there exists a nonsingular matrix $\mathbf{B}(\theta)$ of order $p \times p$ such that

$$n^{1/2}(\hat{\theta}_n - \theta) = n^{1/2} \sum_{i=1}^{n} \mathbf{B}(\theta)\mathbf{A}(\mathbf{X}_i, \theta) + o_p(1), \tag{5.4.4}$$

where

$$\mathbf{A}(\mathbf{X}, \theta)' = \left(\frac{\partial \log f(\mathbf{X}, \theta)}{\partial \theta_1}, \ldots, \frac{\partial \log f(\mathbf{X}, \theta)}{\partial \theta_p} \right). \tag{5.4.5}$$

We now state the main result of this section.

Theorem 5.4.1. Suppose that the regularity conditions (A_0)–(A_6) hold and

(A_7) the matrix $\mathbf{B} + \mathbf{B}' - \mathbf{BJB}'$ is positive definite for all $\theta \in \Theta$. Then

$$T_n \xrightarrow{\mathscr{L}} \sum_{i=1}^{M-p-1} Z_i^2 + \lambda_1 Z_{M-p}^2 + \cdots + \lambda_p Z_{M-1}^2, \tag{5.4.6}$$

where $Z_1, \ldots, Z_{M-1}$ are i.i.d. $N(0, 1)$ and λ_j, $1 \leqslant j \leqslant p$, may depend on θ but $0 \leqslant \lambda_j < 1$.

Proof. Let $I_\sigma(\theta)$ be the σth cell of the partition generated by $\xi_{ij}(\theta)$. N_σ will be the number of observations $\mathbf{X}_1, \ldots, \mathbf{X}_n$ that fall in the cell $I_\sigma(\hat{\theta}_n)$. Let n_σ be the number of observations that fall in $I_\sigma(\theta_0)$ among $\mathbf{X}_1, \ldots, \mathbf{X}_n$ when θ_0 is the true parameter. Let $F_n(\mathbf{x})$ be the empirical distribution function. Note that

(5.4.7)
$$N_\sigma - np_\sigma(\hat{\theta}_n) = n[\Delta_\sigma^{\hat{\theta}} F_n(\mathbf{x}) - \Delta_\sigma^{\hat{\theta}} F(\mathbf{x}, \hat{\theta}_n)]$$

and

(5.4.8)
$$n_\sigma - np_\sigma(\theta_0) = n[\Delta_\sigma^{\theta_0} F_n(\mathbf{x}) - \Delta_\sigma^{\theta_0} F(\mathbf{x}, \theta_0)].$$

Let

(5.4.9)
$$W_n(\mathbf{x}) = n^{1/2}[F_n(\mathbf{x}) - F(\mathbf{x}, \theta_0)].$$

Then

(5.4.10)
$$\begin{aligned} n^{-1/2}[N_\sigma - np_\sigma(\hat{\theta}_n)] &= n^{-1/2}[n_\sigma - np_\sigma(\theta_0)] \\ &\quad + [\Delta_\sigma^{\hat{\theta}} W_n(\mathbf{x}) - \Delta_\sigma^{\theta_0} W_n(\mathbf{x})] \\ &\quad - n^{1/2} \Delta_\sigma^{\hat{\theta}} [F(\mathbf{x}, \hat{\theta}_n) - F(\mathbf{x}, \theta_0)]. \end{aligned}$$

We shall prove that

(5.4.11)
$$\Delta_\sigma^{\hat{\theta}} W_n(\mathbf{x}) - \Delta_\sigma^{\theta_0} W_n(\mathbf{x}) = o_p(1).$$

Note that $\hat{\theta}_n - \theta_0 \xrightarrow{p} 0$ and $\xi_{ij}(\theta)$ are continuous in θ. Hence (5.4.11) holds provided $\boldsymbol{\eta}_n \xrightarrow{p} \mathbf{c}$ implies that $W_n(\boldsymbol{\eta}_n) - W_n(\mathbf{c}) \xrightarrow{p} 0$.

Let $\mathbf{H}: R^k \to [0, 1]^k$ be continuous such that $\mathbf{Y} = \mathbf{H}(\mathbf{X})$ has a uniform marginals in each direction. Let U be the distribution function of $\mathbf{Y}$ and U_n be the empirical distribution function. Let

$$W_n^*(\mathbf{u}) = n^{1/2}(U_n(\mathbf{u}) - U(\mathbf{u})).$$

It is well known that $W_n^* \xrightarrow{\mathscr{L}} W^*$ on D_k, where D_k is the space of functions defined on the unit cube having utmost jump discontinuities [one can define a metric on D_k of Skorokhod type so that D_k is a complete separable metric space; see Section 1.10 and Neuhaus (1971)], where W^* is a Gaussian process and $W^*(u)$ is continuous a.s. Since $\mathbf{H}(\boldsymbol{\eta}_n) \to \mathbf{H}(\mathbf{c})$ a.s., it it easy to see that $(W_n^*, \mathbf{H}(\boldsymbol{\eta}_n)) \xrightarrow{\mathscr{L}} (W^*, \mathbf{H}(\mathbf{c}))$ on the space $D_k \times R^k$ [cf. Billingsley (1968, Theorem 4.1)]. It can now be checked that

$$W_n^*(\mathbf{H}(\boldsymbol{\eta}_n)) - W_n^*(\mathbf{H}(\mathbf{c})) \xrightarrow{p} 0$$

by using the continuous mapping theorem [cf. Moore (1971)] in Billingsley (1968, Theorem 5.1). Hence

$$W_n(\boldsymbol{\eta}_n) - W_n(\mathbf{c}) \xrightarrow{p} 0. \tag{5.4.12}$$

This proves (5.4.11). Therefore

$$\begin{aligned} n^{-1/2}[N_\sigma - np_\sigma(\hat{\theta}_n)] = n^{-1/2}&[n_\sigma - np_\sigma(\theta_0)] \\ &- (\Delta_\sigma^{\theta} \partial F)' n^{1/2}(\hat{\theta}_n - \theta_0) + o_p(1), \end{aligned} \tag{5.4.13}$$

where

$$\partial F' = \left(\frac{\partial F}{\partial \theta_1}, \ldots, \frac{\partial F}{\partial \theta_p} \right)\Bigg|_{\theta = \theta_0} \tag{5.4.14}$$

and

$$\Delta_\sigma^{\theta} \partial F = \left(\Delta_\sigma^{\theta} \frac{\partial F}{\partial \theta_1}, \ldots, \Delta_\sigma^{\theta} \frac{\partial F}{\partial \theta_p} \right). \tag{5.4.15}$$

But

$$(\Delta_\sigma^{\theta} \partial F)' n^{1/2}(\hat{\theta}_n - \theta_0) - (\Delta_\sigma^{\theta_0} \partial F)' n^{1/2}(\hat{\theta}_n - \theta_0) = o_p(1) \tag{5.4.16}$$

since $n^{1/2}(\hat{\theta}_n - \theta_0) = O_p(1)$ by assumption (A_6) and ∂F is continuous. On the other hand,

$$\begin{aligned} \Delta_\sigma^{\theta_0}\left[\frac{\partial F(\mathbf{x}, \theta)}{\partial \theta_s} \right] &= \frac{\partial}{\partial \theta_s} \Delta_\sigma^{\theta_0} F(\mathbf{x}, \theta) \\ &= \frac{\partial}{\partial \theta_s} \int_{I_\sigma(\theta_0)} f(\mathbf{x}, \theta) d\mathbf{x} \\ &= \int_{I_\sigma(\theta_0)} \frac{\partial f(\mathbf{x}, \theta)}{\partial \theta_s} d\mathbf{x} \end{aligned} \tag{5.4.17}$$

by (A_2). Let $\mathbf{W}_\sigma(\theta)$ be the σth column vector of the matrix W defined by assumption (A_5). Applying, assumption (A_6) and relations (5.4.13), (5.4.16), and (5.4.17), we have

$$\text{(5.4.18)}\quad n^{1/2}[N_\sigma - np_\sigma(\hat{\theta}_n)] = n^{-1/2}\left[n_\sigma - np_\sigma(\theta_0) - \mathbf{W}_\sigma(\theta_0)'\mathbf{B}n^{-1/2}\sum_{i=1}^{n}\mathbf{A}(\mathbf{X}_i, \theta_0)\right] + o_p(1)$$

$$= n^{-1/2}\sum_{i=1}^{n}[C_\sigma(\mathbf{X}_i) - \mathbf{W}_\sigma'\mathbf{B}\mathbf{A}(X_i)] + o_p(1),$$

where

$$C_\sigma(\mathbf{x}) = \begin{cases} 1 - p_\sigma(\theta_0) & \text{if } \mathbf{x} \in I_\sigma(\theta_0) \\ -p_\sigma(\theta_0) & \text{if } \mathbf{x} \notin I_\sigma(\theta_0) \end{cases}$$

and θ_0 is suppressed in the last expression in (5.4.18) for convenience. By the multivariate central limit theorem, it follows that

$$\text{(5.4.19)}\quad \{n^{-1/2}[N_\sigma - np_\sigma(\hat{\theta}_n)];\ \sigma = 1, \ldots, M\} \xrightarrow{\mathscr{L}} \mathbf{N}_M(\mathbf{0}, \boldsymbol{\Sigma}(\theta_0)),$$

where $\boldsymbol{\Sigma}(\theta_0)$ is an $M \times M$ matrix with entries

$$\text{(5.4.20)}\quad \Sigma_{\sigma\tau} = E_{\theta_0}[(C_\sigma(\mathbf{X}) - \mathbf{W}_\sigma\mathbf{B}\mathbf{A}(\mathbf{X})), (C_\tau(\mathbf{X}) - \mathbf{W}_\tau\mathbf{B}\mathbf{A}(\mathbf{X}))].$$

Since $p_\sigma(\hat{\theta}_n)/p_\sigma(\theta_0) \xrightarrow{p} 1$ for $\sigma = 1, \ldots, M$, it follows that

$$\text{(5.4.21)}\quad \left\{\frac{N_\sigma - np_\sigma(\hat{\theta}_n)}{(np_\sigma(\hat{\theta}_n))^{1/2}};\ \sigma = 1, \ldots, M\right\} \xrightarrow{\mathscr{L}} \mathbf{N}_M(\mathbf{0}, \mathbf{P}^{-1/2}\boldsymbol{\Sigma}\mathbf{P}^{-1/2}),$$

where $\mathbf{P}$ is the $M \times M$ matrix with entries

$$\text{(5.4.22)}\quad \begin{aligned} P_{\sigma\sigma} &= p_\sigma(\theta_0), & \sigma &= 1, \ldots, M, \\ P_{\sigma\tau} &= 0, & \sigma &\neq \tau. \end{aligned}$$

In view of relation (5.4.21) and the definition of T_n, it follows that

$$\text{(5.4.23)}\quad T_n \xrightarrow{\mathscr{L}} \sum_{i=1}^{M} \lambda_i Z_i^2,$$

where λ_i, $1 \leqslant i \leqslant M$, are the eigenvalues of $\mathbf{P}^{-1/2}\boldsymbol{\Sigma}\mathbf{P}^{-1/2}$. Let $\mathbf{P}$ be column vector with entries $p_\sigma(\theta_0)$, $\sigma = 1, \ldots, M$. It is easy to check that

$$\text{(5.4.24)} \qquad \begin{aligned} \mathbf{\Sigma} &= \mathbf{P} - \mathbf{pp}' - \mathbf{W}'\mathbf{BW} - \mathbf{W}'\mathbf{B}'\mathbf{W} + \mathbf{W}'\mathbf{BJB}'\mathbf{W} \\ &= \mathbf{P} - \mathbf{C}, \end{aligned}$$

where $\mathbf{W}$ is as defined by (A_5) and

$$\text{(5.4.25)} \qquad \mathbf{C} = \mathbf{pp}' + \mathbf{W}'(\mathbf{B} + \mathbf{B}' - \mathbf{BJB}')\mathbf{W}.$$

Note that λ_i are also the eigenvalues of $\mathbf{P}^{-1}\mathbf{\Sigma}$ and they are nonnegative since $\mathbf{P}^{-1/2}\mathbf{\Sigma}\mathbf{P}^{-1/2}$ is a covariance matrix and hence nonnegative definite. On the other hand

$$p_1 + \cdots + p_M = 1 \quad \text{and} \quad \sum_{\sigma=1}^{M} (W_\sigma)_s = 0 \text{ for } s = 1, \ldots, m.$$

In other words, the sum of the columns of $\mathbf{\Sigma}$ is zero. Hence at least one eigenvalue is zero.

Let $r(D)$ denote the rank of a matrix D. Note that $r(\mathbf{P}^{-1}\mathbf{C}) = r(\mathbf{C}) = q$ (say). Since

$$\text{(5.4.26)} \qquad \det[\lambda\mathbf{I} - \mathbf{P}^{-1}\mathbf{\Sigma}] = \pm\det[(1-\lambda)\mathbf{I} - \mathbf{P}^{-1}\mathbf{C}]$$

from (5.4.24), there are exactly $M - q$ of the λ_i equal to 1.

By hypothesis, the matrix $\mathbf{D} = \mathbf{B} + \mathbf{B}' - \mathbf{BJB}'$ is positive definite. Hence the rank of $\mathbf{W}'\mathbf{DW}$ is the same as that of $\mathbf{W}$, which is p by (A_5) of the hypothesis. Since the vectors $\mathbf{p}$ and $\mathbf{W}_\sigma$ are linearly independent, $r(\mathbf{C})$ is $p + 1$ from (5.4.25). Hence exactly $M - p - 1$ of the λ_i are equal to 1. Note that $\mathbf{C}$ is nonnegative definite since $\mathbf{pp}'$ and $\mathbf{W}'\mathbf{DW}$ are. Hence, it follows that all of the λ_i are less than are equal to 1 from (5.4.26).

This completes the proof of Theorem 5.4.1.

The problem of interest is to find sufficient conditions so that the eigenvalues λ_i in the theorem derived above do not depend on the parameter θ. This is necessary in practice in order to obtain the cutoff points for tests of significance. We now derive such a result. Suppose the matrix $\mathbf{B} = \mathbf{J}^{-1}$, where $\mathbf{J}$ is the information matrix. This holds in case a MLE $\hat{\theta}$ is used as an estimator for θ. It is easy to see that $\mathbf{D} = \mathbf{J}^{-1}$ in such a case, and we have the following theorem.

Theorem 5.4.2. Suppose conditions (A_0)–(A_6) hold. If either

(i) $F(\mathbf{x}, \theta) = F(\mathbf{x} - \theta)$ and $\xi_{ij}(\theta) = \theta_i + a_{ij}$, a_{ij} constant

or

(ii) $F(x, \theta, \phi) = F\left(\dfrac{x_1 - \theta_1}{\phi_1}, \ldots, \dfrac{x_p - \theta_p}{\phi_p}\right)$, $\xi_{ij}(\theta, \phi) = \theta_i + a_{ij}\phi_i$, a_{ij} constant,

then the eigenvalues λ_i in Theorem 5.4.1 do not depend on the true parameters.

Proof. Note that in the case of location, $\mathbf{J}$ and $p_\sigma(\theta)$ do not depend on the parameter θ. Furthermore, $W_\sigma(\theta)$ does not depend on θ. But λ_i are eigenvalues of $\mathbf{P} - \mathbf{pp}' - \mathbf{W}'\mathbf{J}^{-1}\mathbf{W}$, and hence λ_i are not dependent on θ. In the second case, the matrices $\mathbf{J}$ and $\mathbf{W}$ depend on the scale parameter $\phi_1, \ldots, \phi_p$ only. It can be shown that $\mathbf{W}'\mathbf{J}^{-1}\mathbf{W}$ is independent of (θ, ϕ). Hence λ_i are independent of (θ, ϕ).

Remarks. If the number of cells are increased, then the λ_i converge to zero and the χ^2 distribution with M-p-1 degrees of freedom can be used as an approximation for the asymptotic distribution of T_n for large M. In fact, if conditions $(\mathbf{A}_0)$–$(\mathbf{A}_6)$ hold with $\mathbf{B} = \mathbf{J}^{-1}$, $M \to \infty$, and $\xi_{ij}(\theta)$ are chosen so that $\xi_{i1}(\theta) \to -\infty$ and $\xi_{i,v_i-1}(\theta) \to +\infty$ for all i and $\theta \in \Theta$ such that $\sup_j |\xi_{ij}(\theta) - \xi_{i,j-1}(\theta)| \to 0$ for all i and $\theta \in \Theta$, then $\lambda_i \to 0$ for $1 \leqslant i \leqslant p$ and all $\theta \in \Theta$. For proof, see Moore (1971).

The ω^2 statistic or the Cramer–von Mises statistic for testing a simple hypothesis about an unknown continuous distribution function F of i.i.d. random variables $X_1, \ldots, X_n$ is defined by

$$\omega_n^2 = n \int_{-\infty}^{\infty} [F_n(x) - F_0(x)]^2 dF_0(x), \tag{5.4.27}$$

where $F_n(x)$ is the empirical distribution function based on $X_1, \ldots, X_n$ and $F(x) = F_0(x)$ by the null hypothesis. Substituting $t = F_0(x)$, it is easy to see that

$$\omega_n^2 = n \int_0^1 [F_n(t) - t]^2 dt = \int_0^1 V_n^2(t)\, dt, \tag{5.4.28}$$

where $V_n(t) = \sqrt{n}[F_n(t) - t]$. It is well known that $V_n(\cdot)$ converges weakly to the Brownian bridge $V(\cdot)$ on $[0, 1]$ in $L_2[0, 1]$ as $n \to \infty$ and hence the distribution of ω_n^2 converges weakly to the distribution of

$$\omega^2 = \int_0^1 V^2(t) dt. \tag{5.4.29}$$

But

(5.4.30) $$V(t)=\sum_{k=1}^{\infty}\frac{1}{\pi k}V_k\sin\pi kt \quad \text{a.s.},$$

where the $V_k \geqslant 1$, are i.i.d. random variables with standard normal distribution. [cf. Hajek and Sidak (1967)]. Hence

(5.4.31) $$\omega^2=\sum_{k=1}^{\infty}\frac{1}{\pi^2k^2}V_k^2$$

and the distribution of this quadratic form gives the asymptotic distribution of ω_n^2.

If the null hypothesis to be tested is composite, that is, $F\in\{F(x,\theta),\theta\in\Theta\}$, then we can consider the statistic

(5.4.32) $$\hat{\omega}_n^2=n\int_{-\infty}^{\infty}[F_n(x)-F(x,\hat{\theta}_n)]^2\,dF(x,\hat{\theta}_n)$$

as discussed in Section 5.1, where $\hat{\theta}_n$ is a suitable estimator of θ. We now study the case when $\Theta\subset R$.

Let there be $\theta_0\in\Theta$ such that the true distribution function $F(x)=F(x,\theta_0)$. Making the transformation $t=F(x,\theta_0)$, we have

(5.4.33) $$\hat{\omega}_n^2=n\int_0^1[F_n(t)-G(t,\hat{\theta})]^2\,dG(t,\hat{\theta}),$$

where $G(t,\theta)=F[F^{-1}(t,\theta_0),\theta]$. Note that $G(t,\theta_0)=t$. Let

(5.4.34) $$U_n(t)=\sqrt{n}[F_n(t)-G(t,\hat{\theta})].$$

It can be shown under some regularity conditions that

(5.4.35) $$U_n(\cdot)\xrightarrow{\mathscr{L}}U(\cdot)$$

in $L_2[0,1]$, where $U(\cdot)$ is a Gaussian process with expansion

(5.4.36) $$U(t)=\sum_{k=1}^{\infty}\mu_kV_ka_k(t),$$

where $a_k(t)$ is an orthonormal basis in $L_2[0, 1]$, the V_k are i.i.d. standard normal random variables, and the μ_k are constants. However, μ_k and $a_k(\cdot)$ depend on the family $\{F(x, \theta), \theta \in \Theta\}$, The asymptotic distribution of $\hat{\omega}_n^2$ can be shown to be that of the quadratic form

$$\hat{\omega}^2 \equiv \sum_{k=1}^{\infty} \mu_k V_k^2, \tag{5.4.37}$$

where μ_k depends on the family $\{F(x, \theta), \theta \in \Theta\}$ as mentioned above.

Khmaladze (1979) considered the problem of finding a kernel $K(t, s)$, which might depend on the family $\{F(x, \theta), \theta \in \Theta\}$ such that

$$\int_0^1 \int_0^1 K(t, s) dU(t) dU(s) = \sum_{k=1}^{\infty} \lambda_k^2 Z_k^2, \tag{5.4.38}$$

where λ_k^2 *does not depend* on the family but coincide with a prespecified sequence. Observe that

$$\hat{\omega}^2 = \int_0^1 \int_0^1 \min(t, s) dU(t) dU(s). \tag{5.4.39}$$

We now describe some results of Khmaladze (1979).

Suppose there exist functions $h_k(x, \theta)$ such that

(5.4.40) (i) $$\int_{-\infty}^{\infty} h_k(x, \theta) h_r(x, \theta) dF(x, \theta) = \delta_{kr},$$

(ii) $$\int_{-\infty}^{\infty} h_k(x, \theta) dF(x, \theta) = 0,$$

(iii) $$\int_{-\omega}^{\infty} h_k(x, \theta) \frac{\partial f(x, \theta)}{\partial \theta} dx = 0$$

for every $\theta \in \Theta$ where δ_{kr} is the Kronecker δ function and $f(x, \theta) = dF(x, \theta)/d\theta$. One can construct functions satisfying (i) and (ii) easily by choosing an orthonormal system $\{h_k\}$ in $L_2[0, 1]$ and defining

$$h_k(x, \theta) = h_k[F(x, \theta)].$$

Condition (iii) needs verification. Suppose that $\sum_{k=1}^{\infty} \lambda_k^2 < \infty$, $\sqrt{n}(\hat{\theta}_n - \theta_0) =$

$O_p(1)$, and furthermore, for every $\theta \in \Theta$ and for every k,

$$\int_{\theta-\varepsilon \leq \theta' \leq \theta+\varepsilon} \left| \frac{\partial}{\partial\theta} h_k(x, \theta) \right|_{\theta=\theta'} \Bigg| dF(x) \to 0 \quad \text{as } \varepsilon \to \infty, \tag{5.4.41}$$

where $F = F(\cdot, \theta_0)$ is the true distribution function. Then it can be seen that

$$\frac{1}{\sqrt{n}} \sum_{i=1}^{n} h_k(X_i, \hat{\theta}) = \frac{1}{\sqrt{n}} \sum_{i=1}^{n} h_k(X_i, \theta_0) + o_p(1) \tag{5.4.42}$$

as $n \to \infty$ under the hypothesis, and furthermore,

$$\sum_{k=1}^{\infty} \lambda_k^2 \left\{ \frac{1}{\sqrt{n}} \sum_{i=1}^{n} h_k(X_i, \hat{\theta}) \right\} \to \sum_{k=1}^{\infty} \lambda_k^2 Z_k^2, \tag{5.4.43}$$

where the Z_k^2 are independent χ^2 random variables with one degree of freedom. Statistics of the form (5.4.41) are similar to the components of ω^2 statistic considered in Durbin and Knott (1972), and they can be used for testing goodness of fit.

5.5 INFERENCE FOR COUNTING PROCESSES

Aalen (1978, 1982) investigated nonparametric inference for a family of counting processes via the modern theory of semimartingales and their properties. We now present a very special aspect of his result. A brief discussion of his results is given in Prakasa Rao (1984).

Suppose we observe the occurrence over time of several events, which may be of different types, say $1, 2, \ldots, k$. Let $N_i(t)$ be the number of events of type i that have occurred up to and including time t. $N_i(t)$ counts the number of events of type i and hence is called a *counting process*.

Let $A_i(t)dt$ be the conditional probability of an event of type i occurring in the time interval $(t, t+dt)$ given the past (all that has happened up to time t). $A_i(t)$ is called the *intensity of the process* $\{N_i(t)\}$. Suppose that $A_i(t) = \alpha_i(t) Y_i(t)$, where $\alpha_i(t)$ is an unknown function and $Y_i(t)$ is another observed process that may depend arbitrarily on the past. Such a model is called a *multiplicative intensity* model.

We now describe some examples where multiplicative intensity model is appropriate for statistical modeling.

Example 5.5.1 (Competing Risks). Suppose a person is prone to one or several of k types of diseases. Let $N_i(t)$ be the number of deaths due to disease i in $[0, t]$ and $Y_i(t)$ be the number of individuals at risk at time t. Clearly $Y_i(t)$ is the same for each i. Let $\alpha_i(t)$ be the mortality rate at time t due to disease i. It is reasonable to assume that the intensity of death due to disease i is $\alpha_i(t)Y_i(t)$ at time t.

Example 5.5.2 (Birth and Death Process). Let $N_1(t)$ and $N_2(t)$ denote the number of births and deaths, respectively in $[0, t]$. Let $Y(t)$ be the number of individuals alive at time t. It is clear that the intensity of births at time t is $A_1(t) = \alpha_1(t)Y(t)$ and the intensity of deaths at time t is $A_2(t) = \alpha_2(t)Y(t)$, where $\alpha_1(t)$ and $\alpha_2(t)$ are the birth and death rates, respectively.

Example 5.5.3 (Epidemiology). Let $N(t)$ be the number of infections in a population at time t. Let $I(t)$ and $S(t)$ be the number of infectives and susceptibles, respectively, at time t. Then the intensity of an individual being infected can be taken to be $\alpha(t)I(t)S(t)$, where $\alpha(t)$ is a measure of infectiousness. Here we have a multiplicative intensity model with $Y(t) = I(t)S(t)$.

For other examples, see Aalen (1982). We saw in Section 1.8 how the theory of point processes fits into the theory of semimartingales. Let us recall few definitions (see Section 1.8).

Let $(\Omega, \mathscr{B}, P)$ be a complete probability space and $\{\mathscr{B}_t, t \in [0, 1]\}$ be an increasing right-continuous family of sub-σ-algebras of $\mathscr{B}$ with the usual property that each $\mathscr{B}_t$ contain all P-null sets.

Definition 5.5.1. A stochastic process $\mathbf{N} = \{(N_1(t), \ldots, N_k(t));\ 0 \leqslant t \leqslant 1\}$ is called a *k-dimensional counting process* if (i) each of the k component processes N_i has a sample function that is a right-continuous step function with a finite number of jumps each of size $+1$ and (ii) two different component processes cannot jump at the same time.

We suppose that $\mathbf{N}(0) = \mathbf{0}$, $E\{N_i(1)\} < \infty$ for $1 \leqslant i \leqslant k$ and that $\mathbf{N}$ is adapted to $\{\mathscr{B}_t\}$.

Since N_i is nondecreasing and integrable,

$$E[N_i(t) - N_i(s) | \mathscr{B}_s] \geqslant 0 \quad \text{if } s \leqslant t, \tag{5.5.0}$$

and hence it is a submartingale. Therefore, by the Doob–Meyer decomposition, there exists a natural continuous nondecreasing process A_i and a square integrable martingale M_i such that

$$N_i = A_i + M_i, \qquad 1 \leqslant i \leqslant k, \tag{5.5.1}$$

where A_i is the compensator of N_i. Suppose A_i is absolutely continuous in the sense that

$$(5.5.2) \qquad A_i(t) = \int_0^t \Lambda_i(s)ds, \qquad 1 \leqslant i \leqslant k,$$

where

$$(5.5.3) \qquad \mathbf{\Lambda} = \{(\Lambda_1(t), \ldots, \Lambda_k(t)), \qquad 0 \leqslant t \leqslant 1\}$$

is adapted to $\{\mathscr{B}_t, 0 \leqslant t \leqslant 1\}$. Then

$$(5.5.4) \qquad M_i(t) = N_i(t) - \int_0^t \Lambda_i(s)ds, \qquad 1 \leqslant i \leqslant k$$

are orthogonal square integrable martingales with quadratic characteristic (see Section 1.8),

$$\langle M_i, M_i \rangle(t) = \int_i^t \Lambda_i(s)ds, \qquad 1 \leqslant i \leqslant k.$$

$\mathbf{\Lambda}$ is the intensity of the process of $\mathbf{N}$. The name can be justified by the following reasoning. Suppose the component A_i is dominated by an integrable random variable. Then

$$\begin{aligned}\lim_{h\downarrow 0} \frac{1}{h} P\{N_i(t+h) - N_i(t) = 1 | \mathscr{B}_t\} &= \lim_{h\downarrow 0} \frac{1}{h} E\{N_i(t+h) - N_i(t) | \mathscr{B}_t\} \\ &= \lim_{h\downarrow 0} \frac{1}{h} E\left\{\int_t^{t+h} \Lambda_i(s)ds \,\middle|\, \mathscr{B}_t\right\} \\ &= E\left(\lim_{h\downarrow 0} \frac{1}{h} \int_t^{t+h} \Lambda_i(s)ds \,\middle|\, \mathscr{B}_t\right) \\ &= E(\Lambda_i(t+0) | \mathscr{B}_t) = \Lambda_i(t+0).\end{aligned}$$

Suppose we have a multiplicative intensity model

$$\Lambda_i(t) = \alpha_i(t) Y_i(t),$$

where $\boldsymbol{\alpha} = (\alpha_1(t), \ldots, \alpha_k(t))$ is unknown and $\mathbf{Y} = (Y_1(t), \ldots, Y_k(t))$ is a random process observable over [0, 1] adapted to $\{\mathscr{B}_t\}$. We can rewrite the relation (5.5.4)

$$(5.5.5)\qquad \begin{aligned} dN_i(t) &= \Lambda_i(t)dt + dM_i(t) \\ &= \alpha_i(t)Y_i(t)dt + dM_i(t), \end{aligned}$$

which is a stochastic differential equation. Given $N_i(s), 0 \leqslant s \leqslant t$, the problem is the estimation of $\alpha_i(t)$ or equivalently

$$(5.5.6)\qquad \beta_i(t) = \int_0^t \alpha_i(s)ds.$$

A natural estimator of $\beta_i(\cdot)$ from (5.5.5) is

$$(5.5.7)\qquad \hat{\beta}_i(t) = \int_0^t Y_i^{-1}(s)dN_i(s)$$

provided Y_i is nonzero. Let $J_i(s) = I[Y_i(s) > 0]$, where $I(A)$ denotes the indicator function of the set A. Let

$$(5.5.8)\qquad \beta_i^*(t) = \int_0^t \alpha_i(s)J_i(s)ds.$$

If $Y_i(t) = 0$, we interpret $J_i(t)/Y_i(t) = 0$, and we assume that there exists $c > 0$ such that $Y_i(t) < c$ implies that $Y_i(t) = 0$ a.s. Define

$$(5.5.9)\qquad \hat{\beta}_i(t) = \int_0^T \frac{J_i(s)}{Y_i(s)} dN_i(s), \qquad 1 \leqslant i \leqslant k, \quad t \in [0, 1],$$

where the integral is a Stieljes integral. We take $\hat{\beta}_i$ as an estimator of β_i^*. Note that

$$(5.5.10)\qquad \hat{\beta}_i(t) - \beta_i^*(t) = \int_0^t \frac{J_i(s)}{Y_i(s)} dM_i(s)$$

and $\{\hat{\beta}_i(t) - \beta_i^*(t), \mathscr{B}_t\}, 1 \leqslant i \leqslant k$, are orthogonal square integrable martingales. Aalen (1978) studied the asymptotic properties of $\hat{\beta}_{in}$ for a sequence of counting processes N_n using the theory of weak convergence. Linear nonparametric tests for comparison of counting processes with applications to censored survival data using the martingale methods described above are discussed in Anderson et al. (1982).

We now discuss an example illustrating this result. For an important class of practical problems when these results are applied, see Aalen (1982).

Example 5.5.4. Suppose $X_1, \ldots, X_k$ are i.i.d. nonnegative random variables with an absolutely continuous distribution function F and failure rate λ. Let

$$N_i(t) = I(X_i \leqslant t), \qquad i = 1, \ldots, k, \quad t \in [0, 1],$$

and $\mathscr{B}_t$ be the σ-algebra generated by $N_i(t)$, $1 \leqslant i \leqslant k$ where $I(A)$ is the indication function of the set A. Then $\mathbf{N} = (N_1, \ldots, N_k)$ is a multivariate counting process with multiplicative intensity

$$\Lambda_i(t) = \alpha_i(t) Y_i(t),$$

where $\alpha_i(t) = \lambda(t)$ and $Y_i(t) = I(X_i \geqslant t)$, $1 \leqslant i \leqslant k$. Let

$$M_i(t) = I(X_i \leqslant t) - \int_0^t \lambda(u) I(X_i \geqslant u) du, \qquad 0 \leqslant t \leqslant 1.$$

Then $\{M_i(t), 1 \leqslant i \leqslant k\}$ are orthogonal square integrable martingales. Note that

$$\begin{aligned}
\bar{N}(t) &\equiv \sum_{i=1}^{k} N_i(t) \\
&= \sum_{i=1}^{k} I(X_i \leqslant t) \\
&= \int_0^t \lambda(u) \left[\sum_{j=1}^{k} I(X_j \geqslant u) \right] du + \sum_{j=1}^{k} M_j(t) \\
&= \int_0^t \lambda(u)(k - \bar{N}(u - 0)) du + \sum_{j=1}^{k} M_j(t).
\end{aligned}$$

The process $\{\bar{N}(t), 0 \leqslant t \leqslant 1\}$ is a univariate counting process satisfying the multiplicative intensity model

$$\bar{\Lambda}(t) = \alpha(t) \bar{Y}(t),$$

where $\alpha(t) = \lambda(t)$ and $\bar{Y}(t) = k - \bar{N}(t - 0)$. $\bar{N}(t)$ is the number of failures in $[0, t]$, and $\bar{Y}(t)$ is the number at risk at time t. If

$$\beta(t) = \int_0^t \alpha(u) du = \int_0^t \lambda(u) du,$$

then $\beta(t)$ is the cumulative failure rate. In this example

$$\begin{aligned} J(t) &= I[\bar{Y}(t) > 0] = I[\bar{N}(t-0) < k] \\ &= I[X_{(k)} \geqslant t], \end{aligned}$$

where $X_{(1)}, \ldots, X_{(k)}$ are the order statistics. An estimator for $\beta(t)$ is

$$\begin{aligned} \hat{\beta}(t) &= \int_0^t \frac{J(s)}{Y(s)} d\bar{N}(s) \\ &= \int_0^t \frac{I(X_{(k)} \geqslant s)}{k - \bar{N}(s-0)} d\bar{N}(s) \\ &= \sum_{j=1}^{\bar{N}(t)} \frac{1}{k-j+1}, \end{aligned}$$

which is known as the Nelson estimator for cumulative failure rate [cf. Nelson (1969)]. If

$$\beta^*(t) = \int_0^t \lambda(u) I(X_{(k)} \geqslant u) du,$$

then

$$\beta^*(t) \xrightarrow{p} \beta(t) \quad \text{as } k \to \infty$$

provided $F(1) < 1$, where F is the common distribution function of $X_1, \ldots, X_n$.

Remarks. The advantage of the general formulation of the problem using the martingale theory is that it allows the study of estimation in the censoring case, too. The censoring need not be adapted to the family $\mathscr{B}_t$ of σ-algebras generated by $N_i(t)$, $1 \leqslant i \leqslant k$. It might depend also on an external random influence in addition to the events that occurred in the counting process up to time t. Statistical inference for counting processes using the theory of martingales is discussed in Jacobsen (1982). The theory of point processes via martingales is studied in Bremaud (1981). For the classical approach to these problems, see Basawa and Prakasa Rao (1980) and Prakasa Rao (1983).

5.6 INFERENCE FROM CENSORED DATA

In order to illustrate the martingale approach to statistical analysis for censored data, we now discuss a special case. For an extensive discussion via this approach, the reader is referred to Gill (1980). A classical approach to the problem is given in detail in Prakasa Rao (1983). Our discussion is based on Gill (1980).

In problems occurring in medical follow-up trials, industrial life testing, and other fields, one is interested in some characteristics of the distributions of n independent positive random variables $X_1, \ldots, X_n$, but is able to observe only the bivariate random vectors $(\tilde{X}_j, \delta_j)$, $1 \leqslant j \leqslant n$, where

$$\text{(5.6.0)} \quad \text{(i)} \quad 0 < \tilde{X}_j \leqslant X_j, \qquad 1 \leqslant j \leqslant n,$$

$$\text{(ii)} \quad \delta_j = \begin{cases} 1 & \text{if } X_j = \tilde{X}_j \\ 0 & \text{otherwise.} \end{cases}$$

Hence, if $\delta_j = 1$, then the observative X_j is *uncensored* and, if $\delta_j = 0$, then $X_j > \tilde{X}_j$.

Censoring might be of several types: type I censoring, type II censoring, random censoring, and so on. For examples that illustrate these several types of censoring, see Prakasa Rao (1983) or Gill (1980).

Let $(\Omega, \mathscr{B}, P)$ be a complete probability space and X_i, $1 \leqslant i \leqslant n$, be independent, positive, possibly infinite-valued random variables with subdistribution functions F_i, $1 \leqslant i \leqslant n$, given by

$$\text{(5.6.1)} \qquad F_i(t) = P(X_i \leqslant t), \qquad F_i(\infty) = P(X_i < \infty) \leqslant 1.$$

Define G_i with values in $\overline{R^+}$ by

$$\text{(5.6.2)} \qquad G_i(t) = \int_0^t (1 - F_i(s-))^{-1} dF_i(s),$$

where $F_i(s-) = F_i(s-0)$. Let

$$\text{(5.6.3)} \qquad \tau_i = \sup\{t : F_i(t) < 1\}.$$

For each i, $F_i(0) = G_i(0) = 0$, G_i is finite on $[0, \tau_i]$, and G_i is constant on $[\tau_i, \infty]$. G_i is called the *cumulative hazard* for the item i. If F_i has a density f_i,

$$\text{(5.6.4)} \qquad \lambda_i = \frac{f_i}{1 - F_i}$$

is called the *hazard rate* or *failure rate* of the item i.

Without loss of generality, we assume that $(\tilde{X}_i, \delta_i)$, $1 \leqslant i \leqslant n$, are defined on the same probability space $(\Omega, \mathscr{B}, P)$. Clearly

$$G_i(\tilde{X}_i) \leqslant G_i(X_i) \leqslant \infty. \tag{5.6.5}$$

Define the stochastic processes N_i, J_i, and M_i by

(5.6.6) (i) $N_i(t) = I[\tilde{X}_i \leqslant t, \delta_i = 1]$,

(ii) $J_i(t) = I[\tilde{X}_i \geqslant t]$,

(iii) $M_i(t) = N_i(t) - G_i(\tilde{X}_i \wedge t)$

$$= N_i(t) - \int_0^t J_i dG_i$$

for $1 \leqslant i \leqslant n$. Gill (1980) (see Theorem 3.1.1) proved the following theorem. We omit the proof.

Theorem 5.6.1. Let $\{\mathscr{B}_t, t \geqslant 0\}$ be a family of sub-σ-algebras of $\mathscr{B}$ satisfying the usual conditions. Suppose that $I[\tilde{X}_j \leqslant t]$ and $\delta_j I[\tilde{X}_j \leqslant t]$ are $\mathscr{B}_t$-measurable for each j and t. Furthermore, assume that, for each t, conditional on $\mathscr{B}_t$, the X_j with $\tilde{X}_j > t$ are independent of one another each having the distribution of X_j given that $X_j > t$. Then N_j, J_j, and M_j are $\mathscr{B}_t$-adapted for each j, M_j is a square integrable martingale with

$$\langle M_j, M_j \rangle = \int J_j(1 - \Delta G_j) dG_j, \qquad \langle M_j, M_{j'} \rangle = 0 \quad \text{for } j \neq j', \tag{5.6.7}$$

and for each $t \in [0, \infty)$, conditional on $\mathscr{B}_{t-}$, $\Delta N_j(t)$, $1 \leqslant j \leqslant n$, are independent zero–one random variables with means $J_j(t)\,\Delta G_j(t)$, $1 \leqslant j \leqslant n$.

Remarks. The main condition stated above can be rephrased as that, at time t, given $\mathscr{B}_t$, if $\tilde{X}_j > t$, then $N_j(t)$ has the same conditional probability of having a jump in a small time interval $(t, t+h)$ as if there had been no censoring.

Let us now assume that $F_1 = \cdots = F_n = F$, F not necessarily a continuous (sub) distribution function (say) and $G = G_j$, $1 \leqslant j \leqslant n$, $\tau = \tau_j$, $1 \leqslant j \leqslant n$. Define

(5.6.8) (i) $N(t) = \sum_{j=1}^n N_j(t) = \#\{j : \tilde{X}_j \leqslant t \text{ and } \delta_j = 1\}$,

(ii) $Y(t) = \sum_{j=1}^n J_j(t) = \#\{j : \tilde{X}_j \geqslant t\}$,

(iii) $M(t) = \sum_{j=1}^n M_j(t) = N(t) - \int_0^t Y dG$.

Let

$$J(t) = I[Y(t) > 0] \tag{5.6.9}$$

and

$$\hat{F}(t) = 1 - \prod_{s \leqslant t} \left(1 - \frac{\Delta N(s)}{Y(s)}\right), \tag{5.6.10}$$

where we apply the convention $0/0 = 0$. Note that N is nondecreasing and right continuous, Y is nonincreasing and left continuous, and both take values in $\{0, 1, \dots, n\}$. Furthermore, $Y(0) = n$ a.s. and $\Delta N(s) \leqslant Y(s)$ for all s. If $\Delta N(s) = Y(s)$ for some s, then for $t > s$, $N(t) = N(s)$ and $Y(t) = 0$. Clearly $Y(\infty) = 0$ a.s.

$\hat{F}(t)$ is the *Kaplan–Meier estimator* of $F(t)$ [cf. Kaplan and Meier (1958)]. It reduces to the empirical distribution function if $\delta_j = 1$ for each j. $\hat{F}$ is a subdistribution function on $[0, \infty)$ assigning mass to the uncensored observations and

$$\frac{\Delta \hat{F}(t)}{1 - F(t-)} = \frac{\#\{j : \tilde{X}_j = t, \delta_j = 1\}}{\#\{j : \tilde{X}_j \geqslant t\}}. \tag{5.6.11}$$

If F is a discrete distribution, then (5.6.11) is a natural estimator of

$$P(X_j = t \mid X_j \geqslant t) = \frac{\Delta F(t)}{1 - F(t-)}. \tag{5.6.12}$$

$\hat{F}(\cdot)$ is a maximum likelihood estimator in a generalized sense [cf. Johansen (1978)]. Observe that

$$(1 - \hat{F}(t)) \prod_{s \leqslant t} \left\{1 - \frac{Y(s) - Y(s+) - \Delta N(s)}{Y(s) - \Delta N(s)}\right) = \frac{Y(t+)}{n} \tag{5.6.13}$$

from (5.6.10), where $Y(s+) = Y(s+0)$. Hence

$$\left(\frac{Y(t+)}{n}\right)(1 - \hat{F}(t))^{-1} \tag{5.6.14}$$

is nonincreasing, nonnegative, and equal to 1 at time zero. Therefore

$$\hat{F}(t) \leqslant 1 - \frac{Y(t+)}{n}. \tag{5.6.15}$$

On the other hand,

$$\frac{N(t)}{n} = 1 - \prod_{s \leqslant t}\left(1 - \frac{\Delta N(s)}{n - N(s-)}\right) \tag{5.6.16}$$

and hence

$$\hat{F}(t) \geqslant \frac{N(t)}{n}. \tag{5.6.17}$$

Combining relations (5.6.15) and (5.6.17), we have the inequality

$$\frac{N(t)}{n} \leqslant \hat{F}(t) \leqslant 1 - \frac{Y(t+)}{n}. \tag{5.6.18}$$

Equation (5.6.10) can also be written in the form

$$\hat{F}(t) = \int_0^t (1 - \hat{F}(s-))\frac{dN(s)}{Y(s)}. \tag{5.6.19}$$

Note that F and G satisfy the relation

$$F(t) = \int_0^t (1 - F(s-))\, dG(s), \tag{5.6.20}$$

and hence

$$\int_0^t \frac{1}{Y(s)}\, dN(s), \tag{5.6.21}$$

the empirical cumulative failure rate, can be considered as an estimator of $G(t)$ [cf. Nelson (1972)].

Before we discuss further properties of $\hat{F}(t)$, we first state and prove a proposition due to Gill (1980).

Lemma 5.6.2. Suppose U and V are right-continuous functions of locally bounded variation on $[0, \infty)$. Then for all $t > 0$,

$$U(t)V(t) = U(0)V(0) + \int_{s\in(0,t]} U(s-)\, dV(s) + \int_{s\in(0,t]} V(s)\, dU(s). \tag{5.6.22}$$

Proof. Obvious.

Remark. Relation (5.6.22) can also be written

(5.6.23) $$d(UV) = U_- dV + V dU$$

in terms of differentials, which in turn implies that

(5.6.24) (i) $$d(U^n) = \left(\sum_{i=1}^{r-1} U^i U_-^{r-1-i} \right) dU, \qquad r \geqslant 1,$$

(ii) $$d(U^{-1}) = -(UU_-)^{-1} dU.$$

Hence, if U is nondecreasing and nonnegative in addition, then

(5.6.25) $$rU_-^{r-1} \leqslant d(U^r) \leqslant rU^{r-1} dU, \qquad r \geqslant 1.$$

Proposition 5.6.3. Suppose A and B are right-continuous nondecreasing functions on $[0, \infty)$, zero at time zero. Suppose $\Delta A \leqslant 1$ and $\Delta B < 1$ on $[0, \infty)$. Then there exists a unique locally bounded solution Z of

(5.6.26) $$Z(t) = \int_0^t \frac{1 - Z(s-)}{1 - \Delta B(s)} (dA(s) - dB(s)),$$

and it is given by

(5.6.27) $$Z(t) = 1 - \frac{\prod_{s \leqslant t}(1 - \Delta A(s)) \exp(-A_c(t))}{\prod_{s \leqslant t}(1 - \Delta B(s)) \exp(-B_c(t))},$$

where A_c is the continuous component of the process A given by

(5.6.28) $$A_c(t) = A(t) - \sum_{s \leqslant t} \Delta A(s).$$

Proof. Clearly $Z(t)$ is locally bounded where $Z(t)$ is defined by (5.6.27). We first show that it is a solution of (5.6.26). Define

(5.6.29) $$U(t) = \prod_{s \leqslant t} \frac{1 - \Delta A(s)}{1 - \Delta B(s)}$$

and

(5.6.30) $$V(t) = \exp(-A_c(t) + B_c(t)).$$

Note that U and V satisfy the properties stated in Lemma 5.6.2. Hence

$$(5.6.31)\quad Z(t) = 1 - U(t)V(t)$$

$$= 1 - U(0)V(0) - \int_{s\in(0,t]} U(s-)dV(s-) - \int_{s\in(0,t]} V(s)dU(s)$$

$$= -\int_{s\in(0,t]} U(s-)V(s)(-dA_c(s) + dB_c(s))$$

$$- \sum_{s\leqslant t} V(s)U(s-)\left(\frac{1-\Delta A(s)}{1-\Delta B(s)} - 1\right)$$

$$= \int_{s\in[0,t]} \frac{1-Z(s-)}{1-\Delta B(s)}(dA_c(s) - dB_c(s))$$

$$+ \sum_{s\leqslant t} \frac{1-Z(s-)}{1-\Delta B(s)}(\Delta A(s) - \Delta B(s))$$

$$= \int_{s\in[0,t]} \frac{1-Z(s-)}{1-\Delta B(s)}(dA(s) - dB(s)),$$

where $(1-\Delta B)^{-1}$ can be written in the integrand since A_c and B_c are continuous. Therefore relation (5.6.26) holds for $Z(t)$ defined by (5.6.27).

Suppose Z' is any other locally bounded solution of equation (5.6.26). Define $\tilde{Z} = Z - Z'$ and $L(t) = \sup_{0\leqslant s\leqslant t}|\tilde{Z}(s)|$. Let

$$\alpha(t) = \int_0^t (1-\Delta B(s))^{-1}(dA(s) + dB(s)).$$

Then, for any $s \leqslant t$,

$$|\tilde{Z}(s)| \leqslant \int_{u\in[0,s]} L(t)\alpha(u-)d\alpha(u) \leqslant \tfrac{1}{2}L(t)\alpha(s)^2$$

by (5.6.25), and, in general, for any $r \geqslant 1$,

$$|\tilde{Z}(s)| \leqslant \frac{L(t)}{r!}\alpha(s)^r \to 0 \quad \text{as } r \to \infty.$$

Proposition 5.6.4. Let

$$(5.6.32)\qquad G(t) = \int_0^t \frac{1}{1-F(s-)}dF(s)$$

for some subdistribution function F with $F(0) = 0$. Define $\tau = \sup\{t: F(t) < 1\}$.

Then

(i) G determines F uniquely and F can be written

$$F(t) = 1 - \prod_{s \leqslant t} (1 - \Delta G(s)) \exp(-G_c(t)). \qquad t \geqslant 0, \tag{5.6.33}$$

(ii) if F has a density f and failure rate $\lambda = f/(1-F)$, then

$$G(t) = \int_{s \in [0,t]} \lambda(s) ds, \qquad t \geqslant 0, \tag{5.6.34}$$

and

(iii) for all t such that $F(t) < 1$,

$$\frac{1 - \hat{F}(t)}{1 - F(t)} = 1 - \int_0^t \frac{1 - \hat{F}(s-)}{1 - F(s)} \left(\frac{dN(s)}{Y(s)} - dG(s) \right). \tag{5.6.35}$$

Proof. Relation (5.6.33) follows from Proposition 5.6.1 by taking $B = 0$ and $A = G$ in (5.6.26) for all t such that $G(t) < \infty$. By taking limits as $t \uparrow \tau$, it can be seen that relation (5.6.33) holds for $t \geqslant \tau$. Clearly (5.6.34) follows from (5.6.33).

Let t be such that $F(t) < 1$. Let

$$A(t) = \int_0^t \frac{1}{Y(s)} dN(s) \quad \text{and} \quad B(t) = G(t).$$

Then

$$Z(t) = 1 - \frac{1 - \hat{F}(t)}{1 - F(t)}$$

is a solution of (5.6.26). But, with A and B as defined above, equation (5.6.26) is equivalent to (5.6.35) since

$$(1 - F(s-))(1 - \Delta G(s)) = 1 - F(s).$$

In view of (5.6.35), it is easy to see that for t such that $F(t) < 1$ and $Y(t) > 0$,

$$\begin{aligned} \hat{F}(t) - F(t) &= (1 - F(t)) \int_0^t \frac{1 - \hat{F}(s-)}{1 - F(s)} \frac{J(s)}{Y(s)} (dN(s) - Y dG(s)) \\ &= (1 - F(t)) \int_0^t \frac{1 - \hat{F}_-}{1 - F} \frac{J}{Y} dM, \end{aligned} \tag{5.6.36}$$

where J, Y, and M are as defined by (5.6.8) and (5.6.9). Define the stopping time T by the relation

$$T = \inf\{t : Y(t) = 0\}. \tag{5.6.37}$$

Observe that $\hat{F}$ and M are constant on $[T, \infty)$ and relation (5.6.36) holds with $t = T$ provided $F(t) < 1$. Therefore, for any t such that $F(t) < 1$,

$$\begin{aligned} \hat{F}(t) - F(t) &= (1 - F(t)) \int_0^t \frac{1 - \hat{F}_-}{1 - F} \frac{J}{Y} dM \\ &\quad + I[T < t] \left\{ \hat{F}(t) - F(t) - (1 - F(t)) \int_0^T \frac{1 - \hat{F}_-}{1 - F} \frac{J}{Y} dM \right\}, \\ &= (1 - F(t)) \int_0^t \frac{1 - \hat{F}_-}{1 - F} \frac{J}{Y} dM \\ &\quad + I[T < t] \left\{ \hat{F}(t) - F(t) - (1 - F(t)) \frac{\hat{F}(T) - F(T)}{1 - F(T)} \right\} \\ &= (1 - F(t)) \int_0^t \frac{1 - \hat{F}_-}{1 - F} \frac{J}{Y} dM \\ &\quad - I(T < t) \frac{(1 - \hat{F}(T))(F(t) - F(T))}{1 - F(T)}. \end{aligned} \tag{5.6.38}$$

Suppose the assumptions started in Theorem 5.6.1 hold. Then the process M is a square integrable martingale and the process

$$\frac{1 - \hat{F}_-}{1 - F} \frac{J}{Y}$$

is bounded on $[0, t]$ for each t with $F(t) < 1$ and is predictable since J, Y, and $\hat{F}_-$ are left continuous, adapted and F is nonrandom. Hence, by the property of stochastic integrals discussed in Section 1.9, it follows that

$$E\hat{F} = F - E\left(I[T < t] \frac{(1 - \hat{F}(T))(F(t) - F(T))}{1 - F(T)} \right). \tag{5.6.39}$$

Here $\hat{F}$ is unbiased estimate of F on $\{t : F(t) < 1\}$ if and only if $\hat{F}(T) = 1$ a.s or F is constant on $\{t : t \geqslant T$ and $F(t) < 1\}$. A sufficient condition for unbiasedness of $\hat{F}(t)$ is that, if the largest observation is less than $\tau = \sup\{t : F(t) < 1\}$, then this observations and all the observations equal to it must be uncensored, as in this case $F(\tau)=1$ implies $\hat{F}(\tau)=1$ a.s. and we have unbiasedness. Note that relation (5.6.39) implies that, for t such that $F(t)<1$,

$$0<F(t)-\hat{F}(t)\leqslant F(t)\,P(Y(t)=0) \tag{5.6.40}$$

giving bounds for the bias of the estimator $\hat{F}$. Using the fact that

$$\langle M,M\rangle=\int Y(1-\Delta G)\,dG,$$

it can be shown that

$$\begin{aligned}(5.6.41)\quad &\operatorname{Var}(\hat{F}(t)-F(t)-B(t))\\ &=(1-F(t))^2\int_0^t E\left(\frac{(1-\hat{F}_-)^2J}{Y}\right)\frac{dF}{(1-F_-)^2(1-F)},\end{aligned}$$

$$B(t)=-I[T<t]\frac{(1-F(T))(F(t)-F(T))}{1-F(t)} \tag{5.6.42}$$

is the *random bias*. In particular, the variance of $\hat{F}(t)-F(t)$ can be estimated by

$$\begin{aligned}\hat{V}(t)&=(1-\hat{F}(t))^2\int_0^t\frac{J}{Y(1-\hat{F})}\,d\hat{F}\\ &=(1-\hat{F}(t))^2\int_0^t\frac{dN}{Y(Y-\Delta N)}.\end{aligned} \tag{5.6.43}$$

Since $Y/n\leqslant 1-\hat{F}_-$, it follows that

$$\hat{V}(t)\geqslant n^{-1}\hat{F}(t)(1-\hat{F}(t)) \tag{5.6.44}$$

from (5.6.24)(ii). Clearly equality holds in (5.6.44) iff there are no censored on observations on $[0,t]$.

We now briefly discuss asymptotic properties of the product limit estimator using the theory of martingales. For extensive discussion, see Gill (1980, Chapter 4). The underlying probability space and hence also the distribution function may be different for each n. We suppose that, for each n, the model for n censored observations specified earlier holds.

Let $F^n(t)$ denote the true distribution function under the model. We denote dependence on n by superscript. Let t and n be such that $F^n(t)<1$. Then, for any t such that $Y(t) > 0$,

$$\frac{\hat{F} - F^n}{1 - F^n} \int \frac{1 - \hat{F}_-}{1 - F^n} \frac{J}{Y} dM \tag{5.6.45}$$

on $[0, t]$. Let

$$H = \frac{(1 - \hat{F}_-)J}{(1 - F^n)Y} \tag{5.6.46}$$

and

$$Z = \int H \, dM. \tag{5.6.47}$$

H is a bounded predictable process and M is a square integrable martingale on $[0, t]$, and hence $Z^2 - \langle Z, Z \rangle$ is a martingale on $[0, t]$ where

$$\begin{aligned} \langle Z, Z \rangle &= \int H^2 \, d\langle M, M \rangle \\ &= \int \frac{(1 - \hat{F}_-)^2}{(1 - F^n)^2} \frac{J}{Y} (1 - \Delta G^n) \, dG^n \end{aligned} \tag{5.6.48}$$

since

$$\langle M, M \rangle = \int Y(1 - \Delta G^n) \Delta G^n. \tag{5.6.49}$$

Note that $\langle Z, Z \rangle$ is a predictable nondecreasing right-continuous process, zero at time zero. By the martingale property and Doob's optional sampling theorem

$$\mathrm{E}(Z(T)^2) = E(< Z, Z > (T)). \tag{5.6.50}$$

We will now prove the weak consistency of the product limit estimator.

Theorem 5.6.5 (Consistency). Let $t > 0$ be such that

(5.6.51) (i) $Y(t) \xrightarrow{p} \infty$ as $n \to \infty$,

(ii) $\overline{\lim_n} F^n(t-) < 1$.

Then

$$(5.6.52)\quad \text{(i)}\quad \sup_{s\in[0,t]} |\hat{F}(s) - F^n(s)| \xrightarrow{p} 0 \quad \text{as } n\to\infty,$$

$$\text{(ii)}\quad \sup_{s\in[0,t]} \left| \int_0^s \frac{dN}{Y} - G^n(s) \right| \xrightarrow{p} 0 \quad \text{as } n\to\infty.$$

Proof. Since $t>0$ such that (5.6.51) holds, it follows that

$$(5.6.53)\qquad P\left(\frac{\hat{F}-F^n}{1-F^n} = Z \text{ on } [0,t]\right) \to 1 \quad \text{as } n\to\infty$$

from (5.6.45) and the definition of Z. Furthermore

$$(5.6.54)\qquad \underline{\lim}_n \inf_{s\in[0,t)} (1-F^n(s)) > 0$$

from (5.6.51). In order to prove (5.6.52)(i), it is sufficient to show that

$$(5.6.55)\qquad \sup_{s\in[0,t)} Z^2(s) \xrightarrow{p} 0.$$

Note that, for any $\varepsilon>0$ and $\eta>0$,

$$(5.6.56)\quad P\left(\sup_{0\leqslant s<t} Z^2(s) \geqslant \varepsilon\right) \leqslant \frac{\eta}{\varepsilon} + P(\langle Z,Z\rangle(t-) > \eta)$$
$$\leqslant \frac{\eta}{\varepsilon} + P\left(\frac{G^n(t-)}{1-F^n(t-)^2\,Y(t)} > \eta\right)$$

by (5.6.48) and Proposition 1.10.25. In view of inequality (5.6.51) and ε and η being arbitrary, it follows that

$$(5.6.57)\qquad \sup_{0\leqslant s\leqslant t} Z^2(s) \xrightarrow{p} 0$$

and hence

$$(5.6.68)\qquad \sup_{0\leqslant s\leqslant t} |\hat{F}(s) - F^n(s)| \xrightarrow{p} 0.$$

On the other hand, on the event $[Y(t)>0]$,

$$(5.6.59)\quad \Delta\hat{F}(t) - \Delta F^n(t) = (1 - \hat{F}(t-))\frac{\Delta N(t)}{Y(t)} - (1 - F^n(t-))J(t)\Delta G^n(t)$$

by relations (5.6.19) and (5.6.20). In order to complete the proof of (5.6.52)(i), it is enough to show that

$$(5.6.60)\qquad \frac{\Delta N(t)}{Y(t)} - J(t)\Delta G^n(t) \xrightarrow{p} 0.$$

Observe that

$$(5.6.61)\qquad \int \frac{dN}{Y} - \int J\, dG^n = \int Y^{-1}\, dM$$

is a square integrable martingale on $[0, t]$ such that

$$(5.6.62)\qquad \left\langle \int Y^{-1}\, dM, \int Y^{-1}\, dM \right\rangle = \int \left(\frac{J}{Y}\right)(1 - \Delta G^n)dG^n,$$

and hence, one can again apply Proposition 1.10.25 to show that

$$(5.6.63)\qquad P\left(\sup_{0 \leqslant s \leqslant t} \left| \int_0^s \frac{dN}{Y} - \int_0^s J\, dG^n \right| \geqslant \varepsilon \right) \to 0$$

as $n \to \infty$ for all $\varepsilon > 0$, proving (5.6.60) and also (5.6.52)(ii). This completes the proof.

Gill (1980) proved the following theorem giving the weak convergence of the product limit estimator. We omit the proof. An alternative proof due to Meier (1975) is given in Prakasa Rao (1983).

Theorem 5.6.6 (Weak Convergence). Suppose that $F^n = F$ for all n and that Y/n converges uniformly on $[0, \infty)$ to a function y in probability as $n \to \infty$. Then

$$(5.6.64)\qquad n^{1/2}(\hat{F} - F) \xrightarrow{\mathscr{L}} (1 - F)Z^\infty \quad \text{as } n \to \infty$$

on $D(\tau)$, where $\tau = \{t : y(t) > 0\}$ and Z^∞ is a zero-mean Gaussian process with independent increments and

$$(5.6.65)\qquad \operatorname{Var}(Z^\infty(t)) = \int_0^t \frac{I_{[0,1]}(\Delta G)}{1 - \Delta G} \frac{dG}{Y}.$$

Here $D(\tau)$ is the space of functions with discontinuities utmost of the first kind on the set τ endowed with the Skorokhod topology.

Remarks. The limiting variance can be consistently estimated by

$$n\int_0^t \frac{I[\Delta N < Y]}{Y - \Delta N}\frac{dN}{Y} = \frac{\hat{V}(t)}{(1-\hat{F}(t))^2} \tag{5.6.66}$$

if $\hat{F}(t) < 1$, where $\hat{V}(t)$ is as defined in (5.6.43). The method used by Gill (1980) is a martingale approach as in the case of his proof of consistency coupled with a central limit theorem due to Rebolledo (1980) and the standard theory of weak convergence [cf. Billingsley (1968)].

Nonparametric inference for a class of semi-Markov processes with censored observations has been studied recently by Voelkel and Crowley (1984) using the theory of semimartingales. They show that a particular class of semi-Markov models fit into the multiplicative intensity model discussed in Section 5.5 after a random time change.

REFERENCES

Aalen, O. (1978). Nonparametric inference for a family of counting processes. *Ann. Statistics* **6**, 701–726.

Aalen, O. (1982). Practical applications of the nonparametric statistical theory for counting processes. Statistical Research Report. Institute of Mathematics, University of Oslo.

Anderson, P. K., Borgan, Ø., Gill, R., and Keiding, N. (1982). Linear nonparametric tests for comparison of counting processes with applications to censored survival data. *Inst. Statist. Rev.* **50**, 219–258.

Basawa, I. V. and Prakasa Rao, B. L. S. (1980). *Statistical Inference for Stochastic Processes*, Academic Press, New York.

Billingsley, P. (1968). *Convergence of Probability Measures*, Wiley, New York.

Bremaud, P. (1981). *Point Processes and Queues: Martingale Dynamics*, Springer-Verlag, Berlin.

Burke, M. D., Csörgö, M., Csörgö, S., and Revesz, P. (1978). Approximations of the empirical process when parameters are estimated. *Ann. Probability* **7**, 790–810.

Chatterjee, S. D. and Mandrekar, V. (1978). Equivalence and singularity of Gaussian measures and applications. *Probabilistic Analysis and Related Topics*. Vol. 1 (Ed. A. Bharucha-Reid), Academic Press, New York.

Chernoff, H. and Lehmann, E. L. (1954). The use of maximum likelihood estimates in χ^2 test for goodness-of-fit. *Ann. Math. Statist.* **25**, 579–586.

Chibisov, D. (1969). *Limit Theorems for Empirical Distribution Functions*, Lecture Notes, Indian Statistical Institute, Calcutta.

Csörgö, S. (1981). The empirical characteristic process when parameters are estimated. *Contribution to Probability* (Ed. J. Gani and V. K. Rohatgi), Academic Press, New York.

Durbin, J. (1973). Weak convergence of the sample distribution function when parameters are estimated. *Ann. Statistics* **1**, 279–290.

Durbin, J. and Knott, M. (1972). Components of Cramer–Von Mises Statistic I. *J. Roy. Statist. Soc. B* **34**, 290–307.

Gill, R. D. (1980). *Censoring and Stochastic Integrals.* Math. Centrum Tracts 124, Amsterdam.

Hajek, J. and Sidak, Z. (1967). *Theory of Rank Tests*, Academic Press, New York.

Hinkley, D. (1970). Inference about the change point in a sequence of random variables. *Biometrika* **57**, 1–17.

Jacobsen, M. (1982). *Statistical Analysis for Counting Processes*, Lecture Notes in Statistics, No. 12, Springer-Verlag, New York.

Johansen, S. (1978). The product limit estimator as maximum likelihood estimator. *Scand. J. Statistics.* **5**, 195–199.

Kaplan, E. L. and Meier, P. (1958). Nonparametric estimation from incomplete observations. *J. Amer. Statist. Assoc.* **53**, 457–481.

Khmaladze, E. V. (1979). The use of ω^2-tests for testing parametric hypotheses. *Theory Probab. Appl.* **24**, 283–301.

Khmaldaze, E. V. (1981). The martingale approach to the theory of nonparametric tests of fit. *Theory Probab. Appl.* **26**, 240–265.

Khmaladze, E. V. (1982). Some applications of the theory of martingales to statistics. *Russian Math. Surveys* **37**, 215–237.

Khmaladze, E. V. (1983). Martingale limit theorems for divisible statistics. *Theory Probab. Appl.* **28**, 530–548.

Liptser, R. and Shiryayev, A. N. (1980). A functional central limit theorem for semimartingales. *Theory Probab. Appl.* **25**, 667–688.

Meier, P. (1975). Estimation of a distribution function from incomplete observations. *Perspectives in Probability and Statistics.* (Ed. J. Gani), Applied Prob. Trust. Sheffield.

Mnatsakanov, R. M. (1983). Martingale limit theorems for additive statistics of selective intervals. *Theory Probab. Appl.* **28**.

Moore, D. S. (1971). A Chi-square statistic with random cell boundaries. *Ann. Math. Statist.* **42**, 147–156.

Nelson, W. (1969). Hazard plotting for incomplete failure data. *J. Quality Tech.* **1**, 27–52.

Nelson, W. (1972). Theory and applications of hazard plotting for censored failure data. *Technometrics* **14**, 945–966.

Neuhaus, G. (1971). Weak convergence of stochastic processes with multidimensional time parameter. *Ann. Math. Statist.* **42**, 1285–1295.

Parthasarathy, K. R. (1967). *Probability Measures on Metric Spaces*, Academic Press, London.

Prakasa Rao, B. L. S. (1983). *Nonparametric Functional Estimation*, Academic Press, New York.

Prakasa Rao, B. L. S. (1984). On some applications of the nonparametric statistical analysis for counting processes. *Gujarat Stat. Rev.* 21–34.

Rebolledo, R. (1980). Central limit theorems for local martingales. *Z. Wahr. und verw Geb.* **51**, 269–286.

Roy, A. R. (1956). On χ^2-statistics with variable intervals. Technical Report No. 1, Department of Statistics, Stanford University.

Shiryayev, A. N. (1981). Martingales: Recent developments, results and applications. *Int. Statist. Rev.* **49**, 199–203.

Voelkel, J. G. and Crowley, J. (1984). Nonparametric interence for a class of semi-Markov processes with censored observations. *Ann. Statistics* **12**, 142–160.

CHAPTER 6

Inference in Nonlinear Regression

6.1 EXISTENCE AND CONSISTENCY OF LEAST-SQUARES ESTIMATORS

In this section and the subsequent sections of this chapter, we shall discuss the asymptotic theory of least-squares estimators from the viewpoint of the theory of stochastic processes. An alternative approach to the discussion of the asymptotic properties of the least-squares estimator may be found in Malinvaud (1970), Jennrich (1969), and Wu (1981), among others. Nonlinear regression models frequently occur in modeling of stochastic phenomena. Several examples are given in Bard (1974).

Our approach to the problem is that of Ivanov (1984) and Prakasa Rao (1984b). We first study some necessary and sufficient conditions for the consistency of a least-squares estimator due to Wu (1981). Some results connected with the rate of convergence of a least-squares estimator are presented later.

Consider a nonlinear regression model

$$X_j = g_j(\theta) + \varepsilon_j, \qquad j \geqslant 1, \tag{6.1.0}$$

where $\varepsilon_j, j \geqslant 1$, are i.i.d. random variables with $E(\varepsilon_1) = 0$, $0 < \sigma^2 = E\varepsilon_1^2 < \infty$, and $\theta \in \Theta \subset R^k$. The problem of interest is the estimation of θ given that $X_1, \ldots, X_n$ are observed. Let

$$Q_n(\theta) = \sum_{j=1}^{n} [X_j - g_j(\theta)]^2 \tag{6.1.1}$$

and $\mathscr{B}^n$ denote the Borel σ-algebra of subsets of R^n. A $\mathscr{B}^n$-measurable mapping

$\hat{\theta}_n: R^n \to \bar{\Theta}$ for which

$$Q_n(\hat{\theta}_n) = \inf_{\theta \in \bar{\Theta}} Q_n(\theta) \tag{6.1.2}$$

is called a *least-squares estimator* (LSE) for the parameter θ based on the observations X_j, $1 \leqslant j \leqslant n$. Here $\bar{\Theta}$ denotes the closure of the set Θ.

Theorem 6.1.1 (Wu, 1981). Let $X_j = g_j(\theta) + \varepsilon_j$, where $\theta \in \Theta \subset R^k$, $g_j(\theta)$, are some functions defined on Θ and ε_j are i.i.d. random variables with a common distribution G that has a positive almost everywhere (with respect to the Lebesgue measure), and absolutely continuous density g with finite Fisher information, that is,

$$\int_{-\infty}^{\infty} \frac{g^{(1)^2}(x)}{g(x)} dx < \infty. \tag{6.1.3}$$

If there exists an estimator $\bar{\theta}_n(X_1, \ldots, X_n)$ such that

$$\bar{\theta}_n(X_1, \ldots, X_n) \xrightarrow{p} \theta \quad \text{as } n \to \infty, \tag{6.1.4}$$

then

$$\phi_n(\theta_1, \theta_2) = \sum_{i=1}^{n} (g_i(\theta_1) - g_i(\theta_2))^2 \to \infty \quad \text{as } n \to \infty \tag{6.1.5}$$

for all $\theta_1 \neq \theta_2$ in Θ.

We first prove a proposition before giving a proof of this theorem.

Proposition 6.1.2 (Akahira and Takeuchi, 1981). Let $(\Omega, \mathscr{B})$ be a measurable space and $(\Omega^{(n)}, \mathscr{B}^{(n)})$ denote its n-fold direct product. Let $P_{\theta,n}$ be a probability measure defined on $(\Omega^{(n)}, \mathscr{B}^{(n)})$, $\theta \in \Theta \subset R^k$, such that

$$P_{\theta,n}(B^{(n)}) = P_{\theta,n+1}(B^{(n)} \times \Omega), \quad B^{(n)} \in \mathscr{B}^{(n)}.$$

Define

$$J_n(\theta_1, \theta_2) = \int_{\Omega^{(n)}} \left| \frac{dP_{\theta_1,n}}{d\mu_n} - \frac{dP_{\theta_2,n}}{d\mu_n} \right| d\mu_n, \tag{6.1.6}$$

where μ_n is a σ-finite measure dominating $P_{\theta_1,n}$ and $P_{\theta_2,n}$. If there exists a weakly consistent estimator θ_n for θ, then

$$\lim_{n\to\infty} J_n(\theta_1, \theta_2) = 2 \tag{6.1.7}$$

for all $\theta_1 \neq \theta_2$ in Θ.

Proof. Let θ_n be a consistent estimator of θ. Let $\theta_1 \neq \theta_2$ in Θ. Choose $\varepsilon = \frac{1}{2}\|\theta_1 - \theta_2\|$. For any $\delta > 0$, there exists an integer $n_0 \geqslant 1$ such that, for every $n \geqslant n_0$,

$$P_{\theta_i,n}(\|\theta_n - \theta\| \leqslant \varepsilon) > 1 - \tfrac{1}{4}\delta, \qquad i = 1, 2. \tag{6.1.8}$$

Let A be the event $[\|\theta_n - \theta_1\| \leqslant \|\theta_n - \theta_2\|]$. Then

$$P_{\theta_1,n}(A) > 1 - \tfrac{1}{4}\delta \quad \text{and} \quad P_{\theta_2,n}(A^c) > 1 - \tfrac{1}{4}\delta. \tag{6.1.9}$$

Note that

$$\begin{aligned} J_n(\theta_1, \theta_2) &= \int_{\Omega^{(n)}} \left| \frac{dP_{\theta_1,n}}{d\mu_n} - \frac{dP_{\theta_2,n}}{d\mu_n} \right| d\mu_n = 2 \sup_{B \in \mathscr{B}^{(n)}} |P_{\theta_1,n}(B) - P_{\theta_2,n}(B)| \\ &\geqslant \int_A \left(\frac{dP_{\theta_1,n}}{d\mu_n} - \frac{dP_{\theta_2,n}}{d\mu_n} \right) d\mu_n + \int_{A^c} \left(\frac{dP_{\theta_2,n}}{d\mu_n} - \frac{dP_{\theta_1,n}}{d\mu_n} \right) d\mu_n \\ &\geqslant 2 - 2P_{\theta_1,n}(A^c) - 2P_{\theta_2,n}(A) \\ &\geqslant 2 - \delta. \end{aligned} \tag{6.1.10}$$

Letting $\delta \to 0$, we obtain that

$$\underline{\lim}_{n\to\infty} J_n(\theta_1, \theta_2) \geqslant 2. \tag{6.1.11}$$

On the other hand, it is clear that

$$\overline{\lim}_{n\to\infty} J_n(\theta_1, \theta_2) \leqslant 2 \tag{6.1.12}$$

from (6.1.10). Relations (6.1.11) and (6.1.12) prove the proposition.

Remarks. Another necessary condition for the existence of a consistent estimator is that

$$\lim_{n\to\infty} I_n(\theta_1, \theta_2) = \infty, \tag{6.1.13}$$

where

$$I_n(\theta_1, \theta_2] = \int_{\Omega^{(n)}} \frac{dP_{\theta_1,n}}{d\mu_n} \log\left(\frac{dP_{\theta_1,n}}{dP_{\theta_2,n}}\right) d\mu_n \tag{6.1.14}$$

is the Kullback–Leibler information provided that the support for each n $\{\omega^{(n)}: dP_{\theta,n}/d\mu_n > 0\}$ does not depend on $\theta \in \Theta$. For proof, see Akahira and Takeuchi (1981, p. 24).

Proposition 6.1.3 (Shepp, 1965). Let $Y = (Y_1, Y_2, \ldots)$ be a sequence of i.i.d. random variables and $a = (a_1, a_2, \ldots)$ be a real sequence. Let μ denote the probability measure corresponding to Y and μ^a denote the probability measure corresponding to $Y + a$. Then the following results hold.

(i) If $\sum_n a_n^2 = \infty$, then $\mu \perp \mu^a$.

(ii) Suppose the distribution function of Y_1 has finite Fisher information I. Then μ is equivalent to μ^a if $\sum_n a_n^2 < \infty$ and $\mu \perp \mu^a$ if $\sum_n a_n^2 = \infty$.

(iii) If μ is equivalent to μ^a for all a with $\sum_n a_n^2 < \infty$, then $I < \infty$.

For proof, see Shepp (1965). For related results on absolutely continuity and singularity of measures, see Section 1.12. The result given above is a consequence of a result of Kakutani given in Theorem 1.12.1.

We now give the proof of Theorem 6.1.1.

Proof. In view of Proposition 6.1.2, if there exists a consistent estimator θ_n, then $\lim_{n\to\infty} J_n(\theta_1, \theta_2) = 2$ for any $\theta_1 \neq \theta_2$ in Θ. Hence the probability measures P_{θ_1} and P_{θ_2} corresponding to θ_1 and θ_2 are mutually singular for any $\theta_1 \neq \theta_2$ in Θ (see Section 1.3). Since the sequence $\{X_j\}$ under θ_1 is a translate of the sequence $\{X_j\}$ under θ_2, it follows from Proposition 6.1.3 that the fact that measures P_{θ_1} and P_{θ_2} are singular with respect to each other implies that

$$\sum_{j=1}^{\infty} [g_j(\theta_1) - g_j(\theta_2)]^2 = \infty$$

for all $\theta_1 \neq \theta_2$ in Θ since ε_j has finite Fisher information by hypothesis. This completes the proof.

Remarks. No conditions on the moments of the random variable ε_i are used in Theorem 6.1.1.

Example 6.1.1. Consider the model

$$X_j = e^{-\alpha j} + \varepsilon_j, \qquad j \geq 1,$$

where $\alpha \in (0, 2\pi)$ and the ε_j are i.i.d. with finite Fisher information. There exists no consistent estimator for α since $\sum_{j=1}^{\infty}(e^{-\alpha j} - e^{-\beta j})^2 < \infty$ for any α and β in $(0, 2\pi)$. Hence the least-squares estimator is inconsistent.

The next result is concerned with the strong consistency of a least-squares estimator $\hat{\theta}_n$.

Proposition 6.1.4. Suppose that, for any $\delta > 0$,

$$\underline{\lim}_{n\to\infty} \inf_{\|\theta-\theta_0\| \geq \delta} \{Q_n(\theta) - Q_n(\theta_0)\} > 0 \quad \text{a.s. } [P_{\theta_0}]. \tag{6.1.15}$$

Then

$$\hat{\theta}_n \to \theta_0 \quad \text{a.s. } [P_{\theta_0}] \text{ as } n \to \infty. \tag{6.1.16}$$

Proof. Note that, if $\hat{\theta}_n \not\to \theta_0$ a.s. $[P_{\theta_0}]$, then there exists $\delta > 0$ such that

$$P_{\theta_0}\left\{\overline{\lim_{n}} \|\hat{\theta}_n - \theta_0\| \geq \delta\right\} > 0,$$

which in turn implies that

$$P_{\theta_0}\left(\underline{\lim}_{n} \inf_{\|\theta-\theta_0\| \geq \delta} \{Q_n(\theta) - Q_n(\theta_0)\} \leq 0\right) > 0.$$

The following theorem gives a necessary and sufficient condition for strong consistency in case the parameter space Θ is finite.

Theorem 6.1.5 (Wu, 1981). Suppose the parameter space Θ is finite. Then

$$\phi_n(\theta, \theta_0) = \sum_{j=1}^{n} (g_j(\theta) - g_j(\theta_0))^2 \to \infty \quad \text{for all } \theta \neq \theta_0 \tag{6.1.17}$$
$$\Rightarrow \hat{\theta}_n \to \theta_0 \quad \text{a.s. } [P_{\theta_0}].$$

Conversely, if the support of ε_i is neither bounded above nor bounded below, then

$$\hat{\theta}_n \to \theta_0 \quad \text{a.s. } [P_{\theta_0}] \Rightarrow \phi_n(\theta, \theta_0) \to \infty \quad \text{for all } \theta \neq \theta_0 \text{ in } \Theta. \tag{6.1.18}$$

Proof. We first prove sufficiency. Since Θ is finite, it is sufficient to verify

(6.1.15) for each $\theta \in \Theta$ in view of Proposition 6.1.4. Note that

$$Q_n(\theta) - Q_n(\theta_0) = -2 \sum_{i=1}^{n} \varepsilon_i [g_i(\theta) - g_i(\theta_0)] + \sum_{i=1}^{n} [g_i(\theta) - g_i(\theta_0)]^2. \tag{6.1.19}$$

But $\phi_n(\theta, \theta_0) = \sum_{i=1}^{n} [g_i(\theta) - g_i(\theta_0)]^2 \to \infty$ for $\theta \neq \theta_0$. Hence

$$Q_n(\theta) - Q_n(\theta_0) \to \infty \quad \text{a.s. } [P_{\theta_0}] \text{ for } \theta \neq \theta_0, \tag{6.1.20}$$

which implies that $\hat{\theta}_n \to \theta_0$ a.s. $[P_{\theta_0}]$ as $n \to \infty$.

Conversely suppose $\hat{\theta}_n \to \theta_0$ a.s. $[P_{\theta_0}]$ but

$$\phi_n(\theta_1, \theta_0) = \sum_{i=1}^{n} [g_i(\theta_1) - g_i(\theta_0)]^2 \to c < \infty \tag{6.1.21}$$

for some $\theta_1 \neq \theta_0$. One can construct a random variable Y with $EY = 0$ and $\text{Var}[Y] = c$ such that

$$\sum_{i=1}^{n} [g_i(\theta_1) - g_i(\theta_0)] \varepsilon_i \to Y \quad \text{a.s. } [P_{\theta_0}]. \tag{6.1.22}$$

Hence

$$Q_n(\theta_1) - Q_n(\theta_0) \to -2Y + c \quad \text{a.s. } [P_{\theta_0}]. \tag{6.1.23}$$

It is sufficient to prove that $P_{\theta_0}(-2Y + c < 0) > 0$. Since the number c is not specified, it is sufficient to show that the support of Y is not bounded above. But this is clear since

$$\sum_{i=1}^{n} [g_i(\theta) - g_i(\theta_0)] \varepsilon_i$$

has unbounded support for any n and hence for Y in the almost sure limit.

Remarks. Jennrich (1969) and Wu (1981) have given sufficient conditions for strong consistency of a least-squares estimator. We shall not discuss them here.

In view of Lemma 3.1.2, it can be checked that, if $g_j(\theta)$ are continuous in θ for all j and $\bar{\Theta}$ is compact, then there exists a least-squares estimator as described above. In the following, we assume that $\hat{\theta}_n$ is such an estimator.

Let $\mathbf{d}_n = \mathbf{d}_n(\theta)$, $\theta \in \Theta$, be a diagonal matrix of order $k \times k$ and d_{in}, $1 \leqslant i \leqslant k$, be the diagonal elements. Define

$$\phi_n(\theta_1, \theta_2) = \sum_{j=1}^{n} [g_j(\theta_1) - g_j(\theta_2)]^2, \qquad \theta_1, \theta_2 \in \bar{\Theta}, \tag{6.1.24}$$

as before and, for any fixed $\theta \in \Theta$, set

$$\psi_n(\mathbf{u}_1, \mathbf{u}_2) = \phi_n(\theta + n^{1/2}\mathbf{d}_n^{-1}\mathbf{u}_1, \theta + n^{1/2}\mathbf{d}_n^{-1}\mathbf{u}_2), \tag{6.1.25}$$

where $\mathbf{u}_1, \mathbf{u}_2 \in \overline{U_n(\theta)}$ and

$$U_n(\theta) = \{\theta^* : \theta + n^{1/2}\mathbf{d}_n^{-1}\theta^* \in \Theta\}. \tag{6.1.26}$$

For any $\tau > 0$, let $B(\tau) = \{\mathbf{u} \in R^k : \|\mathbf{u}\| \leqslant \tau\}$, where $\|\mathbf{u}\|$ is the ordinary Euclidean norm in R^k.

Let $K \subset \Theta$ be compact. Suppose that the following regularity conditions hold.

(R_1) For any $\varepsilon > 0$ and $\rho > 0$, there exists $\delta > 0$ such that for $n > n_0$ (depending on K),

$$\sup_{\theta \in K} \sup_{\substack{\mathbf{u}_1, \mathbf{u}_2 \in B(\rho) \cap \overline{U_n(\theta)} \\ \|\mathbf{u}_1 - \mathbf{u}_2\| \leqslant \delta}} n^{-1}\psi_n(\mathbf{u}_1, \mathbf{u}_2) \leqslant \varepsilon. \tag{6.1.27}$$

(R_2) For some $\rho_0 > 0$ and any $r \in (0, \rho_0]$, there $\Delta > 0$ and $\tau > 0$ such that for $n > n_0$,

$$\text{(i)} \quad \inf_{\theta \in K} \inf_{\mathbf{u} \in B(\rho_0) \setminus B^0(r) \cap \overline{U_n(\theta)}} n^{-1}\psi_n(\mathbf{0}, \mathbf{u}) \geqslant \tau, \tag{6.1.28}$$

$$\text{(ii)} \quad \inf_{\theta \in K} \inf_{\mathbf{u} \in \overline{U_n(\theta)} \setminus B^0(\rho_0)} n^{-1}\psi_n(\mathbf{0}, \mathbf{u}) \geqslant 4\sigma^2 + \Delta. \tag{6.1.29}$$

(R_3) $E|\varepsilon_1|^s < \infty$ for some integer $s \geqslant 3$ and $\sigma^2 = E\varepsilon_1^2$.

Theorem 6.1.6. Suppose conditions (R_1)–(R_3) hold. Then, for any $\rho > 0$,

$$\sup_{\theta \in K} P_\theta\{\|n^{-1/2}\mathbf{d}_n(\theta)(\hat{\theta}_n - \theta)\| \geqslant \rho\} = o(n^{-(s-2)/2}). \tag{6.1.30}$$

Proof. Let

$$\text{(i)} \quad W_n(\theta_1, \theta_2) = \sum_{j=1}^{n} \varepsilon_j(g_j(\theta_1) - g_j(\theta_2)), \tag{6.1.31}$$

$$\text{(ii)} \quad V_n(\theta_1, \theta_2) = \begin{cases} \phi_n^{-1}(\theta_1, \theta_2) W_n(\theta_1, \theta_2), & \theta_1, \theta_2 \in \bar{\Theta}, \quad \theta_1 \neq \theta_2 \\ 0, & \theta_1 = \theta_2. \end{cases}$$

Let

$$V_n(\mathbf{u}) = V_n(\theta + n^{1/2}\mathbf{d}_n^{-1}\mathbf{u}, \theta). \tag{6.1.32}$$

Since $\hat{\theta}_n$ is a LSE,

$$\begin{aligned}\sum_{j=1}^{n} \varepsilon_j^2 = Q_n(\theta) &\geqslant Q_n(\hat{\theta}_n) \\ &= \sum_{j=1}^{n} \varepsilon_j^2 + \phi_n(\hat{\theta}_n, \theta)(1 - 2V_n(\hat{\theta}_n, \theta))\end{aligned} \tag{6.1.33}$$

and hence

$$\phi_n(\hat{\theta}_n, \theta)(2V_n(\hat{\theta}_n, \theta) - 1) \geqslant 0. \tag{6.1.34}$$

Therefore, for any $r > 0$ and $n > n_0$ such that

$$\inf_{\theta\in K} \inf_{\mathbf{u}\in \overline{U_n(\theta)}\setminus B^0(r)} \psi_n(\mathbf{0}, \mathbf{u}) > 0,$$

it follows that

$$\sup_{\theta\in K} P_\theta\{\|n^{-1/2}\mathbf{d}_n(\theta)(\hat{\theta}_n - \theta)\| \geqslant r\} \leqslant \sup_{\theta\in K} P_\theta\left\{\sup_{\mathbf{u}\in \overline{U_n(\theta)}\setminus B^0(r)} V_n(\mathbf{u}) \geqslant \tfrac{1}{2}\right\}. \tag{6.1.35}$$

Let $r > 0$ be fixed and choose ρ_0 and τ as in condition (R_2). Note that, for any $\theta \in K$,

$$\begin{aligned}P_\theta\left\{\sup_{\mathbf{u}\in \overline{U_n(\theta)}\setminus B^0(r)} V_n(\mathbf{u}) \geqslant \tfrac{1}{2}\right\} &\leqslant P_\theta\left\{\sup_{\mathbf{u}\in \overline{U_n(\theta)}\setminus B^0(\rho_0)} V_n(\mathbf{u}) \geqslant \tfrac{1}{2}\right\} \\ &\quad + P_\theta\left\{\sup_{\mathbf{u}\in \overline{U_n(\theta)}\cap(B(\rho_0)\setminus B^0(r))} V_n(\mathbf{u}) \geqslant \tfrac{1}{2}\right\} \\ &= P_1 + P_2 \quad \text{(say)}.\end{aligned} \tag{6.1.36}$$

Observe that

$$\begin{aligned}P_1 &\leqslant P_\theta\left\{\frac{1}{n}\sum_{j=1}^{n} \varepsilon_j^2 \geqslant \tfrac{1}{4} \inf_{u\in \overline{U_n(\theta)}\setminus B^0(\rho_0)} n^{-1}\psi_n(\mathbf{0}, \mathbf{u})\right\} \\ &\leqslant P_\theta\left\{\frac{1}{n}\sum_{j=1}^{n} \varepsilon_j^2 \geqslant \frac{\Delta}{4}\right\} \equiv \pi_n\left(\frac{\Delta}{4}\right)\end{aligned} \tag{6.1.37}$$

by (R_2)(ii). The last term is $o(n^{-(s-2)/2})$ under condition (R_3) by Proposition 1.14.16 [cf. Petrov (1975), Chapter 4, Theorems 27 and 28].

Let us consider the set $B(\rho_0)\backslash B^0(r)$ in R^k. Since this set is compact, it can be covered by a finite number of open sets $F^{(1)},\ldots,F^{(m)}$, each of diameter less than δ given by condition (R_1) for any given ε and ρ. Hence

$$P_2 \leqslant \sum_{i=1}^{m} P_\theta\left\{\sup_{\mathbf{u}\in \overline{U_n(\theta)}\cap F^{(i)}} V_n(\mathbf{u}) \geqslant \tfrac{1}{2}\right\}. \tag{6.1.38}$$

Let $\mathbf{u}^{(i)}\in F^{(i)}\cap \overline{U_n(\theta)}$, $1\leqslant i\leqslant m$. Note that

$$\begin{aligned} P_\theta\left\{\sup_{\mathbf{u}\in \overline{U_n(\theta)}\cap F^{(i)}} V_n(\mathbf{u}) \geqslant \tfrac{1}{2}\right\} &\leqslant P_\theta\{|V_n(\mathbf{u}^{(i)})|\geqslant \tfrac{1}{4}\} \\ &\quad + P_\theta\left\{\sup_{\mathbf{u}_1,\mathbf{u}_2\cap \overline{U_n(\theta)}\cap F^{(i)}} |V_n(\mathbf{u}_1)-V_n(\mathbf{u}_2)|\geqslant \tfrac{1}{4}\right\} \\ &= P_3+P_4 \quad \text{(say)}. \end{aligned} \tag{6.1.39}$$

In view of Proposition 1.14.13, it follows from condition (R_2)(i) that there exist a constant $c_s>0$ such that

$$\begin{aligned} P_3 &\leqslant c_s E|\varepsilon_1|^s[\psi_n(\mathbf{0},\mathbf{u}^{(i)})]^{-s/2} \\ &\leqslant c_s E|\varepsilon_1|^s \tau^{-s/2} n^{-s/2}. \end{aligned} \tag{6.1.40}$$

On the other hand,

$$\begin{aligned} |V_n(\mathbf{u}_1)-V_n(\mathbf{u}_2)| &\leqslant |W_n(\theta+n^{1/2}\mathbf{d}_n^{-1}\mathbf{u}_1,\theta)|\,|\psi_n^{-1}(\mathbf{u}_1,\mathbf{0})-\psi_n^{-1}(\mathbf{u}_2,\mathbf{0})| \\ &\quad + \psi_n^{-1}(\mathbf{u}_2,\mathbf{0})|W_n(\theta+n^{1/2}\mathbf{d}_n^{-1}\mathbf{u}_1,\theta+n^{1/2}\mathbf{d}_n^{-1}\mathbf{u}_2)| \end{aligned} \tag{6.1.41}$$

and

$$\begin{aligned} &|\psi_n^{-1}(\mathbf{u}_1,\mathbf{0})-\psi_n^{-1}(\mathbf{u}_2,\mathbf{0})| \\ &\quad \leqslant 2^{1/2}\psi_n^{1/2}(\mathbf{u}_1,\mathbf{u}_2)\{\psi_n^{-1/2}(\mathbf{u}_1,\mathbf{0})\psi_n^{-1}(\mathbf{u}_2,\mathbf{0}) \\ &\qquad + \psi_n^{-1}(\mathbf{u}_1,\mathbf{0})\psi_n^{-1/2}(\mathbf{u}_2,\mathbf{0})\}. \end{aligned} \tag{6.1.42}$$

Therefore, for $\mathbf{u}_1,\mathbf{u}_2\in \overline{U_n(\theta)}\cap F^{(i)}$,

$$\begin{aligned} |V_n(\mathbf{u}_1)-V_n(\mathbf{u}_2)| &\leqslant \left(\sum_{j=1}^{n}\varepsilon_j^2\right)^{1/2}\psi_n^{1/2}(\mathbf{u}_1,\mathbf{u}_2)\{(1+2^{1/2})\psi_n^{-1}(\mathbf{u}_2,\mathbf{0}) \\ &\quad + 2^{1/2}\psi_n^{-1/2}(\mathbf{u}_1,\mathbf{0})\psi_n^{-1/2}(\mathbf{u}_2,\mathbf{0})\} \\ &\leqslant (1+2^{3/2})\varepsilon^{1/2}\tau^{-1}\left(n^{-1}\sum_{j=1}^{n}\varepsilon_j^2\right)^{1/2} \end{aligned} \tag{6.1.43}$$

and hence

$$P_4 = o(n^{-(s-2)/2}) \tag{6.1.44}$$

by Proposition 1.14.16 if ε is chosen so that $\{\tau^2/16(1+2^{3/2})^2\varepsilon\} > \sigma^2$. Combining the estimates for P_1, P_2, P_3, and P_4, we obtain the result stated in Theorem 6.1.6. This completes the proof of the theorem.

Remarks. A one-dimensional version of a variation of the above result was proved in Ivanov (1976a). This result was extended to the case of dependent errors ε_i in Prakasa Rao (1984a) and an exponential rate for Gaussian errors was given in Prakasa Rao (1984c). The rate of convergence for the multidimensional case was also obtained in Prakasa Rao (1984d) under stronger conditions on the moments of $\{\varepsilon_i\}$, where ε_i are independent but not necessarily identically distributed random variables.

Ivanov (1984) proved that, if the condition $E|\varepsilon_1|^s < \infty$ for s such that $s^2 > s + k$ holds, there exists a constant $c > 0$ such that

$$\sup_{\theta \in K} P_\theta\{\|\mathbf{d}_n(\theta)(\theta_n - \theta)\| \geqslant c(\log n)^{1/2}\} = o(\bar{n}^{(s-2)/2}) \tag{6.1.45}$$

for any compact $K \subset \Theta$ under some additional conditions. We shall not go into the details here.

6.2 ASYMPTOTIC NORMALITY OF A LEAST-SQUARES ESTIMATOR

Consider the nonlinear regression model

$$X_j = g_j(\theta) + \varepsilon_j, \qquad j \geqslant 1, \tag{6.2.1}$$

as in Section 6.1, where $\{g_n(\theta), n \geqslant 1\}$ is a sequence of functions possibly nonlinear in $\theta \in \Theta$, open in R and $\{\varepsilon_n, n \geqslant 1\}$ are i.i.d. random variables with mean zero and finite known variance $0 < \sigma^2 < \infty$. Without loss of generality, we assume that $\sigma^2 = 1$. Let

$$Q_n(\theta) = \sum_{j=1}^{n} w_j(X_j - g_j(\theta))^2, \tag{6.2.2}$$

where $\{w_i\}$ is a sequence of positive numbers called *weights*. An estimator $\hat{\theta}_n$ based on the observations $X_1, \ldots, X_n$ is called a *weighted least-squares estimator* (WLSE) corresponding to the weights $\{w_i\}$ if $\hat{\theta}_n$ is a measurable solution of the equation

(6.2.3)
$$Q_n(\hat{\theta}_n) = \inf_{\theta \in \bar{\Theta}} Q_n(\theta).$$

Define

(6.2.4)
$$\phi_n(\theta_1, \theta_2) = \sum_{i=1}^{n} [g_i(\theta_1) - g_i(\theta_2)]^2 w_i^2.$$

Suppose that the following regularity conditions hold.

(C_1) There exist $0 < k_1 < k_2 < \infty$ such that

(6.2.5)
$$nk_1(\theta_1 - \theta_2)^2 \leqslant \phi_n(\theta_1, \theta_2) \leqslant nk_2(\theta_1 - \theta_2)^2$$

for all θ_1, θ_2 in Θ.

(C_2) $g_i(\theta)$ is differentiable with respect to θ for every $i \geqslant 1$ and, for any $\theta_0 \in \Theta$, there exists a neighbourhood V_{θ_0} of θ_0 in Θ such that, for all $i \geqslant 1$,

(6.2.6)
$$|g_i(\theta) - g_i(\theta_0) - (\theta - \theta_0) g_i^{(1)}(\theta_0)| \leqslant d_i(\theta_0)|\theta - \theta_0|^2$$

for all $\theta \in V_{\theta_0}$. Here $g_i^{(1)}(\theta_0)$ denotes the derivative of $g_i(\theta)$ evaluated at θ_0. Furthermore,

(6.2.7)
$$\overline{\lim_{n}} \frac{1}{n} \sum_{i=1}^{n} d_i^2(\theta_0) < \infty.$$

(C_3) $0 < a < w_i < b < \infty$ for all $i \geqslant 1$.

Assume further that

(C_4) (i) $0 < K^2 = \lim_{n \to \infty} \frac{1}{n} \sum_{i=1}^{n} w_i^2 g_i^{(1)}(\theta_0)^2 < \infty,$

(ii) $0 < K^* = \lim_{n \to \infty} \frac{1}{n} \sum_{i=1}^{n} w_i^2 g_i^{(1)}(\theta_0)^2 < \infty.$

It is clear that, if $\hat{\theta}_n$ is a LSE minimizing $Q_n(\theta)$, it also minimizes $Q_n(\theta) - Q_n(\theta_0)$ for any fixed $\theta_0 \in \Theta$. Let

(6.2.8)
$$\begin{aligned} J_n(\phi) &= Q_n(\theta_0 + n^{-1/2}\phi) - Q_n(\theta_0) \\ &= \sum_{i=1}^{n} w_i[\varepsilon_i + g_i(\theta_0) - g_i(\theta_0 + n^{-1/2}\phi)]^2 - \sum_{i=1}^{n} w_i\varepsilon_i^2 \\ &= 2\sum_{i=1}^{n} w_i\varepsilon_i[g_i(\theta_0) - g_i(\theta_0 + n^{-1/2}\phi)] \\ &\quad + \sum_{i=1}^{n} w_i[g_i(\theta_0) - g_i(\theta_0 + n^{-1/2}\phi)]^2 \\ &= 2Z_n(\phi) + T_n(\phi) \quad \text{(say).} \end{aligned}$$

In the following discussion, we consider ϕ such that $\theta_0 + n^{-1/2}\phi \in V_{\theta_0}$.

Theorem 6.2.1. Under conditions (C_1)–(C_4), the finite-dimensional distributions of the process $Z_n(\phi)$, $\phi \in [-\tau, \tau]$, converge weakly to the corresponding finite-dimensional distributions of the process

(6.2.9)
$$Z(\phi) = K\phi\xi, \qquad -\tau \leqslant \phi \leqslant \tau,$$

where ξ is $N(0, 1)$, that is, ξ has a standard normal distribution.

Proof. We shall first show that

(6.2.10)
$$Z_n(\phi) \xrightarrow{\mathscr{L}} K\phi\xi \quad \text{as } n \to \infty$$

and

(6.2.11)
$$\frac{Z_n(\phi_1)}{\phi_1} - \frac{Z_n(\phi_2)}{\phi_2} \xrightarrow{p} 0 \quad \text{as } n \to \infty$$

for any ϕ_1, ϕ_2. Observe that $E(Z_n(\phi)) = 0$ and

$$\begin{aligned} s_n^2 &= \operatorname{Var}(Z_n(\phi)) \\ &= \sum_{i=1}^{n} w_i^2[g_i(\theta_0) - g_i(\theta_0 + n^{-1/2}\phi)]^2 \\ &= \sum_{i=1}^{n} w_i^2[n^{-1/2}\phi g_i^{(1)}(\theta_0) + n^{-1}\phi^2 h_i(\theta_n')]^2 \end{aligned}$$

where $|\theta_n' - \theta_0| < \phi n^{-1/2}$ and $|h_i(\theta_n')| \leqslant d_i(\theta_0)$ for all $i \geqslant 1$ for sufficiently large n by (C_2). Hence

(6.2.12)
$$s_n^2 = \phi^2 n^{-1} \sum_{i=1}^{n} w_i^2 g_i^{(1)}(\theta_0)^2 + \phi^4 n^{-2} \sum_{i=1}^{n} w_i^2 h_i(\theta_n')^2 + 2\phi^3 n^{-3/2} \sum_{i=1}^{n} w_i^2 g_i^{(1)}(\theta_0) h_i(\theta_n') = K^2\phi^2[1+o(1)]$$

as $n \to \infty$ by conditions (C_1)–(C_4) and the fact that

(6.2.13)
$$\overline{\lim_{n\to\infty}} \frac{1}{n} \sum_{i=1}^{n} |g_i^{(1)}(\theta_0)|^r < \infty, \qquad 0 \leqslant r \leqslant 2,$$

which follows as a consequence of (C_3) and (C_4). Observe that

(6.2.14)
$$Z_n(\phi) = \sum_{k=1}^{n} a_{nk}\varepsilon_k, \qquad a_{nk} = w_k[g_k(\theta_0) - g_k(\theta_0 + n^{-1/2}\phi)]$$

and

(6.2.15)
$$\frac{Z_n(\phi)}{s_n} \xrightarrow{\mathscr{L}} N(0,1)$$

provided

(6.2.16)
$$\max_{1\leqslant k\leqslant n} \frac{a_{nk}^2}{s_n^2} \to 0 \quad \text{as } n \to \infty$$

by the central limit theorem stated in Proposition 1.4.6 (Eicker, 1963) as $\varepsilon_k, k \geqslant 1$, are i.i.d. random variables with mean zero and finite positive variance. But relation (6.2.16) holds if

(6.2.17)
$$\max_{1\leqslant k\leqslant n} a_{nk}^2 \to 0 \quad \text{as } n \to \infty$$

since $s_n \to K\phi > 0$ as $n \to \infty$ by (6.2.12). Note that

(6.2.18)
$$\begin{aligned}\max_{1\leqslant k\leqslant n} a_{nk}^2 &= \max_{1\leqslant k\leqslant n} w_k^2[g_k(\theta_0) - g_k(\theta_0 + n^{-1/2}\phi)]^2 \\ &\leqslant \max_{1\leqslant k\leqslant n} w_k^2[g_k^{(1)}(\theta_0)n^{-1/2}\phi + n^{-1}\phi^2 h_k(\theta_n')]^2 \\ &\leqslant 2n^{-1}\left\{\max_{1\leqslant k\leqslant n} w_k^2 g_k^{(1)}(\theta_0)^2\right\}\phi^2 \\ &\quad + 2n^{-2}\phi^4 b^2 \max_{1\leqslant k\leqslant n} d_k^2(\theta_0).\end{aligned}$$

The last term on the right-hand side of (6.2.18) tends to zero since

$$\overline{\lim_{n}} \frac{1}{n} \max_{1 \leqslant k \leqslant n} d_k^2(\theta_0) < \infty \tag{6.2.19}$$

by (C_2). Furthermore,

$$\frac{1}{n} \max_{1 \leqslant k \leqslant n} w_k^2 g_k^{(1)}(\theta_0)^2 \to 0 \tag{6.2.20}$$

by the following elementary lemma since

$$0 < \lim_{n \to \infty} \frac{1}{n} \sum_{k=1}^{n} w_k^2 g_k^{(1)}(\theta_0)^2 < \infty$$

by (C_4). We state a general version of the lemma.

Lemma 6.2.2 (Wu, 1981). Let $\mathbf{x}_i$, $1 \leqslant i \leqslant n$, be vectors in R^k such that there exists $\tau_n \uparrow \infty$ and $\lim_{n\to\infty} \tau_{n-1}/\tau_n = 1$ with $\tau_n^{-1} \sum_{i=1}^{n} \mathbf{x}_i \mathbf{x}_i'$ converging to a positive definite matrix $\mathbf{\Sigma}$. Then

$$\max_{1 \leqslant i \leqslant n} \mathbf{x}_i' \left(\sum_{i=1}^{n} \mathbf{x}_i \mathbf{x}_i' \right)^{-1} \mathbf{x}_i \to 0.$$

We omit the proof of this lemma and continue with the main proof of Theorem 6.2.1.

Relations (6.2.18)–(6.2.20) together with earlier remarks prove that

$$Z_n(\phi) \xrightarrow{\mathscr{L}} N(0, K^2 \phi^2] \quad \text{as } n \to \infty. \tag{6.2.21}$$

On the other hand,

$$\begin{aligned} E\left[\frac{Z_n(\phi_1)}{\phi_1} - \frac{Z_n(\phi_2)}{\phi_2}\right]^2 &= \frac{1}{\phi_1^2 \phi_2^2} E[\phi_2 Z_n(\phi_1) - \phi_1 Z_n(\phi_2)]^2 \\ &= \frac{1}{\phi_1^2 \phi_2^2} \sum_{i=1}^{n} w_i^2 \{\phi_2 [g_i(\theta_0) - g_i(\theta_0 + n^{-1/2}\phi_1)] \\ &\quad - \phi_1 [g_i(\theta_0) - g_i(\theta_0 + n^{-1/2}\phi_2)]\}^2 \end{aligned} \tag{6.2.22}$$

$$=\frac{1}{\phi_1^2\phi_2^2}\sum_{i=1}^{n} w_i^2\{\phi_2[g_i^{(1)}(\theta_0)n^{-1/2}\phi_1+n^{-1}\phi_1^2 h_{i1}(\theta')]$$

$$-\phi_1[g_i^{(1)}(\theta_0)n^{-1/2}\phi_2+n^{-1}\phi_2^2 h_{i2}(\theta'')]\}^2$$

$$=n^{-2}\left(\sum_{i=1}^{n} d_i^2(\theta_0)\right)O(1)=o(1)$$

by conditions (C_2) and (C_3). Hence

$$\frac{Z_n(\phi_1)}{\phi_1}-\frac{Z_n(\phi_2)}{\phi_2}\xrightarrow{p}0 \quad \text{as } n\to\infty \tag{6.2.23}$$

for any fixed ϕ_1 and ϕ_2. Relations (6.2.21) and (6.2.23) together prove the convergence of the two-dimensional distributions of the process Z_n to that of Z. Convergence of the higher-order finite-dimensional distributions of Z_n to those of Z defined by (6.2.9) can be proved by analogous arguments. This completes the proof.

Proposition 6.2.3. Under condition (C_1),

$$E[Z_n(\phi_1)-Z_n(\phi_2)]^2\leqslant k_2|\phi_1-\phi_2|^2 \tag{6.2.24}$$

for all n, ϕ_1, and ϕ_2 where k_2 is given by (C_1).

Proof. Note that

$$E[Z_n(\phi_1)-Z_n(\phi_2)]^2=E\left[\sum_{i=1}^{n} w_i\varepsilon_i\{g_i(\theta_0+n^{-1/2}\phi_2)-g_i(\theta_0+n^{-1/2}\phi_1)\}\right]^2$$

$$=\sum_{i=1}^{n} w_i^2\{g_i(\theta_0+n^{-1/2}\phi_2)-g_i(\theta_0+n^{-1/2}\phi_1)\}^2$$

$$\leqslant k_2(\phi_2-\phi_1)^2$$

by (C_1) where k_2 is independent of n, ϕ_1, and ϕ_2.

Theorem 6.2.4. Suppose conditions (C_1)–(C_4) hold. Then, for any $\tau>0$, the sequence of processes $\{Z_n(\phi),\ -\tau\leqslant\phi\leqslant\tau\}$ converges in distribution on $C[-\tau,\tau]$ to the process $\{Z(\phi),-\tau\leqslant\phi\leqslant\tau\}$.

Proof. Observe that $Z_n(0)=0$ for all n and $Z_n\in C[-\tau,\tau]$. The result now follows from Theorem 6.2.1 and Proposition 6.2.3 in view of Proposition

1.10.12 [cf. Billingsley (1968)]. Since $T_n(\phi) \in C[-\tau, \tau]$ for all $n \geqslant 1$ and

(6.2.25) $$T_n(\phi) \to K^* \phi^2 \quad \text{as } n \to \infty$$

uniformly in $\phi \in [-\tau, \tau]$, where $T_n(\phi)$ is as defined by (6.2.8), it follows from Slutsky's type theorem for random elements taking values in Polish spaces that the sequence of processes

(6.2.26) $$J_n(\phi) = 2Z_n(\phi) + T_n(\phi), \qquad -\tau \leqslant \phi \leqslant \tau,$$

converges in distribution to the process

(6.2.27) $$J(\phi) = 2K\phi\xi + K^*\phi^2, \qquad -\tau \leqslant \phi \leqslant \tau,$$

where K^* is given by (C_4).

Hence we have the following result.

Theorem 6.2.5. Suppose conditions (C_1)–(C_4) hold. Then, for any $\tau > 0$, the sequence of processes $\{J_n(\phi), -\tau \leqslant \phi \leqslant \tau\}$ converges in distribution on $C[-\tau, \tau]$ to the process $\{J(\phi), -\tau \leqslant \phi \leqslant \tau\}$ defined in (6.2.27).

Remarks. The result proved above is due to Prakasa Rao (1984b). Under conditions (C_1) and (C_3), Ivanov (1976a) proved that

(6.2.28) $$P_{\theta_0}(n^{1/2}|\hat{\theta}_n - \theta_0| > \rho) = O(\rho^{-2}).$$

In fact, he proved that, if $E|\varepsilon_1|^k < \infty$ for some $k \geqslant 2$ and (C_1) and (C_3) hold, then

(6.2.29) $$P_{\theta_0}\{n^{1/2}|\hat{\theta}_n - \theta_0| > \rho\} \leqslant c\rho^{-k},$$

where c is a constant independent of n and ρ. The method of proof is similar to that given in Section 6.1 [see Ivanov (1976a) or Prakasa Rao (1984a)]. It is easy to check that, if $k > 2$, then relation (6.2.29) implies that

$$\hat{\theta}_n \xrightarrow{\text{a.s.}} \theta_0 \quad \text{as } n \to \infty$$

by the Borel–Cantelli lemma. In particular, it follows that

(6.2.30) $$\hat{\theta}_n \in [\theta_0 - \tau n^{-1/2}, \theta_0 + \tau n^{-1/2}]$$

with probability approaching one as $\tau \to \infty$ and $n \to \infty$.

In view of (6.2.8) and Theorem 6.2.5, one can again use Proposition 1.10.12 to conclude that

$$n^{1/2}(\hat{\theta}_n - \theta_0) \xrightarrow{\mathscr{L}} \hat{\phi} \quad \text{as } n \to \infty, \tag{6.2.31}$$

where $\hat{\phi}$ is defined by the relation

$$J(\hat{\phi}) = \min_{\phi} J(\phi). \tag{6.2.32}$$

Observe that

$$\hat{\phi} = \frac{-K}{K^*} \xi \tag{6.2.33}$$

and hence

$$n^{1/2}(\hat{\theta}_n - \theta_0) \xrightarrow{\mathscr{L}} N\left(0, \frac{K^2}{K^{*2}}\right).$$

We have now the following theorem giving the asymptotic distribution of the least-squares estimator.

Theorem 6.2.6. Suppose conditions (C_1)–(C_4) hold. Then

$$P_{\theta_0}(n^{1/2}|\hat{\theta}_n - \theta_0| > \rho) = O(\rho^{-2}) \tag{6.2.34}$$

and

$$n^{1/2}(\hat{\theta}_n - \theta_0) \xrightarrow{\mathscr{L}} N(0, K^2/K^{*2}) \quad \text{as } n \to \infty. \tag{6.2.35}$$

Remarks. In particular, if $w_i = 1$ for all i and the least-squares estimator is the ordinary LSE, then it follows that

$$\hat{\theta}_n \xrightarrow{p} \theta_0 \quad \text{as } n \to \infty \tag{6.2.36}$$

and

$$n^{1/2}(\hat{\theta}_n - \theta_0) \xrightarrow{\mathscr{L}} N(0, K^{*-1}) \quad \text{as } n \to \infty, \tag{6.2.37}$$

where

(6.2.38) $$K^* = \lim_{n\to\infty} \frac{1}{n} \sum_{i=1}^{n} g_i^{(1)}(\theta_0)^2.$$

The regularity conditions (C_1)–(C_4) do not involve the second derivatives of $g_i(\theta)$, as has been the case in the classical approach [cf. Wu (1981) and Jennrich (1969)]. However, Wu (1981) considers a more general growth rate condition than the one given by (C_1). We now discuss an example where the classical approach to asymptotic theory of LSE is not applicable.

Example 6.2.1. Consider the nonlinear regression model

(6.2.39) $$X_i = |\gamma_i - \theta| + \varepsilon_i, \qquad i \geqslant 1, \qquad \theta \in \Theta \subset R,$$

where $\varepsilon_i, i \geqslant 1$, are i.i.d. random variables with mean 0 and variance 1. $\{\gamma_i, i \geqslant 1\}$ is a sequence of real numbers satisfying the conditions:

(6.2.40) (i) $$\sum_{i=1}^{n} (|\gamma_i - \theta_1| - |\gamma_i - \theta_2|)^2 \geqslant nk_1(\theta_1 - \theta_2)^2$$

for some constant $k_1 > 0$ and

(ii) $$\sum_{i=1}^{n} [|\gamma_i - \theta| - |\gamma_i - \theta - n^{-1/2}\phi_1|][|\gamma_i| - |\gamma_i - n^{-1/2}\phi_2|] \to \phi_1\phi_2 \quad \text{as } n \to \infty$$

for all ϕ_1 and ϕ_2 and $\theta \in \Theta$. It can be shown that modified versions of conditions (C_1)–(C_4) hold and

(6.2.41) $$\hat{\theta}_n \xrightarrow{p} \theta_0 \quad \text{as } n \to \infty$$

and

(6.2.42) $$n^{1/2}(\hat{\theta}_n - \theta_0) \xrightarrow{\mathscr{L}} N(0, 1) \quad \text{as } n \to \infty.$$

For details, see Prakasa Rao (1984b). It is easy to check that the sequence $\gamma_i = i$ satisfies the relations (6.2.40)(i) and (ii) if Θ is compact and hence the least-squares estimator $\hat{\theta}_n$ for the model

(6.2.43) $$Y_i = |i - \theta| + \varepsilon_i, \qquad i \geqslant 1,$$

is consistent and asymptotically normal provided Θ is compact and $\{\varepsilon_i\}$ are i.i.d. with mean Θ and variance 1. Furthermore

$$\sup_{\theta} P_\theta(n^{1/2}|\hat{\theta}_n - \theta| > \rho) = O(\rho^{-2}). \tag{6.2.44}$$

A more general version of the Example 6.2.1 is discussed in Section 6.3.

Remarks. Bounds of the Berry–Esseen type and asymptotic expansion for the distribution of LSE are given in Ivanov (1976a, b). We have discussed asymptotic properties of LSE in the one-parameter case. The results in this case are of limited practical interest since most nonlinear models have multiparameters. We now discuss some results in the multiparameter case due to Prakasa Rao (1984c).

Consider the nonlinear regression model

$$X_i = g_i(\theta) + \varepsilon_i, \qquad i \geqslant 1, \tag{6.2.45}$$

where $\{g_i(\theta), i \geqslant 1\}$ is a sequence of functions, possibly nonlinear in $\theta \in \Theta$, open in R^k, and ε_i, $i \geqslant 1$, are independent random variables with mean 0 and finite variances. Let

$$Q_n(\theta) = \sum_{i=1}^{n} [X_i - g_i(\theta)]^2. \tag{6.2.46}$$

An estimator $\hat{\theta}_n$ based on the observations X_i, $1 \leqslant i \leqslant n$, is called a *least-squares estimator* (LSE) if $\hat{\theta}_n$ is a measurable solution of the equation

$$Q_n(\hat{\theta}_n) = \inf_{\theta \in \Theta} Q_n(\theta). \tag{6.2.47}$$

Define

$$\phi_n(\theta_1, \theta_2) = \sum_{i=1}^{n} [g_i(\theta_1) - g_i(\theta_2)]^2. \tag{6.2.48}$$

Suppose the following regularity conditions hold:

(C_1') There exist $0 < k_1 < k_2 < \infty$ such that

$$nk_1 \|\theta_1 - \theta_2\|^2 \leqslant \phi_n(\theta_1, \theta_2) \leqslant nk_2 \|\theta_1 - \theta_2\|^2$$

for all θ_1, θ_2 in Θ.

(C_2') $g_i(\theta)$ has partial derivatives with respect to $\theta = (\theta^{(1)}, \ldots, \theta^{(k)})$ and, for any $\theta_0 \in \Theta$, there exists a neighborhood V_{θ_0} of θ_0 in Θ such that, for all $i \geqslant 1$,

$$|g_i(\theta) - g_i(\theta_0) - (\theta - \theta_0)'\nabla g_i(\theta_0)| \leqslant \alpha_i(\theta_0)\|\theta - \theta_0\|^2$$

for all $\theta \in V_{\theta_0}$. Here $\nabla g_i(\theta_0)$ denotes the gradient vector $(\partial g_i/\partial\theta^{(1)}, \ldots, \partial g_i/\partial\theta^{(k)})$ evaluated at θ_0. Furthermore,

$$\overline{\lim_n} \frac{1}{n} \sum_{i=1}^{n} \alpha_i^2(\theta_0) < \infty.$$

(C_3') $\sup_i E|\varepsilon_i|^m < \infty$ for some $m > k$ and $m \geqslant 4$ and $\inf_i E(\varepsilon_i^2) \geqslant \sigma^2 > 0$. Let $\sigma_i^2 = E\varepsilon_i^2$.

(C_4') The matrices

$$\text{(6.2.49)} \quad \text{(i)} \quad \mathbf{K} = \lim_{n\to\infty} \frac{1}{n} \sum_{i=1}^{n} [\nabla g_i(\theta_0)\nabla g_i(\theta_0)']\sigma_i^2,$$

$$\text{(ii)} \quad \mathbf{K}^* = \lim_{n\to\infty} \frac{1}{n} \sum_{i=1}^{n} [\nabla g_i(\theta_0)\nabla g_i(\theta_0)']$$

are finite and positive definite.

Clearly, if $\hat{\theta}_n$ is a LSE minimizing $Q_n(\theta)$, then $\hat{\theta}_n$ minimizes $Q_n(\theta) - Q_n(\theta_0)$ for any fixed $\theta_0 \in \Theta$. Let

$$\text{(6.2.50)} \qquad \begin{aligned} J_n(\phi) &= Q_n(\theta_0 + n^{-1/2}\phi) - Q_n(\theta_0) \\ &= \sum_{i=1}^{n} [\varepsilon_i + g_i(\theta_0) - g_i(\theta_0 + n^{-1/2}\phi)]^2 - \sum_{i=1}^{n} \varepsilon_i^2 \\ &= 2\sum_{i=1}^{n} \varepsilon_i[g_i(\theta_0) - g_i(\theta_0 + n^{-1/2}\phi)] \\ &\quad + \sum_{i=1}^{n} [g_i(\theta_0) - g_i(\theta_0 + n^{-1/2}\phi)]^2 \\ &= 2Z_n(\phi) + T_n(\phi) \quad \text{(say)}. \end{aligned}$$

We consider ϕ such that $\theta_0 + n^{-1/2}\phi \in V_{\theta_0}$ defined by (C_2').

Theorem 6.2.7. Under conditions (C_1') and (C_3'), given a compact set $K \subset \Theta$, there exists a constant $C_k > 0$ independent of n and ρ such that

$$\text{(6.2.51)} \qquad \sup_{\theta_0 \in K} P_{\theta_0}(n^{1/2}\|\hat{\theta}_n - \theta_0\| > \rho) \leqslant C_k\rho^{-m}.$$

Theorem 6.2.8. Under conditions (C_1')–(C_4'), the sequence of random fields $\{J_n(\phi), \|\phi\| \leqslant \tau\}$ converge in distribution to the random field $\{J(\phi), \|\phi\| \leqslant \tau\}$ on the space C of continuous functions on the sphere $B(\tau) \equiv \{\phi : \|\phi\| \leqslant \tau\}$ for any fixed $\tau > 0$, where

$$J(\phi) = \phi' K^* \phi + 2\phi' \xi \tag{6.2.52}$$

and ξ has a k-dimensional multivariate normal distribution with mean vector 0 and covariance matrix $\mathbf{K}$, hereafter denoted by $N_k(\mathbf{0}, \mathbf{K})$.

Proofs of Theorems 6.2.7 and 6.2.8 are given in Prakasa Rao (1984d, e) and are similar to those given earlier for the one-dimensional parameter. A result similar to Proposition 1.10.12 under the relation (1.10.19) can be obtained to derive the above theorems. Prakasa Rao (1984e) proved Theorem 6.2.8 under conditions (C_1'), (C_2'), (C_4'), and (C_3^*) given below.

As a consequence of Theorem 6.2.7 and 6.2.8 one can obtain the following theorem again by using an analogue of Proposition 1.10.12 for random fields [cf. Prakasa Rao (1984e)].

Theorem 6.2.9. Suppose conditions (C_1')–(C_4') hold. Then, for any $\rho > 0$,

$$\sup_{\theta_0 \in K} P_{\theta_0}(n^{1/2} \|\hat{\theta}_n - \theta_0\| > \rho) = O(\rho^{-m}) \tag{6.2.53}$$

and

$$n^{1/2}(\hat{\theta}_n - \theta_0) \xrightarrow{\mathscr{L}} N_k(\mathbf{0}, \mathbf{K}^{*-1}\mathbf{K}\mathbf{K}^{*-1}) \quad \text{as} \quad n \to \infty, \tag{6.2.54}$$

where $\mathbf{K}$ and $\mathbf{K}^*$ are defined by (C_4').

Remarks. Condition (C_3') is too strong, even though for practical purposes one can assume that errors are bounded, in which case (C_3') holds. It is a technical condition that is necessary because of the method of proof. We conjecture that condition (C_3') can be relaxed to

$$\sup_i E|\varepsilon_i|^{2+\delta} < \infty, \qquad \inf_i E\varepsilon_i^2 \geqslant \sigma^2 > 0 \tag{C_3^*}$$

for some $\delta > 0$ in the Theorem 6.2.9. This condition does not depend on k, the dimension of the parameter space. Theorem 6.2.8 is proved under condition (C_3^*) in addition to some other regularity conditions in Prakasa Rao (1984e). By arguments analogous to those in Inagaki and Ogata (1975) and Prakasa

Rao (1972). An alternative set of sufficient conditions [in addition to (C_3^*)] for consistency and asymptotic normality are given in Prakasa Rao (1986). However, we do not present them here, because they are not yet satisfactory. Wu (1981) has given an alternative set of sufficient conditions for the asymptotic normality of LSE in the multivariate case. These conditions involve conditions on the second-order partial derivatives of the functions g_j.

We now present an example where conditions (C_1')–(C_4') hold.

Example 6.2.2. Consider the following nonlinear regression model

$$X_i = g_i(\theta) + \varepsilon_i, \qquad i \geq 1, \tag{6.2.55}$$

where

$$g_i(\theta) = \exp\{-\theta_1 \lambda_i e^{-\theta_2 \mu_i}\}, \qquad \theta = (\theta_1, \theta_2). \tag{6.2.56}$$

The model arises in the following way as given in Bard (1974, p. 123). Let x_i be the fraction remaining at time y_i of a chemical compound A under going a first-order reaction. Suppose x_i is given by the relation

$$x_i = e^{-\gamma y_i}, \tag{6.2.57}$$

where γ is the rate constant with the initial condition $x_1 = 1$ at $y_1 = 0$. The rate constant γ depends in turn on the absolute temperature t_i through the relation

$$\gamma = \theta_1 e^{-\theta_2/t_i},$$

where θ_1 is the frequency constant and θ_2 is the activation energy measured in some unit. Hence

$$x_i = \exp\{-\theta_1 y_i e^{-\theta_2 t_i^{-1}}\} + e_i, \tag{6.2.58}$$

where e_i is the error and this is the model stated above.

We make the following assumptions about the sequences $\{\lambda_i\}$ and $\{\mu_i\}$ and the parameters θ_1 and θ_2:

(A_1) There exists $a > 0$ and $b > 0$ such that

(i) $0 < a \leq \lambda_i \leq b < \infty, \quad i \geq 1,$

(ii) $0 < a \leq \mu_i \leq b < \infty, \quad i \geq 1$

and there exist $\alpha_i > 0$, $i = 1, 2$, and $\beta_1 > 0$ such that

(iii) $0 < \beta_1 \leqslant \theta_1 \leqslant \alpha_1 < \infty$,

(iv) $0 \leqslant \theta_2 \leqslant \alpha_2 < \infty$.

(A_2) Let

$$h_i(\theta_1, \theta_2) = \exp\{-\theta_1 \lambda_i e^{-\theta_2 \mu_i} - \theta_2 \mu_i\}, \qquad \theta = (\theta_1, \theta_2).$$

Suppose that

$$\frac{1}{n} \sum_{i=1}^{n} \lambda_i^2 h_i^2 \to k_{11}(\theta),$$

$$\frac{1}{n} \sum_{i=1}^{n} \mu_i \lambda_i^2 h_i^2 \to \frac{k_{12}(\theta)}{\theta_1}$$

and

$$\frac{1}{n} \sum_{i=1}^{n} \mu_i^2 \lambda_i^2 h_i^2 \to \frac{k_{22}(\theta)}{\theta_1^2},$$

where $k_{ij}(\theta) < \infty$ for $i = 1, 2$ and $j = 1, 2$ with $k_{12}(\theta) = k_{21}(\theta)$.

(A_3) Let

$$\mathbf{K}(\theta) = \begin{bmatrix} k_{11}(\theta) & k_{12}(\theta) \\ k_{21}(\theta) & k_{22}(\theta) \end{bmatrix}.$$

Suppose the matrix $\mathbf{K}(\theta)$ is positive definite for all $\theta \in \Theta$.

(A_4) Suppose $\{\varepsilon_i\}$ are i.i.d. with mean 0 and $E|\varepsilon_i|^4 < \infty$. Let $\sigma^2 = E\varepsilon_1^2$.

We leave it to the reader to check that conditions (C_1')–(C_4') hold with $m = 4$ under conditions (A_1)–(A_4). Hence, for any compact $K \subset \Theta$, there exists $C_K > 0$ such that

$$\sup_{\theta \in K} P_\theta(n^{1/2} \|\hat{\theta}_n - \theta\| > \rho) < C_K \rho^{-4} \tag{6.2.59}$$

and

$$n^{1/2}(\hat{\theta}_n - \theta_0) \xrightarrow{\mathscr{L}} N_2(\mathbf{0}, \sigma^2 \mathbf{K}(\theta_0)^{-1}) \quad \text{as} \quad n \to \infty \tag{6.2.60}$$

by Theorem 6.2.9. In particular, relation (6.2.59) implies that

$$\hat{\theta}_n \to \theta_0 \quad \text{a.s.} \quad [P_{\theta_0}] \quad \text{as} \quad n \to \infty.$$

6.3 ASYMPTOTIC PROPERTIES OF A LEAST-SQUARES ESTIMATOR IN A NONREGULAR MODEL

In this section, we shall study asymptotic properties of LSE in a nonregular nonlinear regression model. Results in this section are due to Prakasa Rao (1985).

Consider the nonlinear regression model

$$X_i = |\gamma_i - \theta|^\lambda + \varepsilon_i, \qquad i \geqslant 1, \tag{6.3.1}$$

where $0 < \lambda < \frac{1}{2}$ and $\{\varepsilon_i\}$ are i.i.d. random variables with mean 0 and finite positive variance σ^2 (say) $\sigma^2 = 1, \theta \in \Theta$ compact contained in R. Suppose $\{\gamma_i, i \geqslant 1\}$ is a real sequence with the property

$$\sum_{i=1}^{n} \{|\gamma_i - \theta|^\lambda - |\gamma_i - \theta_0|^\lambda\}^2 = 2nC(\lambda)|\theta - \theta_0|^{2\lambda+1}(1 + o(1)) \tag{6.3.2}$$

as $n \to \infty$, where $C(\lambda) \neq 0$ and there exist $0 < k_1 < k_2 < \infty$ such that

$$\begin{aligned} nk_1|\theta_1 - \theta_2|^{2\lambda+1} &\leqslant \sum_{i=1}^{n} \{|\gamma_i - \theta_1|^\lambda - |\gamma_i - \theta_2|^\lambda\}^2 \\ &\leqslant nk_2|\theta_1 - \theta_2|^{2\lambda+1} \end{aligned} \tag{6.3.3}$$

for all θ_1 and θ_2 in Θ. Let $\hat{\theta}_n$ be a LSE of θ obtained by minimizing

$$Q_n(\theta) = \sum_{i=1}^{n} (X_i - |\gamma_i - \theta|^\lambda)^2 \tag{6.3.4}$$

over $\theta \in \Theta$. Clearly $\hat{\theta}_n$ also minimizes

$$\begin{aligned} Q_n(\theta) - Q_n(\theta_0) &= \sum_{i=1}^{n} (X_i - |\gamma_i - \theta|^\lambda)^2 - \sum_{i=1}^{n} (X_i - |\gamma_i - \theta_0|^\lambda)^2 \\ &= 2\sum_{i=1}^{n} \varepsilon_i(|\gamma_i - \theta|^\lambda - |\gamma_i - \theta_0|^\lambda) \\ &\quad + \sum_{i=1}^{n} (|\gamma_i - \theta|^\lambda - |\gamma_i - \theta_0|^\lambda)^2 \end{aligned} \tag{6.3.5}$$

for any fixed $\theta_0 \in \Theta$. In view of (6.3.2), it is easy to see that

$$E_{\theta_0}[Q_n(\theta) - Q_n(\theta_0)] = \sum_{i=1}^{n} (|\gamma_i - \theta|^\lambda - |\gamma_i - \theta_0|^\lambda)^2 = 2nC(\lambda)|\theta - \theta_0|^{2\lambda+1}(1 + o(1)), \tag{6.3.6}$$

$$\text{var}_{\theta_0}[Q_n(\theta) - Q_n(\theta_0)] = 4\sum_{i=1}^{n} (|\gamma_i - \theta|^\lambda - |\gamma_i - \theta_0|^\lambda)^2 = 8nC(\lambda)|\theta - \theta_0|^{2\lambda+1}(1 + o(1)), \tag{6.3.7}$$

and in general, there exists $k_3 > 0$ independent of n, θ, θ_0 such that

$$E_{\theta_0}[Q_n(\theta) - Q_n(\theta_0)] \leqslant nk_3|\theta - \theta_0|^{2\lambda+1} \tag{6.3.8}$$

and

$$\text{var}_{\theta_0}[Q_n(\theta) - Q_n(\theta_0)] \leqslant 4nk_3|\theta - \theta_0|^{2\lambda+1}. \tag{6.3.9}$$

Furthermore,

$$\begin{aligned} \text{cov}_{\theta_0}&[Q_n(\theta_1) - Q_n(\theta_0), Q_n(\theta_2) - Q_n(\theta_0)] \\ &= 4\sum_{i=1}^{n} (|\gamma_i - \theta_1|^\lambda - |\gamma_i - \theta_0|^\lambda)(|\gamma_i - \theta_2|^\lambda - |\gamma_i - \theta_0|^\lambda) \\ &= 4nC(\lambda)[|\theta_1 - \theta_0|^{2\lambda+1} + |\theta_2 - \theta_0|^{2\lambda+1} - |\theta_1 - \theta_2|^{2\lambda+1}](1+o(1)) \end{aligned} \tag{6.3.10}$$

from the relation

$$\|\mathbf{f}\|^2 + \|\mathbf{g}\|^2 - \|\mathbf{f} - \mathbf{g}\|^2 = 2\langle \mathbf{f}, \mathbf{g}\rangle$$

for any $\mathbf{f}, \mathbf{g}$ in R^m. Let

$$J_n(\theta) = Q_n(\theta) - Q_n(\theta_0). \tag{6.3.11}$$

In view of relations (6.3.6)–(6.3.10), it can be shown by arguments similar to those in Prakasa Rao (1968a, Theorem 3.6) (cf. Proposition 3.7.2) that there exists $\eta > 0$ such that

$$\lim_{\tau\to\infty} \overline{\lim_{n\to\infty}} P_{\theta_0}\left[\inf_{|\theta-\theta_0|>\tau n^{-\rho}} \frac{J_n(\theta)}{n|\theta - \theta_0|^{2\lambda+1}} \leqslant \eta\right] = 0, \tag{6.3.12}$$

where $\rho = (2\lambda + 1)^{-1}$. In fact, the same proof shows that, for any $\tau > 0$,

(6.3.13) $$P_{\theta_0}[n^{\rho}|\hat{\theta}_n - \theta_0| > \tau] \leqslant C\tau^{-(2\lambda+1)},$$

where C is independent of n and τ. In view of inequality (6.3.12), it follows that the process $J_n(\theta)$ has a minimum in the interval $[\theta_0 - \tau n^{-\rho}, \theta_0 + \tau n^{-\rho}]$ with probability approaching 1 as $n \to \infty$ and $\tau \to \infty$. For any such $\tau > 0$, let

(6.3.14) $$R_n(\phi) = J_n(\theta_0 + \phi n^{-\rho}), \qquad \phi \in [-\tau, \tau],$$

and let $R(\phi)$ be the Gaussian process with

(6.3.15) (i) $ER(\phi) = 2C(\lambda)|\phi|^{2\lambda+1}$,

(ii) $$\operatorname{cov}[R(\phi_1), R(\phi_2)] = 4C(\lambda)[|\phi_1|^{2\lambda+1} + |\phi_2|^{2\lambda+1} - |\phi_1 - \phi_2|^{2\lambda+1}].$$

It can be shown that the sequence of processes $\{R_n(\phi), -\tau \leqslant \phi \leqslant \tau\}$ converge in distribution on $C[-\tau, \tau]$ to the Gaussian process $\{R(\phi), -\tau \leqslant \phi \leqslant \tau\}$ with mean and covariance defined by (6.3.15) [cf. Prakasa Rao (1985)]. Applying arguments similar to those given in Section 3.7 [cf. Prakasa Rao (1968a)], it now follows that

(6.3.16) $$P_{\theta_0}[n^{\rho}|\hat{\theta}_n - \theta_0| > \tau] \leqslant C\tau^{-(2\lambda+1)}$$

and

(6.3.17) $$n^{1/(1+2\lambda)}(\hat{\theta}_n - \theta_0) \xrightarrow{\mathscr{L}} \hat{\phi} \quad \text{as} \quad n \to \infty,$$

where $\hat{\phi}$ has the distribution of the location of the minimum of the Gaussian process $R(\phi)$, $-\infty < \phi < \infty$ with mean and covariance given by (6.3.15).

Remarks. Observe that $\rho = (2\lambda + 1)^{-1} > \frac{1}{2}$ if $0 < \lambda < \frac{1}{2}$. The asymptotic variance of the LSE here is of the order $O(n^{-2\rho})$, which is small compared to the asymptotic variance of the least-squares estimator in the smooth case as the latter is of the order $O(n^{-1})$ as discussed in Section 6.2 [cf. Prakasa Rao (1984a) and Ivanov (1976a)]. Condition (6.3.2) is not an unreasonable condition since it is known that

(6.3.18) $$\int_{-\infty}^{\infty} (|x - \theta_1|^{\lambda} - |x|^{\lambda})(|x - \theta_2|^{\lambda} - |x|^{\lambda})\, dx = C(\lambda)[|\theta_1|^{2\lambda+1} + |\theta_2|^{2\lambda+1} - |\theta_1 - \theta_2|^{2\lambda+1}],$$

where

$$(6.3.19)\quad C(\lambda) = \Gamma(\lambda+1)\Gamma(\tfrac{1}{2}-\lambda)(2^{2\lambda+1}\pi^{1/2}(2\lambda+1))^{-1}(2-2\cos\pi\lambda)$$

from Prakasa Rao (1966) [cf. Prakasa Rao (1968b)].

REFERENCES

Akahira, M. and Takeuchi, K. (1981). *Asymptotic Efficiency of Statistical Estimators: Concepts and Higher Order Asymptotic Efficiency*, Lecture Notes in Statistics, No. 7, Springer-Verlag, New York.

Bard, Y. (1974). *Nonlinear Parameter Estimation*, Academic Press, New York.

Billingsley, P. (1968). *Convergence of Probability Measures*, Wiley, New York.

Eicker, F. (1963). Central limit theorems for families of sequences of random variables. *Ann. Math. Statist.* **34**, 439–446.

Inagaki, N. and Ogata, Y. (1975). The weak convergence of likelihood ratio random fields and its applications. *Ann. Inst. Statist. Math.* **27**, 391–419.

Ivanov, A. V. (1976a). An asymptotic expansion for the distribution of the least squares estimator of the non-linear regression parameter. *Theory. Probab. Appl.* **21**, 557–570.

Ivanov, A. V. (1976b). The Berry–Esseen inequality for the distribution of the least squares estimate. *Mathematical Notes* **20**, 721–727.

Ivanov, A. V. (1984). Two theorems on consistency of a least squares estimator. *Theor. Probability and Math. Statist.* **28**, 25–34.

Jennrich, R. (1969). Asymptotic properties of nonlinear least squares estimators. *Ann. Math. Statist.* **40**, 633–643.

Malinvaud, E. (1970). The consistency of nonlinear regression. *Ann. Math. Statist.* **41**, 956–969.

Petrov, V. (1975). *Sums of Independent Random Variables*, Springer-Verlag, Berlin.

Prakasa Rao, B. L. S. (1966). Asymptotic distributions in some nonregular statistical problems. Ph.D. Thesis, Michigan State University.

Prakasa Rao, B. L. S. (1968a). Estimation of the location of the cusp of a continuous density. *Ann. Math. Statist.* **39**, 76–87.

Prakasa Rao, B. L. S. (1968b). On evaluating a certain integral. *Amer. Math. Monthly* **75**, 55.

Prakasa Rao, B. L. S. (1972). Maximum likelihood estimation for Markov processes. *Ann. Inst. Statist. Math.* **24**, 333–345.

Prakasa Rao, B. L. S. (1984a). The rate of convergence of the least squares estimator in nonlinear regression model with dependent errors. *J. Multivariate Analysis* **14**, 315–322.

Prakasa Rao, B. L. S. (1984b). Weak convergence of least squares process in the smooth case (preprint), Indian Statistical Institute, New Delhi. Also in: *Statistics* **17** (1986) (to appear).

Prakasa Rao, B. L. S. (1984c). On the exponential rate of convergence of the least squares estimator in the nonlinear regression model with Gaussian errors. *Statist. Probab. Lett.* **2**, 139–142.

Prakasa Rao, B. L. S. (1984d). On the rate of convergence of the least squares estimator in nonlinear regression model for multiparameter (preprint), Indian Statistical Institute, New Delhi. Also in: *J. Ramanujan Math. Soc.* **1** (1986) (to appear).

Prakasa Rao, B. L. S. (1984e). Weak convergence of least squares random field in the smooth case (preprint), Indian Statistical Institute, New Delhi. Also in: *Stat. Decisions* **4** (1986) (to appear).

Prakasa Rao, B. L. S. (1985). Asymptotic theory of least squares estimator in a nonregular nonlinear regression model. *Statist. Probab. Lett.* **3**, 15–18.

Prakasa Rao, B. L. S. (1986). The weak convergence of least squares random fields and its application. *Ann. Inst. Statist. Math.* **38** (to appear).

Shepp, L. A. (1965). Distinguishing a sequence of random variables from a translate of itself. *Ann. Math. Statist.* **36**, 1107–1112.

Whittle, P. (1960). Bounds for the moments of linear and quadratic forms in independent variables. *Theory Probab. Appl.* **5**, 302–305.

Wu, C. F. (1981). Asymptotic theory of nonlinear least squares estimation. *Ann. Statistics* **9**, 501–513.

CHAPTER 7

Von Mises Functionals

7.1 VON MISES FUNCTIONALS

In order to study the asymptotic behavior of statistics that are functionals of the empirical distribution function, von Mises (1936, 1947) proposed a technique. We now discuss some results in this direction due to Fernholz (1983). An earlier approach under different conditions is given in Serfling (1980). Examples of statistics that are functionals of the empirical distribution function include sample moments.

Suppose $X_1, X_2, \ldots, X_n$ are i.i.d. random variables with distribution function F. Let F_n be the empirical distribution function based on $X_1, \ldots, X_n$. If

$$\mu_k' = \frac{1}{n} \sum_{i=1}^{n} X_i^k$$

is the kth sample raw moment, then μ_k' is clearly a functional based on the empirical distribution function F_n since

$$\mu_k' = \int_{-\infty}^{\infty} x^k \, dF_n(x).$$

Another example where a statistic is defined implicitly is the following. Let $\psi(\cdot,\cdot)$ be a function on $R \times R$, and define T_n by $\sum_{i=1}^{n} \psi(X_i, T_n) = 0$. For instance, MLE is such a statistic under some conditions, and it is a functional based on the empirical distribution function.

Let $T(F)$ be a functional based on $F \in \mathscr{F}$, a class of distribution functions. Suppose we estimate $T(F)$ by $T(F_n)$. Heuristically, let us expand $T(F_n)$ around

$T(F)$ using a Taylor expansion given by

$$T(F_n) = T(F) + T_F^{(1)}(F_n - F) + R(F_n - F), \tag{7.1.0}$$

where $T_F^{(1)}$ is some type of derivative of $T(\cdot)$ at F and $R(\cdot)$ is the remainder term. Under some regularity conditions, it is clear that

$$\sqrt{n}\, T_F^{(1)}(F_n - F) \xrightarrow{\mathscr{L}} N(0, \sigma^2) \tag{7.1.1}$$

for some σ^2 since it is the average of i.i.d. random variables. If

$$\sqrt{n}\, R(F_n - F) \xrightarrow{p} 0, \tag{7.1.2}$$

then it follows that

$$\sqrt{n}(T(F_n) - T(F)) \xrightarrow{\mathscr{L}} N(0, \sigma^2). \tag{7.1.3}$$

It is therefore clear that one has to define the notion of derivative of a functional that is appropriate to the present context before proceeding with the asymptotic theory of such functionals.

Let $\mathbf{v}$ and $\mathbf{w}$ be topological vector spaces, and let $L(\mathbf{v}, \mathbf{w})$ be the set of continuous linear transformations from $\mathbf{v}$ to $\mathbf{w}$. Let S be a class of compact subsets of $\mathbf{v}$, and suppose further S contains all singletons. Let A be an open subset of $\mathbf{v}$.

Definition 7.1.1. A function $T: A \to \mathbf{w}$ is said to be *Hadamard differentiable* at $F \in A$ if there exists $T_F^{(1)} \in L(\mathbf{v}, \mathbf{w})$ such that for any $K \in S$

$$\lim_{t \to 0} \frac{T(F + tH) - T(F) - T_F^{(1)}(tH)}{t} = 0 \tag{7.1.4}$$

uniformly for $H \in K$. $T_F^{(1)}$ is called the *Hadamard derivative* of T at F.

Hereafter we assume that the domain and the range of the functionals considered are contained in real Banach spaces.

Proposition 7.1.1 (Chain Rule). Let $\mathbf{v}, \mathbf{w}$, and $\mathbf{z}$ be topological vector spaces with $T: \mathbf{v} \to \mathbf{w}$ and $Q: \mathbf{w} \to \mathbf{z}$. If T is Hadamard differentiable at $F \in \mathbf{v}$ and if Q is Hadamard differentiable at $T(F) \in \mathbf{w}$, then $Q \circ T$ is Hadamard differentiable at F and

$$(Q \circ T)_F^{(1)} = Q_{T(F)}^{(1)} \circ T_F^{(1)}. \tag{7.1.5}$$

Proof. See Yamamuro (1974, p. 11).

If a statistic is defined implicitly through an equation as in the case of the MLE, for instance, Hadamard differentiability of such functionals can be derived by using the following *implicit function theorem.* For proof, see Fernholz (1983, p. 21).

Theorem 7.1.2 (Fernholz, 1983). Let $(G_0, \theta_0) \in \mathbf{v} \times R^p$, and let $\mathcal{N}$ be a neighborhood of G_0 and $\mathcal{M}$ be a neighborhood of θ_0. Let $\psi: \mathcal{N} \times \mathcal{M} \to R^p$ be Hadamard differentiable at (G_0, θ_0). Suppose that

$$\psi(G_0, \theta_0) = 0 \tag{7.1.6}$$

and the second partial derivative $D_2\psi(G_0, \theta_0)$ is nonsingular. Furthermore, suppose there exists a neighborhood $\mathcal{N}_0$ of $\mathbf{0}$ in R^p such that $\psi(G, \theta) = t$ has a unique solution $T(G, t) = \theta$ for $t \in \mathcal{N}_0$ and $(G, \theta) \in \mathcal{N} \times \mathcal{M}$. If, for any compact $K \subset \mathbf{v} \times R^p$, any $\varepsilon_n \to 0$, and any sequence $\{(H_n, t_n)\} \subset K$,

$$\frac{T(G_0 + \varepsilon_n H_n, \varepsilon_n t_n) - T(G, \mathbf{0})}{\varepsilon_n} \tag{7.1.7}$$

is bounded for large n, then the function $\tau: \mathcal{N} \to R^p$ defined by $\tau(G) = T(G, \mathbf{0})$ is Hadamard differentiable at G_0, and the Hadamard derivative of τ is

$$\tau_{G_0}^{(1)} = -(D_2\psi_{(G_0,\theta_0)})^{-1} \circ D_1\psi_{(G_0,\theta_0)}. \tag{7.1.8}$$

In order to study the asymptotic behavior of functionals of type $T(F_n)$, where F_n is the empirical distribution function as defined earlier, we can suppose that the support of F is contained in $[0, 1]$ for the following reason, provided F is a continuous and strictly increasing distribution function. For, in such an event, $Y_j \equiv F(X_j)$, $1 \leqslant j \leqslant n$, are i.i.d. uniform random variables on $[0, 1]$. Let U_n be the empirical distribution function based on $Y_1, \ldots, Y_n$. Then $F_n = U_n \circ F$, and we can defined a functional $\tau(\cdot)$ by the relations

$$\tau(U_n) = T(U_n \circ F) = T(F_n) \tag{7.1.9}$$

and

$$\tau(U) = T(U \circ F) = T(F), \tag{7.1.10}$$

where U is the uniform distribution on $[0, 1]$. In general, for any distribution function G with support contained in $[0, 1]$, we define $\tau(G) = T(G \circ F)$, and the statistic T induces a functional τ on the space of all distribution functions on $[0, 1]$.

Consider the space $C[0, 1]$ with sup norm, and endow the space $D[0, 1]$ *also* with sup norm. Note that $D[0, 1]$ is no longer a separable space; it is complete metric space. The space $C[0, 1]$ can still be considered a closed subset of $D[0, 1]$. Furthermore, the elements of $D[0, 1]$ can be uniformly approximated by step functions.

Proposition 7.1.3. Let $K \subset C[0, 1]$. Then K is compact if and only if it is closed, bounded, and equicontinuous.

Proposition 7.1.4. A bounded set $K \subset D[0, 1]$ has compact closure if and only if for every $\varepsilon > 0$ there exists a partition

$$0 = t_0 < t_1 < \cdots < t_n = 1 \quad \text{and} \quad s_i \in [t_{i-1}, t_i), \quad 1 \leqslant i \leqslant n,$$

such that

$$\sup_{s \in [t_{i-1}, t_i)} |G(s_i) - G(s)| < \varepsilon, \qquad G \in K, \quad i = 1, \ldots, n.$$

For proof of Proposition 7.1.3, see Billingsley (1968) or Parthasarathy (1967). A proof of Proposition 7.1.4 is given in Fernholz (1983, p. 29). Note that, if a set is compact in $C[0, 1]$, it is also compact in $D[0, 1]$.

Let $(\Omega, \mathscr{B}, P)$ be a probability space and $Y_i, 1 \leqslant i \leqslant n$, be i.i.d. random variables defined on Ω with uniform distribution U on $[0, 1]$. Let U_n be the empirical distribution function based on $Y_1, \ldots, Y_n$. Note that U_n is an element in $D[0, 1]$. Let U_n^* be a continuous modification of U_n obtained in the following manner. Denote by $Y_{(1)}, \ldots, Y_{(n)}$ the order statistics corresponding to $Y_i, 1 \leqslant i \leqslant n$. Let $Y_{(0)} = 0$ and $Y_{(n+1)} = 1$. Note that $Y_i \neq Y_j$ for $i \neq j$ with probability 1. U_n^* assigns uniform distribution with mass $1/(n+1)$ to the interval $[Y_{(i-1)}, Y_{(i)}]$, $1 \leqslant i \leqslant n+1$. Note that $U_n^* \in C[0, 1]$ a.s. and

$$\| U_n^* - U_n \| \leqslant 1/n \quad \text{a.s.} \tag{7.1.11}$$

It is well known that

$$Z_n = \sqrt{n}(U_n^* - U) \xrightarrow{\mathscr{L}} B \quad \text{as} \quad n \to \infty \tag{7.1.12}$$

in $C[0, 1]$, where $\{B(t), 0 \leqslant t \leqslant 1\}$ is the Brownian bridge.

Suppose $\tau: C[0, 1] \to R$ is Hadamard differentiable at U. Define

$$R(tH) = \tau(U + tH) - \tau(U) - \tau_U^{(1)}(tH). \tag{7.1.13}$$

Then

$$\lim_{t\to 0}\frac{R(tH)}{t}=0 \tag{7.1.14}$$

uniformly for $H\in K$ for any compact set $K\subset C[0,1]$.

Proposition 7.1.5. Suppose that T is a statistical functional inducing a functional τ from $C[0,1]$ to R. If $\tau(\cdot)$ is Hadamard differentiable at $U\in C[0,1]$, then

$$n^{1/2}R(U_n^*-U)\xrightarrow{p}0. \tag{7.1.15}$$

Proof. Let μ_n be the probability measure induced by the process $\sqrt{n}(U_n^*-U)$ on $C[0,1]$. Since μ_n converges weakly to a probability measure μ on $C[0,1]$ by (7.1.12), the family $\{\mu_n, n\geqslant 1\}$ is tight and hence, given $\varepsilon>0$, there exists a compact set $K\subset C[0,1]$ such that

$$\mu_n(K)>1-\varepsilon,\qquad n\geqslant 1. \tag{7.1.16}$$

In view of (7.1.14),

$$\lim_{n\to\infty} n^{1/2}R(n^{-1/2}H)=0 \tag{7.1.17}$$

uniformly for $H\in K$, and hence there exists n_1 such that, for $n\geqslant n_1$,

$$|n^{1/2}R(n^{-1/2}H)|<\varepsilon,$$

uniformly for $H\in K$. Therefore, by relations (7.1.16) and (7.1.17), it follows that

$$P\{|n^{1/2}R(n^{-1/2}n^{1/2}(U_n^*-U))|<\varepsilon\}>1-\varepsilon$$

for all $n\geqslant n_1$. Since $\varepsilon>0$ is arbitrary, we have

$$n^{1/2}R(U_n^*-U)\xrightarrow{p}0. \tag{7.1.18}$$

It is sometimes important to consider functionals τ defined on $D[0,1]$ endowed with sup norm rather than on $C[0,1]$ with sup norm.

For any $H\in D[0,1]$ and $K\subset D[0,1]$, define

$$d(H,K)=\inf_{G\in K}\|H-G\|. \tag{7.1.19}$$

Lemma 7.1.6. Let $Q:D[0,1]\times R\to R$, and suppose that

(7.1.20)
$$\lim_{t\to 0} Q(H,t)=0$$

uniformly for $H\in K$ for any compact $K\subset D[0,1]$. Let $\varepsilon>0$ and $\delta_n\downarrow 0$. Then, for any compact $K\subset D[0,1]$, there exists n_0 such that

(7.1.21)
$$d(H,K)\leqslant \delta_n \Rightarrow |Q(H,\delta_n)|<\varepsilon$$

for $n\geqslant n_0$.

Proof. Obvious.

Note that $\sqrt{n}(U_n-U)$ is not a random element on $D[0,1]$ with sup norm. Let P_* denote the inner measure corresponding to P.

Lemma 7.1.7. For every $\varepsilon>0$, there exists a compact set $K\subset D[0,1]$ such that

(7.1.22)
$$P_*\{d(\sqrt{n}(U_n-U),K)\leqslant 1/\sqrt{n}\}>1-\varepsilon$$

for all $n\geqslant 1$.

Proof. Define U_n^* as before. Observe that $\| U_n^*-U_n\|\leqslant 1/n$. We have seen that, given $\varepsilon>0$, there exists a compact set $K\subset C[0,1]$ such that

$$P\{\sqrt{n}(U_n^*-U)\in K\}>1-\varepsilon$$

for all $n\geqslant 1$. But K is compact in $D[0,1]$ too. Furthermore, if $\sqrt{n}(U_n^*-U)\in K$ and $\| U_n^*-U_n\|\leqslant 1/n$, then

$$d(\sqrt{n}(U_n-U),K)\leqslant \frac{1}{\sqrt{n}}$$

and hence

$$P_*\left\{d(\sqrt{n}(U_n-U),K)\leqslant \frac{1}{\sqrt{n}}\right\}>1-\varepsilon$$

for all $n\geqslant 1$.

Proposition 7.1.8. Suppose T is a statistical functional inducing a functional τ on $D[0,1]$. Suppose further that $\tau(\cdot)$ is Hadamard differentiable at the uniform distribution $U \in D[0,1]$. Then

$$n^{1/2}R(U_n - U) \xrightarrow{p} 0. \tag{7.1.23}$$

Proof. Let $\varepsilon > 0$. In view of Lemma 7.1.7, there exists a compact set $K \subset D[0,1]$ such that

$$P_*\left\{d(\sqrt{n}(U_n - U), K) \leqslant \frac{1}{\sqrt{n}}\right\} > 1 - \frac{\varepsilon}{2} \tag{7.1.24}$$

for all $n \geqslant 1$. Hence there exists a measurable event E_n such that

$$E_n \Rightarrow \left[d(\sqrt{n}(U_n - U), K) \leqslant \frac{1}{\sqrt{n}}\right] \tag{7.1.25}$$

and

$$P(E_n) > 1 - \varepsilon. \tag{7.1.26}$$

Let $Q(H,t) = R(tH)/t$. Applying Lemma 7.1.6, we obtain that there exists n_0 such that, for $n > n_0$,

$$d(H,K) \leqslant \frac{1}{\sqrt{n}} \Rightarrow |n^{1/2}R(n^{-1/2}H)| < \varepsilon. \tag{7.1.27}$$

It can be shown that $R((U_n - U))$ is measurable (see proof of Theorem 7.1.10 given later). Choosing $H = n^{1/2}(U_n - U)$, it follows that

$$P\{|n^{1/2}R(U_n - U)| < \varepsilon\} \geqslant P(E_n) > 1 - \varepsilon.$$

Hence

$$n^{1/2}R(U_n - U) \xrightarrow{p} 0.$$

Proposition 7.1.9. Suppose T is a statistical functional inducing a functional τ on $D[0,1]$. Suppose that τ is Gateaux differentiable with derivative $\tau_U^{(1)}$ at U in the sense that, for any $H \in D[0,1]$,

$$\lim_{t \to 0} \frac{\tau(U + tH) - \tau(U) - \tau_U^{(1)}(tH)}{t} = 0. \tag{7.1.28}$$

Then the influence curve of T defined by

$$\text{(7.1.29)} \qquad \mathrm{IC}(x; F, T) = \frac{d}{dt} T(F + t(\delta_x - F))|_{t=0}$$

exists, where δ_x denotes the distribution function degenerate at x. Furthermore,

$$\text{(7.1.30)} \qquad \mathrm{IC}(x; F, T) = \tau_U^{(1)}((\delta_x - F)\circ F^{-1}).$$

Proof. Note that

$$\begin{aligned}\mathrm{IC}(x; F, T) &= \lim_{t\to 0} \frac{T(F + t(\delta_x - F)) - T(F)}{t} \\ &= \lim_{t\to 0} \frac{\tau(U + t(\delta_x - F)\circ F^{-1}) - \tau(U)}{t} \\ &= \tau_U^{(1)}((\delta_x - F)\circ F^{-1})\end{aligned}$$

since τ is Gateaux differentiable with derivative $\tau_U^{(1)}$.

Remarks. Note that Hadamard differentiability of a functional implies its Gateaux differentiability. For properties of the influence curve, see Huber (1981).

We now study the limiting properties of statistical functionals that induce functionals on either $C[0,1]$ or $D[0,1]$.

Theorem 7.1.10. Suppose T is a statistical functional and X has a continuous, strictly increasing distribution function F. Let τ be induced by T on $D[0,1]$ by the relation $\tau(G) = T(G\circ F)$ for $G\in D[0,1]$. If τ is Hadamard differentiable at the uniform distribution U on $[0,1]$ and if

$$\text{(7.1.31)} \qquad 0 < \sigma^2 = \mathrm{Var}_F\, \mathrm{IC}(x; F, T) < \infty,$$

then

$$\text{(7.1.32)} \qquad n^{1/2}(T(F_n) - T(F)) \xrightarrow{\mathscr{L}} N(0, \sigma^2).$$

Proof. Observe that

$$\begin{aligned}n^{1/2}(T(F_n) - T(F)) &= n^{1/2}(\tau(U_n) - \tau(U)) \\ &= n^{1/2}\tau_U^{(1)}(U_n - U) + n^{1/2} R(U_n - U).\end{aligned}$$

But

$$\begin{aligned}(7.1.33)\qquad \tau_U^{(1)}(U_n - U) &= \tau_U^{(1)}((F_n - F)\circ F^{-1})\\ &= \tau_U^{(1)}\left(\frac{1}{n}\sum_{i=1}^{n}(\delta_{X_i} - F)\circ F^{-1}\right)\\ &= \frac{1}{n}\sum_{i=1}^{n} \mathrm{IC}(X_i; F, T)\end{aligned}$$

by Proposition 7.1.9. Hence $\tau_U^{(1)}(U_n - U)$ is a random element in $D[0, 1]$. Since $\tau(U_n)$ is measurable, it follows that $R(U_n - U)$ is measurable and hence, by Proposition 7.1.8, we have

$$n^{1/2}R(U_n - U)\xrightarrow{p} 0.$$

The theorem now follows from the central limit theorem for i.i.d. random variables from the representation (7.1.33).

Theorem 7.1.11. Suppose T is a statistical functional and X has a continuous, strictly increasing distribution function F. Let τ be induced by T on $C[0, 1]$ by the relation $\tau(G) = \tau(G\circ F)$ for $G\in C[0, 1]$ satisfying $\tau(U) = T(F)$ and $\tau(U_n) = T(F_n)$, where U_n^* is a continuous version of $U_n = F_n\circ F^{-1}$. Suppose τ is Hadamard differentiable at U. If

$$(7.1.34)\qquad 0 < \sigma^2 = \mathrm{Var}_F\, \mathrm{IC}(x; F, T) < \infty,$$

then

$$(7.1.35)\qquad n^{1/2}(T(F_n) - T(F))\xrightarrow{\mathscr{L}} N(0, \sigma^2).$$

Proof of this theorem is similar to that of Theorem 7.1.10 in view of Proposition 7.1.5. For details, see Fernholz (1983, p. 41).

7.2 APPLICATIONS

As we have seen, Hadamard differentiability of the induced functional on $D[0, 1]$ is a sufficient condition for the asymptotic normality of estimators of the form $T(F_n)$. Fernholz (1983) discusses the asymptotic properties of various estimators like M, L, and R estimators using these results. These were studied under different conditions using different type of derivatives in Boos and Serfling (1980) and Serfling (1980). We do not discuss these results of Fernholz (1983) here. In order to illustrate the theorem, we shall sketch an example due to Fernholz (1983).

Example 7.2.1. Let $X_1, \ldots, X_n$ be i.i.d. random variables with distribution function F, which is continuous and strictly increasing. An L estimator is a linear combination of a function of order statistics. More generally, we consider statistics of the form

$$T_n = \int_0^1 h(F_n^{-1}(x))m(x)\,dx$$

as L estimators, where $m(x)$ is the density of a signed measure M on $[0, 1]$ (weight function) and h is a real-valued function. This estimator is generated by the statistical functional

$$T(F) = \int_0^1 h(F^{-1}(x))m(x)\,dx,$$

which induces the functional τ on $D[0, 1]$ by

$$\tau(G) = \int_0^1 h(F^{-1}(G^{-1}(x))m(x)\,dx$$

for $G \in D[0, 1]$. Fernholz (1983, Proposition 7.2.1) proved that if h is continuous and piecewise differentiable with bounded derivative, if m is square integrable with support $[\alpha, 1-\alpha]$ for some $\alpha > 0$, and if F is an absolutely continuous increasing distribution function, then τ is Hadamard differentiable at U, the uniform distribution in $D[0, 1]$. As a consequence, it follows that

$$n^{1/2}(T(F_n) - T(F)) \xrightarrow{\mathscr{L}} N(0, \sigma^2),$$

where

$$\sigma^2 = \mathrm{Var}_F \, \mathrm{IC}(X_1; F, T)$$

from Theorem 7.1.10. Huber (1981) shows that

$$\mathrm{IC}(x; F, T) = \int_{-\infty}^{x} h^{(1)}(y)m(F(y))\,dy - \int_{-\infty}^{\infty} (1 - F(y))h^{(1)}(y)m(F(y))\,dy.$$

Remarks. It should be interesting to examine the recent work of Serfling (1984) on generalized L, M, and R statistics using Hadamard differentiability instead of Gateaux differentiability as used in Serfling (1980, 1984). For results on almost sure behavior of statistical functionals, see Sen (1974) and Ghosh and Sen (1970).

REFERENCES

Billingsley, P. (1968). *Convergence of Probability Measures*, Wiley, New York.

Boos, D. D. and Serfling, R. J. (1980). A note on differentials and the CLT and LIL for statistical functions with applications to M-estimates. *Ann. Statistics* **8**, 618–624.

Fernholz, L. T. (1983). *Von Mises Calculus for Statistical Functions*, Lecture Notes in Statistics, No. 19, Springer-Verlag, New York.

Ghosh, M. and Sen, P. K. (1970). On the almost sure convergence of von Mises differentiable statistical functions. *Calcutta Statist. Assoc.* **19**, 41–44.

Huber, P. (1981). *Robust Statistics*, Wiley, New York.

Parthasarathy, K. R. (1967). *Probability Measures on Metric Spaces*, Academic Press, London.

Sen, P. K. (1974). Almost sure behaviour of U-statistics and von Mises differentiable statistical functions. *Ann. Statistics* **2**, 387–395.

Serfling, R. J. (1980). *Approximation Theorems of Mathematical Statistics*, Wiley, New York.

Serfling, R. J. (1984). Generalized L-, M- and R- statistics. *Ann. Statistics* **12**, 101–118.

von Mises, R. (1936). Les lois de probabilite pour les fonctions statistique. *Ann. Inst. H. Poincaré* **6**, 185–212.

von Mises, R. (1947). On the asymptotic distributions of differentiable statistical functions. *Ann. Math. Statist.* **18**, 309–348.

Yamamuro, S. (1974). *Differential Calculus in Topological Linear Spaces*. Lecture Notes in Mathematics, No. 374, Springer-Verlag, Berlin.

CHAPTER 8

Empirical Characteristic Function and Its Applications

8.1 ESTIMATION OF PARAMETERS OF STABLE LAWS

Stable distributions are of practical interest and have been used in applications to astronomy and physics [cf. Chandrasekhar (1943)] and more recently in business and economics, for instance, to provide models for changes in stock market prices [cf. Mandelbrot (1963), Fama (1965), and Leitch and Paulson (1975)]. Hence the problem of constructing estimates of parameters involved in them and study of the properties of estimators is of importance. The difficulty in using the methods developed in the earlier chapters is that explicit expressions for the densities of stable laws are known only for a few cases. We first describe some properties of stable laws.

Definition 8.1.1. A random variable X is said to have a *stable law* with parameters $(\alpha, \beta, \gamma, \lambda)$ if its characteristic function $\phi(t)$ is of the form (A):

(8.1.0) $$\log \phi(t) = it\gamma - \lambda |t|^\alpha \omega_A(t, \alpha, \beta),$$

where

(8.1.1) $$\omega_A(t, \alpha, \beta) = \begin{cases} 1 - i\beta \tan\left(\frac{\pi}{2}\alpha\right) \operatorname{sgn} t & \text{if} \quad \alpha \neq 1 \\ 1 + i\beta \frac{2}{\pi} \log|t| \operatorname{sgn} t & \text{if} \quad \alpha = 1 \end{cases}$$

and

(8.1.2) $$0 < \alpha \leqslant 2, \qquad -1 \leqslant \beta \leqslant 1, \qquad -\infty < \gamma < \infty, \qquad \lambda \geqslant 0.$$

An alternative form of representation of stable laws is the following one due to Zolotarev (1957). In this case the characteristic function $\phi(t)$ is given by the form (B):

$$\log \phi(t) = it\gamma - \lambda |t|^{\alpha} \omega_{B}(t, \alpha, \beta), \tag{8.1.3}$$

where

$$\omega_{B}(t, \alpha, \beta) = \begin{cases} \exp\left\{ -i\frac{\pi}{2}\beta K(\alpha) \operatorname{sgn} t \right\} & \text{if} \quad \alpha \neq 1 \\ 1 + i\beta\left(\frac{2}{\pi}\right) \log|t| \operatorname{sgn} t & \text{if} \quad \alpha = 1, \end{cases} \tag{8.1.4}$$

where $K(\alpha) = 1 - |1 - \alpha| = \min(\alpha, 2 - \alpha)$ and α, β, γ, and γ satisfy (8.1.2).

α is called the *index* of the stable law. The parameters in forms (A) and (B) are related by the following formulas where $\alpha_A, \beta_A, \gamma_A$, and λ_A denote the parameters in form (A) and $\alpha_B, \beta_B, \gamma_B$, and λ_B denote the parameters in form (B). Clearly

$$\alpha_A = \alpha_B = \alpha, \qquad \gamma_A = \gamma_B = \gamma. \tag{8.1.5}$$

If $\alpha \neq 1$, then

$$\text{(8.1.6) (i)} \quad \beta_A = \cot\left(\frac{\pi}{2}\alpha\right) \tan\left(\frac{\pi}{2}K(\alpha)\beta_B\right),$$

and

$$\text{(8.1.6) (ii)} \quad \lambda_A = \lambda_B \cos\left(\frac{\pi}{2}K(\alpha)\beta_B\right).$$

If $\alpha = 1$, then

$$\beta_A = \beta_B \quad \text{and} \quad \lambda_A = \lambda_B. \tag{8.1.7}$$

Let us consider the subfamily $\mathscr{F}$ of all stable laws with parameter $(\alpha, \beta, \gamma, \delta)$ where $\gamma = 0$ if $\alpha \neq 1$ and $\beta = 0$ if $\alpha = 1$. This subfamily is a three-parameter family of stable laws. The characteristic function $\phi(t)$ of any stable law belonging to this subfamily can be written in the form (C):

$$\log \phi(t) = -\exp\left\{ \nu^{-1/2}\left[\log|t| + \tau - \frac{i\pi}{2}\eta \operatorname{sgn} t + c(1 - \nu^{1/2}) \right] \right\}, \tag{8.1.8}$$

where

$$(8.1.9) \qquad \nu \geqslant \tfrac{1}{4}, \qquad |\eta| \leqslant \min(1, 2\nu^{1/2} - 1), \qquad -\infty < \tau < \infty,$$

and c is the Euler constant. (ν, η, τ) is related to $(\alpha, \beta, \gamma, \delta)$ in the form (B) by the relations

$$(8.1.10) \quad \text{(i)} \quad \nu = \alpha^{-2}, \qquad \eta = \begin{cases} \dfrac{K(\alpha)}{\alpha}\beta & \text{if } \alpha \neq 1 \\ \dfrac{2}{\pi}\arctan\left(\dfrac{\gamma}{\lambda}\right) & \text{if } \alpha = 1 \end{cases}$$

and

$$(8.1.10) \quad \text{(ii)} \quad \tau = \frac{1}{2\alpha}\log(\lambda^2 + \gamma^2) + c\left(\frac{1}{\alpha} - 1\right).$$

It is known that all stable laws are absolutely continuous. Furthermore, all stable laws belonging to the family $\mathscr{F}$ are continuous with respect to the parameter (ν, θ, τ) in the sense of weak convergence. It is again well known that if X has a stable law with parameters $(\alpha, \beta, \gamma, \lambda)$, then $E|X|^s < \infty$ for $s < \alpha$. The only stable laws for which explicit expressions for densities are known are $\alpha = 2$ (normal law); $\alpha = 1, \beta = 0$ (Cauchy law); and $\alpha = \frac{1}{2}, \beta = 1$ (Levy law). Hence, in order to study inference problems where the underlying distributions came from a stable family, one has to use other functions that have one-to-one correspondence with the densities like the characteristic functions whose forms are, of course, explicitly known. For tables and graphs of the stable probability density function, see Holt and Crow (1973).

In the following discussion, we assume that the stable law is represented in form (B).

Suppose $X_1, \ldots, X_n$ are i.i.d. with a stable law with parameters $\theta \equiv (\alpha, \beta, \gamma, \lambda)$. Let $s_{\alpha,\beta}(x)$ be the density function of a stable distribution with $\lambda = 1$ and $\gamma = 0$. $\hat{\theta}_n = (\hat{\alpha}_n, \hat{\beta}_n, \hat{\gamma}_n, \hat{\lambda}_n)$ is said to be a maximum likelihood estimator of $\theta = (\alpha, \beta, \gamma, \lambda)$ based on the sample $X_1, \ldots, X_n$ if

$$(8.1.11) \qquad L_n(\hat{\theta}_n) = \sup_{\theta} L_n(\theta),$$

where

$$(8.1.12) \qquad L_n(\theta) = \prod_{i=1}^{n} s_{\alpha,\beta}\left(\frac{X_i - \gamma}{\lambda}\right)\frac{1}{\lambda}.$$

Du Mouchel (1973) has proved that, if both α and γ are unknown, then the likelihood function will have no maximum in $0<\alpha\leqslant 2$, $-\infty<\gamma<\infty$, and $L_n(\alpha,\beta,\gamma,\lambda)\to\infty$ as $(\alpha,\gamma)\to(0,X_k)$, where X_k is any one of the observations. However, if L_n is maximized over $\alpha\geqslant\varepsilon>0$, then the maximum likelihood estimators are consistent and asymptotically normal.

Theorem 8.1.1 (Du Mouchel, 1973). The maximum likelihood estimator $\hat{\theta}_n$ for $\theta=(\alpha,\beta,\gamma,\lambda)$ based on an i.i.d. sample of size n from a stable law with parameter θ restricted to the parameter space with $\alpha\geqslant\varepsilon>0$ is consistent and asymptotically normal provided the true parameter θ_0 belongs to the interior of the parameter space (i.e., the cases $\alpha_0<\varepsilon$, $\alpha_0=2$, and $\beta_0=\pm 1$ are excluded) and the additional case $(\alpha_0=1,\beta_0\neq 0)$ is excluded.

The permissible set of values of θ in the above theorem is

$$\Theta=\{(\alpha,\beta,\gamma,\lambda):0<\varepsilon\leqslant\alpha<1 \quad \text{or} \\ 1<\alpha<2, -1<\beta<1, -\infty<\gamma<\infty, \lambda>0\}.$$

This result can be checked using the standard theory of maximum likelihood estimation [cf. Cramer (1946), Wald (1949), or Le Cam (1953)]. Dumouchel (1973) proves the result using Le Cam (1953). We omit the details.

Press (1972) has suggested the method of moments for the estimation of θ. Let $X_1, X_2,\ldots,X_n$ be i.i.d. with characteristic function $\phi(t)$ of a stable law, with index α as before. Note that

$$|\phi(t)|=e^{-\lambda|t|^\alpha}. \tag{8.1.13}$$

The function

$$\hat{\phi}(t)=\frac{1}{n}\sum_{j=1}^{n} e^{itX_j} \tag{8.1.14}$$

is called the *empirical characteristic function.* We shall study the properties of the stochastic process $\{\hat{\phi}(t), -\infty<t<\infty\}$ later in this chapter. It is easy to see that $E[\hat{\phi}(t)]=\phi(t)$ for all t and $\hat{\phi}(t)\xrightarrow{\text{a.s.}}\phi(t)$ as $n\to\infty$ by the strong law of large numbers.

Suppose $\alpha\neq 1$. For any $t_1\neq t_2$, both different from zero,

$$\lambda|t_1|^\alpha=-\log|\phi(t_1)|, \qquad \lambda|t_2|^\alpha=-\log|\phi(t_2)|$$

and hence

(8.1.15) $$\alpha = \frac{\log|\log|\phi(t_1)|/\log|\phi(t_2)||}{\log|t_1/t_2|}$$

and

(8.1.16) $$\log \lambda = \frac{\log|t_1|\log[-\log|\phi(t_2)|] - \log|t_2|\log[-\log|\phi(t_1)|]}{\log|t_1/t_2|}.$$

One can substitute $\hat{\phi}(t)$ as an estimator for $\phi(t)$ to get moment estimators $\hat{\alpha}$ and $\hat{\lambda}$ for α and λ, respectively. Similar estimators can be obtained for β and γ.

Suppose $\alpha = 1$. Then

(8.1.17) $$u(t) \equiv \mathrm{Im}[\log \phi(t)] = \gamma t - \frac{2}{\pi}\lambda\beta t \log|t|,$$

where $\mathrm{Im}[\psi(t)]$ denotes the imaginary part of the complex-valued function $\psi(t)$. Let $t_3 \neq t_4$, both nonzero. Then

(8.1.18) $$\gamma - \frac{2\lambda \log|t_k|}{\pi}\beta = \frac{u(t_k)}{t_k}, \qquad k = 3, 4.$$

Solving these equation for β and γ, we have

(8.1.19) $$\beta = \left\{\frac{u(t_3)}{t_3} - \frac{u(t_4)}{t_4}\right\} \Big/ \left\{\frac{2\lambda}{\pi}\log\left|\frac{t_4}{t_3}\right|\right\}$$

and

(8.1.20) $$\gamma = \left(\frac{\log|t_4|}{t_3}u(t_3) - \frac{\log|t_3|}{t_4}u(t_4)\right) \Big/ \log|t_4/t_3|.$$

One can estimate λ by

(8.1.21) $$\hat{\lambda} = -\log|\hat{\phi}(t_1)|/|t_1|$$

from (8.1.13) and $u(t)$ by

(8.1.22) $$\hat{u}(t) = \mathrm{Im}[\log \hat{\phi}(t)]$$
$$= \arctan\left(\sum_{j=1}^{n} \sin tX_j\right) \Big/ \left(\sum_{j=1}^{n} \cos tX_j\right)$$

and use these estimators to get $\hat{\beta}$ and $\hat{\gamma}$ from (8.1.19) and (8.1.20), respectively.

It is easy to see that these estimators are consistent since $\hat{\phi}(t)$ is a consistent estimator of $\phi(t)$. However, no optimality properties can be stated for these estimators.

Another procedure for estimating the parameter is the minimum distance method suggested by Press (1972). The procedure consists in choosing a $\theta = (\alpha, \beta, \gamma, \lambda)$ that minimizes

$$I(\theta) = \int_{-\infty}^{\infty} |\hat{\phi}(u) - \phi(u)|^2 e^{-u^2}\, du. \tag{8.1.23}$$

In general θ might be chosen to minimize

$$I_w(\theta) = \int_{-\infty}^{\infty} |\hat{\phi}(u) - \phi(u)|^2 w(u)\, du, \tag{8.1.24}$$

where $w(u)$ is a specified weight function. The motivation for this approach is that two distribution functions are equal if and only if their characteristic functions are the same. Asymptotic properties of the estimators so obtained are discussed in Thornton and Paulson (1977) and Heathcote (1977). Paulson et al. (1975) study computational procedures for estimating all the parameters of the stable law simultaneously. We shall not discuss these results here because we shall consider a more general problem of estimation in Section 8.2.

Zolotarev (1980, 1981) studied asymptotic properties of moment-type estimators when the stable law belongs to the three-parameter famly $\mathscr{F}$ defined earlier. We now briefly discuss his results.

Suppose X_i, $1 \leqslant i \leqslant n (n \geqslant 2)$ are i.i.d. with stable law with parameters ν, η, τ as specified in (8.1.10). Let

$$U_j = \operatorname{sgn} X_j \quad \text{and} \quad V_j = \log|X_j|, \qquad 1 \leqslant j \leqslant n. \tag{8.1.25}$$

Let

$$\tilde{\eta}_n = \frac{1}{n}\sum_{i=1}^{n} U_i, \qquad \tilde{\tau}_n = \frac{1}{n}\sum_{i=1}^{n} V_i. \tag{8.1.26}$$

It can be shown that $\tilde{\eta}_n$ and $\tilde{\tau}_n$ are unbiased estimators of η and τ, respectively. Let

$$\tilde{\nu}_n = \frac{6}{\pi^2} s_V^2 - \tfrac{3}{2} s_U^2 + 1 \tag{8.1.27}$$

and

$$\hat{v}_n = \max\{\tilde{v}_n, \tfrac{1}{4}(1 + |\tilde{\eta}_n|^2)\}, \tag{8.1.28}$$

where

$$s_V^2 = \frac{1}{n-1}\sum_{i=1}^{n}(V_i - \bar{V})^2, \qquad s_U^2 = \frac{1}{n-1}\sum_{i=1}^{n}(U_i - \bar{U})^2.$$

It can be shown that $\hat{v}_n$ is an asymptotically unbiased estimator of v. For details, see Zolotarev (1980).

One can extend the method described above to estimate the parameters of multivariate stable distributions. Press (1972) discussed the method of moments for the estimation of parameters in a multivariate symmetric stable law with characteristic function defined by

$$\log\phi(\mathbf{t}) = i\gamma'\mathbf{t} - \tfrac{1}{2}(\mathbf{t}'\mathbf{\Sigma}\mathbf{t})^{\alpha/2}, \tag{8.1.29}$$

where Σ is a positive definite symmetric matrix and $0 < \alpha \leqslant 2$. Zolotarev (1981) studied estimation for a spherically symmetric stable law with characteristic function defined by

$$\phi(\mathbf{t}) = \exp(-\lambda\|\mathbf{t}\|^{\alpha}), \quad 0 < \alpha \leqslant 2, \quad \lambda > 0. \tag{8.1.30}$$

Another parametrization for this family is given by

$$\log\phi(\mathbf{t}) = -\exp[(\log\|\mathbf{t}\| + \tau + c)v^{-1/2} - c], \tag{8.1.31}$$

where $v = \alpha^{-2}$, $\tau = (\log\lambda + c)\alpha^{-1} - c$, and c is the Euler constant. Zolotarev (1981) proved that, if $\mathbf{Z}$ has a spherically symmetric stable law on R^n with characteristic function defined by (8.1.30), then

$$E\|\mathbf{Z}\|^s = \lambda^{s/2}2^s\frac{\Gamma(1-s/\alpha)}{\Gamma(1-s/2)}\frac{\Gamma((n+s)/2)}{\Gamma(n/2)} \tag{8.1.32}$$

for any $0 < s < \alpha$. Let $V = \log\|\mathbf{Z}\|$. Using (8.1.32), one can compute $E(V)$ and $\operatorname{Var}(V)$ in an explicit form. Suppose $\mathbf{Z}_i, i = 1, \ldots, m$, is an i.i.d. sample distributed as $\mathbf{Z}$. Let $V_i = \log\|\mathbf{Z}_i\|$. With the aid of V_i, $1 \leqslant i \leqslant m$, Zolotarev (1981) constructed consistent and asymptotically unbiased estimators of v, τ, α, and λ.

8.2 ESTIMATION BY EMPIRICAL CHARACTERISTIC FUNCTION

We saw in Section 8.1 that the characteristic functions may be used instead of the densities or the distribution functions in problems of inference. This could be the case either because the densities and hence the likelihood function or the distribution functions may not be expressible in terms of known elementary functions, as in the case of stable laws, or the likelihood function is too complicated for computational purpose. The characteristic function may be smoother than the corresponding density even if the form of the density is known.

We now develop a procedure for estimation of parameters of familiar distributions whose characteristic functions are known except for the parameter involved. Our approach is that of Heathcote (1977). A similar approach under different conditions was studied in Thoronton and Paulson (1977).

Let $X_1, X_2, \ldots, X_n$ be i.i.d. random variables with characteristic function $\phi(t, \theta)$, $\theta \in \Theta \subset R^k$. Let $\phi_n(t)$ be the empirical characteristic function based on $X_i, 1 \leqslant i \leqslant n$.

Definition 8.2.1. A statistic $\hat{\theta}_n = \hat{\theta}_n(X_1, \ldots, X_n)$ is said to be an *integrated squared error estimator* (ISEE) corresponding to a weightfunction $G(\cdot)$ if it minimizes

(8.2.0)
$$I_n(\theta) = \int_{-\infty}^{\infty} |\phi_n(t) - \phi(t, \theta)|^2 \, dG(t).$$

In the following discussion, we assume that $G(\cdot)$ is a nondecreasing weight function with finite total variation taken to be unity without loss of generality. We assume that θ is a scalar. The multidimensional case can be studied on similar lines. We shall discuss that case briefly later in this section. Let

(8.2.1)
$$\phi_n(t) = U_n(t) + iV_n(t)$$

and

(8.2.2)
$$\phi(t, \theta) = U(t, \theta) + iV(t, \theta).$$

Let P_θ be the probability measure corresponding to the characteristic function $\phi(t, \theta)$. Assume that $\{P_\theta, \theta \in \Theta\}$ are dominated by a σ-finite measure μ.

Assume that the following regularity conditions hold:

(R_0) *(Identifiability)* $\theta_1 \neq \theta_2 \Leftrightarrow P_{\theta_1} \neq P_{\theta_2}$.

(R_1) $I_n(\theta)$ is twice differentiable with respect to θ under the integral sign.

(R_2) $U^{(2)}(t, \theta)$ and $V^{(2)}(t, \theta)$ are jointly continuous in t and θ and uniformly bounded by a G-integrable function.

Note that

$$I_n(\theta) = \int_{-\infty}^{\infty} |\phi_n(t) - \phi(t, \theta)|^2 \, dG(t) = \int_{-\infty}^{\infty} \{|U_n(t) - U(t, \theta)|^2 + |V_n(t) - V(t, \theta)|^2\} \, dG(t). \tag{8.2.3}$$

Hence the derivative of $I_n(\theta)$ is given by

$$I_n^{(1)}(\theta) = -2 \int_{-\infty}^{\infty} \{U_n(t) - U(t, \theta)\} U^{(1)}(t, \theta) \, dG(t) - 2 \int_{\infty}^{\infty} \{V_n(t) - V(t, \theta)\} V^{(1)}(t, \theta) \, dG(t)$$

by condition (R_1). Clearly

$$I_n^{(1)}(\theta) = 2n^{-1} \sum_{j=1}^{n} \int_{-\infty}^{\infty} \{\cos tX_j - U(t, \theta)\} U^{(1)}(t, \theta) \, dG(t) - 2n^{-1} \sum_{j=1}^{n} \int_{-\infty}^{\infty} \{\sin tX_j - V(t, \theta)\} V^{(1)}(t, \theta) \, dG(t). \tag{8.2.4}$$

Therefore, if $\hat{\theta}_n$ is an ISEE, then

$$I_n^{(1)}(\hat{\theta}_n) = 0 \quad \text{and} \quad I_n^{(2)}(\hat{\theta}_n) > 0, \tag{8.2.5}$$

where $I_n^{(1)}(\theta)$ is given by (8.2.4). Here $I_n^{(j)}(\theta)$ denotes the jth derivative of $I_n(\theta)$. Let

$$K(x, \theta) = \int_{-\infty}^{\infty} \{\cos tx - U(t, \theta)\} U^{(1)}(t, \theta) \, dG(t) + \int_{-\infty}^{\infty} \{\sin tx - V(t, \theta)\} V^{(1)}(t, \theta) \, dG(t). \tag{8.2.6}$$

We assume further that

(R_3) the Fisher information $\mathscr{I}(\theta) = E_\theta[d \log f(X, \theta)/d\theta]^2 < \infty$, where $f(x, \theta)$ is the density function of X with respect to the σ-finite measure μ, where θ is the parameter.

Note that

$$U^{(1)}(t, \theta) = \int_{-\infty}^{\infty} \cos tx \frac{d f(x, \theta)}{d\theta} \mu(dx) = \operatorname{cov}_\theta\left(\cos tX, \frac{d \log f(X, \theta)}{d\theta} \right) \tag{8.2.7}$$

and similarly

$$V^{(1)}(t, \theta) = \operatorname{cov}_\theta\left(\sin tX, \frac{d \log f(X, \theta)}{d\theta} \right). \tag{8.2.8}$$

It is easy to see from (8.2.6) that

$$\operatorname{cov}_\theta\left[K(X, \theta), \frac{d \log f(X, \theta)}{d\theta} \right] = \int_{-\infty}^{\infty} |\phi^{(1)}(t, \theta)|^2 \, dG(t) \equiv \lambda(\theta). \tag{8.2.9}$$

Assume that

(R_4) $0 < \lambda(\theta) < \infty$.

Observe that

$$\begin{aligned} \int_{-\infty}^{\infty} |\phi^{(1)}(t, \theta)|^2 \, dG(t) &\leqslant \mathscr{I}(\theta) \int_{-\infty}^{\infty} [\operatorname{var}_\theta(\cos tX) + \operatorname{var}_\theta(\sin tX)] \, dG(t) \\ &= \mathscr{I}(\theta) \int_{-\infty}^{\infty} \{1 - |\phi(t, \theta)|^2\} \, dG(t), \end{aligned} \tag{8.2.10}$$

where the first inequality is a consequence of the Cauchy–Schwarz inequality. It is easy to check that $E_\theta[K(X, \theta)] = 0$ and

$$\operatorname{var}_\theta(K(X,\theta)) \leqslant 2\mathscr{I}(\theta), \tag{8.2.11}$$

and in fact,

$$|K(x,\theta)| \leqslant 2\mathscr{I}(\theta)^{1/2}. \tag{8.2.12}$$

Hence $\{K(X_j,\theta),\ j \geqslant 1\}$ are i.i.d. random variables with finite variance by (8.2.11) and

$$I_n^{(1)}(\theta) = -\frac{2}{n}\sum_{j=1}^{n} K(X_j,\theta). \tag{8.2.13}$$

Therefore, by the central limit theorem,

$$n^{1/2} I_n^{(1)}(\theta) \xrightarrow{\mathscr{L}} N(0, 4\operatorname{var}_\theta[K(X,\theta)]) \tag{8.2.14}$$

and SLLN implies that

$$I_n^{(1)}(\theta) \to 0 \quad \text{a.s. as } n \to \infty \tag{8.2.15}$$

when θ is the true parameter.

Theorem 8.2.1 (Heathcote, 1977). Suppose that conditions (R_0)–(R_4) hold. Then there exist a root $\hat{\theta}_n$ of the equation $I_n^{(1)}(\theta) = 0$ that is strongly consistent, and, if θ is the true parameter, then

$$\sqrt{n}(\hat{\theta}_n - \theta) \xrightarrow{\mathscr{L}} N\left(0, \frac{\operatorname{var}_\theta(K(X,\theta))}{\lambda^2(\theta)}\right). \tag{8.2.16}$$

Proof. Let θ_0 denote the true parameter. For any $\delta > 0$,

$$\begin{aligned} I_n(\theta_0 \pm \delta) - I_n(\theta_0) &= \int_{-\infty}^{\infty} \{[U(t,\theta_0 \pm \delta) - U(t,\theta_0)] \\ &\quad \cdot [U(t,\theta_0 \pm \delta) + U(t,\theta_0) - 2U_n(t)] \\ &\quad + [V(t,\theta_0 \pm \delta) - V(t,\theta_0)] \\ &\quad \cdot [V(t,\theta_0 \pm \delta) + V(t,\theta_0) - 2V_n(t)]\}\, dG(t). \end{aligned} \tag{8.2.17}$$

Note that $U_n(t) \to U(t,\theta_0)$ a.s. and $V_n(t) \to V(t,\theta_0)$ a.s. Applying the dominated convergence theorem using (R_2), it can be seen that

$$(8.2.18)\qquad \begin{aligned} E_{\theta_0}[I_n(\theta_0 \pm \delta) - I_n(\theta_0)] &= \int_{-\infty}^{\infty} [\{U(t, \theta_0 \pm \delta) - U(t, \theta_0)\}^2 \\ &\quad + \{V(t, \theta_0 \pm \delta) - V(t, \theta_0)\}^2] dG(t). \end{aligned}$$

Hence, by the SLLN,

$$(8.2.19)\qquad I_n(\theta_0 \pm \delta) > I_n(\theta_0)$$

with probability 1 for large n, which implies that $I_n(\theta)$ achieves a minimum at θ_0 locally with probability 1 for large n. However, this is also a global minimum, for otherwise relation (8.2.18) shows that $U(t, \theta_0 \pm \delta) = U(t, \theta_0)$ and $V(t, \theta_0 \pm \delta) = V(t, \theta_0)$ for all t, which in turn implies that $F(x, \theta_0 \pm \delta) = F(x, \theta_0)$ for all x. Since $\delta \neq 0$, this contradicts (R_0). Since $I_n(\theta)$ is differentiable and δ is arbitrary, there exists a solution $\hat{\theta}_n$ of the equation $I_n^{(1)}(\theta) = 0$ that is strongly consistent.

In view of (R_1),

$$(8.2.20)\qquad \begin{aligned} I_n^{(2)}(\theta) = 2\int_{-\infty}^{\infty} &[U^{(1)}(t, \theta)^2 + V^{(1)}(t, \theta)^2 - \{U_n(t) - U(t, \theta)\} U^{(2)}(t, \theta) \\ &- \{V_n(t) - V(t, \theta)\} V^{(2)}(t, \theta)] dG(t). \end{aligned}$$

Hence

$$(8.2.21)\qquad E_{\theta_0}[I_n^{(2)}(\theta_0)] = 2\int_{-\infty}^{\infty} |\phi^{(1)}(t, \theta_0)|^2 \, dG(t) = 2\lambda(\theta_0).$$

Therefore

$$(8.2.22)\qquad I_n^{(2)}(\theta_0) \to 2\lambda(\theta_0) \quad \text{a.s.}$$

Let $\hat{\theta}_n$ be a consistent solution of the equation $I_n^{(1)}(\theta) = 0$. Then

$$(8.2.23)\qquad 0 = I_n^{(1)}(\hat{\theta}_n) = I_n^{(1)}(\theta_0) + (\hat{\theta}_n - \theta_0) I_n^{(2)}(\theta_0 + \eta(\hat{\theta}_n - \theta_0))$$

for some η in $[-1, 1]$. Hence

$$(8.2.24)\qquad n^{1/2}(\hat{\theta}_n - \theta_0) = \frac{n^{1/2} I_n^{(1)}(\theta_0)}{I_n^{(2)}(\theta_0 + \eta(\hat{\theta}_n - \theta_0))}.$$

Applying (8.2.14) and (R_2), we obtain by the usual methods that

$$n^{1/2}(\hat{\theta}_n - \theta_0) \xrightarrow{\mathscr{L}} N\left(0, \frac{\operatorname{var}_{\theta_0}(K(X, \theta_0))}{\lambda(\theta_0)}\right). \tag{8.2.25}$$

This completes the proof of the theorem.

Remarks. Observe that

$$\frac{\operatorname{var}_\theta[K(X, \theta)]}{\lambda(\theta)} \geqslant \frac{1}{\mathscr{I}(\theta)} \tag{8.2.26}$$

since

$$\lambda(\theta) = \operatorname{cov}_\theta\left[K(X, \theta), \frac{d \log f(X, \theta)}{d\theta}\right]. \tag{8.2.27}$$

Example 8.2.1. Consider the family of densities

$$\begin{aligned} f(x, \theta) = {} & \frac{\varepsilon}{\sigma(2\pi)^{1/2}} \exp\left\{-\frac{1}{2}\left(\frac{x-\theta}{\sigma}\right)^2\right\} \\ & + \frac{(1-\varepsilon)}{(2\pi)^{1/2}} \exp\{-\tfrac{1}{2}(x-\theta)^2\}, \qquad \varepsilon > 0. \end{aligned} \tag{8.2.28}$$

Suppose X_i, $1 \leqslant i \leqslant n$, are i.i.d. with density f. Here the maximum likelihood estimator of θ is inconsistent because of the presence of the nuisance parameter σ. However, the characteristic function of f is

$$\phi(t, \theta) = \{\varepsilon e^{-(1/2)\sigma^2 t^2} + (1-\varepsilon)e^{-(1/2)t^2}\}e^{i\theta t} \tag{8.2.29}$$

and it has derivatives of all orders with respect to θ satisfying the conditions stated in (R_0)–(R_4). Hence the ISEE $\hat{\theta}_n$ is strongly consistent and asymptotically normal.

Remarks. For further discussion about the computational aspects of the ISE estimators, see Heathcote (1977) and Paulson et al. (1975). The results can be extended to the multiparameter case in an obvious way. For instance, see Heathcote (1977) and Csörgö (1981a).

Another method that has been suggested for estimation of the parameter θ is a generalized method of moments. Let $A(t)$ be a complex-valued function of bounded variation with $\overline{A(t)} = A(-t)$. Let $\tilde{\theta}_n$ be a solution of the equation

$$\int_{-\infty}^{\infty} [\phi_n(t) - \phi(t, \theta)]\, dA(t) = 0. \tag{8.2.30}$$

$\tilde{\theta}_n$ is called an *integrated error estimator* (IEE). This was proposed by Feuverger and Mc Dunnough (1979) in the univariate case and Csörgö (1981a) considered the multiparameter case. Under some regularity conditions, they show that $\tilde{\theta}_n$ is consistent and asymptotically normal. We omit the details.

8.3 EMPIRICAL CHARACTERISTIC PROCESS

We have seen the role of empirical characteristic function in problems of inference in the preceding sections. We shall now study some further properties of the empirical characteristic function and discuss its application in tests for goodness of fit.

Let $X_1, X_2, \ldots,$ be i.i.d. random variables with distribution function F and characteristic function

$$\phi(t) = \int_{-\infty}^{\infty} e^{itx}\, dF(x). \tag{8.3.0}$$

Let $F_n(x)$ be the empirical distribution and $\phi_n(t)$ be the empirical characteristic function based on $X_1, X_2, \ldots, X_n$. Note that

$$\phi_n(t) = \frac{1}{n} \sum_{j=1}^{n} e^{itX_j} = \int_{-\infty}^{\infty} e^{itx}\, dF_n(x). \tag{8.3.1}$$

Proposition 8.3.1. For fixed $0 \leqslant T < \infty$,

$$P\left[\limsup_{n \to \infty} \sup_{|t| \leqslant T} |\phi_n(t) - \phi(t)| = 0 \right] = 1. \tag{8.3.2}$$

Proof. The Glivenko–Cantelli lemma implies that $F_n \xrightarrow{w} F$ a.s. as $n \to \infty$. Hence, by the Levy continuity theorem, it follows that

$$\limsup_{n \to \infty} \sup_{|t| \leqslant T} |\phi_n(t) - \phi(t)| = 0 \quad \text{a.s.}$$

for any $0 \leqslant T < \infty$.

Remarks. Since $\phi_n(t)$ is a trigonometric polynomial, it is almost periodic. Furthermore, $|\phi_n(t)| \leqslant 1 = \phi(0)$ for all $n \geqslant 1$. Hence $\phi_n(t)$ must approach $\phi_n(0) = 1$ arbitrarily often as $|t| \to \infty$ [cf. Bohr (1947, pp. 31–38)]. However, $\phi(t) \to 0$ as $|t| \to \infty$ if F is absolutely continuous by the Riemann–Lebesgue lemma [cf. Lukacs (1970, p. 19)]. Hence Proposition 8.3.1 cannot hold in general for $T = \infty$. However, the result holds for $T = \infty$ if F is purely discrete.

Proposition 8.3.2. Let $F(\cdot)$ be purely discrete. Then

(8.3.3) $$P\left[\limsup_{n\to\infty} \sup_t |\phi_n(t) - \phi(t)| = 0\right] = 1.$$

Proof. Let X be a discrete random variable with distribution function F taking values $\{a_n\}$ with probability $\{p_n\}$. Clearly, with an obvious definition for $p_{k,n}$,

$$|\phi_n(t) - \phi(t)| = \left|\sum_{k=1}^{\infty} (p_{k,n} - p_k)e^{ita_k}\right|$$
$$\leqslant \sum_{k=1}^{\infty} |p_{k,n} - p_k|,$$

which tends to 0 a.s. by the SLLN and Scheffé's lemma.

Proposition 8.3.3. If $T_n = o((n/\log\log n)^{1/2})$, then

(8.3.4) $$\Delta_n = \sup_{|t|\leqslant T_n} |\phi_n(t) - \phi(t)| \to 0 \quad \text{a.s. as } n \to \infty$$

for an *arbitrary* distribution function F.

Proof. Let $0 < \varepsilon < 1$ and choose $K > 0$ such that $F(-K) < \varepsilon/6$ and $1 - F(K) < \varepsilon/6$. By the Glivenko–Cantelli lemma, there exists n_0 depending on the sample such that

$$F_n(-K) < \varepsilon/6 \quad \text{and} \quad 1 - F_n(K) < \varepsilon/6$$

for $n \geqslant n_0$, and hence

$$|F_n(\pm K) - F(\pm K)| < \varepsilon/6.$$

Hence, with probability 1,

$$\begin{aligned}\Delta_n &\leqslant \varepsilon + \sup_{|t|\leqslant T_n}\left| -it\int_{-K}^{K}(F_n(x)-F(x))e^{itx}\,dx\right| \\ &\leqslant \varepsilon + 2KT_n \sup_{-\infty<x<\infty}|F_n(x)-F(x)| \\ &\leqslant \varepsilon + 3KT_n(n^{-1}\log\log n)^{1/2}\end{aligned}$$

by the law of iterated logarithm for empirical processes for an *arbitrary distribution function* F using Theorem 2.3.4 (Komlos et al., 1975).

Proposition 8.3.4. Let F be an arbitrary distribution function and $T_n = o(n^{p/2})$, where $0 < p \leqslant 2$. Then

$$V_n = \int_0^{T_n} |\phi_n(t) - \phi(t)|^p\,dt \xrightarrow{p} 0 \quad \text{as} \quad n \to \infty. \tag{8.3.5}$$

Proof. It is easy to see that

$$\begin{aligned}EV_n &\leqslant \int_0^{T_n} \{E|\phi_n(t)-\phi(t)|^2\}^{p/2}\,dt \\ &\leqslant n^{-p/2}T_n.\end{aligned}$$

The first inequality is a consequence of the elementary inequality

$$E|Z|^p \leqslant (E|Z|^2)^{p/2}, \qquad 0 \leqslant p \leqslant 2.$$

The proposition now follows from Chebyshev's inequality since $T_n = o(n^{p/2})$.

We now discuss the properties of the empirical characteristic process

$$Y_n(t) = n^{1/2}(\phi_n(t) - \phi(t)), \qquad -\infty < t < \infty. \tag{8.3.6}$$

Note that

(i) $EY_n(t) = 0, \qquad -\infty < t < \infty,$

(ii) $EY_n(t_1)\bar{Y}_n(t_2) = \phi(t_1+t_2) - \phi(t_1)\phi(-t_2), \qquad -\infty < t_1, t_2 < \infty.$

Let $Y(t)$ be a zero-mean complex-valued Gaussian process satisfying $\bar{Y}(t) = Y(-t)$ and having the same covariance structure as $Y_n(t)$.

Theorem 8.3.5 (Feuerveger and Mureika, 1977). Suppose that $E|X|^{1+\delta} < \infty$ for some $\delta > 0$. Then $Y_n \equiv \{Y_n(t), T_1 \leqslant t \leqslant T_2\}$ converges in distribution to the process $Y \equiv \{Y(t), T_1 \leqslant t \leqslant T_2\}$ on $C[T_1, T_2]$ for $-\infty < T_1 \leqslant T_2 < \infty$ where $C[T_1, T_2]$ is the Banach space of continuous complex-valued functions defined on $[T_1, T_2]$ endowed with supremum norm.

Proof. It is easy to see by the multivariate central limit theorem that the finite-dimensional distributions of Y_n converge weakly to the corresponding finite-dimensional distributions of Y. On the other hand,

$$
\begin{aligned}
E|Y_n(t_2) - Y_n(t_1)|^2 &\leqslant 2[1 - \operatorname{Re}\phi(t_1 - t_2)] \\
&\leqslant 2\int_{-\infty}^{\infty} [1 - \cos x(t_2 - t_1)]\, dF(x) \\
&\leqslant 2\int_{\infty}^{\infty} |x(t_2 - t_1)|^{1+\delta}\, dF(x) \\
&= 2|t_2 - t_1|^{1+\delta} E|X|^{1+\delta}
\end{aligned}
$$

for $T_1 \leqslant t \leqslant T_2$. Hence the real and imaginary parts of the process $\{Y_n, n \geqslant 1\}$ form a tight family, which proves the weak convergence stated in the theorem by Proposition 1.10.12.

The above theorem and earlier results are due to Feuerverger and Mureika (1977). Marcus (1981) and Csörgö (1981b) relaxed the moment condition in Theorem 8.3.5. We now discuss their results. Proofs are omitted.

Let us consider the process

$$
\begin{aligned}
(8.3.7) \qquad Y_n(t) &= n^{1/2}(\phi_n(t) - \phi(t)), && -\tfrac{1}{2} \leqslant t \leqslant \tfrac{1}{2} \\
&= n^{1/2}\int_{-\infty}^{\infty} e^{itx}\, d(F_n - F)(x), && -\tfrac{1}{2} \leqslant t \leqslant \tfrac{1}{2}.
\end{aligned}
$$

This can be rewritten

$$
(8.3.8) \qquad Y_n(t) = \sum_{k=1}^{n} \left\{ \frac{e^{iX_k t} - \phi(t)}{n^{1/2}} \right\}, \qquad -\tfrac{1}{2} \leqslant t \leqslant \tfrac{1}{2}
$$

and hence the question of weak convergence of $Y_n(t)$ is that of the validity of the central limit theorem for the random variables $e^{iXt} - \phi(t)$, which satisfy

$$
\begin{aligned}
(8.3.9) \quad &\text{(i)} \quad E[e^{iXt} - \phi(t)] = 0, \\
&\text{(ii)} \quad E[(e^{iXt} - \phi(t))(e^{-iXs} - \overline{\phi(s)})] = \phi(t - s) - \phi(t)\phi(-s).
\end{aligned}
$$

For convenience, let $T_1 = -\frac{1}{2}$ and $T_2 = \frac{1}{2}$. Each process $\{Y_n(t), -\frac{1}{2} \leqslant t \leqslant \frac{1}{2}\}$ induces a measure on the space $C[-\frac{1}{2}, \frac{1}{2}]$. By the finite-dimensional central limit theorem, if $\{Y_n(t), t \in [-\frac{1}{2}, \frac{1}{2}]\}$ converges weakly to a limiting process $\{Y(t), t \in [-\frac{1}{2}, \frac{1}{2}]\}$, then Y must be a complex-valued Gaussian process with covariance

$$\phi(t-s) - \phi(t)\phi(-s).$$

Hence a necessary condition for $Y_n \xrightarrow{\mathscr{L}} Y$ is that Y should be a Gaussian process with continuous sample paths a.s. This process can be represented in the present context in the form

$$Y(t) = \int_{-\infty}^{\infty} e^{itx} \, dB(F(x)), \tag{8.3.10}$$

where B is a Brownian bridge in the light of the representation (8.3.7) for Y_n. Note that

$$\begin{aligned} Y(t) &= \int_0^1 \exp(itF_*^{-1}(y)) \, dB(y) \\ &= \int_0^1 \exp(itF_*^{-1}(y)) \, dW(y) - W(1) \int_0^1 \exp(itF_*^{-1}(y)) \, dy \\ &= \int_{-\infty}^{\infty} \exp(itx) \, dW(F(x)) - W(1)\phi(t), \end{aligned} \tag{8.3.11}$$

where $F_*^{-1}(y) = \sup\{t \mid F(t) \leqslant y\}$ and W is a standard Wiener process. Hence $\{Y(t), t \in [-\frac{1}{2}, \frac{1}{2}]\}$ has continuous sample paths if and only if the complex-valued stationary Gaussian process

$$J(t) = \int_{-\infty}^{\infty} \exp(itx) \, dW(F(x)), \qquad -\tfrac{1}{2} \leqslant t \leqslant \tfrac{1}{2}, \tag{8.3.12}$$

has continuous sample paths. Note that $J(t)$ has continuous sample paths a.s. iff the real-valued stationary Gaussian process

$$\hat{J}(t) = \int_{-\infty}^{\infty} \cos tx \, dW(F(x)) + \int_{-\infty}^{\infty} \sin tx \, dW^*(F(x)), -\tfrac{1}{2} \leqslant t \leqslant \tfrac{1}{2}, \tag{8.3.13}$$

has continuous sample paths a.s., where W^* is an independent copy of W.

Furthermore,

(8.3.14)
$$\begin{aligned}\sigma^2(|t-s|) &\equiv E|J(t)-J(s)|^2\\ &= E(\hat{J}(t)-\hat{J}(s))^2\\ &= 4\int_{-\infty}^{\infty} \sin^2\frac{|t-s|x}{2}\,dF(x).\end{aligned}$$

Note that σ is defined on $[-1,1]$. Let

(8.3.15)
$$m_\sigma(x) = \lambda\{u\in[-1,1]:\sigma(u)<x\},$$

where λ is the Lebesgue measure and

(8.3.16)
$$\bar{\sigma}(u) = \sup\{y\,|\,m_\sigma(y)<u\}.$$

The function $\bar{\sigma}$ is called a nondecreasing rearrangement of σ. It is a nondecreasing function on $[0,2]$. Note that $\bar{\sigma}$ has the same distribution function with respect to the Lebesgue measure on $[0,2]$ that σ has with respect to the Lebesgue measure on $[0,1]$. Let

(8.3.17)
$$I(\sigma) = \int_0^2 \frac{\bar{\sigma}(s)}{s(\log 16/s)^{1/2}}\,ds.$$

By the Dudley–Fernique necessary and sufficient condition for the continuity of stationary Gaussian processes [cf. Jain and Marcus (1978, Theorem 7.6 and Corollary 6.3)], the process $\hat{J}$ and hence J has continuous sample paths a.s. iff $I(\sigma)<\infty$. Hence if $Y_n \xrightarrow{\mathcal{L}} Y$, then $I(\sigma)<\infty$. The following theorem due to Marcus (1981) proves the sufficiency of the condition. We omit the proof.

Theorem 8.3.6 (Marcus, 1981). Let X be a random variable with distribution function F and characteristic function $\phi(t)$. Let

(8.3.18)
$$\sigma^2(t) = 4\int_{-\infty}^{\infty} \sin^2\frac{xt}{2}\,dF(x) = 2(1-\operatorname{Re}\phi(t)).$$

Define $I(\sigma)$ by (8.3.17). If $I(\sigma)<\infty$, then the process Y_n converges weakly to the process Y on $C[-\frac{1}{2},\frac{1}{2}]$, where Y is a Gaussian process with mean 0 and covariance $\phi(t-s)-\phi(t)\phi(-s)$. If $I(\sigma)=\infty$, then the Gaussian process Y with covariance $\phi(t-s)-\phi(t)\phi(-s)$ does not have continuous sample paths and hence Y_n does not converge weakly on $C[-\frac{1}{2},\frac{1}{2}]$.

The following result due to Csörgö (1981b) gives sufficient conditions for the weak convergence of Y_n to Y in $C[T_1, T_2]$. It follows from Theorem 8.3.6.

Theorem 8.3.7 (Csörgö, 1981b). Suppose the distribution F of the random variable X satisfies the condition

(8.3.19) $$x^\alpha F(-x) + x^\alpha(1 - F(x)) = O(1) \quad \text{as} \quad x \to \infty \quad \text{for some} \quad \alpha > 0.$$

Then Y has continuous sample paths and $Y_n \xrightarrow{\mathscr{L}} Y$ in $C[T_1, T_2]$.

Remarks. Condition (8.3.19) is equivalent to the condition

(8.3.19′) there exists $0 < \alpha < 2$ such that

$$1 - \operatorname{Re} \phi(t) = 0(t^\alpha) \quad \text{as} \quad t \to 0+$$

by Kawata (1972, p. 420). Observe that condition (8.3.19′) holds for stable laws. Hence the weak convergence of the empirical characteristic process Y_n to Y in $C[T_1, T_2]$ holds whenever X has a stable law.

Csörgö (1981b) obtained strong approximation to the process Y_n using the results of Komlos et al. (1975) (see Theorem 2.3.4). We now state his result. The proof is omitted.

Theorem 8.3.8 (Csörgö 1981b). Suppose condition (8.3.19) holds for some $\alpha > 0$. Then there exists a sequence of Brownian bridges $\{B_n(y): 0 \leqslant y \leqslant 1\}$ and a Kiefer process $\{K(y, t): 0 \leqslant y \leqslant 1, 0 \leqslant t < \infty\}$ such that

(8.3.20) $$P\left\{\sup_{T_1 \leqslant t \leqslant T_2} |Y_n(t) - Z_n(t)| > C_1 r_1(n)\right\} \leqslant L_1 n^{-(1+\delta)}$$

and

(8.3.21) $$P\left\{\sup_{T_1 \leqslant t \leqslant T_2} |Y_n(t) - K_n(t)| > C_2 r_2(n)\right\} \leqslant L_2 n^{-(1+\delta)},$$

where $\delta > 0$ is arbitrarily large and the positive constants C_1 and C_2 depend only on δ, F, T_1, T_2, whereas L_1 and L_2 depend only on $T_2 - T_1$. Here

(8.3.22) $$Z_n(t) = \int_{-\infty}^{\infty} e^{itx}\, dB_n(F(x)), \qquad T_1 \leqslant t \leqslant T_2,$$

(8.3.23) $$K_n(t) = n^{-1/2} \int_{-\infty}^{\infty} e^{itx}\, dK(F(x), n), \qquad T_1 \leqslant t \leqslant T_2,$$

and

$$r_k(n) = n^{-\alpha(2\alpha+4)}(\log n)^{(k\alpha+1)/(\alpha+2)}; \qquad k = 1, 2. \tag{8.3.24}$$

Remarks. Note that the processes $\{Z_n(t),\ T_1 \leqslant t \leqslant T_2\}$, $\{K_n(t),\ T_1 \leqslant t \leqslant T_2\}$, and $\{Y(t),\ T_1 \leqslant t \leqslant T_2\}$ have the same distribution. Furthermore, it follows from (8.3.20) and (8.3.21) that

$$\Delta_n^{(1)} \equiv \sup_{T_1 \leqslant t \leqslant T_2} |Y_n(t) - Z_n(t)| = O(r_1(n)) \quad \text{a.s.} \tag{8.3.25}$$

and

$$\Delta_n^{(2)} \equiv \sup_{T_1 \leqslant t \leqslant T_2} |Y_n(t) - K_n(t)| = O(r_2(n)) \quad \text{a.s.} \tag{8.3.26}$$

In particular, it follows that $Y_n \xrightarrow{\mathscr{L}} Y$ in $C[T_1, T_2]$. As a corollary to Theorem 8.3.8, we have the following result, which has applications in testing problems using the empirical characteristic functions.

Corollary 8.3.9 (Csörgö, 1981b). Consider the following functionals defined on $C[T_1, T_2]$:

$$\text{(i)} \quad \psi_1(u) = \int_{T_1}^{T_2} |u(t)|^2\, dH(t), \tag{8.3.27}$$

$$\text{(ii)} \quad \psi_2(u) = \int_{T_1}^{T_2} (\operatorname{Re} u(t))^2\, dH(t),$$

$$\text{(iii)} \quad \psi_3(u) = \int_{T_1}^{T_2} (\operatorname{Im} u(t))^2\, dH(t),$$

where H is a distribution function with support $[T_1, T_2]$. Let $\psi_4(u)$ be an arbitrary real-valued functional for which the Lipschitz condition

$$|\psi_4(u) - \psi_4(v)| \leqslant L \sup \| u - v \|, \qquad u, v \in C[T_1, T_2] \tag{8.3.28}$$

holds for some constant $L > 0$. Suppose $\psi_k(Y)$ has density function $f_k(x)$, $k = 1, \ldots, 4$, with respect to the Lebesgue measure and condition (8.3.19) holds. Then

$$\sup_x |P(\psi_k(Y_n) < x) - P(\psi_k(Y) < x)| = O(r_1(n)), \qquad 1 \leqslant k \leqslant 4, \tag{8.3.29}$$

provided that the functions $f_4(x)$ and $x^{1/2} f_k(x)$, $k = 1, 2, 3$, are bounded.

Proof. We consider the case $k = 1$. The proofs for the cases $k = 2, 3, 4$ are similar under the conditions stated in the theorem.

Let $\alpha_n = (\psi_1(Y_n))^{1/2}$ and $\beta_n = (\psi_1(Z_n))^{1/2}$, where Z_n is as defined in Theorem 8.3.8. Note that β_n has the same distributions as $(\psi_1(Y))^{1/2}$. Let f be the density of β_n. Then $f(x^{1/2}) = 2x^{1/2} f_1(x)$ a.e. It is easy to see that

$$|\alpha_n - \beta_n| \leqslant \left\{ \int_{T_1}^{T_2} (Y_n(t) - Z_n(t))^2 \, dH(t) \right\}^{1/2} \leqslant \Delta_n^{(1)}, \tag{8.3.30}$$

where $\Delta_n^{(1)}$ is as given by (8.3.25). Let $r(n) = C_1 r_1(n)$. Relation (8.3.20) implies that

$$P(|\alpha_n - \beta_n| > r(n)) \leqslant L_1 n^{-(1+\delta)}. \tag{8.3.31}$$

Let $D_n(x)$ and $D(x)$ be the distribution functions of $\psi_1(Y_n)$ and $\psi_1(Y)$, respectively. Then

$$\begin{aligned} D_n(x) &\leqslant P(\alpha_n < x^{1/2}, |\alpha_n - \beta_n| \leqslant r(n)) + L_1 n^{-(1+\delta)} \\ &\leqslant P(\beta_n < x^{1/2} + r(n)) + L_1 n^{-(1+\delta)} \\ &= D(x) + \int_{x^{1/2}}^{x^{1/2}+r(n)} f(y) \, dy + L_1 n^{-(1+\delta)} \\ &= D(x) + Mr(n) + o(r(n)) + L_1 n^{-(1+\delta)} \\ &= D(x) + O(r_1(n)), \end{aligned} \tag{8.3.32}$$

where $M = \sup_x \{x^{1/2} f_1(x)\} < \infty$ by hypothesis. Similarly, we can show that

$$D_n(x) \geqslant D(x) + O(r_1(n)). \tag{8.3.33}$$

We now discuss some applications of the results obtained above for testing problems.

Example 8.3.1 (Testing for Goodness of Fit) (Feigin and Heathcote, 1976). Suppose the problem is to test the hypothesis that a random variable X has a specified distribution function $F_0(x)$. Clearly this is equivalent to the problem of testing that the random variable X has the specified characteristic function

$$\phi_0(t) = \int_{-\infty}^{\infty} e^{itx} \, dF_0(x) = U_0(t) + iV_0(t). \tag{8.3.34}$$

Let $X_1, X_2, \ldots, X_n$ be i.i.d. as the random variable X with characteristic function $\phi(t) = U(t) + iV(t)$. Let $\phi_n(t)$ be the empirical characteristic function and

$$\phi_n(t) = U_n(t) + iV_n(t). \tag{8.3.35}$$

It is easy to see that

$$E \cos tX = U(t), \quad E \sin tX = V(t),$$

$$\operatorname{var} \cos tX = \tfrac{1}{2}\{1 + U(2t) - 2U^2(t)\} \equiv \sigma^2(t)$$

and

$$\operatorname{var} \sin tX = \tfrac{1}{2}\{1 - U(2t) - 2V^2(t)\} \equiv \gamma^2(t).$$

Furthermore, it follows by the central limit theorem that

$$\sqrt{n}(U_n(t) - U(t)) \xrightarrow{\mathscr{L}} N(O, \sigma^2(t))$$

and

$$\sqrt{n}(V_n(t) - V(t)) \xrightarrow{\mathscr{L}} N(O, \gamma^2(t)).$$

In order to test the hypotheses H_0: $\phi(t) = \phi_0(t)$, one can use the test based on either $U_n(t)$ or $V_n(t)$. For instance, the following test can be used. Reject the null hypothesis H_0 if

$$\frac{\sqrt{n}|U_n(t) - U_0(t)|}{\sigma_0(t)} > c,$$

where c is determined by the specified significance level and t is chosen so that

$$\sigma_0(t) = \tfrac{1}{2}\{1 + U_0(2t) - 2U_0^2(t)\} > 0.$$

The power of this test can be computed again under a specified alternative.

This test can be used for instance for testing H_0: $\phi(t) = e^{-t^2/2}$ against H_1: $\phi(t) = e^{-(1/2)|t|^\alpha}$ where $0 < \alpha < 2$. For details, see Feigin and Heathcote (1976). As an alternative test, one can use an analog of the Cramer–von Mises statistic, namely,

$$n \int_{T_1}^{T_2} |\phi_n(t) - \phi(t)|^2 \, dH(t),$$

where $H(\cdot)$ is a distribution function with support $[T_1, T_2]$.

We shall discuss testing for goodness of fit for composite hypothesis later in this section.

Example 8.3.2 (Testing for Symmetry). Since the characteristic function of a random variable X is real if and only if the distribution F is symmetric about origin, one can use

$$R_n = \int_{-\infty}^{\infty} \operatorname{Im}[\phi_n(t)]^2 \, dH(t) \tag{8.3.36}$$

as a test statistic, where H is a specified weight function for testing for symmetry. Let H be a distribution function with support $[T_1, T_2]$, where $-\infty < T_1 \leqslant T_2 < \infty$. Note that, under hypothesis H_0, F is symmetric about the origin or equivalently the characteristic function ϕ is real, so we have

$$\begin{aligned} nR_n &= \int_{-\infty}^{\infty} \operatorname{Im}[Y_n(t)]^2 \, dH(t) \\ &= \psi_3(Y_n), \end{aligned}$$

where ψ_3 is as defined in (8.3.27) and Y_n is as defined by (8.3.6). Hence, if F satisfies condition (8.3.19) and $x^{1/2} f_3(x)$ is bounded, where f_3 is the density of $\psi_3(Y)$, then

$$\psi_3(Y_n) \xrightarrow{\mathscr{L}} \psi_3(Y)$$

by Corollary 8.3.9. Hence the asymptotic distribution of nR_n under the null hypothesis is the same as that of

$$\psi_3(Y) = \int_{T_1}^{T_2} \operatorname{Im}[Y(t)]^2 \, dH(t),$$

where Y is a complex-valued Gaussian process with mean 0 and covariance function

$$\phi(t-s) - \phi(t)\phi(-s).$$

Remarks. We now turn to the problem of testing for goodness of fit when the null hypothesis is composite, as in Section 5.1. The problem of goodness of fit here can be rephrased in terms of the characteristic functions as follows. We are required to test the hypothesis that the characteristic function

$\phi(t) \in \{\phi(t, \theta): \theta \in \Theta\}$, $\Theta \subset R^p$. Since θ is unknown, the problem can be reduced to investigating the properties of the *estimated empirical characteristic process*

$$\hat{Y}_n(t) = n^{1/2}(\phi_n(t) - \phi(t, \hat{\theta}_n)), \tag{8.3.37}$$

where $\hat{\theta}_n$ is a suitable estimator of θ, as a first step toward the construction of test statistics and study of their asymptotic properties. This problem is investigated in Csörgö (1981a) when $\hat{\theta}_n$ is either ISEE or IEE as discussed in Section 8.2. The limiting process of $\hat{Y}_n$ depends on the unknown parameter θ and hence is not distribution free and $\hat{Y}_n$ cannot be used for defining test statistics for testing purposes. Techniques similar to those given in Section 5.1 due to Khmaladze (1981) may be useful in this context.

Multivariate empirical characteristic processes and their properties are investigated in Csörgö (1981a). The properties of empirical characteristic process when the observation $X_i, i \geqslant 1$, are under random censorship are studied in Csörgö (1983). Finally, we remark that condition (8.3.19) may be replaced by a more general condition that

(8.3.19*) there exists a continuous function $h(x)$ such that

(i) $\dfrac{h(x)}{x^\alpha} \uparrow \infty$ as $x \uparrow \infty$ for some $\alpha > 0$,

(ii) $h(x)F(-x) = O(1)$, $\quad h(x)(1 - F(x)) = O(1)$,

and the results in Theorem 8.3.8 and Corollary 8.3.9 hold.

8.4 GENERALIZED LEAST SQUARES FOR LINEAR MODELS

We now discuss another application of the empirical characteristic function to the problem of estimation of slope in linear regression. The method of estimation is an extension of the least squares and may be termed *generalized least squares*. Results here are due to Chambers and Heathcote (1981).

Consider the problem of estimation of the vector $\boldsymbol{\beta}$ in the linear model

$$Y_j = \mathbf{X}_j'\boldsymbol{\beta} + \varepsilon_j, \qquad 1 \leqslant j \leqslant n, \tag{8.4.0}$$

where $\mathbf{X}_j$ are known $p \times 1$ vectors, $\boldsymbol{\beta} = (\beta_1, \ldots, \beta_p)'$, ε_i, $1 \leqslant i \leqslant n$, are i.i.d. random variables, and $Y_i, 1 \leqslant i \leqslant n$, are the observed data.

Suppose the errors ε_i are i.i.d. $N(0, \sigma^2)$. Then the least-squares estimator $\hat{\boldsymbol{\beta}}$ of

$\boldsymbol{\beta}$ is the statistic minimizing the function

$$Q_n(\boldsymbol{\beta}) = \frac{1}{n}\sum_{j=1}^{n}(Y_j - \mathbf{X}_j'\boldsymbol{\beta})^2, \tag{8.4.1}$$

which is an estimator of σ^2. Since $E(e^{it\varepsilon_j}) = e^{-(1/2)\sigma^2 t^2}$ for $1 \leqslant j \leqslant n$, we can write

$$\sigma^2 = -t^{-2}\log|E(e^{it\varepsilon})|^2 \equiv L(t) \quad \text{(say)},$$

where ε is distributed as ε_1. Note that $L(t)$ is well defined for all $t \neq 0$ even when ε_1 does not have finite variance.

Motivated by the above remark, we can define

$$L_n(\boldsymbol{\beta}; t) = -t^{-2}\log\left|\frac{1}{n}\sum_{j=1}^{n} e^{it(Y_j - \mathbf{X}_j'\boldsymbol{\beta})}\right|^2, \qquad t \neq 0, \tag{8.4.2}$$

and consider $\hat{\boldsymbol{\beta}}_n$ a *generalized least-squares estimator* (GLSE) if it minimizes $L_n(\boldsymbol{\beta}; t)$. No conditions on the existence of moments of ε_1 are assumed. Clearly $\hat{\boldsymbol{\beta}}_n$ depends on the value of t. Observe that, if $\phi_n(t)$ denotes the empirical characteristic function based on $\varepsilon_1, \ldots, \varepsilon_n$ and $\boldsymbol{\beta}_0$ is the true parameter, then

$$L_n(\boldsymbol{\beta}_0; t) = -t^{-2}\log|\phi_n(t)|^2. \tag{8.4.3}$$

However $L_n(\boldsymbol{\beta}; t)$ is *not* convex in $\boldsymbol{\beta}$. The limiting case $t = 0$ corresponds to the *ordinary least squares.*

We now study some asymptotic properties of the estimator $\hat{\boldsymbol{\beta}}_n(t)$. Let $\mathbf{0} \neq \boldsymbol{\delta} \in R^p$ and $\boldsymbol{\beta}_0$ be the true parameter. It is easy to see that

$$\begin{aligned} &L_n(\boldsymbol{\beta}_0 \pm \boldsymbol{\delta}; t) - L_n(\boldsymbol{\beta}_0; t) \\ &\quad = -t^{-2}\log\left|\frac{n^{-1}\sum_{j=1}^{n}\exp(it(\varepsilon_j \mp \mathbf{X}_j'\boldsymbol{\delta}))}{n^{-1}\sum_{j=1}^{n}\exp(it\varepsilon_j)}\right|. \end{aligned} \tag{8.4.4}$$

By the SLLN, it follows that

$$\frac{1}{n}\sum_{j=1}^{n} e^{it\varepsilon_j} \to \phi(t) \quad \text{a.s. as} \quad n \to \infty,$$

where $\phi(t)$ is the characteristic function of ε_1. Suppose that

$$\phi(t) \neq 0 \quad \text{and} \quad \left|\lim_{n\to\infty} n^{-1}\sum_{j=1}^{n}\exp\{\pm\mathbf{X}_j'\boldsymbol{\delta}\}\right| < 1 \quad \text{for } \boldsymbol{\delta} \neq \mathbf{0}. \tag{8.4.5}$$

Then

$$(8.4.6)\quad L_n(\boldsymbol{\beta}_0 \pm \boldsymbol{\delta}; t) - L_n(\boldsymbol{\beta}_0; t) \xrightarrow{\text{a.s.}} -t^{-2}\log\left|\lim_{n\to\infty} n^{-1}\sum_{j=1}^{n} \exp\{\mp it\mathbf{X}_j'\boldsymbol{\delta}\}\right|^2$$

for every $\boldsymbol{\delta} \neq \mathbf{0}$. Hence the minimum of $L_n(\boldsymbol{\beta}; t)$ is achieved at $\boldsymbol{\beta}_0$ with probability 1. But

$$(8.4.7)\qquad L_n(\hat{\boldsymbol{\beta}}_n; \mathbf{t}) \leqslant L_n(\boldsymbol{\beta}; t)$$

for all $\boldsymbol{\beta} \in R^p$ by the definition of $\hat{\boldsymbol{\beta}}_n$. Hence $\hat{\boldsymbol{\beta}}_n(t)$ is a consistent estimator of $\boldsymbol{\beta}$.

Let $\mathbf{X}$ be the $n \times p$ matrix with $\mathbf{X}_j'$ as the jth row. Let $(\mathbf{X}'\mathbf{X})_\infty$ denote the centered asymptotic form of the matrix $\mathbf{X}'\mathbf{X}$ with (i, j)th element

$$(8.4.8)\qquad \lim_{n\to\infty} n^{-1}\sum_{r=1}^{n} x_{ri}x_{rj} - \left(\lim_{n\to\infty} n^{-1}\sum_{r=1}^{n} x_{ri}\right)\left(\lim_{n\to\infty} n^{-1}\sum_{r=1}^{n} x_{rj}\right).$$

Let

$$(8.4.9)\quad \sigma^2(t) = \frac{U^2(t)(1 - U(2t)) - 2U(t)V(t)V(2t) + 2V^2(t)(1 + U(2t))}{2t^2(U^2(t) + V^2(t))^2},$$

where

$$(8.4.10)\qquad \phi(t) = E(e^{it\varepsilon}) = U(t) + iV(t).$$

The following theorem can be proved by standard techniques using Taylor expansion. For details, see Chambers and Heathcote (1981).

Theorem 8.4.1. Suppose that condition (8.4.6) holds and the limits in (8.4.8) exist. Furthermore, suppose that $(\mathbf{X}'\mathbf{X})_\infty$ is nonsingular and

$$(8.4.11)\qquad \lim_{n\to\infty} n^{-2}\sum_{r=1}^{n} x_{ri}^2 x_{rj}^2 = 0 \quad \text{for} \quad 1 \leqslant i, j \leqslant p.$$

Then there exists a solution $\hat{\beta}_n(\mathbf{t})$ of the equation

$$(8.4.12)\quad l_{nk}(\boldsymbol{\beta}; t) \equiv t^{-1}n^{-2}\sum_{i=1}^{n}\sum_{j=1}^{n} x_{jk}\sin[t\{Y_j - Y_i - (\mathbf{X}_j' - \mathbf{X}_i')\boldsymbol{\beta}\}] = 0, \qquad 1 \leqslant k \leqslant p,$$

and

$$(8.4.13)\qquad n^{1/2}(\hat{\boldsymbol{\beta}}_n(t) - \boldsymbol{\beta}_0) \xrightarrow{\mathscr{L}} \mathbf{N}_p(\mathbf{0}, \boldsymbol{\Omega}(t)),$$

where

$$\boldsymbol{\Omega}(t) = (\mathbf{X}'\mathbf{X})_{\infty}^{-1}\sigma^2(t). \tag{8.4.14}$$

Remarks. Equation (8.4.12) can be derived as follows: To minimize $L_n(\boldsymbol{\beta}; t)$, we can write

$$L_n(\boldsymbol{\beta}; t) = -t^{-2}\log[U_n^2(\boldsymbol{\beta}; t) + V_n^2(\boldsymbol{\beta}; t)], \tag{8.4.15}$$

where

$$U_n(\boldsymbol{\beta}; t) = n^{-1}\sum_{j=1}^{n} \cos\{t(Y_j - \mathbf{X}_j'\boldsymbol{\beta})\} \tag{8.4.16}$$

and

$$V_n(\boldsymbol{\beta}; t) = n^{-1}\sum_{j=1}^{n} \sin\{t(Y_j - \mathbf{X}_j'\boldsymbol{\beta})\}. \tag{8.4.17}$$

Equating the derivative of $L_n(\boldsymbol{\beta}; t)$ with respect to $\boldsymbol{\beta}$ to zero leads to the estimating equation

$$t^{-1}\Bigg[U_n(\boldsymbol{\beta}; t)n^{-1}\sum_{j=1}^{n} x_{jk}\sin\{t(Y_j - \mathbf{X}_j'\boldsymbol{\beta})\} - V_n(\boldsymbol{\beta}; t)n^{-1}\sum_{j=1}^{n} x_{jk}\cos\{t(Y_j - \mathbf{X}_j'\boldsymbol{\beta})\}\Bigg] = 0, \qquad 1 \leqslant k \leqslant p, \tag{8.4.18}$$

which reduces to equation (8.4.12). Observe that, if $t \to 0$, then equations (8.4.12) become

$$n^{-2}\sum_{i=1}^{n}\sum_{j=1}^{n} x_{jk}\{Y_j - Y_i - (\mathbf{X}_j' - \mathbf{X}_i')\boldsymbol{\beta}\} = 0, \qquad 1 \leqslant k \leqslant p,$$

which are the normal equations for least-squares estimation. Asymptotic normality of $\hat{\boldsymbol{\beta}}_n$ is proved by expanding $l_n(\boldsymbol{\beta}; t)$ around the true parameter and using the method of proof for asymptotic normality of MLE, as in Cramer (1946).

Since the asymptotic covariance matrix of the estimator $\hat{\boldsymbol{\beta}}_n(t)$ of $\boldsymbol{\beta}$ depends on t through $\sigma(t)$, a natural way to choose t is so that $\sigma(t)$ is minimum. Chambers and Heathcote (1981) discussed the choice of t when $\phi(u) = \exp(-\frac{1}{2}|u|^{\alpha})$, $0 < \alpha \leqslant 2$. They show that the optimum value of $t = 1.6$ if the errors are Cauchy. Let ζ be the family of all distributions that are normal, or do not have variance, or satisfy the property $E[\varepsilon - E(\varepsilon)]^4 > 3\sigma_\varepsilon^4$, where σ_ε^2 is the variance of ε. Chambers and Heathcote (1981) proved that, of all the error distributions that are in ζ, the normal distribution is the only one for which $\sigma^2(t)$ achieves a global minimum at $t=0$.

REFERENCES

Bohr, H. A. (1947). *Almost Periodic Functions*, Chelsea, New York.

Chambers, R. L. and Heathcote, C. R. (1981). On the estimation of the slope and the identification of outliers in linear regression. *Biometrika* **68**, 21–33.

Chandrasekhar, S. (1943) Stochastic problems in physics and astronomy. *Rev. Mod. Phys.* **15**, 1–89.

Cramer, H. (1946). *Mathematical Methods of Statistics*, Princeton University Press, Princeton, New Jersey.

Csörgö, S. (1981a). The empirical characteristic process when parameters are estimated. *Contribution to Probability* (Ed. J. Gani and V. K. Rohatgi), Academic Press, New York.

Csörgö, S. (1981b). Limit behaviour of the empirical characteristic function. *Ann. Probability* **9**, 130–144.

S. (1983). Estimating characteristic functions under random censorship. *Theory Probab. Appl.* **28**, 615–622.

Du Mouchel, W. H. (1971). Stable distributions in statistical inference, Ph.D. Thesis, Yale University.

Du Mouchel, W. H. (1973). On the asymptotic normality of the maximum likelihood estimate when sampling from a stable distribution. *Ann. Statistics* **1**, 948–957.

Fama, E. (1965). The behaviour of stock market prices. *J. Business* **38**, 34–105.

Feigin, P. D. and Heathcote, C. R. (1976). The empirical characteristic function and the Cramer–von Mises Statistic. *Sankhya Ser. A* **38**, 309–325.

Feuerverger, A. and Mc Dunnough, P. (1979). On the efficiency of empirical characteristic function procedures (preprint).

Feuerverger, A. and Mureika, R. A. (1977). The empirical characteristic function and its applications. *Ann. Statistics* **5**, 88–97.

Heathcote, C. R. (1977). The integrated squared error estimation of parameters. *Biometrika* **64**, 255–264.

Holt, D. R. and Crow, E. L. (1973). Tables and graphs of the stable probability density functions. *J. Res. Nat. Bur. Std. B* **77** *B*(3, 4), 143–198.

Jain, N. C. and Marcus, M. B. (1978). Continuity of subgaussian processes. *Probability on Banach Spaces* (Ed. J. Kuelbs), Advances in Probability, No. 4, Dekker, New York.

Kawata, T. (1972). *Fourier Analysis in Probability Theory*, Academic Press, New York.

Khmaladze, E. V. (1981). The martingale approach to the theory of nonparametric tests of fit. *Theory Probab. Appl.* **26**, 240–265.

Komlos, J., Major, P., and Tusnady, G. (1975). An approximation of partial sums of independent rv's and the sample df. I. Z. *Wahr. verw. Geb.* **32**, 111–131.

Le Cam., L. (1953). On some asymptotic properties of maximum likelihood and related Bayes estimates. *Univ. of California Publ. in Stat.* **1**, 277–330.

Leitch, R. A. and Paulson, A. S. (1975). Estimation of stable law parameters: stock price application. *J. Amer. Statist. Assoc.* **70**, 690–697.

Lukacs, E. (1970). *Characteristic Functions*, Griffin, London.

Mandelbrot, B. (1963). The variation of certain speculative prices. *J. Business* **36**, 394–419.

Marcus, M. B. (1981). Weak convergence of the empirical characteristic function. *Ann. Probability* **9**, 194–201.

Paulson, A. S., Holcomb, E. W., and Leitch, R. A. (1975). The estimation of the parameters of the stable laws. *Biometrika* **62**, 163–170.

Press, S. J. (1972). Estimation in univariate and multivariate stable distributions, *J. Amer. Statist. Assoc.* **67**, 842–846.

Thornton, J. C. and Paulson, A. S. (1977). Asymptotic distribution of characteristic function based estimators for the stable laws. *Sankhya Ser. A* **39**, 341–354.

Wald, A. (1949). Note on the consistency of the maximum likelihood estimate. *Ann. Math. Statist.* **20**, 595–601.

Zolotarev, V. M. (1957). Mellin–Stieltjes transformations in probability theory. *Theory Probab. Appl.* **2**, 433–460.

Zolotarev, V. M. (1980). Statistical estimates of the parameters of stable laws. *Mathematical Statistics, Banach Center Publications* Vol. 6. Polish Scientific Publishers, Warsaw.

Zolotarev, V. M. (1981). Integral transformations of distributions and estimates of parameters of multidimensional spherically symmetric stable laws. *Contributions to Probability* (Ed. J. Gani and V. K. Rohatgi), Academic Press, New York.

Author Index

Note: Numbers in *italics* indicate pages on which full references appear.

Subject Index

Applied Probability and Statistics (Continued)

JUDGE, HILL, GRIFFITHS, LÜTKEPOHL and LEE • Introduction to the Theory and Practice of Econometrics
JUDGE, GRIFFITHS, HILL, LÜTKEPOHL and LEE • The Theory and Practice of Econometrics, *Second Edition*
KALBFLEISCH and PRENTICE • The Statistical Analysis of Failure Time Data
KISH • Survey Sampling
KUH, NEESE, and HOLLINGER • Structural Sensitivity in Econometric Models
KEENEY and RAIFFA • Decisions with Multiple Objectives
LAWLESS • Statistical Models and Methods for Lifetime Data
LEAMER • Specification Searches: Ad Hoc Inference with Nonexperimental Data
LEBART, MORINEAU, and WARWICK • Multivariate Descriptive Statistical Analysis: Correspondence Analysis and Related Techniques for Large Matrices
LINHART and ZUCCHINI • Model Selection
LITTLE and RUBIN • Statistical Analysis with Missing Data
McNEIL • Interactive Data Analysis
MAINDONALD • Statistical Computation
MANN, SCHAFER and SINGPURWALLA • Methods for Statistical Analysis of Reliability and Life Data
MARTZ and WALLER • Bayesian Reliability Analysis
MIKÉ and STANLEY • Statistics in Medical Research: Methods and Issues with Applications in Cancer Research
MILLER • Beyond ANOVA, Basics of Applied Statistics
MILLER • Survival Analysis
MILLER, EFRON, BROWN, and MOSES • Biostatistics Casebook
MONTGOMERY and PECK • Introduction to Linear Regression Analysis
NELSON • Applied Life Data Analysis
OSBORNE • Finite Algorithms in Optimization and Data Analysis
OTNES and ENOCHSON • Applied Time Series Analysis: Volume I, Basic Techniques
OTNES and ENOCHSON • Digital Time Series Analysis
PANKRATZ • Forecasting with Univariate Box-Jenkins Models: Concepts and Cases
PIELOU • Interpretation of Ecological Data: A Primer on Classification and Ordination
PLATEK, RAO, SARNDAL and SINGH • Small Area Statistics: An International Symposium
POLLOCK • The Algebra of Econometrics
PRENTER • Splines and Variational Methods
RAO and MITRA • Generalized Inverse of Matrices and Its Applications
RIPLEY • Spatial Statistics
RIPLEY • Stochastic Simulation
RUBINSTEIN • Monte Carlo Optimization, Simulation, and Sensitivity of Queueing Networks
SCHUSS • Theory and Applications of Stochastic Differential Equations
SEAL • Survival Probabilities: The Goal of Risk Theory
SEARLE • Linear Models
SEARLE • Matrix Algebra Useful for Statistics
SPRINGER • The Algebra of Random Variables
STEUER • Multiple Criteria Optimization
STOYAN • Comparison Methods for Queues and Other Stochastic Models
TIJMS • Stochastic Modeling and Analysis: A Computational Approach
TITTERINGTON, SMITH, and MAKOV • Statistical Analysis of Finite Mixture Distributions
UPTON • The Analysis of Cross-Tabulated Data
UPTON and FINGLETON • Spatial Data Analysis by Example, Volume I: Point Pattern and Quantitative Data
WEISBERG • Applied Linear Regression, *Second Edition*

(*continued from front*)